To those early workers in soils and geomorphology who inspired my generation

John Frye   *Hans Jenny*   *Roger Morrison*   *Gerry Richmond*   *Bob Ruhe*

Contents

Preface   xi

Chapter 1

THE SOIL PROFILE, HORIZON NOMENCLATURE, AND SOIL CHARACTERISTICS   1

Introduction   1
Definition of Soil   2
Soil Profile and the Pedon   2
Soil Horizon Nomenclature   3
Soil Characteristics   9
   Color   9
   Texture   10
   Organic Matter   11
   Structure   12
   Bulk Density   13
   Comparison of Data Sets   13
   Soil Moisture Retention and Movement   15
   Cation-Exchange Capacity, Exchangeable Cations, and Percentage Base Saturation   18
   Soil pH   18
Comparing the Development of Soils: Use of Soil Indices   19
   Color Indices and the Podzolization Index   20
   Profile Development Indices   21
   Index of Profile Anisotropy   22
Soil Micromorphology   23
Recognition of Buried Soils   24

Chapter 2

SOIL CLASSIFICATION   29

Soil Classification at Lower Levels, and Making Geological Maps   29
Soil Classification at the Higher Categories   32
   Classification of Soil Moisture and Soil Temperature   42
   Soil Orders   42
Distribution of Soil Orders and Suborders in the United States   48
Distribution of Soil Orders of the World   49
Soil Landscape Analysis   49

Chapter 3

WEATHERING PROCESSES   53

Physical Weathering   53
Chemical Weathering   59
   Congruent and Incongruent Dissolution   59
   Chelation   65
   Hydration and Dehydration   66
   Ion Exchange   66
Measurement of the Amount of Chemical Weathering That Has Taken Place   66
Chemical Weathering Disguised as Physical Weathering   74
Rate of Present-Day Chemical Weathering   76
Chemical Denudation: Does the Landscape Lower?   81

Chapter 4

THE PRODUCTS OF WEATHERING   85

Clay Minerals   85

Crystalline and Noncrystalline Compounds of Aluminum and Iron   90

Cation-Exchange Capacity of Inorganic and Organic Colloidal Particles   92

Ion Mobilities   93

Genesis of Noncrystalline and Crystalline Products   95

Clay-Mineral Distribution with Depth in Soil Profiles   102

Continental and Global Clay Formation   104

Chapter 5

PROCESSES RESPONSIBLE FOR THE DEVELOPMENT OF SOIL PROFILES   105

Relative Importance of Processes by Soil Order   106

Formation of the A Horizon   107

Translocation of Iron and Aluminum (Podzolization)   108

The Chelate Connection   110

The Inorganic Pathway   111

Translocation of Clay-Sized Particles   112

Field and Laboratory Data   113

Thin-Section and Scanning Electron Microscope (SEM) Analysis   114

Mass-Balance Studies   117

Summary   118

Dispersion and Flocculation   118

Translocation of Silt   120

Fragipan and Its Origin   121

Morphology and Origin of Lateritic (Oxisolic) Profiles   121

The Lateritic Profile   122

Chemistry and Mineralogy of Lateritic Profiles   123

Genesis   124

Climate and Plate Tectonics   127

Origin of $CaCO_3$-Rich Horizons, and Those with More Soluble Salts   127

Stages of Accumulation   129

Chemistry and Mineralogy   130

Source of the Carbonate   131

Silica Accumulation   132

More Soluble Salts   132

Horizons Associated with High Water Table or Perched Water Table, and Redoximorphic Features   134

Phosphorus Transformations   137

Radiocarbon Dating of Soil Horizons   137

Chapter 6

FACTORS OF SOIL FORMATION   141

Factors and the Fundamental Equation   141

Arguments for and Against the Use of Factors, and Monogenetic versus Polygenetic Soils   144

Other Models of Pedogenesis   145

The Approach Used Here   146

Steady State   146

Chapter 7

INFLUENCE OF PARENT MATERIAL ON WEATHERING AND SOIL FORMATION   148

Mineral Stability   148

Rock Chemical Composition   150

Susceptibility of Rocks to Chemical Weathering   151

Influence of Parent Material on the Formation of Clay Minerals   154

Influence of Original Texture of Unconsolidated Sediments on Soil Formation   155

Examples of Extreme Rock-Type Control on Soil Formation   156

Soils Formed on Limestone   156

Soils Formed from Volcanic Ash Deposits   158

Test for Uniformity of Parent Material   159

Cumulative Soil Profiles   165

The Loess Transect: Cumulic Soil Formation on Parent Materials with Different Textures and Different Depositional Rates   167

Chapter 8

WEATHERING AND SOIL DEVELOPMENT WITH TIME   171

The Time Scale   171

Rate of Rock and Mineral Weathering   172

Chronosequences and Chronofunctions in
Different Climates   178
  Classification of Chronosequences   178
  Landscape Evolution versus
  Soil Formation   179
  Chronofunctions   180
  Thresholds   181
  Data Presentation   182
  Climatic Gradient of the Soil
  Chronosequence   182
Chronosequences in Cold Polar Regions and
High Mountains   183
Noncalcic Soils in a Semiarid ⇒
Humid Transect   190
  Tills at Mountain Fronts: Western United
  States   190
  Soils on Dune Sands   191
  River-Terrace Chronosequences: Western
  United States   192
  Marine Terraces of the California Coast   192
  Chronosequences in Wetter Climates   194
  Unusual Chronosequences of Tropical
  Carbonate Ocean Islands   196
Chronosequences Involving Transects from
Noncalcic to Calcic Soils   201
Chronosequences in Aridic Regions with
Gypsic and Calcic Soils   202
  Cold Desert versus Hot Desert Soils   202
  Chronosequences Involving Calcic Soils:
  Western United States   203
  Chronosequences Involving Both Surface
  and Buried Calcic Soils   209
  Chronosequences Involving Gypsic Soils:
  Israel, Western United States, and Peru   210
  Summary of Desert Soil Development with
  Time   214
Leaching of Parent-Material Carbonate   214
Summaries on the Time Required to
Reach Various Levels of Pedological
Development   215
  A-Horizon Organic Matter and Associated
  Properties   215
  Comparative Data on Clay and Iron Buildup
  in Soils of Different Climates   216
  Comparison of PDI Values with Time in
  Different Climates   217
  Summary of Bt (Argillic) Horizon
  Development   218

  Summary of Development of Red Color   220
  Summary of Clay Mineralogy   221
Relationship between Diagnostic Horizons,
Soil Orders, and Time   225
The Steady-State Condition and Time
Necessary to Attain It   226
Regressive Pedogenesis   226
Estimating Ages of Deposits   228

Chapter 9

TOPOGRAPHY–SOIL RELATIONS WITH TIME IN DIFFERENT CLIMATIC SETTINGS   230

Influence of Slope Orientation on
Pedogenesis   232
Soil Catenas   235
Catenas in Polar, Arctic, and
Alpine Areas   241
Catenas for Noncalcic Soils in a Temperate ⇒
Humid Transect   245
  Western United States, Israel, and New
  Zealand   245
  Midcontinent United States and the
  Gumbotil Problem   253
  Warm, Humid Environments   257
Catenas Involving Calcic and
Gypsic Soils   261
Concluding Remarks about Catenas   265

Chapter 10

SUMMARY OF CLIMATE–SOIL RELATIONS, AND PALEOCLIMATIC INTERPRETATIONS   268

Climatic Parameters   269
Regional Soil Trends Related to Climate:
Pedogenic Gradients or Climosequences   273
Variation in Specific Soil Properties with
Climate   278
  Organic Matter Content   279
  Clay Content   282
  Clay Mineralogy   285
  Soil Redness and Pedogenic Fe Minerals   288
  Fragipan Occurrence   289
  Calcium Carbonate, and
  More Soluble Salts   290
  Trends in Iron, Aluminum, and
  Phosphorus   292
Soil-Elevation Transects   292

PDI as One Example of Poor Correlation
between Climate and Soil   293

Reconstruction of Past Climates from
Pedologic Data   293

   Overall Soil Morphology   294

   *Organic Matter Distribution and
Content*   296

   Bt-Horizon Properties   297

   Clay Mineralogy   299

   Mineral Etching   300

   *Position of Horizon of CaCO₃ Accumulation
and Morphology of the K Horizon*   300

   *Position of Gypsic and Salic Horizons, and
Associated Weathering Features*   303

   *Isotopic Evidence in Soils for Climate
Change*   304

Chapter 11

**APPLICATION OF SOILS TO
GEOMORPHOLOGICAL,
SEDIMENTOLOGICAL, AND
ENVIRONMENTAL STUDIES   307**

Use of Soils in Quaternary Stratigraphic
Studies   307

   *Soil Stratigraphy and Soil-Stratigraphic
Units (Geosols and Pedoderms)*   308

   *Soil-Stratigraphic Units as an Aid in
Defining Stratigraphic Units*   308

Determining the Time of Soil Formation   309

   General Concepts   309

   *Concept and Explanation of the
Soil-Forming Interval*   309

Using Soil-Stratigraphic Units for Subdivision
and Correlation of Quaternary Deposits   311

   *U.S. Geological Survey and Soil
Conservation Service Mapping Projects:
Basic Data and Applied Maps*   311

   *Till and Moraine Correlations, and
Comparison with Cosmogenic
Isotope Dating*   312

   *Using Soil Stratigraphy to Date
Mountainous Landscapes and
Assess Plant Nutrient Relations*   319

   Using Soils to Date Mass Movements   319

   *Floodplains, River Terraces, and Flood
Predictions*   321

   *Long-Range Correlation, Including Loess
and Deep-Sea Sediments*   322

Using Soils to Date Tectonic Activity   326

   Dating Faults   326

   Dating Folds   331

   *Soils and Safety of the Yucca Mountain
Repository*   332

Using Soils in Archeological Studies   333

Using Soils in Environmental Studies   337

Paleosols   339

   Definition of the Term Paleosol   339

   Recognizing Paleosols   340

   Laboratory Analyses   340

   Classification of Paleosols   341

   *Criticism of Paleosol Identification and
Classification*   342

   Examples of Use of Paleosols   344

Appendix 1

DESCRIBING SOIL PROPERTIES   347

Horizon Nomenclature   347

Digging and Photographing Soils   347

Soil Properties   348

   Depth   349

   Percentage Estimate   349

   Color   349

   Structure   352

   Gravel Content   353

   Consistence   353

   Texture   354

   pH   354

   Clay Films   355

   Horizon Boundaries   356

   Stages of Carbonate Morphology   356

   Carbonate Effervescence   357

   Salts and Silica Development   357

   Cementation   357

   Silt Caps   359

Appendix 2

**CALCULATION OF PROFILE-
DEVELOPMENT INDEX   360**

   Plotting and Comparing Soil-Index Data   361

References   368

Index   423

Although there are many textbooks on soils, there are few that also serve the needs of geomorphologists, sedimentologists, environmental geologists, and archeologists working in Quaternary research. This book is an attempt to fill that gap. The emphasis is on the study of soils in their natural setting—the field. Such studies are commonly called "pedology." I believe that soils cannot be adequately used in the field for any purpose without understanding the processes and factors that control their formation. Hence, the overall organization of the book is, first, a discussion of soil morphology, weathering, and soil-forming processes and, then, variation in soils with variation in the soil-forming factors (climate, organisms, topography, parent material, and time). Only the U.S. soil classification system is used, and I have tried to keep it painless. The book ends with a chapter on applications.

For those readers beginning to work in soils, I want to stress the importance of sound field work, because pedology is a field science. Laboratory studies are necessary, but they are only as good as the field work and sampling on which they are based.

This book is a revision of two earlier editions. I initially had no plans for a new revision, until it was requested by Joyce Berry of Oxford University Press. The more I thought about it, the more I realized it was an opportunity for me to showcase the creative work of many talented workers in the fields of soils and geomorphology. Many people helped me with the first two editions of this book, and they are duly acknowledged there.

As in other editions, the focus is on the United States, with which I am most familiar; I also am familiar with some of the overseas areas. My experience indicates that to explain something, it helps to have

been there. I have tried to bring many of the topics up to date, to continue to stress the importance of airborne materials on pedogenesis, to use examples from more places, and, in the last chapters, to add more on detecting climate change, on the global loess–soil record, and on the use of soils in environmental studies.

The literature cited includes both the relevant geological and pedological literature in attacking problems common to both disciplines. A more encyclopedic treatment of the literature would have necessitated deleting materials I consider more important in conveying important principles to the reader. Where appropriate, books that cover the materials well are used as references. The advantage of going through three editions over a quarter of a century is that the older pertinent literature is retained, and I am proud to be able to show the younger generation that many useful concepts were first presented in the older literature, and, in some cases, they have not changed much.

Two appendices appear at the end of the book. Appendix 1 is designed to give the reader information on soil data that needs to be collected to describe a soil profile adequately. Appendix 2 takes the reader through the calculation and application of the profile development index.

I am grateful to several individuals for reading parts of the manuscript. Ed Larson reviewed part of Chapter 8, Dave Swanson reviewed Chapter 9, and Margaret Berry, Emmett Evanoff, Vance Holliday, Jay Quade, and Emily Taylor reviewed parts of Chapters 10 and 11. I accept full responsibility for the entire text, however.

Many people helped computer draft new figures. Kathy Haller prepared many of the figures when she co-authored a soil manual with Mike Machette and myself. John Russell, funded by a mentor grant from this department, drafted many others. Additional single ones came from John Drexler, Emmett Evanoff, Edwin Larson, Paula Maat, Dan Muhs, Dennis Netoff, and Jay Noller.

Since the last edition, many people have taken the time to help me better understand certain soil topics, to work on soil projects with me, or to take me into new field areas. For these courtesies and for sharing thoughts, I would like to thank M.E. Berry, R.M. Burke, A.S. Campbell, W. Chesworth, M. Churchward, E.A. Colhoun, O. Conchon, D.R. Muhs, R. Gerson, L. H. Gile, J.B. J. Harrison, J. W. Hawley, V.T. Holliday, E.E. Larson, M. N. Machette, L.D. McFadden, G. Mews, C. D. Miller, J. Noller, P. Patterson, J. Pitlick, D. Rodbell, R. R. Shroba, F. Smith, R.P. Suggate, E.M. Taylor, P. T. Tonkin, P. Walker, and T.R. Walker. Suzanne Larsen and her staff at our new Jerry Crail Johnson Earth Sciences Library patiently helped with many literature search problems.

Several pieces of unpublished research by myself are included in this edition. Funding for the field study of soils on carbonate islands of the Pacific Ocean (Chapter 8) was from the National Geographic Society. John Drexler and Fred Luiszer did the total chemical analyses, Rolf Kihl analyzed one soil, and Edwin Larson arranged for the REE analyses by the Phoenix Memorial Laboratory in Michigan and interpreted the results. People associated with the University of Guam and the USDA Soil conservation Service (Rota), Kenting National Park (Taiwan), and ORSTOM (Mare' and Vanuatu) helped with accommodations and logistics. Funding for the field study of the New Mexico catena (Chapter 9) was from a Gladys W. Cole Memorial Research Award of the Geological Society of America. The University of Colorado funded Paula Maat and Janet Slate to do the laboratory analyses.

Special thanks to Vivian Lu and Connie Ge; my wife Suzanne and I traded time spent with them for computer help for this computer-challenged old timer from their mother and my colleague, Shemin Ge. Miles Waite kindly formatted the book. John Drexler and Fred Luiszer were also a great help with the computers, and questions about them.

I thank all of my thesis students, for I have learned much from them by being involved in their projects.

My family has been especially helpful in all of my scientific endeavors. When the children were younger, we always went into the field as a group, and each of them, Karl and Robin, in their own way, pitched in and helped make each trip a success. My wife Suzanne shared many fun (and some not so) experiences in getting the soil field data in many places (some strange) around the world since the last edition of this book.

Peter W. Birkeland
Boulder, Colorado

Soils and
Geomorphology

Chapter 1

Berry Def: study of soils and their
use in evaluating landform
evolution and age, landform
stability, surface processes,
and past climates.

The Soil Profile, Horizon Nomenclature, and Soil Characteristics

■ Introduction

The topics covered in this book are best described by the title, soils and geomorphology. It is hard to separate the two disciplines because to understand the genesis of a soil fully one has to have an in-depth appreciation of its geomorphic setting. A subdiscipline has even been started, termed soil geomorphology (McFadden and Knuepfer, 1990), and the broad scope of this is covered in a recent book edited by Knuepfer and McFadden (1990). As I left for the meeting that produced the latter book I asked former student Margaret Berry for a definition of soil geomorphology, and she provided this very inclusive one: the study of soils and their use in evaluating landform evolution and age, landform stability, surface processes, and past climates. Recent books on soils and Quaternary geology (Catt, 1986) and on climatic geomorphology (Bull, 1991) demonstrate the use of soils in many diverse climates. Thus, there now is a good integration of soils and geology, an idea that began over 100 years ago in the United States (Amundson and Yaalon, 1995).

My interest in this field started late in my work on my dissertation, when I was introduced to soils and met some of the premier soil geomorphologists of the United States. I was working on the glacial history of part of the Sierra Nevada in the late 1950s, and Rocky Crandell (US Geological Survey), who was writing a review on the topic, visited me and showed me that soils and other weathering features could help in unraveling the history of the area. Soon after (1960s), I received more formal training in soils from Hans Jenny and Rod Arkley at the University of California–Berkeley. As my field work expanded into Nevada, I received help from two prominent soil geomorphologists, Roger Morrison (US Geological Survey) and John Frye (Illinois State Geological Survey). Field trips with these latter workers showed me the potential of using soils in Quaternary stratigraphic studies. At that time, many of us were strongly influenced by the work of two other soil-geomorphologists, mainly because of the field trips they led: Gerry Richmond (US Geological Survey) in the western United States, and Bob Ruhe (Iowa State University, US Soil Conservation Service, University of Indiana) in the midcontinent. Few of us got to go overseas, but when we did, Dan Yaalon (The Hebrew University, Jerusalem) was always there to offer advise. This was an exceptionally talented group, and from them came the present field of soil geomorphology. Finally, knowing the soil geomorphic setting and Quaternary history of regions made studying soil-forming factors and processes the next logical step, toward which many present workers have made significant contributions.

Soils differ from geological deposits generally, but in some places the two are so similar in appearance that they are difficult to tell apart. The focus here is on those aspects of soils of interest to an interdisciplinary group that includes geomorphologists, soil scientists (especially pedologists), sedimentary petrologists, engineering geologists, neotectonists, archeologists, and botanists. Soils will be studied in their natural setting, the field, that part of soil science called pedology (Daniels, 1988). One cannot overstress the importance of sound field work in any soil project, for no amount of laboratory work or statistical treatment can correct improper site selection, soil description, or sampling. Hence, the stress will be on field relations, but we will also use data from other specialty fields of soil science (for example, soil chemistry) to help understand the soil in its natural setting. Finally, because most soils have formed during the Quaternary, an adequate understanding of the geological, climatological, and botanical history of the last 2 million years is needed (Flint, 1971; Bowen, 1978; Bradley, 1985; Wright, 1984; Wright and others, 1993).

■ Definition of Soil

The term "soil" has many definitions, depending on who is using the term. For example, to some engineers "soil" is unconsolidated surficial material, or the regolith. To many soil scientists it is mainly the medium for plant growth, but this definition is too imprecise. For example, some trees grow out of cracks in solid rock with little evidence for soil. A definition of soil that serves our purpose well is a slight modification of that of Joffe (1949): a soil is a natural body consisting of layers (horizons) of mineral and/or organic constituents of variable thicknesses, which differ from the parent materials in their morphological, physical,

chemical, and mineralogical properties and their biological characteristics (Fig. 1.1). Most of these horizon properties are pedogenic, that is they are the result of complex pedogenic processes. Soil horizons generally are unconsolidated, although some contain sufficient amounts of silica, $CaCO_3$, or iron oxides to be cemented. Although geological materials such as colluvium can have properties similar to specific soil horizons, with many colluvial deposits the properties can be inherited from the in situ soil that eroded to produce the colluvium. However, once the colluvium has been deposited and the landscape stablized, pedogenic processes operating over time can produce soil horizons.

■ Soil Profile and the Pedon

A soil profile consists of the vertical arrangement of all the soil horizons down to the parent material. In studying a soil, therefore, the investigator must be able to identify the parent material from which the soil formed. This is no easy task and requires a good deal of experience in geology and pedology. However, once the parent material is recognized, and its original properties estimated, one can begin to determine departures in the properties of the original material and identify these materials as pedogenically altered soil horizons. In an ideal case the parent material (e.g., volcanic ash) is isotropic with uniform properties in all directions. As a soil forms on the ash it becomes anisotropic, that is layered due to pedogenic processes. Because most pedogenic processes operate from the surface downward, the layers (called soil horizons)

A horizon

Bt horizon

Cox horizon

Figure 1.1 Soil formed on marine-terrace deposits near San Diego, California.

will parallel the land or geomorphic surface (Ruhe, 1956). If there are depositional layers in the parent material, the soil horizons may or may not parallel them. However, if the geomorphic surface crosscuts the depositional layers, so also will the soil horizons (Fig. 1.2).

One should distinguish between a soil profile and a pedon. A soil profile is a two-dimensional body, commonly studied in artificial cuts. We scrape away a width of about 1 m, and describe it until (ideally) unaltered material is encountered at depth. In contrast, a pedon is a three-dimensional body, with the same thickness as a profile, but with a surface area that ranges between 1 and 10 m². The reason for the area is that soils vary laterally and so the pedon has to be large enough to represent the repeating lateral variability in the nature and arrangement of soil horizons and in specific soil properties (Soil Survey Division Staff, 1993). In essence, the pedon is the smallest unit of the soil body used in soil mapping. Some feel that the relationship between the pedon and the soil mapping body is akin to that of the unit cell to the crystal (Soil Survey Staff, 1975). Although many published articles list properties for pedons, few demonstrate that the described soil meets the definition criteria of the pedon—that is that the investigator has examined the soil over an area large enough to include the repeatability in variation of properties.

Some geologists distinguish between a soil profile and a weathering profile (Ruhe, 1969). Where this is done, the soil profile is generally considered to make up the upper part of the much thicker weathering profile. If we consider a soil profile formed on granite, the weathering profile would include the weathered granite at greater depth, the saprolite of many workers (Stolt and Baker, 1994). However, everyone recognizes how difficult it is to distinguish between soil and saprolite, and indeed saprolite can have pedological properties (Graham and others, 1994). Because of the difficulty in separating these two profiles on the basis of the processes involved and the resulting properties, I will not make the above distinctions. What some would call the weathering profile beneath the soil profile probably would qualify as a Cox or Cr soil horizon under the horizon nomenclature used here. This, in effect, is the pedo-weathering profile of Tandarich and others (1994); however, rather than use D horizon for unaltered surficial materials, I will continue to use Cu. What all workers are trying to do is identify pedogenesis as well as they can, and so it helps to have a horizon nomenclature that aids in this process. Another useful nomenclature in the midcontinent is the subdivision of the C horizon (Follmer and others, 1979, Appendix 3, Table A). Where possible, the soil extends to such depths that alteration of the presumed parent material, from either field or laboratory data, no longer exists. A problem with thick weathered zones (say thicker than 30 m) could be the separation of the products of soil formation from those of diagenesis. The boundary between the two is difficult to define, and no doubt all gradations exist. Fortunately, most soils are not thick enough for this to be a major problem.

■ Soil Horizon Nomenclature

Two sets of soil horizon nomenclature are in use in the United States, one for the diagnostic horizons used for classification purposes and one for field descriptions. The diagnostic horizons require laboratory analyses, and will be described in Chapter 2. The set used for field descriptions is that of the Soil Survey Division Staff (1993), with some modifications. The modifications are used to help clarify some of the definitions. A comparison between the old and the new field description systems, important for comparing the older and newer literature, is given in Appendix 1 (Table A1.1). Soil profile depth functions of various key properties are helpful in visualizing the more common horizons (Fig. 1.3).

Figure 1.2 Soil formed on dissected alluvial fan deposit, north of Elat, southern Israel. The fan deposit has horizontal stratification, and the soil (s) follows the present topography, cross cutting the strata.

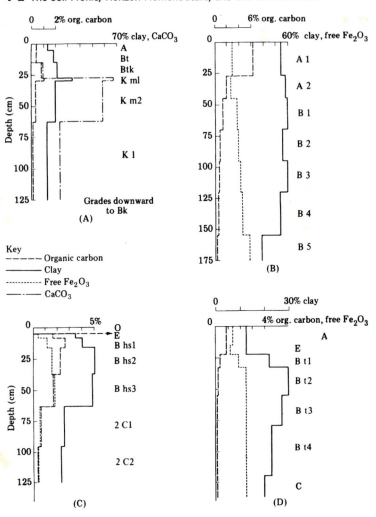

Figure 1.3 Laboratory data on soil profiles that illustrate properties commonly associated with various soil horizons. (A) Parent material is gravelley alluvium, and all data are for the nongravel fraction (Terino soil, Gile et al., 1966, Table 2, ©1966, The Williams & Wilkins Co., Baltimore). (B) Parent material is serpentine (Soil Survey Staff, 1960. profile No. 27). (C) Parent material is glacial outwash (Soil Survey Staff, 1960, profile No. 20). (D) Parent material is loess. (*Soil Survey Staff, 1960, profile No. 11*)

Many properties form by the vertical movement and accumulation of some material, either in solid or dissolved form (e.g., clay and iron, respectively). The result is an eluvial–illuvial relationship. When iron is dissolved and removed from a near-surface horizon, and precipitated in a horizon at greater depth in the profile, the overlying horizon from which the iron was removed is termed eluvial, and the horizon where iron accumulates is termed illuvial. Many other properties have an eluvial–illuvial relationship within the profile. In desert soils carbonate ($CaCO_3$) is removed from the surface horizons and accumulates at depth. In many temperate soils clay particles are moved as a solid from the surface horizon and deposited in a subsurface horizon.

Three kinds of symbols are used in soil horizon nomenclature: capital letters, lower case letters, and Arabic numerals. Capital letters denote master (most prominent) horizons, lower case letters some specific characteristic or subdivision of the master horizon, and Arabic numerals either a further subdivision of the horizon or parent material layering.

Most soil profiles can be divided into several master or most prominent horizons (Table 1.1). Surface or near-surface horizons relatively high in organic matter are designated O and A horizons, the difference between the two being determined by the amount of organic matter present. Beneath the O or A horizon in some environments (usually forested ones), there is a light-colored horizon relatively leached of

Table 1.1 Soil-Horizon Nomenclature

Description of master horizon horizon and subhorizons

O horizon—Surface accumulations of mainly organic material; may or may not be, or has been, saturated with water. Subdivided on the degree of decomposition as measured by the fiber content after the material is rubbed between the fingers.

Oi horizon—Least decomposed organic materials; rubbed fiber content is greater than 40% by volume.

Oe horizon—Intermediate degree of decomposition; rubbed fiber content is between 17 and 40% by volume.

Oa horizon—Most decomposed organic materials; rubbed fiber content is less than 17% by volume.

A horizon—Accumulation of humified organic matter mixed with mineral fraction; the latter is dominant. Occurs at the surface or below an O horizon; Ap is used for those horizons disturbed by cultivation.

E horizon—Usually underlies an O or A horizon, and can be used for eluvial horizons within or between parts of the B horizon (e.g., common above fragipan, x). Characterized by less organic matter and/or fewer sesquioxides (compounds of iron and aluminum) and/or less clay than the underlying horizon. Many are marked by a concentration of sand and silt. Horizon is light colored due mainly to the color of the primary mineral grains because secondary coatings on the grains are absent; relative to the underlying horizon, color value will be higher or chroma will be lower.

B horizon—Underlies an O, A, or E horizon, and shows little or no evidence of the original sediment or rock structure. Several kinds of B horizons are recognized, some based on the kinds of materials illuviated into them, others on residual concentrations of materials. Subdivisions are:

Bh horizon—Illuvial accumulation of amorphous organic matter–sesquioxide complexes that either coat grains or form sufficient coatings and pore fillings to cement the horizon.

Bhs horizon—Illuvial accumulation of amorphous organic matter–sesquioxide complexes, and sesquioxide component is significant; both color value and chroma are three or less.

Bk horizon—Illuvial accumulation of alkaline earth carbonates, mainly calcium carbonate; the properties do not meet those for the K horizon.

B*l* horizon—Illuvial concentrations primarily of silt (Forman and Miller, 1984). Used when silt cap development reaches stages 5 and 6.

Bo horizon—Residual concentration of sesquioxides, the more soluble materials having been removed.

Bq horizon—Accumulation of secondary silica.

Bs horizon—Illuvial accumulation of amorphous organic matter–sesquioxide complexes if both color value and chroma are greater than three.

Bt horizon—Accumulation of silicate clay that has either formed in situ or is illuvial (clay translocated either within the horizon or into the horizon); hence it will have more clay than the assumed parent material and/or the overlying horizon. Illuvial clay can be recognized as grain coatings, bridges between grains, coatings on ped or grain surfaces or in pores, or thin, single or multiple near-horizontal discrete accumulation layers of pedogenic origin (clay bands or lamellae). In places, subsequent pedogenesis can destroy evidence of illuviation. Although Soil Survey Division Staff (1993) does not include this, clay accumulation that lacks evidence for illuviation is included (could have been formed in situ, for example).

Bw horizon—Development of color (redder hue or higher chroma relative to C) or structure, or both, with little or no apparent illuvial accumulation of material.

By horizon—Accumulation of secondary gypsum.

Bz horizon—Accumulation of salts more soluble than gypsum.

K horizon. A subsurface horizon so impregnated with carbonate that its morphology is determined by the carbonate (Gile and others, 1965). Authigenic carbonate coats or engulfs nearly all primary grains in a continuous medium. The uppermost part of a strongly developed horizon is laminated, brecciated, and/or pisolithic (Machette, 1985). The cemented horizon corresponds to some caliches and calcretes.

C horizon—A subsurface horizon, excluding R, like or unlike material from which the soil formed, or is presumed to have formed. Lacks properties of A and B horizons, but includes materials in various stages of weathering.

Cox and Cu horizons—In many unconsolidated deposits, the C horizon consists of oxidized material overlying seemingly unweathered C. The oxidized C does not meet the requirements of the Bw horizon. In stratigraphy, it is important to differentiate between these two kinds of C horizons. Here Cox is used for oxidized C horizons and Cu for unweathered C horizons. Cu is from the nomenclature of England and Wales (Hodson, 1976). Alternatively the Cox can be termed BC or CB.

Cr horizon—In soils formed on bedrock, there commonly will be a zone of weathered rock between the soil and the underlying rock. If it can be shown that the weathered rock has formed in place, and has not been transported, it is designated Cr. Such material is the saprolite of geologists; in situ formation is demonstrated by preservation of original rock features, such as grain-to-grain texture, layering, or dikes. If such material has been moved, however, the original structural features of the rock are lost, and the transported material may be the C horizon for the overlying soil. Those Cr horizons with translocated clay, as shown by clay films, are termed Crt.

R horizon—Consolidated bedrock underlying soil.

(continued)

Table 1.1 (continued)

Description of master horizon horizon and subhorizons

Selected Subordinate Departures

Lower-case letters follow the master horizon designation. Those that are mainly specific to a particular master horizon are given above. Some can be found in a variety of horizons; they are listed below.

b Buried soil horizon with major features formed prior to burial. May be deeply buried and not affected by subsequent pedogenesis; if shallow, they can be part of a younger soil profile.

c Concretions or nodules cemented by accumulations of iron, aluminum, manganeses, or titanium.

f Horizon cemented by permanent ice. Seasonally frozen horizons are not included, nor is dry permafrost material (material that lacks ice but is colder than 0°C).

g Horizon in which gleying is a dominant process, that is, either iron has been removed during soil formation or saturation with stagnant water has preserved a reduced state. Common to these soils are neutral colors, with or without mottling. Most have chroma of 2 or less and many have redox concentrations (see Chapter 5). Strong gleying is indicated by chromas of one or less, and hues bluer than 10Y. Much of the above color is due to the color of reduced iron, or the color of uncoated grains from which iron pigment has been removed. Bg is used for horizons with pedogenic features in addition to gleying; however, if gleying is the only pedogenic feature, the horizon is designated Cg.

j Used in combination with other horizon designation (Btj, Ej) to denote incipient development of that particular feature or property (National Soil Survey Committee of Canada, 1974). A rule for some designations would be to use it for those horizons that do not meet criteria for diagnostic horizons (e.g., Ej for an eluvial horizon that does not meet the criteria of the albic horizon) (Chapter 2).

k Accumulation of alkaline earth carbonates, commonly $CaCO_3$.

m Horizon that is more than 90% cemented. Denote the cementing material (Km, carbonate; qm, silica; Kqm, carbonate and silica; etc.).

n Accumulation of exchangeable sodium.

ss Presence of slickensides.

v Has two uses. (1) One is plinthite, iron-rich, humus-poor, reddish material that hardens irreversibly when dried. (2) If A horizons in arid environments have a vesicular structure (round voids), they are designated Av (McFadden, 1988).

x Subsurface horizon characterized commonly by a bulk density greater than that of the adjacent horizons, firmness and brittleness, and very coarse prismatic structure with bleached vertical faces (fragipan character). An E horizon may overly the fragipan horizon at depth as well as as between the A and Bt horizons higher in the profile. If the E-horizon nomenclature designations are identical, and both are pedogenic, a prime is applied to the lower E horizon. In this example, the profile would be A/E/Bt/E'/Bx/Cox.

y Accumulation of gypsum.

z Accumulation of salts more soluble than gypsum (e.g., NaCl).

(From Soil Survey Division Staff [1993], with modification [all modifications are referenced]).

iron, aluminum, and organic compounds, the E horizon. I suggest using E if the horizon meets the color specifications of the Albic horizon (Chapter 2), and Ej if it does not. The B horizon is beneath the surface horizon or horizons. This horizon encompasses a multitude of soil characteristics relative either to those of the assumed parent material or to the overlying horizon(s). Among the B horizon characteristics are the production of brown to red colors due to the formation of pedogenic Fe compounds (Fe is released from parent minerals), clay accumulation, the accumulation of Fe and Al compounds with or without organic matter, the residual concentration of resistant materials following the removal of more soluble constituents under conditions of intensive weathering and leaching, and accumulation of $CaCO_3$ and more soluble salts (gypsum and halite). The slightly weathered C horizon (Cox) commonly is beneath the B horizon and beneath that is unweathered bedrock, the R horizon, or unconsolidated unweathered material, the Cu horizon (the proposed D horizon of Tandarich and others, 1994). This latter horizon is useful in geomorphological studies, as it forces the investigator to identify the parent material conditions; this is not commonly done in many studies conducted for more agricultural purposes. In desert environments, carbonate buildup plays an important role in soil morphology and genesis, and horizons high in carbonate are designated K (Fig. 1.3A). Although the K-horizon designation is not used by the Soil Survey Division Staff (1993), it is used here because most pedologists and geologists working in arid lands find it an extremely useful term. If a pedogenic carbonate horizon does not meet the criteria for a K horizon, it is designated Bk.

In arid regions, the progressive development of carbonate horizons has been classified into several morphological stages. Gile and others (1965, 1966, 1981) recognized four stages; later, Machette (1985) further subdivided stage IV for a total of six stages. The latter classification is used here (Appendix 1, Fig. A1.5 and Table A1.5). The stages are identified on particular morphological features that correlate well with age of parent material, and should be included in any soil description. In general, stage I and II morphologies are associated with Bk horizons, and stages III through VI with K horizons.

Other materials precipitate out at depth in semiarid to arid environments, and also form a developmental sequence defined by stages. These include pedogenic gypsum and silica. Harden and others (1991) define four stages of such development, which are similar to the first four stages of carbonate development (Appendix 1, Table A1.8). One could devise a similar stage sequence for other salts such as halite. Under the extreme cold and dry conditions of Antarctica, Bockheim (1990) classified the morphology of the water-soluble salts into six stages.

Within the C horizon, numbers are used to denote variation with depth, but specific numbers carry no specific meaning (C1, C2, . . . C8, C9). In the western United States, many workers find the Cox-Cu subdivision of the C horizon useful. Workers in Illinois have proposed specific meaning for numerical subdivision of the C horizon, because variation in horizon properties occurs in a predictable order with depth (Follmer and others, 1979, Appendix 3, Table A).

In the field, many horizons are transitional rather than sharp. In places the transitional material is not described, but is included in the description of the distinctness of the horizon boundary (Appendix 1). However, there may be soils in which the transitional material should be described and sampled. There are two kinds of transitional horizons. In one, the properties of both horizons are mixed, and those of one are dominant; in this case, both capital letters are used and the first letter indicates that properties of that horizon dominate (AB, BA, AC). The other kind of transitional horizon has distinct parts of both horizons; here the two capital letters are separated by a virgule (A/B, E/B, B/R).

Master horizons are further subdivided by use of both lower-case letters (Table 1.1) and Arabic numerals. The lower-case letters follow the capital letter (Bt), and certain rules pertaining to the order of lower-case letters, if more than one suffix is used, are given in Appendix 1. If the horizon is buried, b is usually written last (Btkb); however, if some soil properties are imparted after burial, the symbol denoting the property follows b (Btbk). In general, the lower-case letter sequence is a time sequence. Arabic numerals are used to further subdivide horizons identified by a unique set of letters. Such subdivision can be based on slight changes in color, structure, or any other property. The numbering starts at 1 denoting the uppermost subdivided horizon (e.g., Bt1, Bt2, Bt3). Numbering is restarted if the lower case letters change. Horizons in a particular profile could be A, AB, Bt1, Bt2, Btk, Bk1, Bk2, Bk3, and Btb. If there are several buried soils stacked vertically, place a number after the b, with 1 for first buried soil, counting down from the surface, 2 for the second buried soil, and so forth. Thus Bt2b3 is a subdivided Bt horizon of the third buried soil beneath the surface at the site of interest.

The numbering system creates a problem with the K horizon because the numbers (1, 2, and 3) were defined on specific amounts of carbonate materials enclosing grains (Gile and others, 1966). To be consistent with the numbering system used for the other horizons, however, it seems best to number consecutively, with depth, each subdivision of the K horizon.

Many unconsolidated deposits of Quaternary age consist of depositional layers of contrasting texture and/or lithology, and the soil profile extends through more than one layer. Examples are loess/till, floodplain silt/gravelly outwash, and colluvium/outwash. Such primary, or inherited, differences in texture and/or lithology are important in any soil-profile description, because of the difficulty of differentiating pedogenic from geologic layering. Each different geological layer is so noted by an Arabic numeral, counting from the top down. The numerals precede the master horizon designation, and the numeral for the uppermost layer (1) is omitted (Fig. 1.4). The details of recognizing parent material layers are covered later

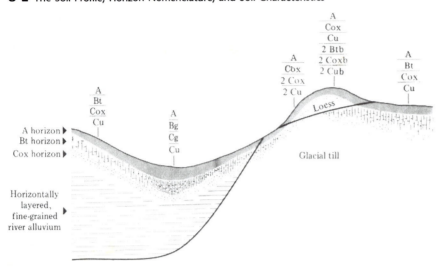

Figure 1.4 Lateral variation in soil profiles due to lateral variation in environmental conditions and lithology and age of parent materials. Soil horizons parallel the land surface at the time of their formation. Those horizons that formed from river alluvium truncate depositional layering; hence many soil properties are independent of the properties of the depositional layers. The soil formed on glacial till was partly truncated by erosion before burial by loess; it grades laterally to the right from a buried soil to an exhumed soil, because the loess has been removed by erosion in that direction.

in this chapter and in Chapter 7. However, a simple test is to compare the orientation of the soil horizons with that of both the parent material layering and the land or geomorphic surface. Soil horizons will more or less parallel the land surface and ideally crosscut depositional layering (Figs. 1.2 and 1.4). Furthermore, soil-horizon properties may change laterally, but this change is usually recognized because it is gradational and related to changing environmental conditions. In contrast, lateral changes in parent material (e.g., cutting and filling channel structures in a fluvial deposit) can be quite abrupt.

Horizon nomenclature for cumulative and noncumulative profiles has not received much attention. These are profiles in which, because of hillslope erosion and deposition associated with topographic position, a particular horizon alters to another horizon (Fig. 7.10). For example, a B horizon can convert to an A horizon or vice versa. If the properties of the previous horizon are still detectable, it might be useful to so indicate this with the horizon nomenclature. One suggestion is to use an arrow; hence, A ⇒ B would indicate a former A horizon converting to a B horizon in a cumulative soil, and B ⇒ A would indicate a former B horizon converting to an A horizon in a noncumulative soil. In contrast, if the A horizon is overthickened due only to parent material additions (eolian or hillslope deposits), Harrison (1982) suggests using

the notation Acum to denote the cumulative nature of the horizon; this could be shortened to Ac.

New Zealand soil scientists have devised an extremely useful classification scheme for soil horizons that have gley features formed in soils with high water content (Clayden and Hewitt, 1993). The definitions are more detailed than those used in the United States, and they are based on reduction–oxidation segregations and both the percentage and location of low-chroma colors. People working with wetland soils will find this classification quite useful.

Soils can be classified by their position in a stratigraphic section and in the landscape due to the geological and pedological history of the area (Ruhe, 1965; Morrison, 1978). Three soils are recognized (Fig. 1.4). Relict soils are those that have remained at the land surface since the time of initial formation; they may or may not have acquired most of their properties some time in the past. Buried soils are those that were formed on some ancient land surface, were subsequently buried by a younger deposit, and generally are far enough below the present land surface not to be affected by present pedogenic processes. Exhumed soils were formerly buried but subsequently exposed to current pedogenesis with erosional removal of the overlying material. The letter "P" preceding the master horizon designation can be used to denote exhumed soil horizons (Ruhe and Daniels, 1958).

■ Soil Characteristics

An undisturbed soil sample consists of a matrix of inorganic and organic solid particles in association with interconnected voids. Depending on local conditions, varying amounts of soil water and gases occupy the voids. I will discuss the main physical and chemical characteristics of the soil, along with an outline of water movement. Only those properties most important to field studies will be dealt with here. These and other properties are treated in more detail in introductory soil science texts, such as Singer and Munn (1987) or Buol and others (1997). Laboratory quantification of some of these properties is desirable, and one should consult soil laboratory manuals such as Jackson (1973), Page (1982), Klute (l986), and Singer and Janitsky (1986) for appropriate analytical methods. Definitions of many of the properties recognized in the field are given in Appendix 1.

Color

Color is a valuable aid, if used with caution, in qualitatively recognizing soil-horizon materials and processes that are or have been operating in a soil (Bigham and Ciolkosz, 1993). Indeed, with buried soils color is the property that first attracts attention. All colors are quantified using the Munsell notation. Consult a soil color book for the details of the system. Basically the dominant color (red, yellow, green, etc.) is the hue (2.5YR is red, 10YR is various combinations of yellow and brown, and 2.5Y is mainly yellow), and there is a page devoted to each main hue. As soils age and they change from yellow ⇒ brown ⇒ red, the hue changes from 2.5Y ⇒ 10YR ⇒ 7.5YR ⇒ 5YR. The common reddest hue is 10R. On each hue page the colors are arranged vertically in terms of the relative lightness of the color, a property called value from 1 (dark) to 8 (light). Also on the same hue page, the colors are arranged horizontally in terms of the relative purity or strength of the spectral color, a property called chroma from 1 (least vivid) to 8 (most vivid); 0 chroma is neutral. Thus, 10YR 5/6 is 10YR hue, 5 value and 6 chroma. This is yellowish brown in the U.S. system of color names (Soil Survey Staff, 1975).

Dark brown to black colors (low value and chroma) in near-surface horizons reflect an accumulation of humified and/or nonhumified organic matter. Value decreases with greater contents of organic matter (Schulze and others, 1993). Dark colors may also result from the accumulation of MnO_2, but these usually have a bluish cast and are not always close to the surface.

White or light-gray colors can have several origins. If these occur between the O or A horizon and the B horizon, an E horizon is suggested. In this case there is enough leaching by vertically or laterally moving water so that most of the grains are free of colloidal coatings of oxides and hydroxides of aluminum and iron, and of organic matter; these materials might be found at depth in a B horizon. In humid climates local concentrations of gibbsite in the B horizon can be white. White color below a Bw or Bt horizon in aridic areas is usually due to concentrations of carbonate or gypsum.

Brown to red colors generally denote some kind of pedogenic iron. In unaltered rock or minerals, Fe is usually in the +2 form. Under weathering in oxidizing conditions, several common iron oxide minerals form. Goethite (FeOOH) is characterized by hues of 7.5YR–2.5Y and hematite (Fe_2O_3) by hues of 7.5R–5YR (Schwertmann, 1993; see also Table 4.3). However, coarse-grained hematite (several micrometers diameter) can be reddish purple (5RP–10RP; see also Blodgett and others, 1993). Under ideal conditions one can estimate proportions of the minerals making up the dominant Fe oxide pigments. A relatively new Fe oxide is ferrihydrite ($Fe_5HO_8\cdot4H_2O$), formed in Bs horizons (Schwertmann, 1993); its distinguishing color is 5YR–7.5YR, with a value ≤6.

Alternating oxidizing and reducing conditions (redoximorphic or gleying conditions or features) result in soils that reflect the removal of Fe pigments under reducing conditions and the precipitation of Fe pigments during oxidizing conditions. If the entire pigment is removed, the color of the remaining soil is seen (gray, etc.), but where Fe oxides are precipitated the color mainly of lepidocrocite (FeOOH; 5YR–7.5YR and value ≥6 according to Schwertmann, 1993) is seen. Commonly under these conditions two or more colors occur in an intricate pattern, termed mottled. The formation of these color patterns is discussed more fully in Chapter 5, and they are described in Appendix 1.

Intensity of color gives a measure of the amount of pigmenting material present, but not a very accurate one. The reason for this is that the texture of the material (particle-size distribution) greatly influences the amount of surface area that has to be coated to impart a certain color on the material. For example, coarse-grained material has a much lower surface area per unit volume than does fine-grained material (Table 1.2), so it would take much less pigment to impart a certain color to a coarse-grained soil than it would to a fine-grained soil. This effect is seen well in the relation between value and organic matter content, because for a given organic matter content the finer-grained soils have a higher value (lighter color) than the coarse-grained soils (Fig. 5.4 of Schulze and others, 1993).

Texture

Soil texture depends on the proportion of sand, silt, and clay sizes, based on the inorganic soil fraction (organic and cementing materials removed) that is less than 2 mm in diameter. Particle-size classes used in pedology are given in Table 1.2; the U.S. system will be used here. However, this is not the only system in

use; some investigators use 4.9 μm as the upper limit for clay. Specific combinations of sand, silt, and clay define soil textural classes (Fig. 1.5). Mechanical analysis is needed for textural classification, but an approximate classification can be made by the use of simple field tests (Appendix 1).

The variation in particle size originates in several ways. One is inheritance from the parent material. This can include all particle sizes. Second is mechanical weathering, in which any particle in a particular class is produced by weathering of the next-larger class. Examples are sand derived from the weathering of fine gravel, silt derived from the weathering of sand, and clay from the weathering of silt. Third is the atmospheric additions of solids to the soil. Sand could be so delivered if close to a source, but more commonly silt and clay are added to the soil surface and these can be from distant source areas hundreds to thousands of kilometers away. Fourth is neoformation, usually clay-size material formed by the precipitation of the constituent elements. The clay-size fraction consists of both

Table 1.2 Particle Size Classes Used in Pedology and Some of Their Properties

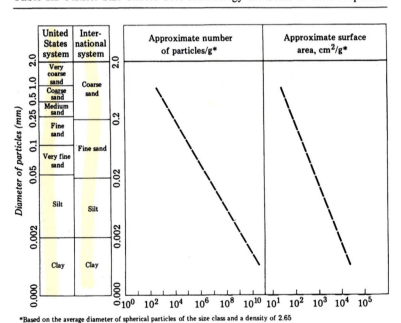

*Based on the average diameter of spherical particles of the size class and a density of 2.65

Modified from Black (1957), © John Wiley & Sons.

Figure 1.5 Soil textural classes plotted on a triangular diagram (Soil Survey Division Staff, 1993, Fig. 3–16). Dashed lines are values for moisture equivalent calculated from the equation, moisture equivalent = 0.023 sand +0.25 silt +0.61 clay. Moisture equivalent approximates field capacity; thus, the plot gives a general relation between soil-texture classes and field capacity for soils low in organic matter. *(Redrawn from Bodman and Mahmud, 1932, Fig. 3, ©1932, The Williams & Wilkins Co., Baltimore, and Jenny and Raychaudhuri, 1960, Fig. 5.)*

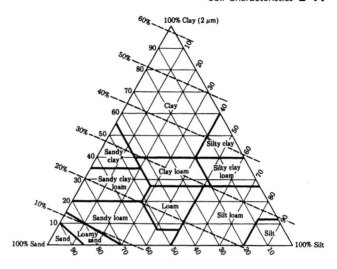

layer-lattice clay minerals and other crystalline and amorphous materials. Because clay refers to both a size fraction and a suite of minerals, the term "clay" will be used here for all material that is less than 0.002 mm (2 μm) in diameter, and "clay mineral" will be used for the layer-lattice clay minerals (Chapter 4). Because some workers use other size limits for clay (4.9 μm), the size used is usually stipulated in the data tables.

Texture is one of the more important characteristics of a soil profile. The variation in texture from horizon to horizon can be used to decipher the pedogenic and geological history of a soil and associated geomorphic surface. The down-profile variation in clay is one of the more important properties used in estimating ages of soils. The fine-grained fraction also affects many processes operating within a soil because the surface area per unit volume increases markedly as particle size decreases (Table 1.2). Soils with large internal surface areas are more chemically active than are soils with low surface areas because of their greater charge per unit volume and their capacity to hold greater amounts of water by adsorption. For these and other reasons, fine-grained materials weather rapidly. Many other soil properties, such as organic matter content, nutrient content, and degree of aeration, are closely related to soil texture.

Organic Matter

Organic matter is found in varying amounts in mineral soils and is almost always most concentrated in the uppermost horizon. A wide spectrum of materials makes up the soil organic matter, which ranges from undecomposed plant and animal tissue to humus, the latter being defined as "colloidal soil organic matter that decomposes slowly and colors soils brown or black" (Singer and Munn, 1987; see Stevenson, 1982, for details). Humus commonly makes up the bulk of the soil organic matter. Large amounts of CO_2 are evolved during its formation. Organic carbon makes up over one-half of the organic matter, and organic carbon content is commonly used to characterize the amount of organic matter in soils. Generally, the percentage of organic matter in a soil is considered to be approximately 1.724 times the percentage of organic carbon. The C:N ratio of a soil is a rough measure of the amount of decomposition of the original organic material and the steady-state values are related to environmental conditions. The ratio varies from 5 to 45 in animals and microbes, 8 to 80 in herbaceous materials, and 45 to 400 in trees (Table 6.2 of Singer and Munn, 1987). A ratio for the A horizon near 20 seems to separate forested soils (>20) from nonforested soils (<20) (Chapter 10).

Soil organic matter is important to many soil properties, particularly to the formation of surface soil structures and to reactions that go on during pedogenesis (Ugolini and Sletten, 1991). It considerably increases both the water-holding capacity of mineral soils and the cation-exchange capacity. The organic acids that are produced promote weathering and form chelating compounds that increase the solubility of some ions in the soil environment. The CO_2 that is evolved builds up to reach concentrations much higher than those in the atmosphere (Brook and others, 1983); this results in the formation of abundant carbonic acid, which lowers the soil pH and promotes weathering.

Structure

Structure involves a bonding together into aggregates of individual soil particles (see review in Australian Journal of Soil Research, v. 29, no. 6). Individual aggregates (peds) are classified into several types on the basis of shape (Table 1.3), and the surfaces of the peds persist through cycles of wetting and drying. Structure type has an association with particular soil horizons.

Organic matter is important in the formation of spheroidally shaped structures, as is clay content in the formation of blocky, prismatic, and columnar structures. Some soils in tropical regions, however, are low in organic matter, yet they are well aggregated; this appears to result from cementation by pedogenic iron. Finally, high gravel content seems to impede the development of structure.

Structure is important to the movement of water through the soil and to surface erosion. A-horizon structures, although they vary from soil to soil, tend to produce larger-sized pores than would be the case for a structureless surface soil. These larger pores allow the soil to take up large amounts of rainwater over a short period of time (high infiltration rate), and thus

Table 1.3 Description and Probable Origin of Soil Structure

Type	Sketch[a] and Description	Probable Origin[b]	Usual Associated Soil Horizon
Granular	Spheroidally shaped aggregates with faces that do not accommodate adjoining ped faces	Colloids, mainly organic, bind the particles together; clay and Fe and Al hydroxides may be responsible for some binding, and flocculating capacity of some ions, such as Ca^{2+}, may be helpful; periodic dehydration helps form more stable aggregates	A
Angular blocky / Subangular blocky	Approximately equidimensional blocks with planar faces that are accommodated to adjoining ped faces; face intersections are sharp with angular blocky, rounded with subangular blocky	Many faces may be intersecting shear planes developed during swelling and shrinkage that accompany changes in soil moisture	Bt
Prismatic / Columnar	Particles are arranged about a vertical line, and ped is bounded by planar vertical faces that accommodate adjoining faces; prismatic has a flat top, and columnar a rounded top.	Faces develop as a result of tensional forces during times of dehydration; rounded column tops may be due to some combination of erosion by percolating water and greater amounts of upward swelling of column centers on wetting	Bt / Bn
Platy	Particles are arranged about a inherited horizontal plane	May be related to particle size orientation from parent material or induced by freeze–thaw processes	E, or those with fragipan
		May be related to layering in cementing material, induced during its precipitation(carbonate, silica, Fe hydroxides)	Km, Bqm, Bs

[a]Taken from Soil Survey Staff (1975).
[b]From Baver (1956), Black (1957), Rode (1962), and White (1966).

reduce the possibility of runoff and surface erosion. Remove the A horizon, however, and erosion due to greater runoff may ensue if the infiltration rate in the exposed B horizon is less than that of the former A horizon. The fact that many structural aggregates are water stable is important because it means that the percolating waters are fairly free of clay particles. However, aggregates unstable in water can break down and contribute clay to the percolating soil water. This tends to plug some of the pores, with a concomitant decrease in infiltration rate. B-horizon structures provide avenues for the translocation of water and any contained solids along the ped interfaces. Indeed, this is where most clay films are located.

Vesicular structure is not shown in Table 1.3, although it is common to fine-grained (silty) desert and arctic A horizons. This structure is characterized by a large volume of near-round small voids that in appearance are not unlike the vesicles of some basalts. In hot deserts, they appear to form when soil–air is entrapped and expands as the soil is wetted (Springer, 1958; Evenari and others, 1974); but a different origin is postulated for those in arctic regions (Bunting, 1977).

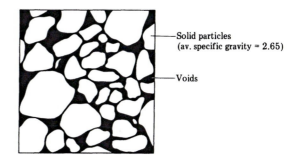

Figure 1.6 Sketch of soil sample to show solid particle and void space distribution. The mineral grains are quartz and feldspar, so 2.65 is used for the average particle density. Bulk density and porosity are calculated as follows:

Weight of oven-dry soil: 63 g
Volume of soil in field: 35 cm^3

$$\text{Bulk density} = \frac{\text{weight}}{\text{volume}} = \frac{63}{35} = 1.8 \text{ g/cm}^3$$

$$\text{Porosity (\%)} = \left(1 - \frac{\text{bulk density}}{\text{particle density}}\right) \times 100$$

$$= \left(1 - \frac{1.8}{2.65}\right) \times 100 = 32$$

Bulk Density

Bulk density is a measure of the weight of the soil per unit volume (grams per cubic centimeter), usually given on an oven-dry (110°C) basis (Fig. 1.6). Variation in bulk density is attributable to the relative proportion and specific gravity of solid organic and inorganic particles and to the porosity of the soil. Retallack (1990, Fig. 4.1) has reviewed much data on bulk density and provides these approximate values: most minerals and rocks are between 3.3 and 2.5; A horizons range from 2.0 to 0.7 and average 1.4; Bw, Bk, and Bt horizons range from 2.0-1.0 and average 1.6. Bulk densities are important in quantitative soil studies, and measurement should be encouraged. Such data are necessary, for example, in calculating soil moisture movement within a profile and rates of clay formation and carbonate accumulation. Even when two soils are compared qualitatively on the basis of their development for purposes of stratigraphic correlation, more accurate comparisons can be made on the basis of total weight of clay formed from (1) 100 g of parent material or (2) in 1 cm^2 of soil (surface area) 1 m thick, than on percentage of clay alone (Fig. 1.7). To convert percentage of a constituent to weight per

unit volume, multiply the weight fraction by bulk density.

Workers have taken large data sets and derived equations to estimate bulk density. One for California shows a close relationship with the organic fraction of the soil and is (Alexander, 1980)

1/2 Bulk density = 1.66 − 0.308 (% organic carbon)

Gravelley soils present problems when it comes to bulk density measurements (Vincent and Chadwick, 1994).

Comparison of Data Sets

There are several ways in which we can portray profile data so that we can compare soils from different environments (Fig. 1.8). One is graphic, the plot of the horizon data with depth. Rather than plot the property value at the position of the middle of each horizon, and draw a smoothed line through these points, I prefer a plot in which the horizon value is extended

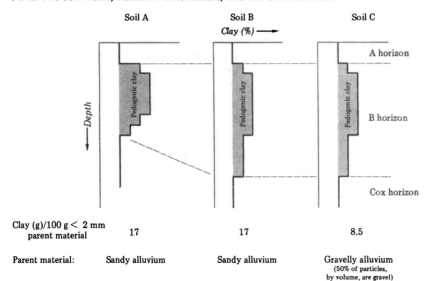

Figure 1.7 Comparison of three soil profiles based on both percentage and mass of clay formed. Assume no clay in the parent material. Soils A and B have identical parent materials and mass of pedogenic clay, but clay is concentrated in a thinner horizon in A than in B; soil A probably would be classed as the more developed soil, therefore. Soils B and C have identical profiles of percentage clay, but soil C has formed in alluvium with 50% gravel (particle-size data commonly are for the <2 mm fraction). Because gravel contributes little to clay formation, yet makes up 50% of the volume, the mass of clay in soil C is one-half that in soil B (see text for calculation of gravel-corrected values).

vertically as a straight line for the depths of each horizon, and these connected by a horizontal line to give a box-shaped plot. These are the common plots used, but usually they do not take into account the gravel content. It is difficult to compare many soils in this manner or to formulate regression equations. A second method, and the preferred one, is to calculate the mass using the bulk density, taking into account both the parent material amounts and the gravel content. A third way is to calculate the accumulation index (Levine and Ciolkosz, 1983). To do this first identify the parent material amounts of the constituent for each horizon, then subtract this from the present horizon value. The latter is the pedogenic amount and is multiplied by the horizon thickness, and all the horizon data summed for a profile value. Gravel content should be accounted for. Accumulation index allows one to compare large data sets where bulk density data are not available. A fourth way is to calculate the weighted mean values of the constituent. In one study

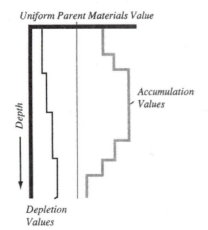

Profile Laboratory Data Indices

(a) Weighted Mean Percent:
$$\frac{\Sigma\,(\%\times \text{Thickness}) \text{ for each horizon}}{\text{profile thickness}}$$

(b) Accumulation Index:
   (i) Σ (% for each horizon − Parent material %) × Horizon Thickness
   (ii) Sum Positive Values

(c) Depletion Index:
   (i) Same as (b,i)
   (ii) Sum negative values

(d) Mass Accumulation:
   (i) (% constituent × bulk density) −
      (% Parent material consituent × bulk density) for each horizon
   (ii) Multiply times horizon thickness
   (iii) Sum horizon data for profile value

Figure 1.8 Several different ways to portray soil data.

in Mexico on the buildup of pedogenic carbonate with time (Slate and others, 1991), the fit is the nearly the same for weighted mean percentage carbonate, mass of carbonate using an estimate of the bulk density, and my calculation of the accumulation index of carbonate. Hence, any of these values could be used in quantitative pedogenic comparison studies.

The gravel content will alter the true value of each of the above portrayals of soil data and should be taken into consideration in any comparative studies. We use percentage gravel volume estimated in the field, as that is the space occupied by the gravel. For example, if the profile mass accumulation for the <2 mm fraction and the field thickness is 25 g/cm^2, and the gravel percentage is 75, the true mass is (25)(1 − 0.75), or 6.25 g. These will be called gravel-corrected values.

Soil Moisture Retention and Movement

Various amounts of water occupy the pore spaces in a soil. This water is held in the soil by adhesive forces between organic and inorganic particles and water molecules and cohesive forces between adjacent water molecules. Thin films of water are held tightly to the particle surfaces and are relatively immobile, whereas thick films are more mobile and the outer part of the water film can migrate from particle to particle both laterally and vertically.

Several soil moisture states are recognized (Fig. 1.9). One can start with a soil devoid of water and begin to add water from the top. Initially, the soil may have its pores saturated with water, but because the outer part of the water film is under low surface tension, this more loosely held water migrates downward, under the influence of gravity, as a wetting front that more or less parallels the ground surface. After 2 or 3 days, redistribution of water ceases for the most part, and the forces that hold water films on the particle surfaces equal the force of downward gravitational pull. At this point, the soil is said to have a water content at field capacity (Fig. 1.9). At field capacity, water can be removed from the soil by evaporation from the surface and transpiration through vegetation, the latter probably being the more important mechanism for water removal from the soil profile. As roots remove water from the pores, the water film becomes thinner and is held by ever stronger forces of attraction until a point is reached at which the water is held so tightly by the particles that the roots can no longer extract it. The water content under these conditions is the permanent wilting point. Under field conditions many soils seldom obtain a water content less than that of permanent wilting point. Available water-holding capacity is the difference between field capacity and permanent wilting point; if given in water depth units, it is the amount of water required to wet a given thickness of soil from permanent wilting point to field capacity.

It is difficult to determine the above moisture contents in the field. Joint laboratory and field studies, however, suggest that water held at 15-atmospheres tension approximates permanent wilting point for a variety of broad-leaved plants, and that held at one-third-atmospheres tension approximates field capacity, although this is not always the case (Salter and Williams, 1965). These are the usual moisture contents reported in the soil literature, and the values are given as percentage water (P_w):

$$P_w = \frac{H_2O \ (g)}{\text{Oven-dry soil (g)}} \times 100$$

Using moisture percent and bulk density data, one can estimate the depth to which water will wet a soil (Fig. 1.10). Arkley (1963) developed a method for calculating annual water movement within a soil, taking into account the water-holding properties of the soil and the seasonal distribution and amount of precipitation and potential evapotranspiration (Fig. 10.3). One positive test of the appropriatness of this approach is that he used the data to calculate long-term movement of carbonate and obtained reasonable ages for the soils. These are important data because percolating water redistributes clay and more soluble constituents downward in the soil or through the entire soil. Thus, many soil properties relate to the depth of wetting. McFadden and Tinsley (1985) have used this approach to calculate the depth of carbonate accumulation with time in a changing climate, and obtained results that mimic the field distribution of carbonate (Chapter 10). Finally, in working with large data sets one is hindered because of the lack of water movement data for many soils. For comparisons of these one might use the leaching index of Arkley (1963; see Table 11.1).

Soil-moisture retention and movement are strongly related to the surface area per unit volume of the soil mass, and this, in turn, is related to the clay and organic matter content. Here are approximate available water-holding capacities for a 10-cm-thick layer of

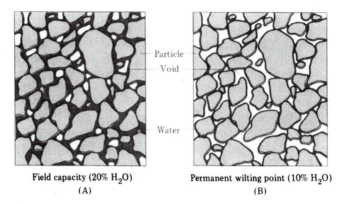

Field capacity (20% H_2O)   Permanent wilting point (10% H_2O)

(A)   (B)

Figure 1.9 Diagram of water in soil at field capacity (A) and permanent wilting point (B) and an example of how to calculate the amount of water present at these two moisture states. The calculation for the amount of water held in 100 cm of soil at the two moisture states is as follows (bulk density = 1.5; P_w = moisture %; D = soil horizon thickness; d = amount of water held in soil (equal to vertical thickness in centimeters): At field capacity (20 g H_2O/100 g soil or 20% H_2O)

$$d = \frac{P_w}{100} \times \text{bulk density} \times D = \frac{20}{100} \times 1.5 \times 100 = 30 \text{ cm } H_2O$$

(30 cm of water are held in 100 cm of this soil at field capacity)

At permanent wilting point (10 g H_2O/100 g soil or 10% H_2O)

$$d = \frac{10}{100} \times 1.5 \times 100 = 15 \text{ cm } H_2O$$

(15 cm of H_2O are held in 100 cm of
this soil at permanent wilting point)

Available water-holding capacity is the difference between the above two water contents, or 15 cm of H_2O.

soil: 1.4 cm for clay, 1.7 cm for silt loam, and 1 cm for sandy loam. Thus, for a given rainfall, sandy soils are wetted to greater depths than are more heavily textured soils. Gravelly sands wet even more deeply than sands because gravel-sized particles have a low surface area per unit volume and basically just take up space. Thus, if a sandy soil wets to a particular depth, a soil with a similar texture but with 50% gravel should wet to twice that depth.

Because of this close association of moisture content with content of colloid-size material, many workers have tried to correlate the two with varying success (Nielsen and Shaw, 1958; Jenny and Raychaudhuri, 1960; Salter and Williams, 1967; Clayton and Jensen,

1973). One could use some of these data to approximate soil-moisture conditions (Fig. 1.5).

Because of the need for water movement data for the gravelly soils of the U.S. desert, Harden (1988) undertook a study of this under controlled conditions. The soils started out dry, and known amounts of water were applied either with an infiltrometer or during the first rain of the season, and the depth of wetting noted. She then calculated the volumetric water-holding capacity (vWHC), defined as the amount of water (in centimeters) applied to a dry soil, divided by the depth of water penetration (in centimeters) after the downward movement of water has ceased or decreased significantly. For soils with ≥50 weight

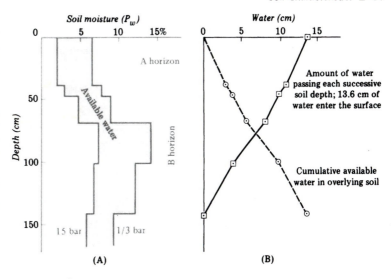

Figure 1.10 (A) Variation in soil-moisture data with depth in a Snelling sandy clay loam (data from Arkley, 1964). (B) Plot of cumulative available water and the amount of water passing through the soil for a 13.6 cm rainfall, calculated by the method in the caption for Figure 1.9. The soil starts at permanent wilting point, 13.6 cm of rainfall is added at the surface, and the water redistributes at depth until the soil to 142 cm is at field capacity. Downward migration would be as a diffuse, near-horizontal wetting front, with the soil above near field capacity and that below near permanent wilting point.

percentage sand and ≤40 weight % gravel, the relation is

$$\% \text{ vWHC} = 16 - 0.09 \text{ (sand\%)} - 0.05 \text{ (gravel\%)}$$

So, for a soil with 20% sand and 30% gravel, vWHC is 12.7, meaning that 1 cm of rainfall will wet to 12.7 cm depth.

Abrupt changes in texture between sedimentary layers have a marked effect on the vertical movement of water. Obviously, water movement is impeded where coarse material overlies fine material. However, if fine material overlies coarse material (for example, loam over sandy gravel), water movement also is impeded and the water content must increase to greater than field capacity before it moves into the coarser material (Aylor and Parlange, 1973; Stuart and Dixon, 1973; Clothier and others, 1977). Water hanging up in this manner has important implications for soil morphology. For example, carbonate can accumulate at these boundaries, or gleying features can be produced.

Some features within the soil can inhibit water movement. Where soil surfaces are exposed to rainfall, perhaps in a desert with low plant cover, a thin crust or seal may form (Mualem and others, 1990). This crust can reduce the infiltration rate and promote runoff. This could foul water-movement calculations based on climatic data (e.g., Fig. 10.3), because the latter require that all of the rainfall enters the soil column. Seals are not restricted to aridic soils, however (Chiang and others, 1994). Gypsum-enriched horizons can alter the water movement in another way (Reheis, 1987). Gypsum is hydroscopic, so absorbs water in greater amounts than the usual mineral grains. This can greatly alter the water movement with depth. As she points out, it is possible that if the gypsum content is sufficiently high, and climate changes to a wetter mode, the gypsum horizon could absorb water and prevent its own leaching.

Capillary rise from a shallow water table may deliver water and soluble salts to the overlying horizons. In theory, the smaller the pores the higher the water will rise, but irregularly shaped soil pores hardly make ideal capillary tubes. Rode (1962) suggests such rise may be 1 m or less in sands and 3 to 4 m in clays. For soils low in clay content, he gives an empirical formula that closely matches the observed data:

$$\text{Capillary rise (mm)} = \frac{75}{\text{particle diameter (mm)}}$$

Although his data are for restricted particle sizes, perhaps they would work for the mean size of a soil. Parts of the soil profile at distances above the water table greater than 3–4 m above the water table should not be influenced by capillary rise. Soluble salts or iron or manganese compounds might accumulate at the top of either the water table or the capillary fringe. Few criteria have been developed, however, to determine if such accumulations are produced by upward- or downward-moving water. If the accumulations are parallel with the ground surface and bear some consistent relationship with the other soil horizons, downward-moving water is suggested. If, however, the accumulations cut across soil horizons, and there is no reasonable depth relationship with the ground surface, derivation from a water table is suggested. If salts of different solubilities are present, the positions of their maximum concentrations may be used to determine the direction of water movement. For example, in a system characterized by downward-moving water, the maximum concentrations of the more soluble salts occur at progressively greater depths; a Bk horizon would overlie a By, which, in turn, would overlie a Bz. Perhaps these horizon positions would be reversed if the waters were moving upward by capillary action.

It is important to note that field capacity and permanent-wilting-point moisture conditions describe the movement of water and not necessarily the amount of water involved in weathering reactions within the soil. Weathering occurs at all water contents, because a thin water film is always in contact with mineral grains. If field capacity is often reached, the ions in the water film released by weathering may be constantly flushed from the soil. If, however, the film is thin and field capacity is seldom reached, ionic concentrations in the water film may approach saturation, and this would inhibit further weathering unless periodic flushing occurs to lower ionic concentrations.

Cation-Exchange Capacity, Exchangeable Cations, and Percentage Base Saturation

Most soil colloids, both inorganic and organic, have a net negative surface charge, the origin of which is discussed in Chapter 4. Cations are attracted to these charged surfaces. The strength of cation attraction varies with the colloid and the particular cation, and some cations may exchange for others. The total negative charge on the surface is called the cation-exchange capacity, and it is expressed in milliequivalents (meq) per 100 g oven-dry material (cmol/kg in SI units). The exchangeable cations are those that are attracted to the negatively charged surfaces. Base saturation is expressed as the percentage of base ions (nonhydrogen) that make up the total exchangeable cations. Thus, there generally is a close relationship between base saturation and pH. Exchangeable

sodium percentage is the amount of sodium relative to the total exchangeable cations, and this is important in defining a natric horizon.

Soil pH

Soil pHs have an extreme range of 2 to 11, but most soil pHs range from 5 to 9. Soil pH is dependent on the ionic content and concentration in both the soil solution and the exchangeable cation complex adsorbed to the surfaces of colloids. In general, the ionic concentration increases from a low in the soil solution to a high at the colloid surface (Fig. 5.7). Furthermore, there is an equilibrium between the ions in solution and the exchangeable ions, and the ions are present in about the same proportions in both environments. Ions commonly present are Ca^{2+}, Mg^{2+}, K^+, Na^+, H^+, Cl^-, NO_3^-, SO_4^{2-}, HCO_3^-, CO_3^{2-}, and OH^-. The relative proportion of these ions helps determine the soil pH (Fig. 1.11; Singer and Munn, 1987).

Hydrogen ions are derived from rainfall and from organic and inorganic acids produced within the soil (see discussion in Ugolini and Sletten, 1991). Rainfall pH varies from 3.0 to 9.8 (Carroll, 1962). Pure water, in equilibrium with atmospheric CO_2 at 25°C, should have a pH of 5.7. Values lower than this are thought to be due partly to atmospheric pollution, whereas higher values are attributed to salts derived both from windblown seawater along coasts or from windblown dust. Carbonic acid is formed within the soil by the combination of CO_2 and water. A pH below 5.7 is possible because CO_2 content up to 10 or more times greater than atmospheric is possible because of respiration by plant roots and by microorganisms (Brook and others, 1983). An important source of hydrogen is the wide variety of organic acids produced within the soil. Still other sources are the exchangeable Al^{3+} and Al-hydroxy ions, which release H^+ during hydrolysis

$$Al^{3+} + H_2O \Rightarrow Al(OH)_2^+ + H^+$$
$$\text{(under very acidic conditions)}$$
$$Al(OH)_2^+ + H_2O \Rightarrow Al(OH)_3 + H^+$$
$$\text{(under less acidic conditions)}$$

The exchange complex in acidic soils is dominated by H^+, Al^{3+}, and Al-hydroxy ions, and base content is low. As cation content increases, however, they replace H^+ and Al-hydroxy ions, and OH^- concentration and pH increase. The alkalinity will depend on the strength of the base formed. For example, $Ca(OH)_2$ is formed in the presence of $CaCO_3$ and the resulting pH can approach 8.5. In contrast, $NaCO_3$ and $NaHCO_3$ form $NaOH$, a stronger base, and results in pH values over 8.5.

One final point on pH is that the pH of a given soil is not uniform throughout. There may be slight variations from place to place due to slight variations in CO_2 or organic acid type or concentration, the content and composition of the exchangeable bases, or the presence of nearby roots, since these commonly contain adsorbed H^+. Thus, although most work on weathering and mineral stabilities uses soil pH values, it should be remembered that these values reflect average conditions and may not reflect conditions where minerals are being weathered or synthesized.

Soil scientists obtain pH values in several ways (Janitsky, 1986). One uses a field kit, and these provide reliable values. The other uses a pH meter in the laboratory. Several soil:H_2O ratios are used, but if the water content is very high the pH will be higher because the solution is more dilute. Sometimes the solution added to the soil for pH measurement contains a salt of known concentration, such as KCl or $CaCl_2$. This is done to mimic concentrations in natural waters and to replace some of the H^+ on the exchange complex. pH measurements using the salt solution are always lower than those measured in water.

In Soil Taxonomy (Chapter 2) base saturation is used to as an aid in identifying the diagnostic horizons and certain taxonomic units. For example, the most common one is 50% base saturation for the mollic epipedon (approximately equal to the A horizon). Base saturation data are difficult to obtain, but because of the relation between base saturation and pH, the latter can be used as a surrogate for the former. In general 50% base saturation occurs at a pH between about 5 and 6 (Fig. 1.11). For each region one could establish a general pH-base saturation relation, and thereafter use pH for classification.

■ Comparing the Development of Soils: Use of Soil Indices

Degree of soil-profile development is used as a qualitative measure of the amount of pedologic change that has taken place in the parent material. It has been used in Quaternary stratigraphy where soils were used to correlate unconsolidated deposits (Richmond, 1962; Morrison, 1964). The ranking is generally on a relative scale, based on the properties of a sequence of soils in an area. A more quantitative scale would be useful, however, because soil of the same age may vary in its development from place to place due to variations in soil-forming factors. The following scheme is modified from Retallack (1990), and although he uses it for paleosols (Chapter 11) it could be used for surface soils.

A very weakly developed soil is one with an A horizon and perhaps a Cox horizon. If pedogenic carbonate is present, carbonate morphology is stage 1 (see Appendix 1).

A weakly developed soil profile is one with an A/Bw/Cox and/or Bk horizon sequence. Accumulation of pedogenic constituents is not sufficient to qualify the horizons as argillic, spodic, or calcic (see Chapter 2). If carbonate horizons are present, morphology is probably stage I.

A moderately developed soil profile is one with an A/Bt/Cox, A/E/Bs/Cox, or A/Bt/Bk/Cox horizon sequence. The Bt, E, Bs, and Bk horizons meet the cri-

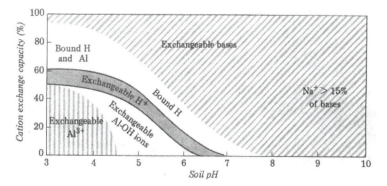

Figure 1.11 General relation between pH and exchangeable cations. Precise values vary from soil to soil for many reasons. Bound H^+ is that H^+ held so tightly to colloidal surfaces that little of it is exchangeable. (*Redrawn mostly from Buckman and Brady, 1969, Fig. 14.1, ©1969, The Macmillan Co.*)

teria for argillic, albic, spodic, and calcic horizons. Carbonate accumulations would display stage II morphology.

A strongly developed soil profile is similar to a moderately developed one with these exceptions: the B horizon in the strongly developed profile is generally thicker and redder, contains more clay, and has a more strongly developed structure (could be kandic horizon, Chapter 2); if carbonates are present, they would form a K horizon (stage III and higher morphology).

The distinctions between a moderately and a strongly developed profile are qualitative, but they could be quantified on color and texture. However, it is difficult to compare soil development on a loess with 20% primary clay with that on a gravelly outwash with 5% primary clay. As regards color, there is no fast rule. Each development rank might be accompanied by a different hue. Thus, if a moderately developed soil has a 10YR hue, a strongly developed soil would have a redder color, perhaps a 5YR hue.

Soil development can be better described by use of quantitative color and profile indices.

Color Indices and the Podzolization Index

Four color indices have been proposed and all are similar to the extent that the Munsell notation is recalculated to a single number. One was proposed by Buntley and Westin (1965); hue is converted to a number (7.5YR = 4; 10YR = 3; 2.5Y = 2; 5Y = 1), which is multiplied by the chroma. The second index, devised by Hurst (1977), also converts hue to a number (5R = 5, 7.5R = 7.5, 10R = 10, 2.5R = 12.5, 5YR = 15, 7.5YR = 17.5, and 10YR = 20), and this is multiplied by the product of the fraction, value/chroma. Third, the rubification index of Harden (1982) compares the color of each horizon with that of the parent material, and a shift in hue (change 1 hue) and of chroma are each worth 10 points (details in Appendix 2). With all indices, values are calculated for each horizon, and these can be multiplied by the horizon thickness. The fourth index is used in Europe and is $(10 - H) \times$ chroma/value, with H being the number that precedes the YR of the hue designation (Torrent and others, 1980, 1983). This latter index shows a good linear relationship with hematite percent. Rubification is the color index used most commonly in the western United States.

For the profile color index value, there are several choices. One is to sum the horizon values. However, most soils in a developmental sequence vary in thickness, and this should be taken into account by multiplying each horizon value by the horizon thickness, and summing the latter. A third way is to divide the profile sum by the thickness of the described profile, to obtain a weighted mean value. A fourth way is to artificially increase the thickness of the lowest horizon in all soils so that the total thickness of all soils is uniform. These latter two choices are preferred over the others. Whatever method is used, the Buntley–Westin and rubification indices increase with soil age, whereas the Hurst index decreases (Fig. 1.12).

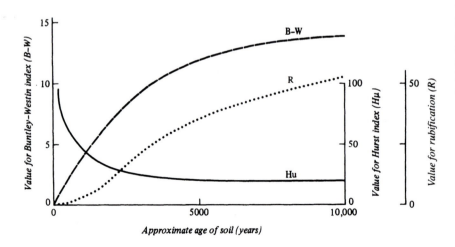

Figure 1.12 Development of color index values with time for a sequence of soils in the alpine region of the Wind River Range, Wyoming. (*data from Miller and Birkeland, 1974.*)

The podzolic development index (POD index) of Schaetzl and others (1988) is an example of using the Munsell system to rank the development of specific soils (well developed podzols would be Spodosols in Soil Taxonomy; see Chapter 2). They subtract the value of the B horizon from that of the E horizon and multiply that by a factor that accounts for the hue difference between the two horizons. The POD index increases with the color contrast, and this could be a function of factors such as age or topographic setting. Similar indices could be devised for other profile properties.

Profile Development Indices

Two indices have been developed to assess profile development based on field properties (Bilzi and Ciolkosz, 1977; Harden, 1982, with later additions by Harden and Taylor, 1983). In both schemes points are assigned for the buildup of particular pedologic properties (e. g., with Bilzi–Ciolkosz, a change in textural class is worth 1 point, a change in structure grade 1 point, and an abrupt boundary 3 points). Although the Bilzi–Ciolkosz index compares the properties of a horizon with those of the deepest C horizon, the index of Harden compares horizon properties with those of the assumed parent material. Because parent material properties might not be represented at depth in the profile, one might have to seek them elsewhere. For example, in working with a river–terrace soil chronosequence, one could use the properties of sediments of the present active floodplain as parent material. Or, if one is working with alpine tills, the parent material proxy could be unweathered till of the Little Ice Age moraines, high in the cirque.

Each of the above profile indices presents the data differently. Under the Bilzi–Ciolkosz scheme, the values of each horizon are plotted versus depth and the results are presented graphically. However, one could sum the horizons or repeat the calculations given above for the color indices to obtain a meaningful number to represent the entire profile. Calculations for the index of Harden (hereafter called profile development index or PDI; also referred to as soil development index or SDI) are more involved (Appendix 2). After obtaining a number for each property for each horizon, property values are normalized by dividing the latter by the maximum possible value for that property. This provides a scale ranging from 0 to 1. All the individual property values per horizon are then summed, the total is divided by the number of properties used, and the latter is multiplied by the horizon thickness.

One can then sum the horizon data for each profile, the usual stopping point. However, because described profiles in a sequence (e.g., chronosequence or toposequence) are usually of different thicknesses, one can adjust for this in the ways suggested for the color indices. If the profile data are divided by either the described thickness or if all profiles are adjusted to a uniform depth (e.g., artifically extended to the depth of the thickest soil in the sequence) and then divided by that depth value, the weighted mean PDI is obtained. These latter values range from 0 (no development) to 1 (maximum development). One advantage of the weighted mean PDI with a uniform artificial depth is in comparing thin, young soils with thick, old soils. Let us say the old soil has 3 m described thickness (Cu horizon not reached) and has a value of 0.65. The young soil could be a 0.5 m thick to the Cu horizon, and when the PDI is divided by the described thickness, have a weighted mean PDI of 0.20. However, if the Cu horizon is artifically extended to 3 m, one then includes 2.5 m of zero values, for a final weighted mean PDI value of 0.03. Only in this way can one compare soil development over the same depth interval. The danger in doing this is that one assumes that the properties of the deepest horizon extend to the artificial depth used, an assumption that may not be correct.

Whichever profile index or method is used, higher values denote greater pedologic development (Table 1.4; also for the Great Valley of California, compare PDI data of Harden, 1982, with Bilzi–Ciolkosz data of Meixner and Singer, 1981). These data also could be used to quantitatively define stages of profile development. Finally, if one wishes to use soils to correlate between deposits in two areas using these indices, one could normalize the totals in each area (divide the profile totals by the maximum value in each area). This may produce similar values for soils of roughly similar ages.

Reheis and others (1989) discuss the variability in PDI both with a greater number of soils described per geomorphic surface and with different investigators. To keep descriptions uniform, we usually have the same person describe all the soils within one study area.

Table 1.4 Profile Indices and mIPA Data for Holocene Soils, Ben Ohau Range, New Zealand

			mIPA values[a]				
Approximate Age (Years)	PDI		pH	Organic Carbon	Al	Fe	P
100	0.02[b]	0.01[c]	0.10	0.1	0.8	0.3	0.04
3000	0.28	18.3	0.16	13.8	2.0	0.3	0.47
4000 (range of three profiles)	0.24–0.37	17.7–27.5	0.06–0.17	12.9–21.5	2.9–4.9	0.6–1.6	0.55–0.93
9000 (range of two profiles)	0.28–0.41	20.7–30.6	0.17–0.19	16.8–49.5	5.2–16.0	1.3–6.8	0.42–0.79

[a]Organic carbon is by the Walkley–Black method; Al and Fe are for the oxalate extract; and P is the acid-extractable fraction (Birkeland, 1984).
[b]Profile sum divided by profile thickness.
[c]Profile sum for a fixed thickness.

Index of Profile Anisotropy

Walker and Green (1976) developed an index (IPA) that provides a single number to depict the anisotropy of a profile for laboratory data. The concept is that, ideally, at time = 0 year, soil properties are isotropic, that is, values are the same irrespective of direction (here, however, we are concerned with the variation in properties with depth). In time, soil properties change with depth, and the degree of such anisotropy increases. They defined the IPA as

$$\text{IPA} = D\,\frac{100}{M}$$

where D is the mean deviation of the sampled horizon from the overall weighted mean value (M) for a particular property for the profile. Horizon totals are summed for a profile value. I suggest using a modification of this (mIPA), which takes into account variation from the parent material:

$$\text{mIPA} = \frac{D}{PM}$$

where D is the numerical deviation of the particular property from that of the parent material (PM). The PM values are obtained as mentioned for the PDI. Val-

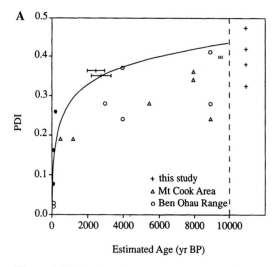

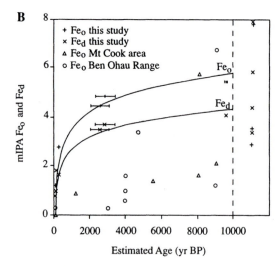

Figure 1.13 (A, B) Index data versus age of Holocene soils, Southern Alps, New Zealand. (*Redrawn from Rodbell, 1990, Figs. 4 and 6.*)

ues are calculated for each horizon and multiplied by the horizon thickness; the data for each horizon are summed to give a profile value and further recalculated as outlined for the color indices. Because anisotropy should increase with time, so should the value of the index (Table 1.4). However, with alluvial soils that are made up of contrasting depositional layers of mineral and organic particles, it is possible to have a relatively high mIPA at $t = 0$ year, for mIPA to initially decrease as pedologic mixing occurs, followed by an increase as the pedologic products overprint and replace the depositional layers.

Rodbell (1990) compared mIPA with weighted mean and PDI data in a study of Holocene soil development on moraines in the Southern Alps of New Zealand. He calculated mIPA for organic carbon, pedogenic iron, and pH, and found that they plot as logarithmic curves when plotted against time. Similar curves were derived for the weighted means of the same constituents and for PDI. All of the curves are initially steep and flatten markedly by 10 ka (Fig. 1.13; see Chapter 8 for definitions of age designations: ka, ky, Ma, My).

■ Soil Micromorphology

The soil properties listed above and in Appendix 1 are mostly macroscopic; however, much can be learned of soil development and processes by the study of soils in thin section, in much the same way that geologists study rocks (Fig. 1.14). This branch of pedology, soil micromorphology, was pioneered by Kubiëna (1970), and there are many excellent books on the topic (e.g., Brewer, 1964; Bullock and Murphy, 1983; Bullock and others, 1985; Kemp, 1985; and FitzPatrick, 1993; the latter has especially high-quality photographs of these features). This aspect of pedology is not routinely done in the United States; perhaps more emphasis should be placed on it. One reason why micromorphology is not commonly done could be that soil research is laboratory intensive. In reading though this book, one should be impressed with the enormous amount of laboratory data used to back up field-based interpretations. Doing micromorphology adds to an already long list of laboratory work, but this does not excuse it not being done.

Researchers in micromorphology have debated the best ways to describe soil features in thin section. Brewer (1964) was the basic text for years. He recognized four main pedologic units. One is voids, which owe their origin to the original sediment characteristics, faunal activity, freezing and thawing, and the entrapment of air bubbles in the surface of desert soils during rainfall (Evenari and others, 1974). A second unit is cutans, which are concentrations of soil constituents (e.g., clay or sesquioxides) by deposition on surfaces such as grains and peds. A third unit is pedotubules, which are tubular in shape, and are formed, perhaps, by faunal activity or roots and later

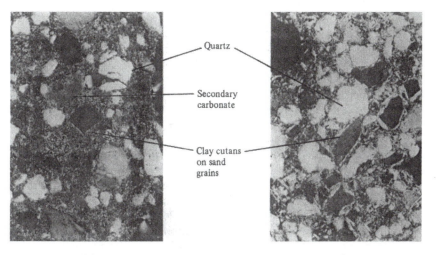

Figure 1.14 Thin sections of soil samples from the Lubbock Lake area, Texas. (A) B horizon that is 100 years old. (B) Btk horizon 3500 years old, showing greater development of clay films (cutans) than in (A). Vertical dimension is 86 μm. (*Photomicrographs by V.T. Holliday.*)

Quartz

Secondary carbonate

Clay cutans on sand grains

(A)   (B)

filled with soil material. The fourth unit is glaebules, which are more or less rounded bodies of constituents; concretions would be a familiar example.

In a 15-year international group effort, Bullock and others (1985) produced a well-illustrated handbook for the description of soil features in thin sections. The emphasis is on description, rather than interpretation. They recognize five major features. One is the groundmass, which is that part of the soil unaffected by pedological features. Examples are unaltered poorly sorted till and unaltered well-sorted dune sand. A second feature is basic mineral components. These are subdivided into a coarse fraction identifiable in thin section (≥ 10 μm), and a fine fraction difficult to identify in thin section. All original minerals and rock fragments are identified, as well as their alteration. Pedogenic minerals are also identified as well as their spatial relation with the groundmass materials. A third feature is the basic organic components, again from the unaltered to the highly altered. A fourth feature is microstructure, that structure seen at the microscope scale. Some of the more important features are the pedofeatures, those features resulting from pedogenesis. The six pedofeatures are (1) textural, such as mechanically transported silt and clay; (2) depletion, such as the loss of a chemical component from the adjacent matrix; (3) growth of crystals, such as calcite in aridic soils; (4) amorphous materials derived from nonmechanical transport, such as the precipitation of iron oxyhydroxides from solution; (5) fabric, or the spatial arrangements of soil constituents; and (6) excremental features. For the most part, jargon is avoided. For example, clay concentrations now are called coatings rather than cutans.

The first task of micromorphology is to separate inherited features from those that are pedological in origin, not an easy task. Such a study may help solve problems of parent material variation and the time and space relationship between several translocated constituents (clay coatings in relation to carbonate, as they are mutually exclusive), or to determine whether ped surfaces have clay coatings due to translocated clay or have oriented clay related to shrink–swell during dry–wet intervals. Another important piece of information would be to differentiate clay due to illuviation from that due to in situ weathering. Such studies can help in paleoclimatic studies; for example, during some climatic intervals, clay coatings might form in specific places in the soil, only to be disturbed during a subsequent climatic interval. These latter features

could be so small that identification could come only from microscopic examination.

A good illustration of the application of micromorphology to soil genesis studies was done by Bullock and Murphy (1979) on interglacial soils in England. They recognized certain features that formed during soil formation under interglacial climatic conditions. Periglacial conditions during the following glacial conditions disrupted some of these features. Although some disruption features can be seen in the field, in places thin-section evidence holds the best clues for changing pedogenic conditions.

The application of thin-section data to the interpretaion of pedological features and conditions will be incorporated into the book where appropriate.

■ Recognition of Buried Soils

Some buried soils are so obvious that few people would argue as to their pedogenic origin (Fig. 1.15); others, however, are quite difficult to differentiate from geological deposits. In field conferences one person's soil may be another's geological deposit. Part of the problem might be that some people have not spent enough time studying surface soils to really understand their properties and profile characteristics. Still another problem is that the pedological features of some thick, red colluvial soils are difficult to differentiate from the inherited features. Much of the red color and clay, for example, could be inherited.

The same criteria used to recognize and describe surface soils should be used for buried soils (Yaalon, 1971a; Valentine and Dalrymple, 1976; Catt, 1990). The first test is to trace the material laterally in outcrop to be sure it is a soil and not a deposit. The relationship between the soil horizons and bedding should be deciphered because soil horizons will truncate geological bedding if the geomorphic surface intersects the bedding at an angle (Figs. 1.2 and 1.4). An additional clue is that pedogenesis in time eradicates inherited bedding due to processes that mix soils. Also, there are predictable lateral changes in soils related to topographic position that should not be inherited (Chapter 9). A problem in places is that the soil forms from a surface that parallels geological bedding, such as the soils formed on alluvial fan deposits in the semiarid southwestern United States. Whenever deposition stopped for a long enough time, a soil could form, and it could be buried later during renewed de-

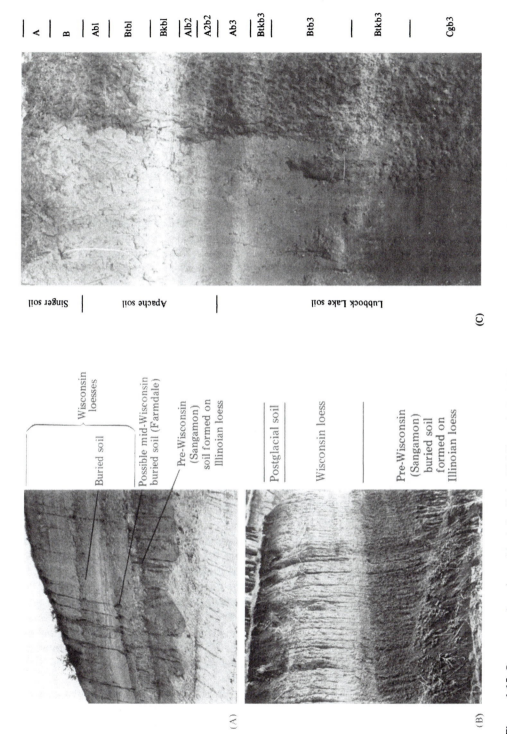

Figure 1.15 Quaternary deposits and buried soils in (A) Missouri, (B) Nebraska, and (C) Texas (latter by V.T. Holliday).

position. In this case, the soil horizons would parallel the bedding in the fan deposit. Some depositional layers in fan deposits can be poorly sorted, however, and resemble soils in their thickness, color, and texture. Thus, one has to look for pedologic features in the zone to be sure soil formation has taken place. Soil features used to recognize buried soils should persist after burial (Table 1.5; Yaalon, 1971b), and the duration of persistence of some features can be given broad age limits (Fig. 11.22). In general, the organic matter in the A horizon does not persist long after burial (Holliday, 1988), but the mineral part of the A horizon may still be present and recognizable by a slightly lower clay content than that of the B horizon, and perhaps by colors that are not so red. Generally, the buried B horizon is the most important horizon for recognizing buried soils. If it is a Bt, it will have a greater clay content, clay films, redder or browner colors, and a better developed structure than the C horizon. In drier regions, the presence of a carbonate-enriched horizon beneath the Bw or Bt horizon is helpful in identifying a buried soil. If a Km horizon with a laminated subhorizon is present, it should be laminated in its uppermost part. Moreover, carbonate horizons should have a distinct relationship to the buried land surface so that a groundwater origin can be ruled out.

One other criterion that is helpful in the recognition of a buried soil is the abruptness of the horizon boundaries (Fig. 1.3; see boundary definitions in Appendix 1). Quite commonly, the upper boundary of a well-developed soil horizon is sharper than the lower boundary, and this need not be the case in geological layering. For example, with depth in a well-developed soil profile the transition from a low-clay-content A horizon to a high-clay-content Bt horizon may be abrupt, whereas with greater depth in the B horizon there is a more gradual decrease in clay content toward the C horizon, perhaps classifying as a diffuse boundary. The same criterion holds for pedogenic carbonate horizons, that is, the upper horizon boundary is sharp and marked by a thin transition zone between the overlying noncarbonate material and the carbonate-enriched horizon, and the lower boundary is less sharp.

Many times the property that defines the horizon has the highest values in the upper part of the horizon. If a Bw horizon is present, the color is redder or browner in the upper part of the horizon, and chroma gradually diminishes with depth. In contrast, postburial diagenetic alteration within a deposit may result in a gradual decrease in color both upward and downward from the zone of maximum alteration. The same argument can be used with the distribution of carbonate with depth; highest values are usually in the upper part of Bk and K horizons.

In most soils, with experience, workers find a relationship between the development of adjacent horizons in a profile. For example, in the aridic soil-moisture regime, red, high-clay Bt horizons are associated with stage III or IV carbonate morphology. This can be used to help one decide if certain features are buried soils. However, in places wind erosion can strip away the B and leave only the K horizon, or younger sediment deposited on the K horizon may have a much less developed B horizon.

Another criterion that is helpful in the recognition of buried soils is the mineralogical characteristics of the profile (Chapter 8). Some nonclay silicate minerals become weathered and/or etched on weathering. If the zone studied is a B horizon, weathering and etching may be greater there than in the underlying C horizon and overlying deposit. Mineral depletion during weathering may be reflected by resistant to less-resistant mineral ratios that have a consistent relationship with depth. Clay minerals may also provide information on the pedogenic origin of a horizon. Quite commonly, the clay minerals that form during soil formation vary in type with depth in a profile in a predictable manner.

A more difficult problem in the recognition of buried soils comes when the upper part of the soil has been removed by erosion, leaving only that part of the profile that was below the B horizon. Here, oxidation colors help in identifying the material as part of a

Table 1.5 Relative Persistence of Soil Horizons and Properties as Possible Indicators of Former Pedologic Conditions

Least Persistent	Relatively Persistent	Persistent
A horizon	O horizon	Bo, Bt, Bn,
By, Bz horizons	A horizon	Bq horizons
pH, base saturation	E horizon	K horizon
Mottles	Bw, Bt, Bs, Bk horizons	Plinthite
	Fragipan	Duripan
	Fe content	Fe content
	$CaCO_3$ content	$CaCO_3$ content
	Clay mineralogy	Clay mineralogy

Modified, in part, from Yaalon (1971b, Table 1).

buried soil, as long as postburial alteration can be ruled out. In the midcontinent, evidence for such a history is shown by the carbonate content of superimposed loesses. The older loess may have been leached of carbonate during an interval of soil formation, the soil B horizon may have been removed during a subsequent period of erosion, and the leached loess may then have been buried by carbonate-bearing loess (Ruhe, 1968). Thus, the major remaining evidence for an unconformity and an interval of soil formation is the presence of carbonate-bearing loess overlying loess that had been leached of its carbonate prior to burial.

Stone lines may help in the location of buried soils. These are thin, more or less planar layers of stones, buried under a younger deposit (Ruhe, 1959). Several origins that are related to their positions in soils have been suggested for them (Johnson, 1990; Johnson and Watson-Stegner, 1990; Wells and others, 1990). Here those due to postdepositional processes, rather than inheritance from layered materials, are discussed. Geological and pedological origins include surface stone concentrations on eroded areas as finer materials are winnowed out, or during freeze–thaw cycles. In many desert environments such surface concentrations form desert pavement, and this is an integral part of the soil-forming processes (McFadden and others, 1987). Shrink–swell of clay-rich soils also produces stone lines, but these are at depth within the soil. Biological activity can produce stone lines at the surface where treethrow takes place in heterogeneous materials, and in the subsurface where animals such as mammals and termites move large quantities of the finer-grained material. Whatever their origin, stone lines could be part of a soil at shallow depth and serve as an indicator of a nearby surface and have a soil associated with it. With careful field work one should be able to determine whether a particular stone line formed at the surface or at shallow depth, and therefore decide if burial by a younger deposit is required. Their importance for use in locating buried soils might be best in tropical areas where deep weathering might obscure individual profiles associated with one or more stone lines. In Madagasgar, stone lines have been used to help recreate the evolution of deeply weathered landscapes (Wells and others, 1990). Some tropical stone lines have been used to infer climate change (Fairbridge and Finkl, 1984).

The depth of buried soils is important in interpreting their properties (Schaetzl and Sorenson, 1987). Depending on the soil-moisture regime, deeply buried soils can be isolated from soil-forming processes related to the present-day surface, and therefore retain many properties from the time when it was at the surface. In contrast, if a buried soil is shallow enough to be within reach of some soil-forming processes (Fig. 1.16), properties related to the surface soil can be superimposed on the buried soil. Ruhe and Olson (1980) propose the term soil welding (composite soils of Morrison, 1978) for such occurrences and discuss the physical and mineralogical properties that can be used to help identify welded soils. Welding materials will differ with the environmental setting. In a wet environment, some translocated clay can be common to both profiles (e.g., a Btbt horizon) and be the welding

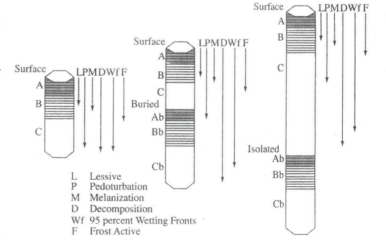

Figure 1.16 Depths to which various soil-forming processes reach in a surface soil (left), shallow buried soil (middle), and soil deep enough to be isolated from most contemporary processes (right). *(Redrawn from Schaetzl and Sorenson, 1987. The concept of "buried" vs "isolated" paleosols: Examples from northeastern Kansas. Soil Sci. 143, 426–435, Fig. 1 ©1987, The Williams & Wilkins Co., Baltimore.)*

L   Lessive
P   Pedoturbation
M   Melanization
D   Decomposition
Wf 95 percent Wetting Fronts
F   Frost Active

component; in a dry environment with shallow burial, pedogenic carbonate could be the welding material (e.g., a Btbk horizon). In places, soil-texture-related water movement can inhibit the postburial alteration of some shallow buried soils (Mausbach and others, 1982).

Soil chemical data can help with the identification and interpretation of buried soils. Constituents measured, however, should be those that persist after burial (Table 1.5). Extractable iron, aluminum, and phosphorus trends probably would persist, but pH and exchangeable cations could be altered soon after burial and give no information of the preexisting soil values. In arid regions, calcium carbonate could be dissolved and translocated to greater depth, but only in those places where soil-moisture regime increased greatly after burial. In many buried Quaternary soils, however, the preburial carbonate content and morphology below the Bt horizon are preserved.

Thin sections of buried soil horizons should be studied. In places in the field, it may be difficult to differentiate parent material features from pedologic ones, and the only real clues that the materials are soils could come from thin-section analysis. In addition, one might be able to differentiate pedofeatures formed when the soil was at the surface from those formed during burial.

Finally, features formed during diagenesis have to be dealt with when studying buried soils. Diagenesis includes the postburial chemical and mechanical processes acting on sediments and rocks during burial (Boggs, 1992). These processes take place far below the depths of soil–water penetration, but above the depths at which metamorphism takes place. These processes produce red-colored sediments, etched mineral grains, and clay minerals. Because these features are common to soils, one could argue that soils are merely a case of shallow diagenesis. The study of Walker (1967) on the diagenetic origin of red beds in desert environments is a classic, and demonstrates the similarity between many diagenetic and pedogenic features. The longer a particular material remains buried, the more difficult it will be to differentiate products of pedogenesis from those of diagenesis (Fig. 11.22).

More details on buried soils (paleosols) are given in Chapter 11.

Chapter 2

Soil Classification

We classify soil by common properties for the purposes of systematizing knowledge about soils and the processes that control similarity within a group and dissimilarities among groups. Here we will focus on Soil Taxonomy (Soil Survey Staff, 1975), the U.S. system; Yaalon (1995) discusses this and other recent systems. The number of individual soils in a group is a function of the limits allowed in the defining properties. Thus, at the lowest category of classification (series) there are many members, because many restrictions are imposed by the limits of the diagnostic properties. Different series can be grouped together at higher levels of classification, and these, in turn, can be regrouped until one ends up at the highest category (order) with a few members, each with wide limits on allowable variations in differentiating properties. In spite of the large number of members, soils in each order should share many properties in common, because they have been formed under somewhat similar soil-forming factors and pedogenic processes. To be most useful, classification should be based on soil properties, but in Soil Taxonomy climate plays a major role at certain steps in the classification. Maps of soils at any category of classification are useful for geomorphic interpretation.

■ Soil Classification at Lower Levels, and Making Geological Maps

Soil maps are prepared after the study of many soil profiles, commonly by augering. With time, it becomes apparent that the soils can be grouped on the basis of similar profile characteristics and that these characteristics can change with changes in the soil-forming factors (e.g., either different materials, or positions on a slope, or vegetation). In general, boundaries between mapping units, although commonly gradational, will be governed by one or more of the factors (Jenny, 1946; Hole and Campbell, 1985; Buol and others, 1989; Soil Survey Division Staff, 1993). The mapping of soils is complex, because as mapping proceeds a multitude of factors must be kept in mind when predicting and drawing boundaries around the mapping units. Most mapping is done with the aid of aerial photographs; these increase mapping effectiveness because subtle tonal changes are often the clue to lateral changes in soil properties and therefore the mapping unit. Many other modern techniques are becoming available for use in soil mapping (Amen and Foster, 1987; Reybold and Peterson, 1987; Mausbach and Wilding, 1991).

The soil series is the basic classification unit used in soil mapping (Soil Survey Staff, 1975; Soil Survey Division Staff, 1993), and in a sense it is similar in concept to geological formations. A soil series is a group of soil profiles with somewhat similar profile characteristics, such as the kind, thickness, arrangement, and properties of soil horizons. Series, therefore, are conceptual and are defined according to permissible ranges in the properties. Series names are derived from a local place name, such as a town or a county. Many times, changes in parent material lithology and texture are reflected in basic soil profile differences, and these have been used to differentiate series. Soil series are subdivided into soil phases, which are the mapping units used in detailed soil surveys. Phase differentiation is based on texture of the surface layer, slope steepness, physiographic position, thickness of soil profile or individual horizons, amount of erosion, stoniness, properties related to soil water (e.g., high water table, well drained), and salinity.

In the United States, soil maps are commonly published on a vertical air photograph base, and each phase is represented by a number (Fig. 2.1). Although the map gives the impression that each mapping unit is pure, and includes just that phase, in reality phases of other series commonly are included in the mapping unit, and the percentage of these other units usually is specified. These are treated as inclusions if they are not mapped as separate series at the scale used. If there

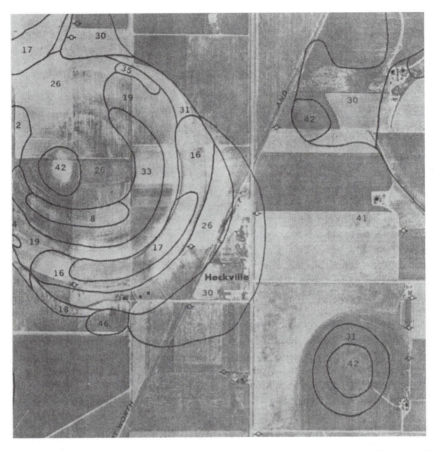

Figure 2.1 Standard soil map of Heckville, Texas. *(Reprinted from Blackstock, 1979.)* See Holliday (1995) for regional geological setting. The numbered phases are as follows: The flat High Plains surface is underlain by Pleistocene eolian sediments in which the Pullman (41), Olton (30, 31), Acuff (2), and Posey (35) soils have formed. The high plains is pockmarked by numerous circular depressions (playas), formed by wind deflation. The rest of the soils are formed in lacustrine clays and calcareous loamy material on the playa floors or in calcareous eolian sediments derived from the lacustrine materials by prevailing westerly winds; the eolian sediments occur on the east side of the playas. The Randall soil (42) has formed in the lacustrine clays, whereas the soils formed in calcareous lacustrine loams include the Midessa (26), Arch (8), and Portales (33) soils. Soils formed in the calcareous eolian sediments include the Drake (16, 17), Midessa (26), and Zita (46) soils. Although the Midessa is formed in both lacustrine and eolian sediments, in geological studies it might be better to have a different name for each parent material. The Pleistocene eolian sediments of the High Plains crop out along the gently sloping margins of the playas, and the Estacado (18, 19) soil is formed in them. This latter soil is probably the uppermost soil in the Blackwater Draw Formation.

is some order or geomorphic prediction to the inclusions, it would be helpful if these were described. For example, are shallow soils of another series present on ridgetops, or are cumulic soils of another series present along small drainageways. Swanson (1990a,b) and Hudson (1992) have emphasized the importance of geomorphology in the design and field recognition of soil mapping units. Another thing one will notice is that all contacts are solid lines. If wanted, contacts could be drawn with different kinds of lines to depict the confidence of the mapper in the true placement of the line. For example, a solid line for high degree of confidence, a long-dash line for moderate degree, and a short dash for low degree.

Soil maps in the United States are made with many purposes in mind. Agriculture, of course, is the main reason for such maps. Other uses, however, are planning for urban areas, highways, and recreational uses, and for these purposes derivative maps can be made (Bartelli and others, 1966; Olson, 1984). In making a derivative map, the soil phase names are deleted and a new, much simplified explanation is written for that specific map. For example, from the soil map of Malibu, California, derivative maps of several kinds were published (Soil Conservation Service, 1967). One, a soil erosion map, depicts areas of slight, moderate, and high soil-erosion hazard. Another outlines areas of soil shrink–swell behavior (low, moderate, and high). The latter map is important to foundation and road work. Such maps are obviously useful for regional or local land-use considerations. People knowledgeable in soils and soil patterns in the landscape are much better at predicting troublesome areas than are geologists or engineers not trained in soils. Soil maps and accompanying text, before derivitative maps are made, should be detailed and include all the pertinent soil and soil-forming-factor data; only in this way can accurate derivative maps be made without additional mapping at considerable cost. In contrast, if a map were made for only one purpose, say soil erosion, it would be very difficult to derive other useful maps from it.

A detailed soil map is useful in interpretating the geology of an area. In Iowa, for example, the soils mapped at the surface are a complex mosaic that can best, and perhaps only, be understood by mapping at the series and phase levels and by knowing the local geology (Fig. 2.2). There, the different series and phases elucidate parent material differentiation, the presence of buried soils and their effect on overlying soils, and could provide a qualitative measure of the rate of hillslope erosion. Another application of the latter is that clay-enriched buried soils, if exposed in new roadcuts, can be the locus of landslides.

Geological maps can be made directly from soil maps. For years I have had students study soil and topographic maps of the same area and combine them into geological maps (Holliday and others, in press). This is best done in areas of widespread uniform Quaternary surficial units (river–terrace deposits, loesses, sand dune deposits) that vary in age. Because of the difference in the parent material textures of the above materials, each is assigned to a different series. Furthermore, the ages of each kind of deposit can be approximated by the degree of soil development. One of the better maps for this purpose is that of Arkley (1964) for a portion of the Great Valley of California, combined with the Oakdale, California 7.5′ topographic map. Alluvial fan deposits of granitic composition dominate the area, and they form a stepped river–terrace sequence. Study of topographic maps puts the terraces into proper geomorphic sequence. Geomorphic expression (degree of preservation of surface, height of terrace above present river; see Fig. 8.10) helps with broad age discrimination (see also Christenson and Purcell, 1985). Soil maps on an air-photograph base are best for deciphering geomorphic expression. One can then group the units into geological mapping units (Fig. 2.3). The area can be expanded with additional soil and topographic maps to produce a regional map (Fig. 2.4).

Another area that students have made useful maps of is Antelope Valley in the eastern California desert. The soil survey is by Woodruff and others (1970), also on an air-photograph base, and the pertinent 7.5' topographic maps are Valyermo and Lovejoy Buttes, California. Geologic maps can be made of the alluvial deposits, and tectonics evaluated, as some terraces are deformed due to proximity to the San Andreas Fault. The best maps turned in are nearly as good as those made by Ponti (1985 and cited Open-File maps). Students can also attempt a correlation with the Great Valley sequence, as Ponti (1985) gives PDI values.

At the very least, if one is making a geological map of an area and there is a soil map, examination of the latter should help identify the map units. After making such maps one should follow it up with field checking.

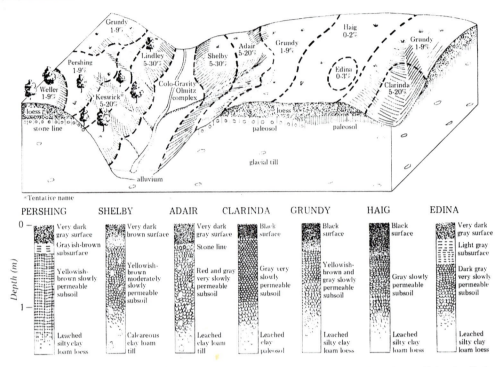

Figure 2.2 Relation of slope (%), vegetation, and parent material to soil series in southern Iowa. Upland soils have formed from loess (O-isotope stage 2 age), and valley side soils have formed either from reexposed older soils (paleosol) or from unweathered Kansan Till. *(Reprinted from Oschwald and others, 1965, Fig. 15.)*

■ Soil Classification at the Higher Categories

Many classification schemes have been proposed over the years, and the history of some of these is reviewed elsewhere (Soil Survey Staff, 1975). There is still no worldwide agreement on soil classification; in general, each country has its own (New Zealand has created one of the more recent ones), although many of the systems now in use have some features in common. There has been an effort by workers from many parts of the world to put together a soil classification scheme for use with a soils map of the world [Food and Agriculture Organization (FAO/UNESCO), 1972, and more recent editions; see Yaalon, 1995]. Until all nations agree on a uniform classification, one must either know several classifications or have access to conversion tables. One especially good conversion chart, with color photographs of the main soil classification units, is that compiled by Lof (1987). In it the systems of the United States, Canada, England and Wales,

France, Germany, and Australia are compared with the FAO system. For the Russian system, see Soil Survey Staff (1975).

Soil Taxonomy is the classification scheme in use in the United States, and it is necessary to learn at least some of it. Many international journals containing articles on soils will ask the author(s) to include the Soil Taxonomy units. The classification is based mainly on observable properties and not some genetic concept, is quantitative, the original publication contained over 700 pages, and carries such exotic tongue-twisting combinations of Latin and Greek as Histic Cryaquoll, Aquic Petroferric Kandiperox, and Glossic Naduralf. The original document has been altered several times (usually by adding more names) and the latest is Soil Conservation Service (1994). To use the classification in detail requires a good working knowledge of it and long experience. To its advantage, however, is the organization as a key, so one just has to follow the directions to key out the soil of interest. In many instances, field observations are not enough to classify

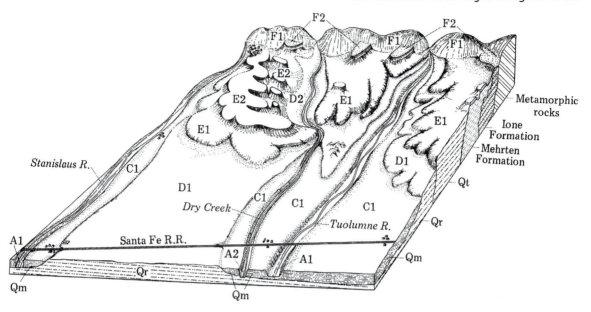

Figure 2.3 Relation of deposits, landscapes, and development of soils in the eastern Stanislaus area of the Great Valley, California. The following table gives data for soils formed on Quaternary sedimentary deposits, mainly of glacial outwash origin. Although many soil series, types, and phases are mapped on each stream–terrace deposit, soil development is fairly consistent for any one deposit and is a strong clue to the geologic age of the deposit. F, soils formed on pre-Quaternary rocks. *(Reprinted from Arkley, 1964, Fig. 21.)*

Data for Soils Formed on Stream Terrace Deposits

Age	Landscape Symbol[a]	Landscape Position	Soil Development	Geologic Formation
Holocene	A	Present-day floodplain	None to weak	
Pleistocene	C	Low-level stream terraces	Weak; some moderate	Modesto (Qm)
Pleistocene	D	Intermediate-level stream terraces	Strong	Riverbank (Qr)
Pleistocene	E	High-level stream terraces	Strong	Turlock Lake (Qt)

[a]The number following the letter in the figure refers to the lithology of the parent material: 1 = granitic; 2 = andesitic. Parent-material texture of the terrace deposits is generally sand and sandy loam.

a soil; these must be supplemented by laboratory quantification of some properties, as well as by information such as the number of wet or dry days per year or the mean annual soil temperature. These climatic parameters may not appear that useful when, indeed, we are supposed to be classifying a soil. However, soil properties often correlate with these climatic parameters, even though it is not explicately stated so in Soil Taxonomy. In places, it is not at all clear why certain decisions were made in the classification, but most of these are explained by Smith (1986), the person most responsible for the system.

To learn Soil Taxonomy, it is suggested that one first scan Soil Survey Staff (1975) to see what the task is going to entail. This publication also has many pages of soil descriptions and analyses, as well as color pho-

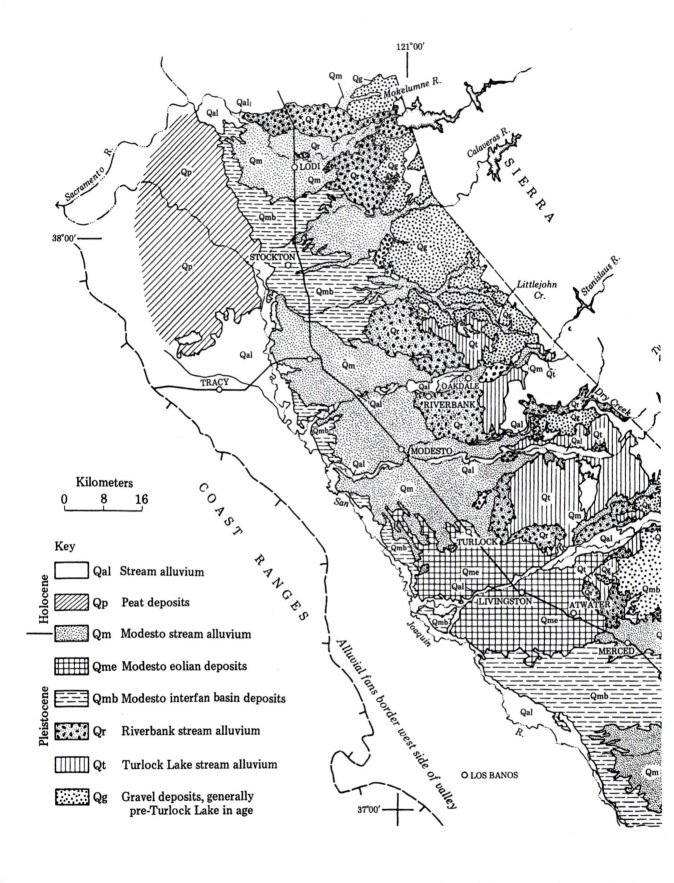

Figure 2.4 Generalized geologic map of part of the San Joaquin Valley, California, compiled from soil surveys of the area (Weir, 1952; Arkley, 1962, 1964; Ulrich and Stromberg, 1962), along with field checking by R.J. Janda and myself. The published surveys also give generalized soil maps, very useful in making this map. The topographic position of the deposits is given in Figure 2.3, a part of this area. Topographic and sedimentary characteristics of the Modesto deposits allow for the identification of the various facies.

—— 38°00'

tographs, so one can become familiar with the classification criteria. However, to avoid becoming overwhelmed, one should read Buol and others (1997). This readable text covers all the orders, chapter by chapter, and each is subdivided into the central concept and geographic distribution of the order, the setting, pedogenic processes, uses, and classification. The reader should be aware, however, that the classification has gone through many revisions, so consult Soil Conservation Service (1994) for the latest (watch for still more recent keys).

The new classification has much appeal for geomorphologists and ecologists, at least down to the suborder level. This is because soil profile development is included in the classification, as well as base saturation, amount of organic matter, and properties indicative of relative wetness and dryness. Most geomorphological soil studies deal with climatic and time factors, and the above properties will later be shown to be related to both.

Eleven orders are now recognized in the new classification; these are subdivided into 55 suborders, and the latter into over 200 great groups. One should be impressed with the magnitude of this task as it begins at the lowest level of classification, the approximately 12,000 soil series in the United States. The orders are basically differentiated by a particular horizon or horizon combinations that occur in the soil profile. These usually can be recognized in the field without recourse to laboratory analysis. One criticism of Soil Taxonomy, from a geomorphological point of view, is the overemphasis on classification at the order level by surface horizon (e.g., Mollisol). In contrast, horizons beneath the A horizon commonly are more important to geomorphologists. Classification into suborders requires an increasingly quantitative knowledge of soil properties and soil-moisture and soil-temperature regimes. However, it is often not necessary to take these measurements because, with experience, soil classification can be estimated from properties recognizable in the field (e.g., thin Av horizon with a calcic horizon at depth points toward aridic moisture regime).

To classify a soil at the order and suborder level, one must be able to identify the diagnostic horizons as well as the soil-moisture and -temperature regimes. The reader will find that some of these definitions are exceedingly complicated, but in time they will make some sense. Diagnostic horizons are so named because they are essential to classify the soil. Epipedons are the surface diagnostic horizons. The diagnostic horizons are somewhat similar to the field-designated soil horizons, although in places they can encompass several different field-designated soil horizons. In an extreme example of the latter, the mollic epipedon can include both the A and B horizons, as long as mollic properties are obtained. For some diagnostic horizons, the criteria are so complex that one has to read the defining criteria in detail using both the field and laboratory data (e.g., spodic horizon). Only the main discriminating criteria are given here (Table 2.1); Soil

Table 2.1 Common Diagnostic Horizons Used in Soil Taxonomy

Diagnostic Horizon	Defining Criteria	Probable Field Horizon Equivalent
Epipedons		
Mollic epipedon	Must be 10 cm thick if on bedrock, otherwise a minimum of 18 or 25 cm thick depending on subhorizon properties and thicknesses; color value darker than 3.5 (moist) and 5.5 (dry); chroma less than 3.5 (moist); organic carbon content at least 0.6%; structure developed and horizon not both massive and hard; base saturation ≥50%	A, A + E + B, A + B
Umbric epipedon	Meets all criteria for mollic epipedon, except base saturation <50%	A
Ochric epipedon	Epipedon that does not meet requirements of either mollic or umbric epipedons	A
Histic epipedon	Complex thickness requirements, but >20 cm thick; >12% organic carbon, with some adjustment for percent clay; saturated with water for 30 consecutive days or more per year, or artificially drained	O
Subsurface Horizons		
Albic horizon	Light colored with few to no coatings on grains—light color is that of grains; if color value (dry) is 7 or more, or color value (moist) is 6 or more,	E

(continued)

Table 2.1 (continued)

Diagnostic Horizon	Defining Criteria	Probable Field Horizon Equivalent
	chroma is 3 or less; if color value (dry) is 5 or 6, or color value (moist) is 4 or 5, chroma is closer to 2 than to 3	
Argillic horizon	Complex thickness requirements, but at least 7.5 or 15 cm thick depending on texture and thickness of overlying horizons; must have these greater amounts of clay relative to overlying eluvial horizon(s) or underlying parent material: (a) if the latter horizons have <15% clay, argillic horizon must have a 3% absolute increase (10 vs. 13%); (b) if the latter horizons have 15 to 40% clay, the ratio of clay in argillic horizon relative to them must be 1.2 or more, and (c) if the latter horizons have >40% clay, the argillic horizon must have an 8% absolute increase (42 vs. 50%); in most cases, evidence for translocated clay should be present (clay as bridges between grains or clay films in pores or on ped faces)	Bt
Kandic horizon	Minimum thickness is either 15 or 30 cm; complex clay increase requirements relative to overlying eluvial horizon(s) or underlying parent material: if the latter horizons have <20% clay, the kandic horizon must have a 4% absolute increase; if the latter horizons have 20–40% clay, the kandic horizon must have at least 20% more clay; if the latter horizons have >40% clay, the kandic horizon must have at least 8% absolute increase; complex depth-texture relations; CEC <16 meq/100 g clay	Bt
Natric horizon	In addition to properties of argillic horizon: prismatic or columnar structure; 15% or more exchangeable sodium; exchangeable magnesium and sodium exceed exchangeable calcium and exchange acidity	Btn
Spodic horizon	Minimum thickness is 2.5 cm, and contains >85% spodic materials: the latter are amorphous materials composed of organic matter and Al, with or without Fe; usually beneath an albic or Ej horizon	Bh, Bs, Bhs
Cambic horizon	Base usually at least 25 cm deep; stronger chroma or redder hue relative to underlying horizon; soil structure or absence of rock or sediment structure; weatherable minerals present; carbonates removed if originally present; no cementation or brittle consistence	Bw
Oxic horizon	At least 30 cm thick; >15% clay and sandy loam or finer; cation exchange capacity ≤16 meq/100g soil; few weatherable minerals	Bo
Calcic horizon	At least 15 cm thick; 15% $CaCO_3$; relative to underlying horizon, has at least 5% more $CaCO_3$, or at least 5% by volume secondary carbonate	Bk
Petrocalcic horizon	Horizon continuously cemented with $CaCO_3$	Km
Gypsic horizon	At least 15 cm thick; at least 5% more gypsum than underlying horizon; product of thickness(cm) times content (%) is 150 or more	By
Petrogypsic horizon	Strongly cemented gypsic horizon, commonly with greater than 60% gypsum	Bym
Salic horizon	At least 15 cm thick; at least 2% salts more soluble than gypsum; product of thickness (cm) times content (%) is 60 or more	Bz
Duripan	Silica cementation is strong enough that fragments do not slack in water	Bqm
Fragipan	Horizon of high bulk density relative to overlying horizons; formed in loamy material; although seemingly cemented with a brittle appearance, slacks in water; slowly permeable to water, so usually mottled; very coarse prismatic structure, usually with some bleached faces	Bx, Cx

Taken from Soil Conservation Service (1994).

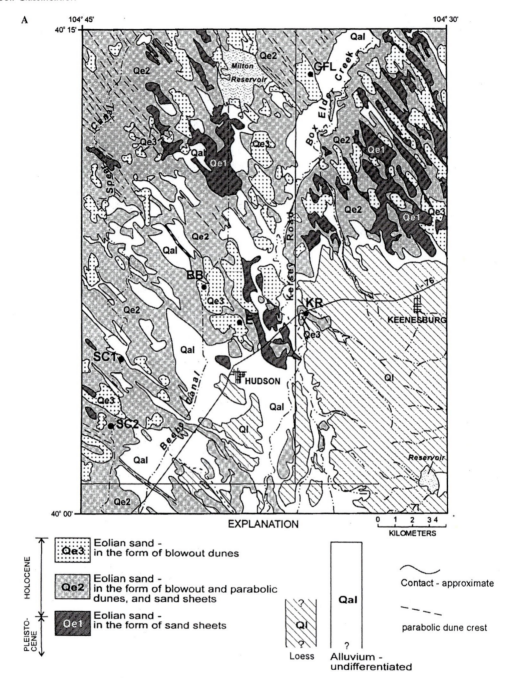

Figure 2.5 (A) Map of eolian deposits south of the South Platte River in eastern Colorado. Soil maps (Crabb, 1980) were used extensively in making this map. Dune sands lie closest to the river, and loess farther downwind, in accord with the prevailing winds from the NW.

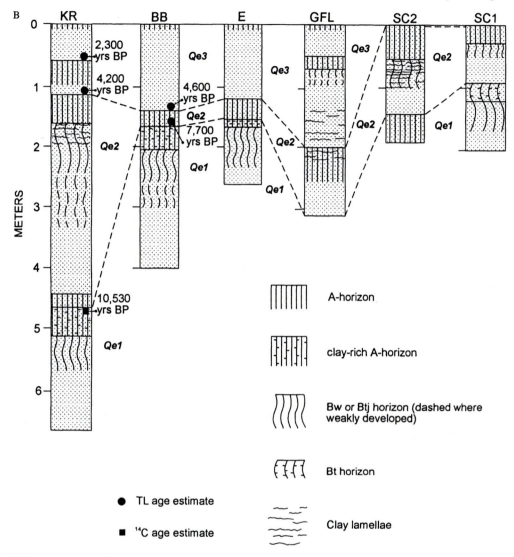

Figure 2.5 (continued). (B) Stratigraphic sections, with letters at the top for keying to localities in (A). Soil development was a major key in age descrimination of the dune deposits (Qe1, Qe2, etc.). *(Reprinted from Maat, 1992.)*

Conservation Service (1994) must be consulted for the details, however. One detail of Soil Taxonomy that has bothered geologists and some pedologists is the restriction of the cambic horizon to textures of very fine sand, loamy very fine sand, or finer texture (see Bockheim, 1990). This restriction excludes some sandy B horizons from being cambic, influences the classification of the soil in an important way, and might be disregarded when soils are studied for geological and other nonagronomic purposes. For example, color B horizons are important in the development of soil sequences in

sand dunes of different ages, and this fact should be reflected in the soil taxonomic name given the units. An example is the younger soils in the dune-fields of eastern Colorado (Jorgenson, 1992; Maat, 1992). These latter studies are yet another example of the use of soil maps to help make geological maps (Fig. 2.5) and undertake geological studies. Jorgenson's project was for my class, and the use of the soil maps helped him get it in by the deadline! Soil Taxonomy would group the younger soils into Psamments, yet there is a developmental sequence within them.

Table 2.2 Generalized Key to Soil Orders and Suborders Based on One or More Diagnostic Horizons

Diagnostic Soil Horizons	Order					
	Entisol	**Inceptisol**	**Aridisol**	**Mollisol**	**Alfisol**	**Spodosol**
Histic		Aquept		Aquoll	Aqualf	Aquod
Mollic		Aquept Tropept		Rendoll Boroll Undoll Ustoll Xeroll		
Umbric		Aquept Umbrept Tropept			a	
Ochric	a	Aquept Ochrept Tropept	a		a	
Albic				Alboll	Boralf Udalf	a
Cambic		a	Cambid			

Argillic
  Kandic for
  some
  Ultisols
			Argid	Base
				saturation
				>50% in
				some part
					Base
					saturation
					>35%	

Natric				Ustoll	Xeralf	
Spodic						Orthod Humod
Oxic						
Petrocalcic			Calcid	Ustoll Xeroll	Ustalf Xeralf	
Calcic			Calcid			
Gypsic			Gypsid			
Salic			Salid			

^aCommon horizon association.

		Order		
Ultisol	**Oxisol**	**Andisol**	**Vertisol**	**Histosol**

Aquult	Aquox	Aquand
		Suborders recognized
		  mainly on temperature or
		  or moisture regime
			May or may not have
			  diagnostic horizons;
			  has >30% clay and
			  shrinks and swells with
			  moisture variation to form
			  cracks that extend to the
			  surface; can have
			  slickensides and gilgai
			  microrelief; suborders
			  partly based on
			  frequency that cracks
			  form and duration
			  they remain open
				Organic soils,
				  many of which
				  are water saturated;
				  suborders defined
				  partly by degree of
				  decomposition of
				  organic matter

a				
a				
a	Torrox			
a				
Base saturation <35%[a] Humult Udult Ustult Xerult				
	Humox Perox Torrox Udox Ustox			

Classification of Soil Moisture and Soil Temperature

Soil-moisture regimes are difficult to summarize because they are based on estimates of depths to which a known amount of water will wet in a certain amount of time, and there are temperature restrictions. The depths of interest vary from 10 to 90 cm, depending on the texture. Water balance data are used to estimate the amount of water in the soil at designated times (Table 11.1). Only the basic concepts can be given here. The soil-moisture regime of a particular area can be estimated by study of a regional map as the regime is an integral part of the classification. For example, the Udalfs of Indiana indicate a udic moisture regime, and the Ustolls of South Dakota indicate an ustic moisture regime. One problem with the above is that temperature can override moisture in the classification. As an example, the soils of northern Montana are Borolls to indicate that the soils are cold; for agronomic purposes the cold is more important than the moisture, so moisture regime cannot be read from the taxonomic term. Five moisture regimes are recognized, based on the length of time a soil is either moist (water content more than that held at 15 atmospheres) or dry (water content is that held at 15 atmospheres). From drier to wetter, the regimes are as follows:

Aridic (and torric) moisture regime: Usually soils in arid climates; the soil is moist for less than 90 consecutive days when the temperature is above 8°C, and dry more than one-half the time (cumulative) when the temperature is above 5°C.

Ustic moisture regime: Soils are dry for 90 or more cumulative days; depending upon the temperature, the soil is (1) moist for at least 180 cumulative or 90 consecutive days or (2) not dry in all parts over one-half the time, and does not meet xeric moisture regime requirements.

Xeric moisture regime: Soils in Mediterranean climates with moist winters and dry summers.

Udic moisture regime: Associated with humid climates with enough summer precipitation that precipitation plus stored moisture exceeds evapotranspiration; it is not dry for as long as 90 cumulative days.

Perudic moisture regime: In all months, water moves through the soil because precipitation exceeds evapotranspiration.

Aquic moisture regime: Soils are saturated with water for at least a few days per year, the level of water can fluctuate with the season, and water is essentially free of dissolved oxygen and conditions are reducing.

Finally, some of driest soils on Earth (parts of Antarctica that are not covered by ice; see Bockheim, 1990) are excluded from the aridic moisture regime classification because the soils are too cold. Geomorphologists and pedologists could overlook this distinction and classify the soils as Aridisols. However, a new order has been proposed for such soils, and, if accepted, they will be called Gelisols (Bockheim et al., 1994). Bockheim and others (1997) and Bockheim and Tarnocai (1998) define gelic materials, the basis for the order, as well as clues for recognizing cryoturbation in soils.

Soil-temperature regimes are based on the mean annual temperature at 50 cm depth. These values can be estimated, as Soil Survey Staff (1975) points out that the mean annual soil temperature of a site is approximately equal to the mean annual air temperature + 1 (in degrees Celsius). Six regimes are recognized (all values are mean annual values): pergelic, below 0°C; Cryic, 0–8°C and summer temperature restrictions depending on whether an O horizon is present and if the soil is saturated with water; frigid, 0–8°C and warmer in summer than the cryic regime; mesic, 8–15°C; thermic, 15–22°C; hyperthermic, more than 22°C. In the frigid, mesic, thermic, and hyperthermic regimes, the difference between mean summer and mean winter temperatures is more than 5°C. If the differences in the latter temperatures are less than 5°C, the prefix iso is added to the name (e.g., isofrigid).

Although at times in Soil Taxonomy it seems as though we are classifying soils more on climatic parameters than on pedologic features, many of the latter features are related to climate (Chapter 10). For example, the calcic horizon is common in the aridic regime, common but at greater depth in the ustic regime, and rare or present only at still greater depth in the udic regime.

Soil Orders

Table 2.2 lists the horizons most diagnostic for classification of the most common soils at the order and suborder level. Only those suborders (54 total) thought to be most useful to geomorphology and Quaternary research are included. To name a suborder, one builds a name in which the suffix indicates the order. For example, if the soil is a Spodosol the suffix is "od" (Table

2.3). If the moisture regime is aquic, the suborder is Aquod. It should be stressed that because of the extremely complex nature of the classification any such simplification (Table 2.2) is bound to contain some errors, particularly at the suborder level. It is an attempt to simplify the system so that it can be used by workers who have neither the necessary time nor desire to learn the system in detail. If the latest key is used (Soil Conservation Service, 1994), it is not difficult to get close to the correct term.

The general concept behind each of the orders follows (more details in Buol and others, 1997; photographs of many in Fig. 2.6). These are presented in the order in which they are listed in the key. As with any key, one continues through the list of classification units until the soil in question meets the criteria listed. The same is true for keying out soils at the suborder level, great group level, and subgroup level. All percentage values are by weight.

Histosols are organic soils. If never saturated with water for more than a few days it must have ≥20% organic carbon, however if saturated for long periods of time the organic carbon content varies from 18% (if clay content is ≥60%) to 12% (no clay). Anaerobic conditions are required to accumulate these materials over long periods of time. These are common to low-lying bogs in cool, humid climates. Four suborders are based on the degree of preservation of the original plant materials.

Spodosols can have the greatest color contrast down-profile of any order, with a blackish O horizon overlying a whitish horizon (albic), which, in turn, overlies a brownish horizon (spodic). Although a soil is a spodosol if it has both albic and spodic horizons, it can be so classified if it has only the latter horizon. Spodosols are most common in cool, humid climates with sandy parent materials. However, they are also common in sandy materials in tropical climates where they were originally termed giant podzols. In general, if the albic horizon is sandy and the spodic horizon is at >2 m depth, the soil is not classified as a Spodosol. This depth restriction excludes some of the thicker and more colorful podzols from being classified as Spodosols, such as those of eastern Australia (Chapter 8). Perhaps for geomorphic purposes, the depth restrictions can be overlooked. Four suborders are recognized, based on the presence of an aquic moisture regime, cold temperatures, and the amount of organic carbon in the spodic horizon.

Andisols are soils with andic properties, but no specific diagnostic horizon. Basically these are soils that

Table 2.3 Nomenclature Key for Common Suborders for Soil Taxonomy

Prefix	Diagnostic Property
alb-	Albic horizon
aqu-	Associated with prolonged wetness
arg-	Argillic horizon
bor-	Cool temperature; relatively high organic matter
cal-	Calcic horizon
camb-	Cambic horizon
cry-	Cryic or pergelic temperature
dur-	Duripan
fluv-	Recent river deposit
gyp-	Gypsic horizon
hapl-	Minimum horizon expression
hum-	High organic matter
natr-	Natric horizon
ochr-	Ochric horizon
orth-	Soils that best typify the order
pale-	Old or maximum development
psamm-	Sandy
rend-	Calcareous parent material (limestone)
sal-	Salic horizon
torr- ⎫ ust- ⎬ xer- ⎭	Associated with dryness for various lengths of time; low organic matter
trop-	Continually warm temperature
ud- ⎫ per- ⎭	Associated with humid climate
umbr-	Umbric epipedon

Suffix indicates Order (key soil out in this order)

-ist	Histosol
-od	Spodosol
-and	Andisol
-ox	Oxisol
-ert	Vertisol
-id	Aridisol
-ult	Ultisol
-oll	Mollisol
-alf	Alfisol
-ept	Inceptisol
-ent	Entisol

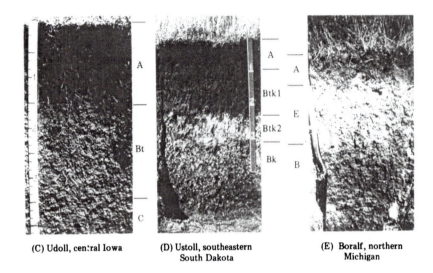

(A) Inceptisol, Searles Lake, California

(B) Argid, Las Cruces, New Mexico

(C) Udoll, central Iowa

(D) Ustoll, southeastern South Dakota

(E) Boralf, northern Michigan

Figure 2.6 Characteristic soil profiles of several soil suborders. All photographs are in color originally, and all but A and B are from the Marbut Memorial Slide Collection, prepared and published by the Soil Science Society of America (Madison, Wisconsin) in 1968. Scale is in feet and inches.

have formed by the weathering of volcanic ash. Andic soil properties are complicated, but basically the material has some combination of <25% organic carbon, a bulk density ≤0.90 g/cm³, high amounts of extractable Al and Fe (oxalate method), and high P retention. These properties are met under most humid conditions. Seven suborder are recognized, based mainly on soil-moisture and -temperature regimes.

Oxisols have either an oxic horizon or a surface horizon with ≥40% clay that overlies a highly weathered kandic horizon with few weatherable minerals remaining. Plinthite may be present. These soils form in warm, humid conditions in tropical areas. Landscapes associated with some of these soils are exceedingly old, measured in millions of years, and thicknesses can be 10s of meters. Five suborders are recognized, based only on soil-moisture regime. One unusual suborder is Torrox, for Oxisols presently in an aridic moisture

regime. This is almost impossible, because high rainfall is required to form Oxisols. Obviously, such soils have experienced a dramatic decrease in rainfall since they formed. Examples of this are the Tertiary lateritic soils of Australia that are now located in arid and semiarid areas (Hubble and others, 1983), a classic example of major climatic change (Chapter 10).

Vertisols do not require a diagnostic horizon. They have a high clay content (≥30%), generally high in smectite, with surface cracks that open and close, commonly seasonally, and slickensides and/or the wedge-shaped peds associated with shrink–swell soils (Yaalon and Kalmar, 1978). The six suborders are based on climate and the percentage of years the cracks are open as well as the time per year they are open. When the cracks are open, material from the walls can fall to the bottom of the crack. On swelling, the cracks close and because the cracks are partially

Figure 2.6 (continued).

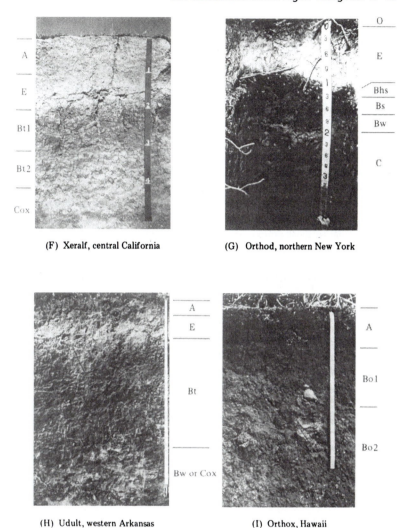

(F) Xeralf, central California

(G) Orthod, northern New York

(H) Udult, western Arkansas

(I) Orthox, Hawaii

filled, the soil expands, producing the wedge-shaped peds. Over time the soil churns, with surface material falling to the bottom of the cracks and deep material moving back to the surface (Yaalon and Kalmar, 1978). Such shrink–swell behavior is important to any engineering projects, as it commonly breaks foundations.

Aridisols are usually characterized by an aridic moisture regime and at least one diagnostic horizon marked by the accumulation of carbonate, gypsum, or silica, or a cambic horizon. In places in arid areas where the water table is near the surface for short periods of time, the associated soils are Aridisols if there is a salic horizon. In the original classification only two suborders were

recognized, but more were needed, so there are now seven suborders, depending on the horizon sequence. There is a developmental sequence at the suborder level that is important to geomorphological studies as both clay and carbonate accumulation relate well to the duration of soil formation. If a cambic horizon is absent in the youngest soil of a sequence, an age sequence would be as follows:

Torriorthent ⇒ Cambid ⇒ Haplargid ⇒ Paleargid

Ultisol are soils with either an argillic or kandic horizon, less than 35% base saturation at depth, with or without a fragipan horizon. Although not required, E horizons are not uncommon in Ultisols. The intent of this order was to split out those well-developed soils in humid climates with low base saturation. This is useful in geomorphological studies because long durations of time might be required to deplete the soil of bases to meet this criterion. Five suborders are recognized, one based on high organic matter (Humult), and the rest on soil-moisture regime.

Mollisols require either a mollic epipedon or a surface horizon with mollic properties, except for thickness, plus an albic horizon and the upper part of the underlying argillic, kandic, or natric horizon. In addition, subsurface horizons must have ≥50% base saturation. Most of these soils have formed under grasslands, and so the A horizons are dark, thick, and fertile. Most of the grains of the world are produced from Mollisols. Seven suborders are recognized, five based on soil moisture or temperature, one on parent rock (Rendoll is common to limestone), and one unusual one that has an albic horizon (see Mollisol requirement, above). Within the Borolls (cool climate Mollisol) one great group subdivision could be confusing for geomorphologists. "Pale" usually indicates features associated with old soils. With the Paleborolls, however, a criterion is that the top of the argillic horizon is ≥60 cm depth. I do not know the rationale for this depth restriction, but this sounds like a cumulic soil with overthickened A or A plus E horizons. Perhaps this will be redefined in the future, as most of the other uses of "pale" have old age properties.

Alfisols require either an argillic, kandic, or natric horizon, or fragipan with clay films. Many of these soils will have ochric epipedons and albic horizons, and have formed under forest. Five suborders are recognized, based mainly on soil climate. Paleboralf appears at the great group level, but it suffers from the same 60-cm-depth restriction as the Paleboroll; I would hope for a better definition in the future.

The criteria for Inceptisols are rather complicated, but in general they will have one of the following: cambic, calcic, petrocalcic, gypsic, petrogypsic, placic, oxic, or sulfuric horizon, or duripan or fragipan. Basically they possess some diagnostic horizon. They are important in an age sequence of soils because they have to be sufficiently old to have formed a cambic horizon, but younger than those soils with argillic horizons. Five suborders are recognized based on soil climate, but one common one requires an ochric epipedon (Ochrept).

Entisols have the simplest definition, the other soils. The reason for this is that all major properties of developed soils have been keyed out already. These are poorly developed soils for reasons of young age, high erosion rate, or resistant rock type. Five suborders are recognized, and the main ones are differentiated on the aquic moisture regime, sandy parent material, or a floodplain setting.

Geologists and geographers might prefer to read through the key in a way that follows soil development. This can be done by use of a flow chart with increasing development to the right (Fig. 2.7).

Suborder names are formed by combining two syllables to indicate the order and some distinguishing characteristic of the suborder (Table 2.3). Unfortunately, the distinguishing characteristic commonly is soil climate. With practice, however, one soon learns that morphological characteristics follow soil climate. For example, in the Mollisols, Udolls will have thicker and more organic-rich mollic epidons than will Ustolls. In the Alfisols, we could predict calcic horizons in the Ustalfs, but not in the Udalfs. In the Ultisols one expects properties related with greater degree of leaching (lower base saturation percentage) in Udults than in Ustults. Soils formed under aquic moisture regimes almost always are keyed out first (Aquod, Aquox, etc.).

Suborders can be subdivided into great groups, and commonly one has to classify at that level to find the best match for the soil in question. There are several hundred of these in the United States. It would be too lengthy to go into these taxonomic units in this book, but once one gets acquainted with Soil Taxonomy, it is not too difficult to classify soils approximately to the great group level. Much information can be gained by classifying to that level, as shown by the following examples.

Several trends can be deciphered from the great groups. An age sequence in the Ustolls would be as follows (Table 2.4):

Haplustoll ⇒ Argiustoll ⇒ Paleustoll,
        or   Calciustoll ⇒ Paleustoll

with the age-related soil features better expressed to the right of each sequence. Leaching conditions, important to agronomists and botanists, can be read in some of the Boralfs for

Glossoboralf ⇒ Eutroboralf ⇒ Natriboralf

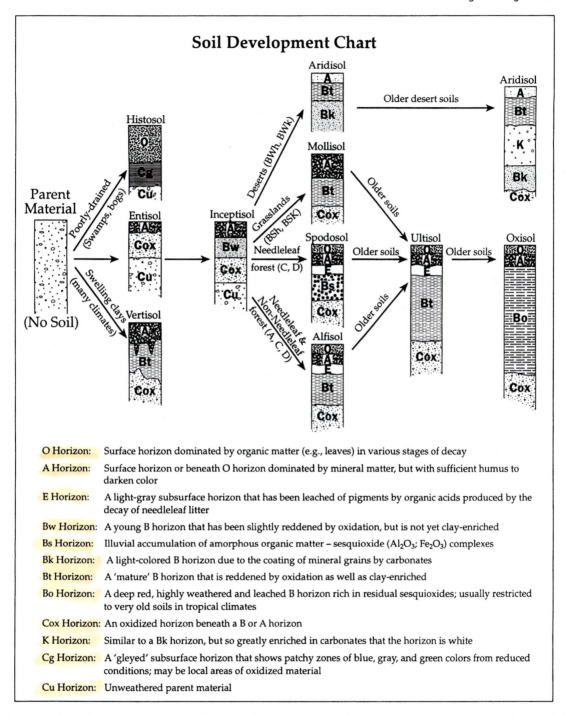

Soil Development Chart

O Horizon: Surface horizon dominated by organic matter (e.g., leaves) in various stages of decay

A Horizon: Surface horizon or beneath O horizon dominated by mineral matter, but with sufficient humus to darken color

E Horizon: A light-gray subsurface horizon that has been leached of pigments by organic acids produced by the decay of needleleaf litter

Bw Horizon: A young B horizon that has been slightly reddened by oxidation, but is not yet clay-enriched

Bs Horizon: Illuvial accumulation of amorphous organic matter – sesquioxide (Al_2O_3; Fe_2O_3) complexes

Bk Horizon: A light-colored B horizon due to the coating of mineral grains by carbonates

Bt Horizon: A 'mature' B horizon that is reddened by oxidation as well as clay-enriched

Bo Horizon: A deep red, highly weathered and leached B horizon rich in residual sesquioxides; usually restricted to very old soils in tropical climates

Cox Horizon: An oxidized horizon beneath a B or A horizon

K Horizon: Similar to a Bk horizon, but so greatly enriched in carbonates that the horizon is white

Cg Horizon: A 'gleyed' subsurface horizon that shows patchy zones of blue, gray, and green colors from reduced conditions; may be local areas of oxidized material

Cu Horizon: Unweathered parent material

Figure 2.7 Soil orders arranged into a development scheme, with greater development (and age, in most cases) to the right. Parent material controls the first step in the development, and time and bioclimate beyond that. Letters in parentheses depict climate classification of Köppen. *(Courtesy of Dennis Netoff, 1997.)*

Table 2.4 Some Great Groups of Ustolls and Boralfs

Order	Suborder	Great Group[a]
Mollisol	Ustoll	Durustoll (duripan)
		Natrustoll (natric horizon)
		Paleustoll (high clay content, red, petrocalcic)
		Calciustoll (calcic horizon)
		Argiustoll (argillic horizon)
		Haplustoll (the other Ustolls; cambic horizon)
Alfisol	Boralf	Paleboralf (upper boundary of argillic horizon deeper than 60 cm)
		Fragiboralf (fragipan)
		Natriboralf (natric horizon)
		Cryoboralf (cryic temperature regime)
		Eutroboralf (>60% base saturation in argillic horizon)
		Glossoboralf (<60% base saturation in some part of argillic horizon)

[a]In Soil Taxonomy, one always reads down lists such as these, so that if the first name encountered does not apply to the soil being classified, one goes down to the next name, and so forth.

are in the direction of increasing base saturation (Table 2.4). In the Orthids, increasing aridity might be suggested in the sequence

$$\text{Cambid} \Rightarrow \text{Calcid} \Rightarrow \text{Gypsid} \Rightarrow \text{Salid}$$

Finally, Petrocambid (petrocalcic horizon but no argillic horizon) suggests an old soil to produce the petrocalcic horizon, but the lack of the expected argillic horizon indicates that it has been either eroded or masked by the carbonate accumulation.

■ Distribution of Soil Orders and Suborders in the United States

A soils map of the world shows some regional trends in the United States that can be roughly related to the soil-forming factors, mainly climatic and vegetation patterns and geology (Fig. 2.8; for a more detailed map of the United States, see the National Atlas of the United States of America, published by the U.S. Geological Survey, 1970). One major trend is that, generally, the soil distribution east of the Rocky Mountains

seems less complex than that to the west. No doubt this results from the western United States being more tectonically complex, and with much larger ranges in elevation locally, relative to the eastern United States. For example, in many western U.S. ranges one can go from Aridisol $\Rightarrow$ Mollisol $\Rightarrow$ Alfisol $\Rightarrow$ Inceptisol with increasing altitude in a short horizontal distance. A similar transect in the eastern United States might require traversing half of the nation.

The soil pattern east of the Rocky Mountains generally follows the gradual regional climatic gradient, although there is some variation probably due in part to erosion in mountainous areas and to age of the landscape. Just east of the Rockies is a wide expanse of Ustolls. Entisols are interspersed with the Ustolls and are related to the erosive shales of eastern Montana and to the dune sands of Nebraska. Udolls lie east of the Ustolls, but in part of the glaciated region to the north these give way to Borolls and Aquolls, the latter being mainly associated with the relatively impermeable sediments of glacial Lake Agassiz. Proceeding eastward, there is a large area of Udults formed on fairly old landscapes south of the glacial boundary that extends almost to the Aquults of the East Coast. Florida does not follow this regional trend and is covered mainly with Aquods, Entisols, and Histosols. Inceptisols are in close proximity with the Ultisols, as shown by Aquepts on the Mississippi River floodplain deposits and Ochrepts in the Appalachian and Ouachita Mountains. Just east of the Mississippi River is a belt of Udalfs, parallel to the river, that mainly are associated with loess deposits of glacial age. Their presence, and the Ultisols to the east, suggests that a predictable age progression will be Alfisol $\Rightarrow$ Ultisol. In the glaciated region of the central and northeastern United States, Udalfs are common to the south, and these grade into Orthods to the north. The Spodosols could indicate sandy parent materials. The only major exceptions are Udolls over much of Illinois and Boralfs in northern Minnesota.

West of the eastern front of the Rocky Mountains there is an intricate mosaic of climatic, vegetation, topographic, and geological patterns. The topography consists of a multitude of mountain ranges separated by intermontane valleys. Bedrock makes up most of the mountains, whereas unconsolidated alluvium of various ages underlies the valleys. In almost all places, the climate and vegetation are closely associated with the topography, with the valleys being the driest and the mountains receiving an increasing amount of moisture relative to altitude; vegetation follows these trends. The one major exception to these climatic trends is the lowland regions west of the crests of the northern Sierra Nevada and the Cascade Range; these areas receive abundant moisture from air masses moving inland from the Pacific Ocean. Still, even here the general climate–altitude relationship holds.

Soil patterns in the Cordilleran mountain ranges follow the overall climatic trends. Boralfs are common in the eastern parts of the Rocky Mountains, whereas the western parts are dominated by Ochrepts, Xerolls, and Orthents, from north to south. Andepts (now called Andisols) are found in parts of the northern Rockies and in the Cascade Range, where they are associated with volcanic rocks. The Sierra Nevada are dominated by Xerults, although soils south of Lake Tahoe might be closer to Xeralfs. Entisols occur in the ranges south of the Sierra Nevada. The ranges along the West Coast grade from Umbrepts in the north to Humults in the central sector to Xeralfs in the south.

The basins show a soil variation from north to south. Those in eastern Oregon and Washington are dominated by Xerolls, with Orthids in the drier parts of the Columbia Plateau. Orthids are common also in the Snake River Plain of southern Idaho and in the northwestern Basin and Range Province, and they grade into Argids to the south in that province. This gradation can be explained, at least in part, by age of landscape, because the widespread alluvial fan and pediment deposits to the south seem to be older than those to the north. Entisols and Orthids are the major soils of the Colorado Plateau.

Humults dominate in the Hawaiian Islands. Oxisols are present there to a limited extent. Most of Alaska has Inceptisols, with Spodosols more common in the forested southern parts of the state.

■ Distribution of Soil Orders of the World

The world distribution of soils has many associations with the soil-forming factors. (Fig. 2.8). In Europe, Asia, and Africa one sees the classic north–south transect, the northern part of which was first identified by the early Russian soil scientists. Inceptisols are present in the tundra that flanks the Arctic Ocean. Immediately south are the Spodosols of the boreal forest. Climate warms to the south to produce the Alfisols of the forests, which give way to the Mollisols of the

grasslands, which, in turn, give way to the Aridisols of the vast desert that extends from China on the east to the Atlantic coast of Africa on the west. Alfisols and Ultisols are common in the humid climate and old landscapes of India, southeast Asia, and the western Pacific islands. Africa and South America demonstrate soil patterns peripheral to the hot, humid climate along the Equator. Oxisols are present in the old landscapes of both continents, and they give way, usually both north and south, to Alfisols and Ultisols.

The approximate ranking of the Orders in terms of land area is as follows (in terms of decreasing area):

Aridisols > Alfisols > Inceptisols = Entisols =
Oxisols = Mollisols > Ultisols > Spodosols >
Vertisols > Andisols = Histisols

In the rest of the book I will use Soil Taxonomy where appropriate so those readers that want to can become more familiar with it. However, I hope to design the book so that it is not necessary to have an in-depth knowledge of Soil Taxonomy to be able to read it.

■ Soil Landscape Analysis

Much information of interest to geomorphologists can be derived from soil maps. Hole (1978; see also Hole and Campbell, 1985) describes some of these, and a brief review will be given here. Basically, the exercise is a quantification of the soil bodies across the landscape.

All soil bodies will have a pattern as shown by the contacts. This pattern can be simple (extensive terraces, with each terrace having a different soil, such as a series) or complex (glaciated bedrock terrain with soils on bedrock and a large variety of glacial materials and terrains). The latter units could consist of hilly topography underlain by unsorted gravelly till, gently sloping river terraces underlain by well-sorted sand, and flat-lying lake terraces underlain by silt. If the deposits are of the last glacial age, the soil pattern and the orientation of the soil bodies (preferred or random) are strongly controlled by the geological materials. In time, however, pedological process will help determine the map pattern. In the extreme case the original landscape can be highly dissected, having lost all traces of its original characteristics, loess blankets the landscape, and Ultisols are the dominant soil. Soil patterns then can be due to position in the slope (catena, Chapter 9).

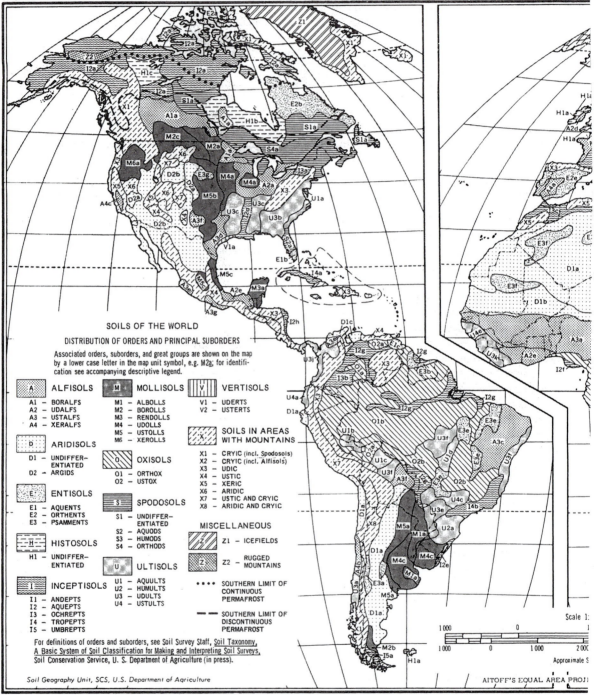

Figure 2.8 Soil map of the world (1972), showing orders and principal suborders. *(Courtesy of the U.S. Department of Agriculture Natural Resources Conservation Service, Soils Division/World Soil Resources.)*

HE WORLD

.nd Principal Suborders

SOIL CONSERVATION SERVICE

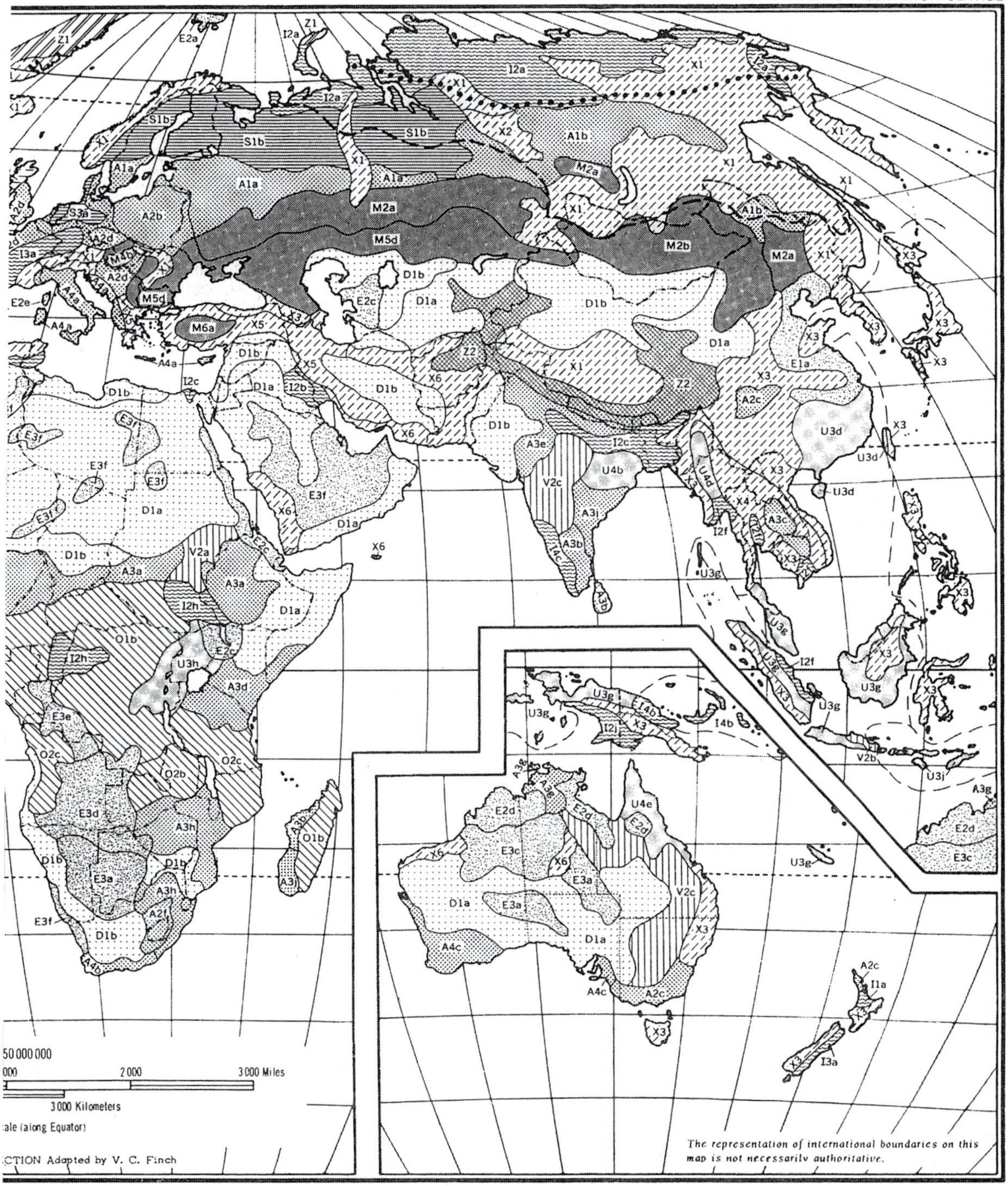

50 000 000
000
2 000
3 000 Miles
3 000 Kilometers
:ale (along Equator)
:CTION Adapted by V. C. Finch
The representation of international boundaries on this map is not necessarily authoritative.

MAY 1972

USDA-SCS-HYATTSVILLE, MD 1972

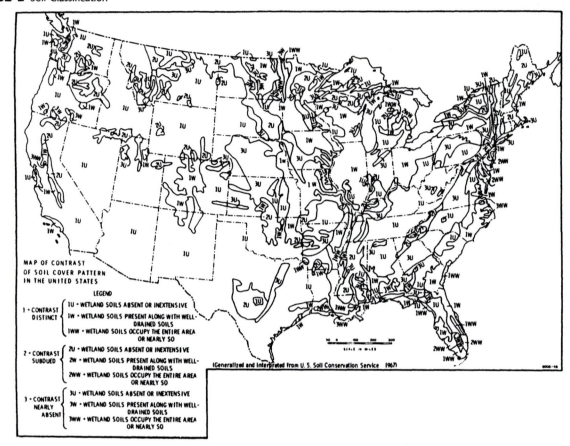

MAP OF CONTRAST
OF SOIL COVER PATTERN
IN THE UNITED STATES

LEGEND

1 - CONTRAST DISTINCT
- 1U - WETLAND SOILS ABSENT OR INEXTENSIVE
- 1W - WETLAND SOILS PRESENT ALONG WITH WELL-DRAINED SOILS
- 1WW - WETLAND SOILS OCCUPY THE ENTIRE AREA OR NEARLY SO

2 - CONTRAST SUBDUED
- 2U - WETLAND SOILS ABSENT OR INEXTENSIVE
- 2W - WETLAND SOILS PRESENT ALONG WITH WELL-DRAINED SOILS
- 2WW - WETLAND SOILS OCCUPY THE ENTIRE AREA OR NEARLY SO

3 - CONTRAST NEARLY ABSENT
- 3U - WETLAND SOILS ABSENT OR INEXTENSIVE
- 3W - WETLAND SOILS PRESENT ALONG WITH WELL-DRAINED SOILS
- 3WW - WETLAND SOILS OCCUPY THE ENTIRE AREA OR NEARLY SO

(Generalized and interpreted from U. S. Soil Conservation Service 1967)

Figure 2.9 Soil contrast map of the United States. *(Reprinted from Hole and Campbell, 1985, Fig. I.2.)*

Quantification of the soil boundaries is possible. One measure they propose is the mean density of the soil bodies, or number of soil bodies per unit area, and the mean soil boundary length (kilometers per square kilometer). In a transect along the 11°C isotherm in the United States, data from a soil survey in Kansas give a much lower mean density relative to the glaciated terrain of the midcontinent (Hole, 1978, Fig. 18). This contrast also is seen across ground moraine of different ages in Iowa (Hole and Campbell, 1985, Fig. 4.13).

Another quantification of the soil boundary is to rank it on the basis of taxonomic unit on both sides of the boundary. Examples of some are phases given a rank of 1, series 4, and order 9. Cumulative length of the ranked boundaries is incorporated into the latter to give the average degree of complexity of soil boundaries. In comparing two watersheds in Wisconsin, a glaciated watershed gave a higher degree of

complexity than did one in the driftless area. If we could track these through time, geological factors might influence soil complexity early in the pedological history of an area, but the complexity later on could result from pedological processes. One interesting outcome of this is a provisional soil contrast map of the United States (Fig. 2.9). It was made by noting the number of soil orders in each delineation in the soil map of the United States (National Atlas of the USA, mentioned above). The ranking ranges from 1 = distinct contrast (many orders) to 3 = contrast nearly absent (few orders). Subdivision is based on the extent of wetland soils. Areas of uniform soils due to uniform conditions (deposits, age, bioclimate) make up large areas of the southeastern United States, whereas areas of distinct contrast are common to the mountainous western part of the country. These analyses can be done at all soil map scales.

Chapter 3

Weathering Processes

Weathering is the physical and chemical alteration of rock or minerals at or near the earth's surface. Most rocks and minerals exposed at and immediately beneath the earth's surface are in an environment quite unlike that under which they were formed. This is particularly true for igneous or metamorphic rocks that were formed under high temperatures and, with the exception of some volcanic rocks, at great confining pressures. Weathering can be defined as the process of rock and mineral alteration to more stable forms under the variable conditions of moisture, temperature, and biological activity that prevail at the surface.

Two main types of weathering are recognized (excellent quantitative review in Winkler, 1975). One is physical weathering (also termed mechanical), in which the original rock disintegrates to smaller-sized material, with no appreciable change in chemical or mineralogical composition. The other is chemical weathering, in which the chemical and/or mineralogical composition of the original rock and minerals are changed. In nature, physical and chemical weathering occur together, and it may be difficult to separate the effects of one from the effects of the other. Traditionally, physical weathering is thought to predominate in relatively cold, dry climates, and chemical weathering in warm, humid climates, as shown in a commonly used figure last published by Pope and others (1995, Fig. 1a).

The emphasis here is on chemical weathering, as it has a closer link to soil processes. Chemical weathering produces many of the materials formed and moving within soils. In contrast, physical weathering can produce clasts within soils, important because they occupy space, as well as alter the particle-size distribution of the <2-mm fraction. Finally, weathering features of clasts on a geomorphic surface are important in assessing the stability of the surface. If the surface has been stable for a long period of time, the soil should be well developed and the form of the surface clasts should show evidence of a high degree of alteration, if a subsequent process has not eradicated the form (e.g., spalling by fire). These weathering features constitute a type of Quaternary dating of a geomorphic surface or deposit called relative age (Colman and others, 1987; Birkeland and Noller, 1998). These methods, including soils, are important to exposure dating and many other Quaternary dating studies (Sowers and others, 1998), as they indicate how long the clasts being dated have been exposed, and whether they had a later history not conducive to the particular dating method (e.g., cosmogenic dating).

There are many excellent books on weathering, and I will draw heavily on these. Good coverage and reference lists on physical weathering are included in current texts on geomorphology (Selby, 1985; Summerfield, 1991; Easterbrook, 1993; Cooke and others, 1993; Abrahams and Parsons, 1994; Robinson and Williams, 1994; Thomas, 1994; Ritter and others, 1995; Bloom, 1998).

■ Physical Weathering

The mechanism common to all processes of physical weathering is the establishment of sufficient stress within the rock so that the rock breaks. If the rock is ruptured along fracture planes, blocks or sheets of varying size are produced. If, however, the lines of weakness are along mineral grain boundaries, physical weathering can produce materials whose size is determined by the size of the grains in the original rock; smaller sizes are possible if the minerals are crosscut by small-scale fractures. The processes that are reported to be most common to physical weathering are unloading by erosion, expansion in cracks or along grain boundaries by freezing water or crystallizing salts, fire, and possibly thermal expansion and contraction of the constituent minerals.

Rock bodies that are either homogeneous or layered can have numerous fracture planes or joints, nearly parallel to the ground surface, that divide the rock into a series of layers or sheets (Fig. 3.1). The spacing between the joints generally increases with depth, and they can be observed for several tens of meters below the surface. The origin of the fractures seems to be the release of stresses contained within the rock (Bradley, 1963). While buried, the rock is under high confining pressures. With erosion of the overlying rock mass, however, the rock has less overburden pressure, so it can expand. If it is close enough to the surface, expansion can be only upward or toward the valley wall in any direction in which the rock body is not confined. This expansion can lead to rupture of the rock along fracture planes oriented at right angles to the direction of the pressure release and, thus, to development of sheeting parallel to the surface. Although the most common sheeting is parallel to the

valley walls, in some places it continues down and conforms to the the bottom of rock canyons. Bradley (1963) shows how these contribute to the arches so common on the canyon walls in the southwestern United States.

The role of freeze–thaw in physical weathering has been debated a long time. The traditional concept is as follows. Water, upon freezing, can set up pressures sufficient to disintegrate most rocks. At 0°C, the increase in volume with the conversion from water to ice is 9%. At localities in which there is a sufficient supply of moisture and a low enough temperature, the moisture contained in the rock can freeze, and the accompanying internal pressures are sufficiently great to exceed the strength of the rock, and the rock ruptures. Even though the water in the system might not be confined, pressures in an unconfined system are probably great enough to rupture most rock types. The direction of the fractures produced could be determined by minute, preexisting planes of weakness, such as joints, or along fractures produced by unloading. In some places this process might be responsible for granular disintegration, provided water has access to voids or cracks between the grains. This process could be most effective in environments in which surface temperatures fluctuate across 0°C many times each year.

Recent work casts doubt on the primary role of freeze–thaw in physical weathering; instead, a process in which thin films of adsorbed water might be the

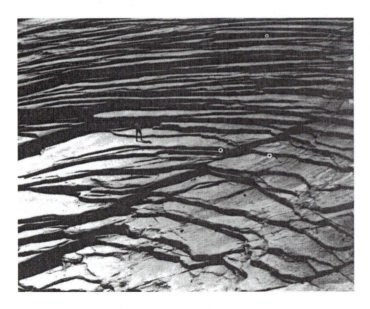

Figure 3.1 Sheeting joints developed in granitic rock, Sierra Nevada, California. *(Photograph by W.C. Bradley.)*

Figure 3.2 Weathered granitic stones in Baja California. The undersides of many stones are virtually unweathered. Salt crystallization in small cracks may be a major factor in the surface weathering of the stones.

agent that does the weathering seems to be favored (White, 1976; Thorn, 1979; Walder and Hallet, 1985). Thin films of water can be adsorbed so tightly that they cannot freeze. Pressures exerted during the migration of these films along microfractures could pry the rock apart. There is still a connection with freezing, as such migration can take place toward existing ice bodies in cracks during times of freezing. Walder and Hallet (1985) believe that the process is most effective at temperatures between -4 and $-15°C$, and a slow cooling rate (0.1–0.5°C/hour).

Saline solutions, if they gain access to fractures in the rock or to the boundaries between grains, can bring about the disintegration of rocks either into blocks or individual grains (reviewed in Cooke and others, 1993; Abrahams and Parsons, 1994) (Fig. 3.2). Salt-produced weathering forms are common in both hot and cold deserts (Watts, 1983; Campbell and Claridge, 1987). Several processes are recognized as important to the breakup of the rock by salts. One is the internal pressures set up during the growth of crystals from solution. A common pedologic example demonstrating this effect is the development of K horizons. In many places, the K horizon contains over 70% $CaCO_3$, and the original silicate grains are no longer in contact with each other. Although there is some evidence that calcite has replaced the original grains in some places (Reheis, 1988; Boettinger and Southard, 1991),

in many places it appears that the grains were pushed aside during the crystallization of the $CaCO_3$ from the soil solution, and much of the Ca may have had an external source (Chapters 5 and 8). Precipitation of gypsum from a solution has similar effects. Another important process is the thermal expansion of salts on heating; this occurs because many common salts have thermal expansion coefficients higher than those of some common rocks. This process might be important in many of the hot deserts that are characterized by large changes in daily temperatures. A final process that is important to rock disintegration is the stresses set up by the volume increase that accompanies hydration of some salts. Winkler (1965), for example, believes that the main cause of exfoliation of Cleopatra's Needle in New York City is hydration of salts that accumulated in the monument while it resided in Egypt; these salts hydrated and expanded once the monument was moved to the humid climate of New York City. Hydration of clay minerals may have similar disruptive effects on rocks.

The salts responsible for the weathering can come from several sources. The obvious one is the sea. Other sources are salt deposits in deserts (from bedrock or playa deposits), which are transported as dust, dissolved in rain, or both. Still another, more nebulous source is salt contained in fluid inclusions in rock minerals. This latter source was invoked by Bradley and others (1978, 1980) in making a case for a salt-weathering origin of cavernous weathering in Australia, when no other process seemed plausible. For cavernous weathering in sandstones, Winkler (1980) hypothesizes case hardening of the more resistant parts, and Young (1987) calls on quartz dissolution of the cavernous part, with the rate of dissolution sped up by the presence of NaCl in the dissolving solution.

A common feature of gravelly soils in hyperarid areas is the occurrence of shattered clasts at depth (Fig. 10.26). When best developed, the clasts maintain their original shapes and disintegrate into small pieces when barely touched with the tip of a knife. In a study in southern Israel, Amit and others (1993) isolated the following conditions as necessary for such shattering:

1. any salt can produce the shattering;
2. the origin seems to be pressure buildup during crystal growth;
3. the clasts must have original cracks, however microscopic, for the salt solutions to enter; and
4. the conditions for clast shattering seem to be best early in the evolution of the soil.

Extreme and rapid variations in soil temperature and moisture at shallow depths result in salt crusts around the gravel surface. These crusts plug the openings of the cracks so that the high pressures attendant with crystal growth shatter the clasts. In time, however, the soils evolve via dust influx to soils with a gravel-free B horizon composed mainly of infiltrated dust. This alters the locus of rainfall–water penetration (greater surface area of the fines in the B horizon results in shallower wetting depth), mutes the temperature variation, and the zone of maximum shattering shifts to shallower depths. Relatively constant moisture conditions at greater depth essentially shuts off the shattering process there; interestingly, some clasts at depth shattered in the early evolution of the soil can become recemented at a later stage of salt accumulation.

In sequences of desert alluvial surfaces it is commonly observed that bar and swale topography made up of large clasts gives way, in time, to a smooth desert pavement with smaller clasts (McFadden and others, 1989; Bull, 1991, Figs. 2.12 and 2.13). Salt weathering no doubt contributes to this reduction in clast size. The rate at which this weathering takes place is impressive, and some of the most rapid I have observed is in southern Israel and northern Sinai.

Fire also is an important factor in the physical weathering of rock (Allison and Goudie, 1994). Rapid temperature fluctuations (to >1000°C) can accompany a fire and, because of the speed with which it might sweep through an area, each subarea might be affected for only a short time. Because rock is a poor conductor of heat, a thin surface layer of the rock attains a high temperature during a fire, and there is a rapid decrease in temperature with depth. The heated surface layer will expand more rapidly than will rock at greater depth, and this expansion can lead to rupture of the rock into thin sheets that eventually fall to the ground. Grain dislodgement also can occur because of the variation in modulus of elasticity of different mineral grains making up the rock. Rapid temperature fluctuations can accompany a fire, and the fire itself may occupy a site only a short time. An example of the weathering effects of fire is seen on Mt. Sopris in western Colorado (Fig. 3.3). Rock-glacier deposits on Mt. Sopris do not have a forest cover. Stones on the surfaces of the rock-glacier deposits and of nonvegetated blockfields have thick oxidation weathering rinds, stone surfaces are oxidized to a reddish color, and the corners of some stones, even after several thousands of years of weathering, still are not too rounded (Birkeland, 1973). However, stones adjacent to or in the forest that burned several decades ago have more rounded corners, thin or no weathering rinds, and lack the pronounced surface oxidation. The differences in corner rounding and in weathering are attributed to fire, in both historic and prehistoric times.

These observations on the effect of fire on weathering are important to Quaternary stratigraphic studies, because commonly the weathered condition of the surface of a stone is one criterion used for age differentiation (Chapter 8). In parts of the Colorado Front Range and the Sierra Nevada, for example, stones in young tills above treeline have more highly pitted surfaces than stones in much older tills within the present forest zone (Fig. 3.4). Fire may explain the difference in weathering of individual stones because, in the forest zone, the stone surfaces can be continually renewed by spalling during a fire. The evidence for ancient fire is indirect, but surely lightning was an important cause. Blackwelder (1927) emphasized that fire, in some places, might be the main weathering process in physical weathering, and I think the stratigraphic and boulder weathering data from some places would support him. If present stone weathering can indeed be related to the presence or absence of fire, and thus to the treeline, stone-weathering studies

Figure 3.3 Comparison of stone weathering in (A) a forested and (B) an adjacent nonforested environment, western Colorado (see Birkeland, 1973). Note the abundance of fresh fracture surfaces in (A) due to expansion on heating by fire, and their absence in (B), which does not experience fires.

(A)

(B)

might be used as one criterion to help estimate the expansion of the upper and lower limit of past forests.

Recent work indicates that boulders also spall during range fires in the brush vegetation of the semiarid valleys of the western United States (Bierman and Gillespie, 1991; Zimmerman and others, 1994). This should result in spalling similar to that in the nearby forests. However, moraines in one Sierra Nevada valley display less weathered boulders in the forests than in the adjacent sagebrush, a difference considered to

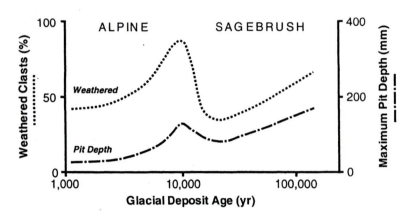

Figure 3.4 Trends in average values for percentage weathered clasts and maximum pit depths for glacial deposits of different ages, Sierra Nevada, California. Deposits between 1 and 10 ka are above the upper treeline, and those between 20 and 140 ka are in sagebrush vegetation, generally below the lower treeline. (*Reprinted from Birkeland and Noller, 1998.*)

be due to past fires (Burke and Birkeland, 1979). Natural range fires should be as frequent as natural forest fires. Perhaps once a boulder attains a certain weathered state, it can withstand stresses imposed by fire better and not spall, whereas less weathered boulders spall. This would explain why, even in some forested moraines, there is a difference in the weathered state of boulders of much different ages (20 vs 140 ka).

Diurnal fluctuations in surface temperatures, if great enough, are thought to bring about surface fracturing or granular disintegration of some rocks (Blackwelder, 1933). This idea is based on repeated observations of weathered rocks in deserts, weathering that seemed best ascribed to physical processes. In theory, the surface of a rock, where temperature fluctuations are greatest, should expand and contract the most (Peel, 1974; Smith, 1977, 1994). The effect is not unlike that of fire, that is, the rock is eventually weakened to the point where a part of the surface spalls off or mineral grains are dislodged. The process envisaged is that each different mineral will expand and contract a different amount at a different rate with surface-temperature fluctuations. With time, the stresses produced are sufficient to weaken the bonds along microfractures or grain boundaries, and thus flaking of rock fragments or dislodging of grains occurs.

The importance of this process in rock weathering has been debated over the years. In an often-cited laboratory experiment, Griggs (1936) heated and cooled rock samples over large fluctuations in temperature to simulate 244 years of weathering and showed little disaggregation of the rocks in the absence of water. Ollier (1969) argued, however, that the time factor was

not taken sufficiently into account in Griggs's study and that small stresses applied over long periods of time might lead to permanent strain. In another laboratory study (Warke and Smith, 1994) it was shown that short-term temperature fluctuations, perhaps due to the passage of a cloud, can contribute to fatigue and rock failure. This process might explain some weathering in arid regions. In most desert regions, however, some moisture is present, as are salts, and therefore other weathering processes could be operative (Smith, 1994).

Plant growth contributes to some physical weathering. Pressures exerted by roots during growth are sometimes able to rupture the rock or to force blocks apart. That such pressures can do this is shown by the common observation of cracked and heaved concrete sidewalks adjacent to tree roots. Lichens growing on rock surfaces also contribute to physical weathering. Loose mineral and rock fragments become attached to the undersides of the lichens and are pulled free of the surface when the lichen contracts during a dry spell (Fry, 1927). If the lichen is removed from the rock surface, it takes rock material with it. One would suspect, however, that the grains would have had to be loosened by some other process prior to removal from the rock surface with the lichen.

Many of the above processes are still hotly debated. Experimental work is not always conclusive, perhaps partly due to the problems of both time and scale. Instead of focusing attention on only one process perhaps several processes acting in concert should be considered. As an example, Rice (1976) cautioned workers against abandoning insolation as a weathering agent because of the importance of both sample size and number of temperature fluctuations in artificial weathering experiments. Finally, subtle chemical weathering might be a factor in weathering phenomena normally attributed to physical weathering.

A recent study by Netoff and others (1995) demonstrates the difficulty in identifying specific weathering processes. They have studied some of the largest weathering pits on Earth (Fig. 3.5). Formed in sandstone in the Colorado Plateau, they have maximum dimensions of 38 m wide and 16.5 m deep, and many walls are vertical to near vertical. They call on several physical weathering processes to weather the sandstone (spalling, growth of salt crystals, hydration of clay minerals, dissolving of the carbonate cement of the rock). The main challenge is how to rid the closed pits of the sediment formed by the above

Figure 3.5 Pit in sandstone in the vicinity of Lake Powell, Utah. Width is 22.6 m and depth of closure is 15.5 m. *(Photograph by D.I. Netoff.)*

processes, and they cite wind removal and piping, and perhaps some chemical weathering (dissolution of the grains).

■ Chemical Weathering

Chemical weathering occurs because rocks and minerals are seldom in equilibrium with near-surface water composition, temperatures, and pressures. Products form that are more stable in near-surface environments. However, if the soil environment changes with time, so too can the rates or products of weathering. That chemical weathering has occurred is shown by several field and laboratory criteria including (1) change in color due to oxidation of iron-bearing minerals, (2) depletion of original minerals (nonclay and clay), (3) alteration of original clay minerals or neoformation of clay minerals, (4) neoformation of iron or aluminum oxides and oxyhydroxides, (5) changes in major-element chemistry versus that of the assumed parent matrial, and (6) the chemistry of the waters that move through the soil and that of the streams draining a particular basin. The change on weathering can be very slight and involve nothing more than the oxidation state of Fe, or it can be quite intense and result in the massive leaching of the more soluble components to produce a product much different from the assumed parent material, such as the formation of an Oxisol from a basalt. There are several processes involved in chemical weathering of the common rocks and minerals.

Good coverage and reference lists on chemical weathering are given in some books (Garrels and MacKenzie, 1971; Curtis, 1976; Bohn and others, 1979; Drever, 1982, 1985, 1997; Colman and Dethier, 1986; Berner and Berner, 1987; Nahon, 1991), from which much of this material comes. An excellent quantitative review is in White and Brantley (1995).

Congruent and Incongruent Dissolution

Chemical weathering can be subdivided into congruent and incongruent dissolution. Congruent dissolution is when the mineral goes into solution completely with no precipitation of other substances. In contrast, with incongruent dissolution, all or some of the ions released by weathering precipitate to form new compounds (clay minerals, oxyhydroxides, etc.). Put another way, when considering incongruent weathering of a mineral the amounts of substances in solution do not correspond to the formula of the weathering mineral.

The soluble salt minerals, halite and gypsum, and silica minerals are examples of congruent dissolution. The respective reactions are

$$NaCl \Rightarrow Na^+ + Cl^-$$
$$\quad c \qquad aq \qquad aq$$

$$CaSO_4 \cdot 2H_2O \Rightarrow Ca^{2+} + SO_4^{2+} + 2H_2O$$
$$\qquad c \qquad aq \qquad aq \qquad \qquad l$$

$$SiO_2 + 2H_2O \Rightarrow H_4SiO_4$$
$$\quad c \qquad l \qquad \quad aq$$

where c is the crystalline, l is the liquid, and aq is the aqueous species. The solubilities of these minerals vary markedly, being about 260 to 350 g/liter of water for halite and 1.6–2 g/liter for gypsum, whereas silica has a range from 7 ppm for quartz to 120 ppm for silica gel (1000 ppm = 1 g/liter). The solubilities of other forms of silica lie between these latter two values. Higher temperatures slightly increase the solubilities of gypsum and silica species. Furthermore, the solubility of silica is somewhat pH dependent; above pH 9 it increases dramatically (Fig. 4.6) according to the reaction

$$H_4SiO_4 \Rightarrow H_3SiO_4^- + H^+$$
$$\quad aq \qquad \quad aq \qquad aq$$

Quartz is a resistant mineral in most environments, and on weathering increases in content relative to most other minerals. An unusual situation occurs along the Brazil–Guayana–Venezuela borders where high mesas underlain with quartzite have weathered for a long time in the tropical climate to form a karst topography not unlike the karst topography related to limestone (Stallard, 1992; Wray, 1997).

The weathering of calcium carbonate is another example of congruent dissolution. It is quite soluble under surface conditions and dissolves according to the equation

$$CaCO_3 + CO_2 + H_2O \Rightarrow Ca^{2+} + 2HCO_3^-$$
$$\quad c \qquad g \qquad l \qquad aq \qquad \quad aq$$

There are many discussions of carbonate equilibria (e.g., Arkley, 1963; Krauskopf, 1979; White, 1988) and all point out that the solubility of $CaCO_3$ varies with differences in CO_2 pressure and in H^+ concen-

tration. Increases in either CO_2 pressure or in H^+ concentration will increase the rate at which $CaCO_3$ dissolves. The partial pressure of CO_2 in the soil atmosphere under vegetation is greater than atmospheric (latter $PCO_2 = 10^{-3.5}$ or 0.035%), and therefore $CaCO_3$ solubility is greater under vegetated surfaces than under surfaces that lack vegetation. Berthelin (1988) gives CO_2 concentrations in soils at up to 100 times atmospheric (0.3–3%, and maybe as high as 5–10%), and Brook and others (1983) calculate a predicted maximum content of 4%. It is also known that CO_2 partial pressure in water is temperature dependent, with colder waters able to contain more CO_2 than warmer waters. Thus, $CaCO_3$ dissolves more readily in cooler climates than in warmer climates. These relationships are diagrammed in Arkley (1963) and White (1988, Fig. 5.2). In addition, chelating agents combine with Ca^{2+} and in this way increase the rate of solution of $CaCO_3$ and the mobility of Ca^{2+}.

Soils formed from limestone should be related to the content of insoluble residue. Pure limestone should not produce a soil because only Ca^{2+} and HCO_3^- result from the reaction, and these are removed from the surface in solution. An impure limestone, however, should produce a soil consisting of the insoluble residue and the properties should reflect this. Not uncommonly, however, eolian additions seem to play a major role in the composition of silicate soils that overlie limestone (Chapter 7).

In arid regions we see the effect of soil solution saturation with respect to limestone clast weathering. In southern Israel limestone clasts sticking above the surface dissolve at a rapid rate and produce a cavernous surface morphology (Fig. 3.6). In contrast, clasts below the surface do not weather as the soil solutions are saturated with respect to $CaCO_3$. Large clasts that are embedded in the soil but rise above the surface essentially get planed off at the ground surface in time.

The congruent weathering of limestone in the humid climate of southeast Asia is responsible for the exceptional examples of karst throughout the region (Fig. 3.7).

The chemical weathering of the aluminosilicate minerals is an example of incongruent dissolution. A general reaction is

$$\text{Aluminosilicate} + H_2O + H_2CO_3 \Rightarrow \text{clay}$$
$$\qquad \quad c \qquad \qquad l \qquad \quad aq$$

$$\text{mineral} + \text{cations} + OH^- + HCO_3^- + H_4SiO_4$$
$$\quad c \qquad \quad aq \qquad \quad aq \qquad \quad aq \qquad \quad aq$$

Figure 3.6 Weathering of a carbonate clast on an alluvial surface in southern Israel.

The usual reaction is that of water and acid on the mineral. The acid shown here is H_2CO_3. Other acids, such as organic acids resulting from the decay of organic matter, or sulfuric or nitric acid, also are important H^+ sources (Ugolini and Sletten, 1991). Although depicted as H^+, in reality in solution the form is H_3O^+ (Curtis, 1976). The common by-products are H_4SiO_4, HCO_3^-, and OH^-, along with clay minerals if aluminum is present in the decomposing minerals and if certain chemical conditions are met. More detailed weathering equations are given in Table 3.1; these reactions are commonly termed hydrolysis. If in-

stead, conditions are acid leaching and no solid products form, the weathering of an aluminosilicate mineral is congruent.

In the above reactions CO_2 is consumed, meaning that atmospheric levels can decrease if the C becomes sequestered during, say, the formation of recent carbonate sediments (Berner, 1995). This has important ramifications for global C cycles and climate change (Raymo and Ruddiman, 1992). In places mountain uplift is thought to be associated with declining CO_2 levels as high relief strips soils, and this and the high degree of physical weathering at high altitude expose more mineral surfaces to chemical weathering.

The fate of the by-products of weathering varies, and this will be discussed in more detail later. Cations can remain in the soil either in the soil solution, as part of the crystal lattices of the clay minerals, or as exchangeable ions adsorbed to the surfaces of the colloidal particles. Some ions can be cycled through the biosphere from the soil and back again. Some cations can be removed from the system, along with HCO_3^-, with the percolating waters; indeed, one measure of the rate of chemical weathering of a region can be gained from the composition of the waters draining the region as long as one can target the sources of the elements. Silica is quite soluble over the normal soil pH range (Fig. 4.6), and it is almost always present in the parent minerals in higher amounts than are necessary to form most clay minerals; therefore some is re-

Figure 3.7 Karst topography near Guilin, China.

Table 3.1 Equations Representing the Hydrolysis of Orthoclase and Albite with Various Clay Minerals as a By-Product

---

$$2KAlSi_3O_8 + 2H^+ + 9H_2O = H_4Al_2Si_2O_9 + 4H_4SiO_4 + 2K^+$$
$$\quad c \qquad aq \qquad\ 1 \qquad\qquad c \qquad\quad aq \qquad aq$$
(orthoclase) (kaolinite)

$$3KAlSi_3O_8 + 2H^+ + 12H_2O = KAl_3Si_3O_{10}(OH)_2 + 6H_4SiO_4 + 2K^+$$
$$\quad c \qquad aq \qquad\ 1 \qquad\qquad c \qquad\qquad aq \qquad aq$$
(orthoclase) (illite)

$$2NaAlSi_3O_8 + 2H^+ + 9H_2O = H_4Al_2Si_2O_9 + 4H_4SiO_4 + 2Na^+$$
$$\quad c \qquad aq \qquad\ 1 \qquad\qquad c \qquad\quad aq \qquad aq$$
(albite) (kaolinite)

$$8NaAlSi_3O_8 + 6H^+ + 28H_2O$$
$$\quad c \qquad aq \qquad\ 1$$
(albite)

$$= 3Na_{0.66}Al_{2.66}Si_{3.33}O_{10}(OH)_2 + 14H_4SiO_4 + 6Na^+$$
$$\qquad (smectite) \qquad\qquad\qquad c \qquad\quad aq \qquad aq$$

---

moved in solution. Aluminum is not very soluble over the normal soil pH range (Fig. 4.6), and so it generally remains near the site of release by weathering to form clay minerals or hydrous oxides. Iron also remains near the point of release in most soils because it also is generally insoluble over the pH range of most soils (Fig. 4.6), and Fe precipitates give the soil or weathered rock the commonly observed oxidation colors.

The integrated effect of all of the above factors, including water movement through the soil, on element distribution can be shown by Polynov's ion mobility series. This is the ease with which ions are leached from rocks by weathering and introduced into rivers. It is done by comparing the average composition of rocks with the chemical composition of river waters. Hudson (1995) has updated the method and come up with this ranking:

$$Cl > SO_4 > Na > Ca > Mg > K > Si > Fe > Al$$
$$—phase\ I \rightarrow\leftarrow II \rightarrow\leftarrow\!\!\!— III \longrightarrow\leftarrow IV \rightarrow\leftarrow V—$$

All of the elements are compared with the relative mobility of Cl to come up with groups of similar mobility, listed as phase I (very mobile) through V (relatively immobile). More will be said on this in Chapters 4 and 9, but in simplistic fashion, in a highly leaching environment only phase V elements remain in the weathering profile and, as conditions get progressively drier, the lower numbered phases remain in the soil,

until under the driest conditions halite and gypsum accumulate (phases I, II, and Ca).

One effect of the hydrolysis reaction is that hydrogen ion is consumed, hydroxide is produced, and the solution becomes more basic. This effect is particularly noticeable when the various silicate and aluminosilicate minerals are ground in distilled water, and the pH of the solution, called the abrasion pH, is taken (Stevens and Carron, 1948). The pH resulting from this grinding and initial hydrolysis is a function of the rapidity at which cations are released to the solution and the strengths of the bases formed (Table 3.2). In any weathering environment, the leaching of the cations and the production of hydrogen ion offset this tendency for most reactions to become basic as weathering proceeds. Grant (1969) has shown that the abra-

Table 3.2 Abrasion pH Values for Some Common Minerals

Mineral	Formula	Abrasion pH
Silicates		
Olivine	$(Mg,Fe)_2SiO_4$	10, 11
Augite	$Ca(Mg,Fe,Al)(Al,Si)_2O_6$	10
Hornblende	$Ca_2Na(Mg,Fe)_4(Al,Fe,Ti)_3$ $Si_6O_{22}(O,OH)_2$	10
Albite	$NaAlSi_3O_8$	9, 10
Oligoclase[a]	$Ab_{90-70}An_{10-30}$	9
Labradorite[a]	$Ab_{50-30}An_{50-70}$	8, 9
Biotite	$K(Mg,Fe)_3AlSi_3O_{10}(OH)_2$	8, 9
Microcline	$KAlSi_3O_8$	8, 9
Anorthite	$CaAl_2Si_2O_8$	8
Hypersthene	$(Mg,Fe)_2Si_2O_6$	8
Muscovite	$KAl_3Si_3O_{10}(OH)_2$	7, 8
Orthoclase	$KAlSi_3O_8$	8
Montmorillonite	$(Al_2,Mg_3)Si_4O_{10}(OH)_2 \cdot nH_2O$	6, 7
Halloysite	$Al_2Si_2O_5(OH)_4$	6
Kaolinite	$Al_2Si_2O_5(OH)_4$	5–7
Oxides		
Boehmite	$AlO(OH)$	6, 7
Gibbsite	$Al(OH)_3$	6, 7
Quartz	SiO_2	6, 7
Hematite	Fe_2O_3	6
Carbonates		
Dolomite	$CaMg(CO_3)_2$	9, 10
Calcite	$CaCO_3$	8

Taken from Stevens and Carron (1948), by permission from the Mineralogical Society of America.

[a]Ab, albite; An, anorthite.

sion pH of weathered material that includes some clay is less than the pH of the original rock, because some cations have been removed, and abrasion pHs of clay minerals commonly are lower than those of the common rock-forming minerals. Abrasion pH, therefore reflects the degree of leaching of cations in the weathered zone, as well as the presence of secondary minerals.

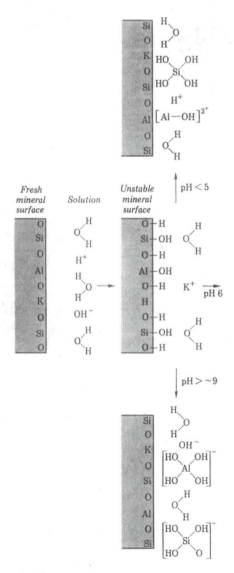

Figure 3.8 Scheme of an orthoclase surface reacting with water at various pH values. Thin surface layer leached of cations is expected but not shown. *(Data partly from Jenny, 1950 and Wollast, 1967.)*

Several processes are involved in the weathering of aluminosilicates (Berner and Berner, 1987; Nahon, 1991; White and Brantley, 1995), and some of this is diagrammed by Curtis (1976) (Fig. 3.8). Water molecules and H^+ react with Si–O–Si of the mineral surfaces and release silica as H_4SiO_4 in solution. The same mechanism should release Al from mineral surfaces and form the various forms of AlOHs in solution; the valency of the AlOH species would be related to the pH of the solution (Nordstrom and Munoz, 1994, Fig. 9.10). As the acids attack the mineral surface, H^+ replaces cations to produce a layer leached of cations. Weathering is not even along the surface as photomicrographs indicate that dissolution is selective and etch pits appear on the mineral surface (Fig. 3.9). This selective dissolution is thought to take place best at dislocations, places where rows of atoms are slightly out of place (Berner and Berner, 1987). Comparing minerals of the same kind on degree of weathering, those with greater densities of dislocations will be more weathered.

The weathering of iron-bearing minerals in oxygenated waters is also an example of incongruent dissolution; it is commonly termed oxidation. Oxidation is the process by which an element loses an electron. This loss results in an increase in positive valency for the element. Iron is the element most commonly oxidized in a soil or weathering environment, and the oxidation products give the altered material the characteristic yellowish brown to red colors. In soils and in many other weathering environments, the common oxidizing agent is oxygen dissolved in the water involved in the weathering reactions. Reducing conditions are important in the mobilization of Fe in the +2 form, and this is treated in Chapter 5 where redoximorphic features are discussed.

Weathering of iron-bearing minerals commonly releases Fe^{2+}, which, if in contact with oxygenated waters, is oxidized to form an oxide or hydrous oxide of iron. An example of this reaction for fayalite (Fe olivine) is (Krauskopf, 1979)

$$Fe_2SiO_4 + 4H_2CO_3 \Rightarrow 2Fe^{2+} +$$
$$c aq aq$$

$$4HCO_3^- + H_4SiO_4 + 2OH^- \text{ (hydrolysis)}$$
$$aq aq aq$$

$$2Fe^{2+} + 4HCO_3^- + 1/2O_2 +$$
$$aq \phantom{2Fe^{2+} +} aq g$$

$$2H_2O \Rightarrow Fe_2O_3 + 4H_2CO_3$$
$$1 c aq \text{ (oxidation)}$$

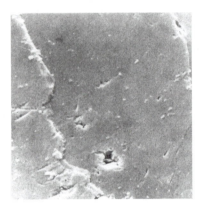 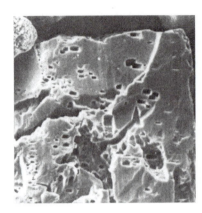

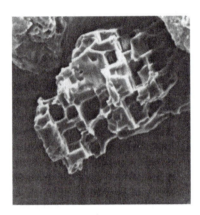

Figure 3.9 Scanning electron microscope photomicrographs showing increasing amounts of etching of feldspars (clockwise from left) in the A horizon of a mountain soil in New Mexico. The etching is considered to be selective along crystal lattice dislocations. *(Taken from Berner and Holdren, 1977, Figs. 2 and 3.)*

where g denotes the gaseous phase. Here the Fe^{2+} is released by weathering, and it enters an oxidizing environment and forms a precipitate, in this case Fe_2O_3. Depending on the prevailing environment, various iron compounds can form (Chapter 4).

The rate at which oxidation takes place and is noticeable in many soils seems to depend on the rate of release of iron on weathering. Stratigraphic studies of unconsolidated materials indicate that the depth and intensity of oxidation increase with age, and that this increase is a fairly slow process on a geological time scale. For example, chronosequences of soils in the western United States that span tens of thousands of years commonly show differences in the depths of oxidation that are a function of soil age. Percolating waters in these well-drained soils no doubt are well oxygenated below the zone of noticeable oxidation (Cox horizon) into seemingly unweathered materials (Cu horizon). It would seem, therefore, that the slow extension of oxidation with depth in the individual members of a soil chronosequence is directly related to the rate of release of iron during weathering because oxygenated waters are passing through materials that appear to be unweathered by most field criteria.

Oxidation of iron in a mineral can bring about the alteration of a mineral (Barshad, 1964). Iron exists as Fe^{2+} in the common rock-forming minerals. Oxidation to Fe^{3+} disrupts the electrostatic neutrality of the crystal, such that other cations leave the crystal lattice to maintain neutrality. These latter cations leave vacancies in the crystal lattice that either bring about the collapse of the lattice or render the mineral more susceptible to attack by other weathering processes. The alteration of biotite to vermiculite is one example of weathering due to oxidation.

Oxidizing and reducing conditions could influence the rate at which iron-bearing minerals weather. It has been suggested that under oxidizing conditions, the Fe that is released precipitates as an oxide on the mineral and that this could protect the mineral from further weathering, or at least slow down the rate of weathering (Siever and Woodward, 1979; Schott and Berner, 1985). In contrast, a protective coat would not form in a reducing environment, and could result in relatively higher rates of weathering. Theoretically, reducing conditions during part of the Precambrian, when atmospheric oxygen levels were low, could have accelerated weathering of iron-bearing minerals.

Chelation

Evidence suggests that chelating agents are responsible for a considerable amount of weathering (Berthelin, 1988; Ugolini and Sletten, 1991; Tan, 1993); in fact, in some places the amount of weathering by this process might exceed that brought about by hydrolysis alone. Chelating agents are formed by biological processes in soil and excreted by lichens growing on rock surfaces. Their structure is varied and complex and can be described as "the formation of more than one bond between the metal and a molecule of the complexing agent and resulting in the formation of a ring structure incorporating the metal ion" (Lehman, 1963, p. 167; see Fig. 3.10). Hydrogen ion is released from the organic molecule during the reaction and can participate in hydrolysis reactions. Once in solution, the chelate may be stable at pH conditions under which the included cations would ordinarily precipitate out, a topic to be discussed later. Laboratory and field work have demonstrated that chelating agents in contact

Table 3.3 Comparison of Weathering of Hawaiian Basalt Erupted in 1907 under Two Surface Weathering Environments: Lichen-Covered and Lichen-Free[a]

	Lichen-Covered Rock	Lichen-Free Rock
Mean thickness of weathering rind (mm)	0.142 (color: 10R 3/4–4/6)	<0.002
Concentration ratio of weathered crust: fresh rock for these elements		
Fe	6.36	1.21
Al	0.58	0.47
Si	0.21	1.20
Ti	0.27	0.965
Ca	0.004	1.24

Taken from Jackson and Keller (1970).

[a]The lichen is *Stereocaulon volcani.*

with rocks or minerals can bring about a significant amount of weathering. Schalascha and others (1967) ground up various minerals and granodiorite and allowed these materials to react with solutions containing chelating agents. Cations were released to the solution at rates greater than would be predicted by a hydrogen ion effect alone. They concluded that the weathering of these materials is a combination of the effect of chelating agents and hydrogen ion. In another laboratory study involving various minerals, Barman and others (1992) found that there was no unique stability of the minerals in different complexing agents, and that the release of the cations from the minerals is related to their positions within the crystal structure as well as the structure itself. For example, Ca^{2+} in interchain spaces in hornblende are the most soluble cations, and Al^{3+} in octahedral sheets of biotite are the least soluble.

Lichens growing on rock surfaces or on soils can bring about substantial amounts of weathering. Lichens excrete chelating agents (Schatz, 1963) and thus are important to the understanding of weathering of rocks and minerals. Jackson and Keller (1970) studied the weathering of a recent basalt in Hawaii under a lichen cover and in the absence of such a cover (Table 3.3). They found good evidence that weathering rinds are thicker and chemical alteration is more extensive beneath a lichen cover than in lichen-free areas of the same rock. Data for the lichen-free rock show slight enrichment in iron, sil-

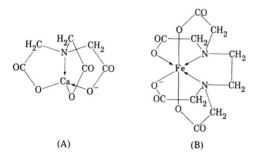

(A) (B)

Figure 3.10 Structure of two chelating agents. (A) Aminotriacetic acid binding Ca; (B) ethylenediaminetetracetic acid (EDTA) binding Fe. *(Redrawn from Ponomareva, 1969.)*

icon, and calcium, little change in titanium, and some depletion of aluminum. In contrast, the lichen-covered rock showed a sixfold enrichment of iron and depletion in varying amounts of all other elements. This study is unique in that it provides a comparison of weathering with and without biological input. These workers concluded that chelating agents in the presence of high hydrogen ion concentration, due to respiratory CO_2 and organic acids, increased weathering beneath lichens.

Chelating agents render substances more soluble as chelates under certain pH conditions. A common example is aluminum, which may be soluble as a chelate over pH values at which it is insoluble as an ion (Fig. 4.6).

Hydration and Dehydration

Hydration and dehydration are processes by which water molecules are added to or removed from a mineral. The result is the formation of a new mineral. These processes probably are not too important in overall chemical weathering because few minerals are affected, and they are not too common. An example of hydration–dehydration is the formation of gypsum and anhydrite by adding or removing water

$$CaSO_4 \; 2H_2O = CaSO_4 + 2H_2O$$
$$\text{c} \qquad\qquad \text{c} \qquad\; \text{l}$$
$$\text{(gypsum)} \qquad\quad \text{(anhydrite)}$$

A marked increase in volume accompanies the reaction anhydrite to gypsum, and, if this takes place within a rock, physical disintegration can occur.

Ion Exchange

Some weathering from one mineral to another can occur through the exchange of ions between the solution and the mineral. The most readily exchangeable cations are those between the layers of the phyllosilicates, such as sodium and calcium. During the exchange, the basic structure of the mineral is unchanged, but interlayer spacing may vary with the specific cation. Because this mechanism is important in the alteration of one clay mineral to another, it is discussed more fully in Chapter 4.

■ Measurement of the Amount of Chemical Weathering That Has Taken Place

Total chemical analysis is the most widely used way of determining the amount of chemical weathering that has taken place in a rock. There are several things one must know, or decide, when using such analyses. First, one must decide if the analysis is to be done on the whole material or on fractions thereof (clay or clay plus silt). This decision will be based on the purpose of the study, that is, is it most important to focus on total change or on the changes within the more reactive finer-grained fraction? Second, all data are presented as oxides, not because they appear that way but because the main balancing anion usually is oxygen. Third, sometimes one or two water percentages are given. The H_2O^- is that loosely held water that is lost at temperatures less than 105°C and probably should not be reported. In contrast, H_2O^+ is termed crystal lattice water, but it really is the weight loss at higher temperatures due to the conversion of crystal lattice OHs to water vapor and oxygen gas. The amount of weight loss varies with clay mineral species, percentage of clay minerals present, and temperature (Jackson, 1973). Fourth, because analyses are given in percentages, we are dealing with relative increases and decreases (e.g., a gain in one oxide must be balanced by a loss in another oxide for the analysis to achieve 100%).

Contents of the individual oxides vary as a function of many soil-forming processes (Chapter 5), and whether they are essential to the products formed, so only broad statements can be made, and all are relative to the parent rock (see chemical compositions of some common rocks in Chapter 7). Three main factors help control the fate of the oxides in leaching environments (e.g., udic moisture conditions). One is pH, as the solubilities of some elements are pH dependent (Fig. 4.6). Silica solubility is relatively constant over the normal soil pH values and increases rapidly with increasingly alkaline pH. Under oxidizing conditions, Fe^{3+} and Al are also insoluble over the normal pH values, but Al solubility increases dramatically at alkaline pH values. A second factor is ionic potential (Fig. 4.5, and discussion thereof). Briefly, K, Na, and Ca remain in the solution, Fe^{3+} and Al^{3+} precipitate as hydroxides, and Si forms a soluble hydroxy acid, H_4SiO_4. A third factor is the redox conditions of the weathering environment (Chapter 5). Under reducing conditions, both Fe and Mn are +2 valency and mobile, so can be removed from the site of release by weathering with percolating waters.

With the above in mind we can make some generalities about the fate of the oxides relative to the parent material (see Table 7.2 for chemical analyses of the common rocks) as shown by major element oxide analyses (Table 3.4). Silicon dioxide is almost always present in the parent material in amounts greater than needed to form clay minerals, so it commonly decreases relative to the parent material. Al_2O_3 has very low solubility over the usual pH range and is essential to most clay minerals, so it commonly shows a relative increase. Iron in most rock-forming minerals is present in the Fe^{2+} form (reported as FeO), and on weathering in an oxidizing environment converts to various Fe^{3+}-bearing minerals or makes up part of the clay minerals; the latter is reported in analyses as Fe_2O_3. Commonly, the Fe_2O_3/FeO ratio increases on weathering in an aerobic environment. A reducing environment will complicate the interpretation of iron

Table 3.4 Chemical Analyses, Molar Ratios of Oxides, and Abrasion pH Values for Two Andesite Rocks and Their Weathered Products, Southern Cascade Range, California

| | Hypersthene Andesite | Saprolite Samples (Increasing Weathering →) | | | | $4a^a$ | $4a - R^b$ | Olivine Andesite | Saprolite Samples (Increasing Weathering →) | | | | $4a^a$ | $4a - R^b$ |
		1	2	3	4				1	2	3	4		
SiO_2	57.0	45.7	41.3	38.9	39.9	20.9	−36.1	53.8	42.6	38.3	36.3	36.8	20.2	−33.6
Al_2O_3	16.7	22.3	29.1	31.6	31.8	16.7	0	16.9	21.7	30.3	30.5	30.7	16.9	0
Fe_2O_3	2.0	5.2	9.8	11.1	11.3	5.9	+3.9	2.3	6.7	15.2	16.0	15.6	8.6	+6.3
FeO	4.7	4.2	1.1	0.50	0.40	0.21	−4.49	5.8	2.9	0.36	0.24	0.10	0.06	−5.74
TiO_2	1.0	1.0	1.2	1.3	1.3	0.68	−0.32	1.2	1.1	1.5	1.5	1.5	0.83	−0.37
MnO	0.12	0.24	0.26	0.27	0.26	0.14	+0.02	0.12	0.15	0.09	0.17	0.10	0.06	−0.06
P_2O_5	0.15	0.07	0.03	0.03	0.03	0.02	−0.13	0.15	0.25	0.04	0.03	0.04	0.02	−0.13
CaO	7.2	4.8	1.2	0.62	0.43	0.23	−6.97	8.4	5.2	0.79	0.20	0.12	0.07	−8.33
MgO	6.1	7.2	2.7	1.4	0.27	0.14	−5.96	7.4	9.7	0.22	0.15	0.15	0.08	−7.32
Na_2O	3.1	1.2	0.86	0.79	0.62	0.33	−2.77	2.5	0.89	0.19	0.09	0.08	0.04	−2.46
K_2O	1.2	0.24	0.28	0.11	0.15	0.08	−1.12	0.60	0.31	0.27	0.15	0.15	0.08	−0.52
H_2O^+	0.35	7.4	11.8	12.6	12.9	6.8	+6.45	0.71	9.1	13.1	13.9	13.7	7.5	+6.79
Total	99.6	99.9	99.6	99.2	99.4	52.13	−47.48	99.9	100.6	100.3	99.3	99.0	54.44	−45.44
Molar sa ratio^c	5.79	3.47	2.40	2.09	2.13			5.40	2.88	2.15	2.03	2.04		
Molar sa ratio saprolite / Molar sa ratio rock		0.60	0.41	0.36	0.37				0.53	0.40	0.38	0.38		
Molar ba ratio^d ($\times 10^{-3}$)	3.36	2.81	1.03	0.59	0.26			3.72	3.44	0.25	0.11	0.09		
Molar ba ratio saprolite / Molar ba ratio rock		0.84	0.31	0.18	0.08				0.92	0.07	0.03	0.02		
Abrasion pH	8.9	5.5	5.1	4.9	4.9			8.6	6.3	5.6	5.4	5.5		

Taken from Hendricks and Whittig (1968, Table 1).

[a]Weight of each oxide (grams) assuming that Al_2O_3 content remains constant on weathering. Total weight is an approximation of what remains from the weathering of 100 g of rock.

[b]Gains and losses by weight (grams) of each oxide obtained by subtracting the rock analysis from that of column 4a.

[c]Molar sa ratio = SiO_2/Al_2O_3.

[d]Molar ba ratio = $(CaO + MgO + Na_2O + K_2O)/Al_2O_3$.

values, however, as Fe^{2+} is soluble and is removed with the percolating waters. One of the more insoluble constitutents, TiO_2, should show a relative increase. Of the major remaining elements (MnO_2, CaO, MgO, Na_2O, and K_2O), Mg is an essential part of a weathering product (i.e., Mg-chlorite); part of the K is tightly held in the interlayer position by illite; Mn can form bluish black mottles; and Mg, Ca, K, and Na are the major exchangeable cations. Most of the latter are depleted in leaching environments, but in aridic environments CaO will increase at depth if $CaCO_3$ has accumulated.

An example of chemical losses and gains is a highly weathered volcanic rock in northern California. Table 3.4 compares analytical data for fresh rock with data for various stages of weathered rock (saprolite), rated from 1 (least weathered) to 4 (most weathered). By stage 3 most of the original minerals have been altered. Note that all data are in percent, and therefore only the relative changes that take place upon weathering, not absolute gains and losses, are given. The obvious relative weathering trends are the loss of silicon and most bases, and gains of iron and aluminum. It is commonly assumed that Al_2O_3 content does not change on weathering because it is relatively insoluble at normal pH values, and much of it is tied up in the clay minerals that form. Thus, if one assumes a constant Al_2O_3 content, all constituents can be recalculated by multiplying by the factor

$$\% \text{ Al}_2\text{O}_3 \text{ in fresh rock}/\% \text{ Al}_2\text{O}_3 \text{ in weathered material}$$

This is done in Table 3.4 (columns 4a) for saprolites weathered to stage 4. Gains and losses can be determined by subtracting data in columns 4a from those of the fresh rock, and these are shown in columns 4a − R. The totals of columns 4a and 4a − R can be interpreted to indicate that of an original 100 g of rock, about 48 g has been removed by weathering and about 52 g is left. The main error in this method is that the gains and losses depend on Al_2O_3 content remaining constant on weathering, a condition not always attained. One can check on the stability of Al_2O_3 in a particular weathering environment by calculating a ratio of Al_2O_3/stable oxide vertically through the weathering zone. In this case, one could use the Al_2O_3/TiO_2 ratio. The ratio for the hypersthene andesite is 16.7, whereas those for the saprolite samples vary from 22.3 to 24.5. This implies that either Al_2O_3 has been added

to the weathered materials or TiO_2 has been depleted, or both.

This identification of the relatively stable oxide is critical to the success of the project. Al_2O_3 has been shown to be depleted from saprolites when compared to their bulk densities (Gardner, 1992) and from weathering rinds when compared to TiO_2 (Colman, 1982). Hence, if Al_2O_3 is used as the stable element in these weathering studies, the amount of elemental losses would be minimal. For any study one should examine all the "stable" oxides for mobility, and use the least mobile one for comparison with the mobile oxides.

A more accurate way of determining chemical weathering is by gains and loses in weight of material on a volume basis. Many times this cannot be done, but the two examples from northern California serve to demonstrate the usefulness of the method (Table 3.5). Because rock structure is retained in saprolite, it could be shown that there is little volume change in going from rock to saprolite; in this case, weathering is said to be isovolumetric. Hence, from data from the chemical analyses (Table 3.4) and the bulk densities for fresh material and for material of all weathering stages, the actual gains and losses in weight can be calculated (Table 3.5). It can be seen from this analysis that both Al_2O_3 and TiO_2 are depleted, and thus the assumption of a constant contents must be reevaluated. It does appear, however, that by assuming constant Al_2O_3 one can show the minimum changes that have occurred on weathering (compare relative losses in column 4a − R of Table 3.4 with those in column 4 − R of Table 3.5).

In regional comparisons of many soils or rock weathering by chemical data, it is cumbersome to use tabulated total chemical analyses, or to compare graphic displays of oxides versus depth (see tables and figures in Loughnan, 1969). Workers generally recalculate the data to a single number for these comparative studies. Molar ratios (percentage oxide divided by molecular weight; gives the relative number of atoms, so compensates for the different molecular weights, and is the number used to calculate a formula) provide good data. Common molar ratios are as follows (Jenny, 1941; Colman, 1982):

- Silica: alumina SiO_2/Al_2O_3
- Silica: iron SiO_2/Fe_2O_3
- Silica:sesquioxides $SiO_2/(Al_2O_3 + Fe_2O_3)$
- Silica:R_2O_3 $SiO_2/(Al_2O_3 + Fe_2O_3 + TiO_2)$

Table 3.5 Weight Change in Oxide Content on Weathering for Two Andesite Rocks of the Southern Cascade Range, California[a]

	Hypersthene Andesite	Saprolite (Increasing Weathering →)				Olivine Andesite	Saprolite (Increasing Weathering →)			
		1	*2*	*3*	*4*		*1*	*2*	*3*	*4*
SiO_2	157	77	48	43	42	143	82	51	47	46
Al_2O_3	46	38	34	35	34	45	42	40	39	39
Fe_2O_3	5.5	8.8	11	12	12	6.1	13	20	20	20
FeO	13	7.1	1.3	0.6	0.4	15	5.6	0.5	0.3	0.1
TiO_2	2.8	1.7	1.4	1.4	1.4	3.2	2.1	2.0	1.9	1.9
MnO	1.3	0.4	0.3	0.3	0.3	0.3	0.3	0.1	0.2	0.1
P_2O_5	0.4	0.1	0.03	0.03	0.03	0.4	0.5	0.05	0.05	0.05
CaO	20	8.1	1.4	0.7	0.5	22	20	1.1	0.3	0.2
MgO	17	12	3.1	1.6	1.0	20	19	0.2	0.3	0.1
Na_2O	8.5	2.0	1.0	0.9	0.4	6.6	1.7	0.3	0.1	0.1
K_2O	3.3	0.4	0.3	0.1	0.2	1.6	0.6	0.4	0.2	0.2
H_2O	1.0	13	14	14	14	1.6	18	17	18	17
Total Fe as Fe_2O_3	20	17	12	13	13	23	19	21	20	20

	Differences between Saprolite (1, 2, 3, 4) and Parent Rock (R)							
	Hypersthene Andesite				Olivine Andesite			
	1 − R	**2 − R**	**3 − R**	**4 − R**	**1 − R**	**2 − R**	**3 − R**	**4 − R**
SiO_2	−80	−109	−114	−115	−61	−92	−96	−97
Al_2O_3	−8	−12	−12	−11	−3	−5	−6	−6
Fe_2O_3	+3.3	+5.5	+6.5	+6.5	+7	+14	+14	+14
FeO	−5.9	−11.7	−12.4	−12.6	−9	−15	−15	−15
TiO_2	−1.1	−1.4	−1.4	−1.4	−1.1	−1.2	−1.3	−1.3
CaO	−12	−19	−19	−19	−12	−21	−22	−22
MgO	−5	−14	−15	−16	−1	−20	−20	−20
Na_2O	−6.5	−7.5	−7.6	−8.1	−5	−6	−7	−7
K_2O	−2.9	−3.0	−3.2	−3.1	−1.0	−1.2	−1.4	−1.4
H_2O	+12	+13	+13	+13	+16	+15	+16	+15
Total Fe as Fe_2O_3	+3.2	−7.5	−7.6	−8.1	−3.4	−2.0	−2.3	−2.5

Taken from Hendricks and Whittig (1968, Tables 2 and 3).

[a]All values in centigrams per cubic centimeter. The volume of the rock was shown to not change upon weathering to the various stages of saprolite.

- Bases:alumina $(K_2O + Na_2O + CaO + MgO)/Al_2O_3$
- Bases:R_2O_3 ($K_2O + Na_2O + CaO + MgO)/Al_2O_3 + Fe_2O_3 + TiO_2$)
- Parkers weathering index 100 $(K_2O/0.25 + Na_2O/0.35 + CaO/0.7 + MgO/0.9)$
- Reiche's weathering potential index (WPI) 100 (sum bases − H_2O)/(sum bases + SiO_2 + R_2O_3)
- Reiche's product index (PI) 100 $(SiO_2)/(SiO_2 + R_2O_3)$
- Iron species ratio Fe_2O_3/FeO

The first six ratios should decrease on weathering in a leaching environment, and some examples of these are given in Table 3.4. In some of the ratios, R_2O_3 is used as the sum of the most stable components in a weathering environment. The above ratios can be used for the parent material and for weathered materials or soil horizons to calculate ratios that express differences between the two. One such ratio is what Jenny (1941) calls the leaching factor:

Leaching factor =
$$\frac{(K_2O + Na_2O)/SiO_2 \text{ of weathered horizon}}{(K_2O + Na_2O)SiO_2 \text{ of parent material}}$$

These ratios have the advantage of expressing six analytical values as a single number. To illustrate this method, the $SiO_2:Al_2O_3$ ratio and $(CaO + MgO +$

$Na_2O + K_2O):Al_2O_3$ ratio for saprolite:rock are presented in Table 3.4.

Not all ratios need to be calculated as molar ratios. The molar SiO_2/Al_2O_3 (sa) ratio makes sense as both constituents are essential to the products formed (layer-lattice silicate minerals, Al-oxides); thus one gains an idea as to how far weathering has progressed toward a particular alteration product. For example, in Table 3.4 the molar sa ratio declines from >5 (rock values) to near 2 for saprolites 3 and 4. According to the sa molar ratios for clay mineral species (Table 4.2), ratios close to 2 indicate nearly complete alteration of the original rock to the 1:1 clay minerals, kaolinite and halloysite. For the saprolites in Table 3.4, halloysite is the most common layer-lattice clay mineral (Hendricks and others, 1967). In contrast, most of the bases are exchangeable cations and not essential for layer-lattice clay minerals, so oxide ratios, not molar ratios, might be sufficient. In Table 3.4 the low bases/alumina ratio is an indication of the extreme alteration of the original rock.

An example of data reduction using chemical analyses and the above ratios is the study of Colman (1982) on weathering rinds formed on volcanic clasts contained in tills in the western United States. Most of the rinds are up to several millimeters thick, and the clasts were collected from soil depths of 20–50 cm to avoid spalling by fires. For a rind representing about

Table 3.6 Weathering Data on a Weathering Rind (0–0.5 mm Depth) Compared to That of Fresh Rock (>2 mm Depth) for a Clast in Till Near McCall, Idaho

Weights assuming TiO₂ constant[a]

Interval (mm)	SiO_2	Al_2O_3	Fe_2O_3	FeO	MgO	CaO	Na_2O	K_2O	TiO_2	MnO	H_2O^+	Sum
0–0.5	16.3	8.1	7.3	2.0	1.3	1.5	0.1	0.2	2.2	0.5	4.4	43.9
>2	53.7	14.1	1.9	9.8	4.2	8.1	2.8	1.5	2.2	0.1	1.2	99.6

Normalized weights assuming TiO₂ constant[b]

0–0.5	0.3	0.6	3.9	0.2	0.3	0.2	0.1	0.2	1.0	0.5	3.6	
>2	1.0	1.0	1.0	1.0	1.0	1.0	1.0	1.0	1.0	1.0	1.0	

Molecular ratios

Interval (mm)	SiO_2/R_2O_3	Bases/R_2O_3	Fe_2O_3/FeO	Parker's index	WPI	PI
0–0.5	1.6	0.4	1.6	11.7	−35.7	62.0
>2	3.6	1.3	0.1	32.3	16.8	78.4

Normalized molecular ratios

0–0.5	0.5	0.3	18.8	0.4	−2.1	0.8
>2	1.0	1.0	1.0	1.0	1.0	1.0

Taken from Colman (1982, Tables 11, 14, 15, and 16).
[a]Calculated by multiplying weight percentage by the ratio % TiO_2 (fresh rock)/% TiO_2 (weathered sample).
[b]Calculated by dividing the value for each interval by the value for that oxide in the fresh rock (here taken as the >2-mm-depth material).

140 ky of weathering, most oxides are depleted; the main increase is in H_2O+ (Table 3.6). In this study TiO_2 is assumed constant, as it seems to be the most insoluble of all the elements released on weathering. These values and sums in Table 3.6 are equivalent to those remaining from the weathering of 100 g of rock. However, for comparison of many samples of different age, it might be best to use the normalized weights calculated by Colman. The ratios, molecular or normalized, display the expected trends with depletions of SiO_2 and bases and an increase in the iron ratio.

Parker's weathering index is a bit different from many indices in that each oxide amount is divided by a number related to the estimated strengths of the cation–oxygen bond. Loss of bases results in a decrease in the index. The weathering potential index and product index of Reiche (1943) incorporate all the major elements into two numbers. Colman considers them the most useful indices, especially when they are plotted together on one graph (Fig. 3.11). Both indices decrease on weathering, and the decrease of the former index is more rapid than the decrease of the lat-

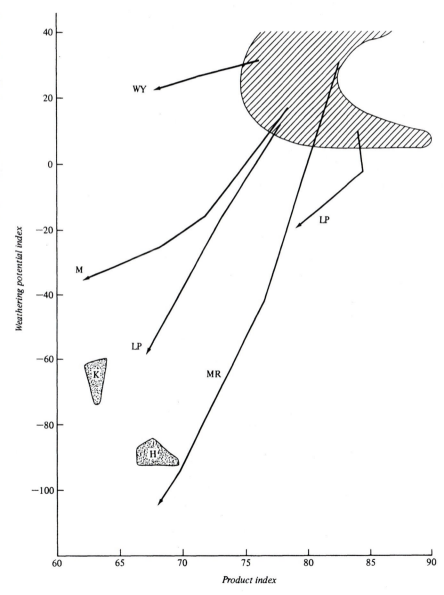

Figure 3.11 Weathering-rind data plotted with respect to both weathering index and product index: lined area, field for igneous rocks; K, kaolinite; H, halloysite. *(Taken from Reiche, 1943.)* Trends from unaltered rock to the outermost parts of weathering rinds are shown by the position of the arrows. All rinds are 140 ka or older. Rocks from West Yellowstone, Montana (WY) and McCall, Idaho (M) are basalt; those from Mt. Rainier, Washington (MR) and Lassen Peak, California (LP) are andesites. *(Redrawn from Colman, 1982, Fig. 24.)*

ter because the former reflects not only the leaching of bases, but also the increase in crystal-lattice water on weathering. Although many of the indices that have been mentioned were initially intended for use in rock weathering studies, they can also be used to depict trends in total chemical analyses of soils.

Still more indices are given in Wakatsuki and others (1977). One that they suggest using is a triangular plot, and a similar one by Chesworth (1992) is interesting because clay minerals and soil orders are added to it (Fig. 3.12). Thus, as stable clay–mineral assemblages change to lower sa values, the soil orders change to more leached soils.

In the humid leaching environment of New Zealand, several workers use the eluvial–illuvial coefficient (EIC) of Muir and Logan (1982). The equation is

$$EIC_h = \frac{S_h/X_h}{S_p/X_p} - 1 \times 100$$

where S = element oxide of interest, %
     X = stable element oxide, %
     h = data for horizon of interest
     p = data for parent material

Positive values reflect a gain, negative values a loss, and the data are relative to 100%. Data are usually presented in graphic form, and those of Knuepfer (1988) indicate depletion of many oxides from shallow depths in perhaps as little as 20 ky (Fig. 3.13). If this equation was

used for aridic soils, it would show gains in CaO and $CO_3{}^{2-}$. Somewhat similar equations have been used by Harden (1988) and Reheis (1990) in chronosequence studies in aridic areas. To compare many profiles or for regression analysis, the weighted mean percent gain and/or loss values for each profile could be used.

Brimhall and colleagues (1991) devised a rather involved method for assessing chemical change during weathering, and it gives values for volume dilution and collapse as well as losses or gains of mass. The basic equation is

$$\frac{V_p \rho_p C_{j,p}}{100} + m_{j,\text{flux}} = \frac{V_w \rho_w C_{j,w}}{100} \quad (3.1)$$

where V is volume, ρ is bulk density, C_j is the concentration of element j in percent, p is parent material, w is the weathered mass or soil, and $m_{j,\text{flux}}$ is the mass of material introduced into (positive) or out of (negative) the parent material. In the equation, therefore, the first element (j in the left-hand side of the equation) is the mass (grams) of the particular element in the parent material, the second element j is the mass of the flux, and the third element j is the resulting mass in the weathered material. Volume changes during weathering are determined by calculating strain

$$\epsilon_{i,w} = \frac{V_w - V_p}{V_p} = \frac{V_w}{V_p} - 1 \quad (3.2)$$

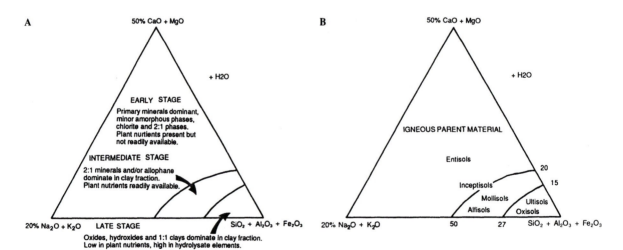

Figure 3.12 Triangular plots showing (A) dominant clay minerals and (B) soil orders for igneous rocks in a humid climate. In a leaching environment with time, both data sets move to the lower right corner of the diagrams. *(Reprinted from Chesworth, 1992, Fig. 2.5, Weathering systems, p. 19–40. In Martini and Chesworth, eds. Weathering, soils & paleosols, © 1992, with kind permission from Elsevier Science, The Netherlands.)*

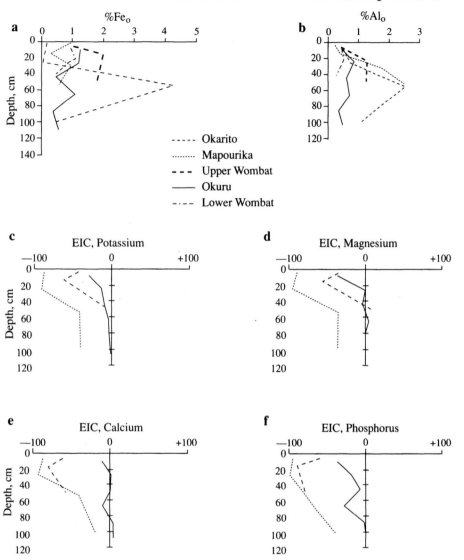

Figure 3.13 Oxalate extracts for Fe and Al (a,b), and EIC data for four elements (c–f), for soils on the humid west coast of the South Island, New Zealand. The soils are arranged in order of age, with Lower Wombat the youngest and Okarito the oldest. Negative EIC indicate losses relative to the parent material. Although the EIC elemental data all show loses with time, Fe_o and Al_o accumulate, thus reflecting their different behavior in the same pedological environment. *(Redrawn from Kneupfer, 1988, Fig. 7, from the Geological Society of America Bulletin 100, 1124–1136 © 1988, with permission from the Geological Society of America, Boulder, Colorado.)*

where i refers to strain determined by an immobile element in an insoluble mineral. Rather that assume isovolumetric weathering, they solve Eq. (3.1) for strain ($m_{i,flux} = 0$ because i is immobile; i can be Ti or Zr) and get

$$\epsilon_{i,w} = \frac{\rho_p C_{i,p}}{\rho_w C_{i,w}} - 1 \qquad (3.3)$$

Positive strains are dilations, and negative strains denote collapse. Incorporation of organic matter into the surface horizons is an example of dilation, and removing soluble minerals from an initial volume and leaving the immobile element more concentrated is an example of collapse. They can then calculate mass gains or loses

$$\delta_{j,w} = \frac{m_{j,\text{flux}}}{V_p} = \frac{\rho_w C_{j,w}(\epsilon_{i,w} + 1) - \rho_p C_{j,p}}{100} \quad (3.4)$$

and the mass fractions relative to the mass of element j in the parent material

$$\tau_{j,w} = \frac{100 \, m_{j,\text{flux}}}{V_p C_{j,p}\rho_p} = \frac{\rho_w \, C_{j,w}}{\rho_p \, C_{j,p}}(\epsilon_{i,w}+1) - 1 \quad (3.5)$$

Vidic (1994) studied the weathering of a chronosequence of soils in Slovenia, and was going to use Eq. (3.5). However she did some substituting and found that EIC and τ are identical, and proved it by getting identical results from both methods. We use EIC because it is easier to calculate.

An example of a soil chronosequence of marine terraces from the northern coast of California demonstrates how the above method works (Brimhall and others, 1992; Merritts and others, 1992). The younger soils (<40 ka) are characterized by dilation relative to Zr, due to incorporation of quartz eolian sands and pedogenic organic carbon near the surface (Fig. 3.14). Collapse is dominant in older soils, and the average amount of negative strain increases with age, due to the loss of Si and other cations. For a soil near 240 ka, approximate average losses are <50% Si, 50–75% Na, and ~75% Ca.

The abrasion pH of weathered materials can be used as a rough measure of the weathering that has taken place. Grant (1969) has shown that

$$\text{Abrasion pH} = f \frac{\text{Na} + \text{K} + \text{Ca} + \text{Mg}}{\text{Clay minerals}}$$

These are compared with various molar ratios in Table 3.4, and it is seen that as the various molar ratios decrease, so too does the abrasion pH. One could also combine abrasion pHs into ratios of weathered material:rock, in much the same way that various oxides can be combined.

Oxide ratios can be calculated on the basis of either weathered rock:parent material or weathering by-product:parent material. The former ratio gives the overall changes in the parent material. The latter, however, gives the direction in which the weathering reactions are going, as reflected in the composition of the by-product. This is particularly true for the SiO_2:Al_2O_3 ratio because this ratio differs with the clay–mineral species that forms.

■ Chemical Weathering Disguised as Physical Weathering

In many field situations, it is difficult to determine quantitatively the amount of weathering due to phys-

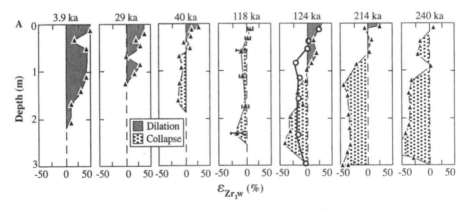

Figure 3.14 Strain plotted with depth for soils of different ages, northern California coast. *(Redrawn from Brimhall and others, 1991, Fig. 1a, Deformational mass transport and invasive processes in soil evaluation. Science 255, 695–702, © 1991, with permission from the American Association for the Advancement of Science.)*

ical processes relative to that due to chemical processes. Indeed, if visual evidence for chemical weathering is lacking, one usually is tempted to look for evidence supporting some physical weathering process. The weathering may be chemical, however, and the evidence for it is either quite subtle or has been removed from the site of weathering.

One common example is the formation of granitic grus, meters or tens of meters thick. In some localities, there is little visual evidence for chemical weathering, such as clay formation or iron oxidation, and the problem confronting many workers is just how this deep weathering is accomplished. The answer lies in the examination of the rock and constitutent minerals under a petrographic microscope or by X-ray. Wahrhaftig (1965) studied the origin of grus in the Sierra Nevada and was able to show that the granitic rock was shattered to considerable depth by the expansion of several minerals on alteration. Some biotite and some plagioclase had altered to clay. Volume increase accompanying the alteration was sufficient to fracture the surrounding minerals, as seen in thin section, and this brought about the disaggregation of a great thickness of rock. Other work confirms this origin for grus (Isherwood and Street, 1976). Biotite alteration was found to be important in the formation of grus in Wyoming (Eggler and others, 1969). In this case, however, the biotite was initially altered during Precambrian, high-temperature oxidation. Later weathering in a near-surface environment exploited the already altered biotite grains, causing them to alter further, expand, and internally shatter the rock to grus. In Australia, hydrothermal cracking, followed by volume increase during biotite and plagioclase weathering seems to account for the formation of grus at least 13 m thick (Dixon and Young, 1981). Grus formation in southern California is also attributed to biotite alteration (Nettleton and others, 1970). Although biotite alteration was not always seen with the petrographic microscope, it could be deciphered by X-ray. As an example of the amount of expansion possible, it was pointed out that the complete alteration of biotite to vermiculite can be accompanied by a 40% volume increase, and data were presented to indicate that this had happened in places. A note of caution should be injected here, as the review by Gerrard (1994) suggests that the above process is not accepted by all workers.

If when working in areas of deep weathering of granite, the weathering classes of Clayton and oth-

ers (1979) might be useful. Seven classes are recognized:

1. Unweathered rock—with bulk density ~ 2.7 g/cm^3.
2. Very weakly weathered—some iron stains from biotite weathering, feldspars have some opacity, junctions of joint planes are angular, and bulk density ~ 2.5.
3. Weakly weathered—hammer blow gives dull ring and results in hand-size fragments, feldspars are opaque, junctions of joint planes are subangular, there is no root penetration, and bulk density ~ 2.3.
4. Moderately weathered—no ring from hammer blow, biotites have yellow sheen, joint sets are indistinct and junctions are rounded, and bulk density ~ 2.2.
5. Moderately well weathered—break into small fragments with hands, roots penetrate along fractures, and bulk density ~ 2.2.
6. Well weathered—fragments disaggregate by hand to sand-size particles (grus), roots penetrate between grains, and bulk density ~ 2.1.
7. Very well weathered—original rock fabric is preserved but feldspars are weathered to clay, material is plastic when wet, and bulk density ~ 1.8.

In the ideal case, the classes should be arranged vertically from 1 at the base of a weathering zone to 7 at the top. Classes 2 and 3 result from initial chemical and physical weathering, 4 through 6 from intense physical and moderate chemical weathering, and class 7 from intense chemical weathering.

Mapping a landscape, using these class distinctions, can be useful in indicating the relative stability of the landscape (7 = very stable and low erosion rate), or with land use decisions. Permeability and ease of excavation are important to many construction projects and it would be helpful to be able to make predictions ahead of time. Also, landslides seem best correlated with class 7. Pierce and Schmidt (1975) have produced a prototype rock-weathering map for the foothills of the Rocky Mountains, but it could be improved using these seven classes.

Another common example of weathering that might be attributed to physical processes is weathering that goes on at high altitudes. For example, stones lying on the surface of Holocene deposits above timber line in the Colorado Front Range are deeply pitted (Fig. 3.15). Commonly, the felsic mineral bands of coarse-grained gneisses form depressions, and the mafic mineral bands stand in relief. The climate is very rigorous, with long cold winters and short cool summers. Because many of the rock surfaces appear fresh, and

Figure 3.15 Pitted stone on the surface of Holocene till above treeline in the Colorado Front Range. Such features could take 5–10 ky to form. (See Birkeland and others, 1987)

there is little visual evidence for chemical weathering, one might suspect that most of the weathering is due to physical processes. However, fine-grained igneous and metamorphic rocks in the Colorado Front Range and in other parts of the Colorado Rockies, presumably of the same age or older, show considerable development of weathering rinds that result

from chemical weathering (Fig. 3.16). Thus, chemical weathering of exposed rocks goes on under the rigorous climatic conditions of high altitudes. This relationship suggests that the coarse-grained, pitted rocks also are undergoing chemical weathering, that this weathering loosens the bonds between mineral grains, and that once the grains are loosened, physical processes remove them from the rock surface. One such process would be the high winds that sweep the tundra in the winter. The evidence, therefore, for chemical weathering does not remain on the rock surface, which can always have a fresh appearance. In contrast, chemical weathering of fine-grained rocks results in a weathering rind that remains intact on the rock surface and thus provides the basic evidence for chemical weathering.

■ Rate of Present-Day Chemical Weathering

It would be helpful to have a measure of regional rates of chemical weathering in near-surface environments. Tombstone weathering studies can provide data for above-ground processes, and air pollution has accelerated the rate of weathering (Meierding, 1981, 1993). Experiments can be devised to determine rates (Trudgill, 1976). For example, packets of crushed rock can be placed in the ground in different environments and weight loss determined over various time spans (Caine, 1979). Clay formation in dated soils could also be used, but clays form so slowly in some environments that thousands of years may have to elapse before clay formation is noticeable. Another complication is that the clays are a by-product of weathering and thus represent only part of the material released

Weathered rind

Unweathered rock

Weathered rind

Figure 3.16 Weathering rind, 1-cm thick, in granitic rock on the surface of a late Pleistocene rock glacier, western Colorado. (See Birkeland, 1973.)

by weathering. Furthermore, some clay can be eolian and not related to weathering (Chapters 5, 7, and 8).

One approach to estimating the present-day rate of chemical weathering is to examine the chemistry of surface waters, because many of the ions in water come from weathering reactions. The system we are dealing with is exceedingly complex (Fig. 3.17), as shown by many workers (e.g., Clayton, 1986, and Mather, 1997). The main sources of ions in the soil solution are eolian input, that contributed by through-fall during passage of rainfall through the vegetation, and release upon mineral weathering or organic matter decomposition. Rainfall will add some ions, but this amount can be determined and subtracted from the overall ion content. Some ions combine to form weathering products (Chapter 4). Once the layer-lattice clay minerals form, they provide a negative template to which cations in the soil solution are attracted, and there is a general order of replacement of one cation by another. Some ions will cycle through the biosphere, and it is probable that the amounts gained from and lost to the soil soon reach a steady state. Analyses of waters that have moved through the soil should provide the best data (e.g., Wolff, 1967). Of the water that moves through the soil, part goes to

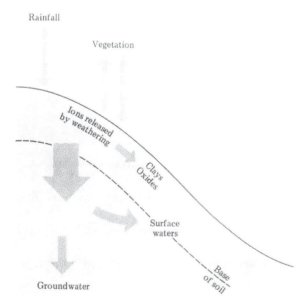

Figure 3.17 Hypothetical flow of ions in a near-surface environment. Arrow width approximates the ion content involved in the reaction or transfer. Atmospheric dust could be added to the influx at the surface.

groundwater reservoirs and part becomes surface-water runoff. Assuming that the dissolved load in surface water is proportional to the rate of chemical weathering, one might be able to use stream dissolved-load data to roughly estimate chemical weathering in weight loss per unit area per year. However, it is highly likely that ions will be added to both ground and surface waters from nonsoil sources—weathering of the aquifer minerals by the ground water is one way. The best indicator of chemical weathering of a watershed has been sought, and SiO_2 is the choice of Yuretich and Batchelder (1988) as it is derived from weathering and less prone to the recycling that the cations go through. We can view this approach to chemical weathering several ways and at several different scales.

Lysimeter studies provide good data on processes within a particular soil. Lysimeters are recepticles placed in the ground at various levels to collect representative soil water samples of that level (Litaor, 1988). A good example comes from a spodosolic soil formed from granitic parent material under forest in British Columbia, Canada (MAP = 220–270; MAP is mean annual precipitation in centimeters) (Feller, 1977). A characteristic trend is that water pH increases as waters progress through the system, expected when waters react with minerals and the latter weather. Rainfall has a pH of 4.5, and that of the throughfall water (percolates through the vegetation canopy) is 4.4. That at the surface of the forest floor (O horizon) is 5.5–6.0, that at the base of the O horizon is 6.0–6.5, and the adjacent stream water is 6.5–7.0.

Chemical analyses of the lysimeter waters for the latter study show differences between the base cations and Fe + Mn (Table 3.7). Base cations are present in rainfall, but Fe + Mn are absent. Base concentrations increase between 2 and 18 times by the time waters have percolated through the canopy. Concentrations for most elements continue to increase to the base of the forest floor, and in general decrease to the base of the soil. Of particular interest is Fe, which is present only at low levels except for the base of the forest floor; this being in a spodosolic environment, this probably reflects the short distance over which Fe is soluble and moves. In contrast, most of the base cations remain at fairly high levels, but the amounts decrease at the base of the soil. The specific connection between the lysimeters and ground and surface waters is less certain, but excess bases move into both water masses, whereas Fe and Mn do not.

Table 3.7 Data on Elements in Solution at Various Places in the Ecosystem (mg/liter)

	K	Na	Mg	Ca	Fe	Mn
Precipitation	0.06	0.3	0.06	0.2	0	0
Throughflow	1.1	0.64	0.3	1.1	0.01	0.19
Surface	1.2	0.56	0.39	1.8	0.02	0.17
Base, O horizon	1.7	1.04	1.82	2.9	0.39	0.26
Base soil	0.35	0.84	0.65	1.8	0.01	0.01
Groundwater	0.09	0.87	0.25	1.3	0	0.01
Stream water	0.09	1.02	0.31	1.5	0	0

Taken from Feller (1977).

Ugolini and others (1991) present work on solutions of several soils in the northwestern United States. For a Spodosol they show the charge balance, and most ions are in greater abundance in the E horizon than at depth. Organic acids in the E and Bhs horizons have chelating properties, form Al and Fe complexes, and allow these metals to move vertically. Still deeper in the profile the solution chemistry indicates that carbonic acid is the main proton donor, incongruent weathering takes place, and Al and Fe precipitate out of solution.

At the next larger scale, we can examine weathering rates of catchments (Drever and Clow, 1995). A clear example of weathering of a small watershed (38.5 ha) is the mass-balance study of Cleaves and others (1970) in Maryland (MAP ~100). Data were collected for a 2.25-year period in the late 1960s. Ideally, the inputs of dissolved solids (precipitation plus that from weathering) should equal the outputs of dissolved solids (runoff, storage, and biomass) at the steady state (Table 3.8). Such equivalence could not be proven because biomass transfers were not measured, but instead taken as the algebraic sum of the other four quantities. Probable sources for dissolved solids in precipitation are pollution (for example, the large amount of SO_4^{2-}) and the sea. Dissolved solids from weathering were estimated from knowledge of the minerals being weathered and the weathering products (they used equations such as those in Table 3.1). Chemical denudation is about 31 kg/ha/year (Table 3.8, cations + SiO_2). They then estimated the dissolved solids released by each of the weathering reactions and the amounts of silicate minerals consumed and formed for that period of time (Table 3.9). One

Table 3.8. Geochemical Balance for 2.25 Years for a Small Watershed in Maryland[a]

	Precipitation		Weathering		Runoff		Storage		Biomass
SiO_2	1.04	+	23.54	=	19.00	+	5.58	+	0
Na^+	2.13	+	2.62	=	3.51	+	1.24	+	0
K^+	1.68	+	2.30	=	1.95	+	0.54	+	1.48
Ca^{2+}	3.16	+	1.28	=	2.99	+	0.35	+	1.11
Mg^{2+}	1.04	+	1.14	=	1.83	+	0.35	+	0
HCO_3^-	3.16	+	28.65	=	16.72	+	3.73	+	11.36
Cl^-	6.30	+	—	=	4.42	+	2.00	−	0.12
SO_4^{2-}	17.83	+	—	=	3.93	+	0.69	+	27.05
H^+	0.27								

Taken from Cleaves and others (1970, Table 8).

[a]Units, kg/hectare/year.

Table 3.9 Amounts of Dissolved Solids Produced and Minerals Consumed and Produced by Weathering in a Small Watershed in Maryland[a]

Reaction	Dissolved Solids (kg)					
	SiO_2	Na^+	Ca^{2+}	Mg^{2+}	K^+	HCO_3^-
Oligoclase to kaolinite	1195	229	112	—	—	948
Biotite to vermiculite	408	—	—	61	199	620
Vermiculite to kaolinite	122	—	—	37	—	915
Kaolinite to gibbsite	314	—	—	—	—	—
Total	2039	229	112	98	199	2483

Reaction	Silicate Minerals Consumed (Minus Sign) or Formed (Plus Sign) (kmol)				
	olig	bio	vermic	ka	gb
Oligoclase to kaolinite	−12.8	—	—	+7.8	—
Biotite to vermiculite	—	−2.8	+1.7	—	—
Vermiculite to kaolinite	—	—	−1.0	+1.5	—
Kaolinite to gibbsite	—	—	—	−2.6	+5.2
Total	−12.8	−2.8	+0.7	+6.7	+5.2

Taken from Cleaves and others (1970, Table 9).

[a]Data are for the entire watershed for the 2.25-year period.

important aspect of this is that as clay minerals form at a particular rate, knowing the clay mineral species provides data on the negative charge buildup in the soils—these are the sites that attract the base cations. Finally, although data for the short term indicate that chemical denudation is five times that of mechanical denudation, they suggest that both are equal in the long term. At the calculated rates, the present landscape could have formed in about 1.5 My.

In 110- to 190-ha watersheds in granitic rock in Idaho (MAP = 100), Clayton and Megahan (1986) report chemical denudation rates of 95–115 kg/ha/year, or three times that above. Plagioclase weathering supplies most of the mass with $SiO_2 >> Ca = Na$.

A multiyear study in a larger watershed (3076 ha), the Hubbard Brook Experimental Forest, New Hampshire, provides detailed information of fluxes of all elements (Likens and others, 1977) (MAP = 130; MAT 5; MAT is mean annual temperature in °C). Bedrock is granitic and metamorphic rocks, and the area has been glaciated. The bulk ionic composition of the incoming precipitation is characterized mainly by H_2SO_4, HNO_3, and HCl. These acids are not sufficient to explain the cations leached from the system, so the

difference (approximately an equal amount) is internally generated H^+. The water leaving the system is much different, being characterized by $CaSO_4$, Na_2SO_4, and $Mg(NO_3)_2$. This changing chemistry is a reflection of the chemical weathering reactions. The amounts of the major elements released can be approximated (Table 3.10), and because 21.1 kg of Ca^{2+} is contained in 1500 kg of rock, this may be the lower limit of the amount of rock weathered per year. Assuming that all of the Ca^{2+} in the rock is released by weathering, the percentage of the other elements released from the rock is calculated and the ranking in the last column of the table is obtained. In general, the Ca-minerals are the most susceptible to weathering, and the K-minerals the least. Although a large amount of Si^{4+} is released by weathering, so much is contained in the rock that the percentage released is low. In contrast, Al^{3+} is released in low quantities, the amount in the rock is moderate, and the percentage released matches that of Si^{4+}. Chemical denudation is about 58 kg/ha/year, and it would be higher if the elements were expressed as oxides.

With all the data, Likens and others (1977) were able to calculate annual budgets for the various ele-

Table 3.10 Differential Chemical Weathering at Hubbard Brook

Element	Abundance in bedrock (%)	A Annual release from bedrock by weathering[a] (kg/ha)	B Amount contained in 1500 kg of bedrock (kg)	Differential weathering ratio[b] = (A/B) × 100 (%)
Ca^{2+}	1.4	21.1	21.1	100
Na^+	1.6	5.8	24.1	24
Mg^{2+}	1.1	3.5	16.5	21
K^+	2.9	7.1	43.6	16
Al^{3+}	8.3	1.9	124.8	2
Si^{4+}	30.7	18.1[c]	461.7	2

Taken from Likens et al. (1977, Table 18).

[a]Based on net output of dissolved substances plus living and dead biomass accumulation, which for Ca^{2+}, Na^+, Mg^{2+}, and K^+ is 9.5, 0.17, 0.9, and 6.1 kg/ha/year, respectively.

[b]Normalized to calcium, i.e., assuming the complete extraction of calcium from 1500 kg of bedrock.

[c]Assuming that dissolved silica is in the form of SiO_2, and 0.1 kg Si per hectare is exported in stream water as organic particulate matter.

ments, Ca being one example (Fig. 3.18; see also Likens and others, 1998). Over six times the amount of Ca leaves the studied ecosystem as enters by precipitation. The soil and rock have the largest reservoir of Ca (to 45 cm depth), and this is released at an estimated annual rate of 21.1 kg/ha/year (Table 3.10). Uptake, however, is three times the release rate, and the reason for this is that the biomass is still accumulating in the ecosystem. It is obvious that in these young postglacial soils there is abundant Ca; this would not be true for well-developed, leached soils of older landscapes.

Incorporating biomass data into the calculations is important because the ions involved are derived from mineral weathering and soils. Taylor and Velbel (1991) point out that by not doing so, one can underestimate calculated rates of mineral weathering and soil formation by up to a factor of 4. In contrast, many studies assume a biomass steady state.

Finally, we can look at one more global view between weathering and soil formatuion. Wakatsuki and Rasyidin (1992) used a mass-balance approach to suggest global rates of soil formation. They first calculated the mean composition of major elements in the earth's crust (they used igneous rocks) and soil (Table 3.11). The results, although general, collaborate what has been said in this chapter, namely on a global scale Si, Fe, and Al either remain the same or are concen-

trated. Of the bases, Mg and K, because they are sequestered in secondary clay minerals, show less loss than does Ca. They then solved the global mass balance equation:

$$R + G = S + W$$

where R is the average rock, G is the atmospheric input in grams to produce the solutes in 1 liter of surface water (W), and S is the grams of soil produced. For the globe, the mean rate of rock weathering is 1100 kg/ha/year, and this produces a mean of 700 kg/ha/year soil. Although the latter is less than the global mean rate of soil erosion (900 kg/ha/year), the range of the latter figure is large and and does not preclude the notion that there is a dynamic equilibrium between soil formation and soil erosion on a global basis. This is only one definition of soil, and perhaps not the best one as it is only a mass. One could think of others. For example, one could calculate global buildup of C, of pedogenic Fe_d, or pedogenic clay, or sum all the pedogenic accumulations into one figure.

To test the above method on a local scale, Wakatsuki and Rasyidin (1992) made similar calculations for the Hubbard Brook ecosystem in New Hampshire, and small watersheds in granite and pyroclastic materials in Japan. The results were that Hubbard Brook values are 0.7 and 0.8 of the global averages for weath-

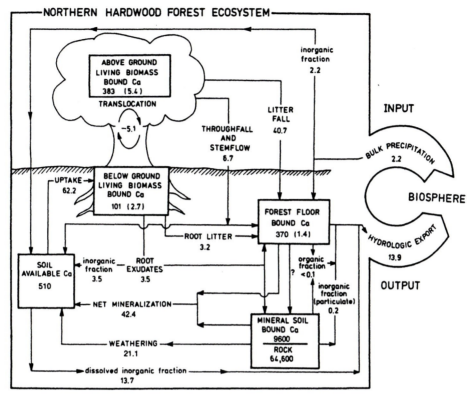

Figure 3.18 Annual Ca budget for an aggrading forested ecosystem at Hubbard Brook Experimental Forest. Standing crop values are in kg/ha, Ca fluxes in kg/ha/year, and values in parentheses are accretion rates/year. *(Reprinted from Likens and others, 1977. Biochemistry of a forested ecosystem, Fig. 30, Springer-Verlag, New York.)*

ering and soil formation, respectively, and the values for Japan are 3.5 and 3.9 times greater than global weathering and soil formation in the granitic terrain, respectively, and 7.2 and 8.1 times greater for the py-

Table 3.11 Mean Composition of Major Elements for Igneous Rocks (R) of the Crust and Soils (S)

Element	R (mg/g)	S (mg/g)
Si	277	292
Al	82	93
Fe	41	45
Ca	41	13
Mg	23	12
Na	23	5
K	21	16

Taken from Wakatsuki and Rasyidin (1992, Table 1).

roclastic watershed. Higher precipitation and temperatures probably account for the greater values in Japan, and parent material susceptibility to weathering for the differences within Japan.

■ Chemical Denudation: Does the Landscape Lower?

Both physical and chemical denudation lower landscapes (Stallard, 1995), and much work has been done on the relative importance of each for individual catchments and on a global scale. The focus here will be on chemical denudation.

Chemical denudation is commonly calculated from dissolved load data of streams, corrected for atmospheric contributions (Berner and Berner, 1987; Summerfield, 1991). The annual load is multiplied by the annual discharge, and divided by the basin area.

Berner and Berner (1987) calculate a world average of 23 tons/km^2/year (230 kg/ha/year), which is 15% of the total (physical + chemical) denudation. On a continental scale, Garrels and MacKenzie (1971, Table 5.1) give this ranking of chemical denudation: Europe > North America = Asia > South America > Africa >> Australia. For contrasts, some of the Russian rivers that flow north to the Arctic Ocean have a dissolved load that is over 70% of the total, whereas that of the Colorado and Brahmaputra Rivers is only 5–7% of the total. Of the world's large rivers, the Yangtze and Brahmaputra have the highest rates of chemical denudation (~100 tons/km^2/year).

The Amazon River Basin is an example of weathering and chemical denudation at a grand scale (6.15 million km^2) in a humid environment (Stallard, 1985, 1992). The high Andes rim the upper part of the basin, and low-relief shields make up the lower basin. Erosional regimes are subdivided into weathering limited and transport limited. Weathering limited means that erosion is limited by the rate at which weathering can deliver materials to the streams, and transport limited means that there is abundant material to move, but low relief inhibits such removal. The mountains, with their steep slopes and thin or no soils, are weathering limited, and chemical denudation will be strongly correlated with the susceptibility of the rock to weathering. In contrast the lowlands, with low relief, low-angle slopes, and thick soils are transport limited, and weathering rate is independent of lithology. Chemical denudation is an order of magnitude higher in the mountains than in the lowlands (Fig. 3.19). This difference is partly explained by the soils. Rainfall has ready access to the unweathered minerals in the fractured rocks of the mountains, and chemical weathering proceeds rapidly. In contrast, the lowlands, with a thick soil, end-point weathering products, and few unweathered minerals are characterized by low rates of weathering, despite the high potential for such weathering (high precipitation and temperature).

Many factors are involved in explaining chemical denudation rates from place to place. First, however, atmospheric influx has to be known, biomass has to be assessed as to whether it is in a steady state, and all elemental data should include the mass of oxygen. Still another source for error is the amount of elements contributed to the stream by minerals weathering during their transport as stream solid load. Several factors influencing chemical denudation are susceptibility of rocks and minerals to weathering, area of

unweathered mineral surfaces for water to react with, amount of stable weathering products in the weathering profile, permeability of the rocks and soils, steepness of slope or relief, tectonic regime, and climate (see discussion in Walling, 1987 and Pinet and Souriau, 1988).

Taiwan and the Southern Alps of New Zealand have high rates of total denudation, probably because both are fast-rising mountains on plate boundaries, in areas of high rainfall. Taiwan has one of the highest denudation rates in the world: 13×10^4 kg/ha/year of physical denudation and 65×10^2 kg/ha/year of chemical denudation (Li, 1976). If there were no uplift, these rates would flatten the Central Range in ~0.5 My. The Southern Alps could be an example of a range at a steady state; the uplift rate (5–10 m/ky) seems balanced by the denudation rate (4–11 m/ky) because of the exceptionally high rainfall (~12 m/year) at the crest (Whitehouse, 1988).

When denudation rates are presented, one envisages a landscape lowering at that particular rate. This is because commonly the data are converted to millimeters or m/unit time. Of course, denudation is an average and indicates nothing about where it is taking place. Take the above example of the Andes. Figure 15.12 of Summerfield (1991) suggests high rates of chemical denudation. We have worked in the headwaters of some of the drainages; they are glaciated, U-shaped, flat-floored valleys, with prominent A/B/C soil profiles across the moraines and steep slopes (Rodbell, 1993; Miller and others, 1993). There is no reason to suspect that dissolved load is not leaving this part of the area, and much of it may be derived from the soils. Landscape lowering, however, may not be great, primarily because solid load removal does not seem to be great in some of these valleys. In contrast, barren, steep, V-shaped valleys downvalley of the glaciated areas could be delivering the greater amount of the total solid and dissolved load from the high Andes; because of the solid load removal, these may be the loci of actual landscape lowering.

In contrast to landscape lowering with solid load removal, chemical denudation does not require landscape lowering (see also discussion in Summerfield, 1991). Rather, removal of ions from a weathering mass of rock might be reflected in a change in bulk density. If a parent material of bulk density 2.0 g/cm^3 converts via volume expansion (common in soil formation) to one of bulk density 1.5, the geomorphic surface can actually go up! If, however, there is a loss

Figure 3.19 Estimated denudation rates and dissolved-solids concentrations of rivers for a lowland shield, upland shield and mountain belt from samples from the Orinoco and Amazon River basins. *(Redrawn from Stallard, 1992, Fig. 6–8, Tectonic processes, continental free-board, and the rate-controlling step for continental denudation, p. 93–121. In Butcher et al., eds., Global geochemical cycles, © 1992, by permission of Academic Press Ltd., London.)*

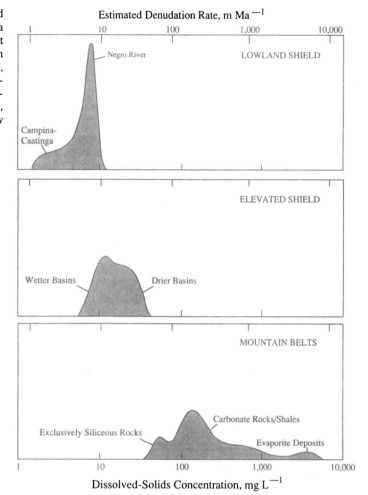

of dissolved load, bulk density decreases, but the surface can remain unchanged. Dethier (1986, Fig. 9) illustrates how the calculation can be made. In any calculation it is difficult to isolate the impact of physical versus chemical denudation. One scenario is that chemical denudation lowers the rock bulk density, which weakens the rock to the point that mechanical denudation takes over and is responsible for the lowering of the landscape.

Many workers have speculated on erosion surfaces, and whether they are in a steady state or not. Pavich (1986) provides some calculations on saprolite and soil formation formed on granitic terrain in the Appalachian Piedmont of Virginia that addresses the issue (MAP = 104; MAT = 10). Soil (<1 m thick) overlies saprolite (average ~12 m thick), which, in turn, overlies rock. Making various assumptions, he calculates a saprolite production rate of 4 m/My. The most active zones of chemical processes are the lower half of the saprolite and the soil. The former produces saprolite and the latter consumes it. In the time required to convert 12 m of rock to saprolite, 12 m of saprolite can be altered to 3 m of soil. But, the soil thickness stays constant at about 1 m. This can be explained by the long-term removal of silica and other elements in solution, and of fines from flat surfaces. Nearby old fluvial gravels help constrain the rate of downcutting, and the calculations show that both the upland surface and the weathering front are descending at about the same rate. The dissected upland landform, therefore, is unchanging over time and in a steady state.

The above examples illustrate the integration of several disciplines to help understand the complex physical, chemical, and biological processes operating in a watershed. Many assumptions had to be made, however, so the results are only approximations. Other studies are reviewed by Drever (1982). Clayton (1983) also reviewed these and other methods as he wanted data on the rate of nutrient supply from rock weathering so that forests can be harvested without depleting the nutrients required by the trees. The point to this discussion is to show that one might be able to make rough estimates on regional rates of chemical weathering and somehow link these to soil formation. The important factors of soil formation and of weathering have to be taken into consideration, however, so that one is fairly certain of the influence of each. For example, the influence of rock type can be held constant by studying only those basins underlain with granitic rocks. The time factor can be held reasonably constant by restricting the study to areas characterized only by weakly or moderately developed soils, or to basins all deglaciated at the same time. One could devise ways of keeping some of the other factors reasonably constant. Dissolved-load data then can be compared with variation in regional climate or tectonic regime to determine the influence of both.

The Products
of Weathering

Materials released during weathering either are removed from the system in leaching water or react in the system to form a variety of crystalline and amorphous products. The most commonly observed reaction products are the clay minerals and hydrous oxides of aluminum and iron. These products can occur alone or in combination, and their distribution with depth can be uniform or variable. Characterization of these products is important because of all the properties of a soil, they probably best reflect the long-term effect of the chemical and leaching environment of the soil, and they influence many soil properties (Nettleton and Brasher, 1983). Their genesis is varied; it can range from relatively simple ion exchange reactions to the more complex combination of aluminum and silicon from solution or a gel to form a crystalline clay mineral.

These minerals are generally so finely divided that their identification is difficult; it can be accomplished, however, by a variety of chemical, thermal, X-ray, and electron microscope techniques (Brindley and Brown, 1980; Dixon and Weed, 1989; Moore and Reynolds, 1989; Amonette and Zelazny, 1994). Only those products commonly found in soils will be dealt with here.

■ Clay Minerals

The common clay minerals are hydrated silicates of aluminum, iron, and magnesium arranged in various combinations of layers (Newman and Brown , 1987; Dixon and Weed, 1989; Velde, 1992). They are called layer silicates or phyllosilicates. Two kinds of sheet structures, the tetrahedral and the octahedral, make up the clay minerals, and variations in combinations of these structures and in their chemical makeup give rise to the multitude of clay minerals. The basic difference between these two sheets is in the geometric arrangement of Si, Al, Fe, Mg, O, and OH. The arrangements differ with the cation because it is the size of the cation that determines how many O or OH ions surround it.

The silicon tetrahedron is the basic structural unit of the tetrahedral sheet. It consists of one Si^{4+} surrounded by four O ions (Fig. 4.1). A sheet is formed when the basal O ions are shared between adjacent tetrahedra and the remaining unshared O ions all point in the same direction. These sheets have a net negative charge.

In the octahedral sheet, the basic unit consists of the octahedral arrangement of six O or OH ions about a central Al, Mg, or Fe ion (Fig. 4.1). In the octahedral sheet, the O and OH ions are shared between adjacent octahedra. If a trivalent ion is the central cation, two-thirds of the available cation sites are filled and the structure is said to be dioctahedral (sometimes called gibbsite structure); in contrast, divalent cations will fill all available sites and exhibit a trioctahedral structure (sometimes called brucite structure). These sheets have a net positive charge.

The variety of clay minerals results in part from different arrangements of the tetrahedral and octahedral sheets. The two adjacent sheets are bound together tightly by the mutual sharing of the O ions of the tetrahedral sheet. Thus in the octahedral sheet of a clay min-

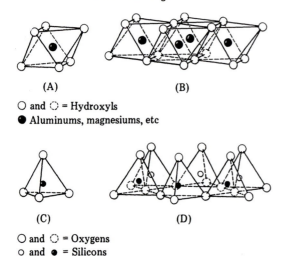

(A)   (B)

○ and ◌ = Hydroxyls
● Aluminums, magnesiums, etc

(C)   (D)

○ and ◌ = Oxygens
o and ● = Silicons

Figure 4.1 Diagram showing (A) octahedral unit, (B) octahedral sheet, (C) tetrahedral unit, and (D) tetrahedral sheet. *(Redrawn from Grim, 1968, Figs. 4.1 and 4.2.)*

eral, the six ions surrounding the central cation include both O and OH ions. Three structural groupings of the two basic sheets are recognized. The first consists of a tetrahedral sheet attached to one side of an octahedral sheet to form a 1:1 layer phyllosilicate. In the tetrahedral–octahedral join, the basic rule is that if a place in space could be occupied by either an O of the tetrahedral sheet or an OH of the octahedral sheet, the O has preference. This explains why Sis are surrounded by Os and Als by both Os and OHs. The second grouping is the symmetrical arrangement of two tetrahedral sheets about a central octahedral sheet to give a 2:1 layer phyllosilicate. A third arrangement is the presence of an octahedral sheet between adjacent 2:1 layers; although these are sometimes known as 2:1:1 layer phyllosilicates, common practise is to include them with the 2:1 clay minerals. In a clay mineral, each layer combination extends for varying distances in the direction of the a and b crystallographic axes.

The 1:1 and 2:1 layers are stacked in the c-axis direction, and bonds between the layers are formed by more than one mechanism. One mechanism involves a bonding of the external O ions of one layer with the external OH ions of the adjacent layer. Bonding probably is due to the mutual sharing of the H ion between the O ions of both sheets to form a hydrogen bond. Bonding can also occur with polar water molecules forming a hydrogen bond between the two layers. The bonding

here is as follows: OH ion (one sheet) to H ion (positive pole of water molecule) to OH ion (negative pole of water molecule) to H ion (adjacent sheet). Another mechanism for bonding involves isomorphous substitution within the clay lattice. Although silicon and aluminum are the more common cations, other cations of near similar size can replace them. The common substitutions are Al^{3+} for Si^{4+} in the tetrahedral sheet, and Fe^{3+}, Fe^{2-}, and Mg^{2+} for Al^{3+} in the octahedral sheet. These substitutions, with the exception of Fe^{3+} for Al^{3+}, are marked by an ion of lower valency substituting for one of higher valency, and the net result is a negative charge in the layer. This charge is balanced by some combination of substituting OH^- for O^{2-}, filling vacant cation sites in the octahedral sheets, and adsorbing hydrated cations at the edges of layers and between layers. Cations between layers can bond the layers together, especially if the opposing layers both have external O ions. The kind of interlayer bonding varies from clay mineral to clay mineral and can be correlated with some properties of the clay minerals.

The classification of the clay minerals is based on both layer type (1:1 vs. 2:1) and charge per unit formula (Bailey, 1980). The charge per unit formula is determined by the extent of substitution within the mineral structure. As an example, take a simple montmorillonite, a 2:1 mineral, with the following formula:

$$Na_{0.33} (Al_{1.67} Mg_{0.33}) Si_4O_{10} (OH)$$

Mg substitutes for Al in the tetrahedral sheet, occupying one-sixth of the sites. This creates a net charge deficit of 0.33, which, in this case, is balanced by Na.

Table 4.1 Classification of the Clay Minerals

Layer type	Group (X = charge per Si_4O_{10})	Subgroup or species
1:1	Kaolinite-serpentine ($X = 0$)	Kaolinite, halloysite
2:1	Smectite ($X = 0.2–0.6$)	Montmorillonite, beidellite, nontronite
	Vermiculite ($X = 0.6–0.9$)	Dioctahedral, trioctahedral
	Mica ($X = 1$)	Muscovite, biotite, illite
	Chlorite (X = variable)	Dioctahedral, trioctahedral, or combination

Modified from Bailey (1980).

A variety of ions could satisfy that charge deficit, so instead of $Na_{0.33}$, we could use $X_{0.33}$. In the classification, the charge per formula unit (or Si_4O_{10}) is used, and in the above case that is $X = 0.33$. The highest level of classification is the layer type (Table 4.1). These are subdivided into groups, with charge deficit being the determining factor. Kaolinite is balanced so $X \cong 0$. Within the 2:1 minerals, subdivision is on relative charge deficit: low charge (smectite), medium (vermiculite), high (mica = illite in places), and variable (chlorite).

Except where noted, the following is taken from various chapters in the book edited by Dixon and Weed (1989), and that by Velde (1992).

Kaolinite and halloysite are the common 1:1 layer clay minerals, and there is little substitution in their structures (Fig. 4.2). Kaolinite has a platy morphology. In contrast, halloysite commonly has a layer of water between successive layers, and has a unique structure of rolled sheets. This comes about because the ideal repeat distance in the crystal lattice of the tetrahedral sheet is slightly greater than that of the octahedral sheet (Drever, 1982). The presence of water results in a relatively loose bond, which allows the rolling to occur. Hydrogen bonding with a water interlayer (10Å halloysite), or without a water interlayer (kaolinite, 7Å halloysite), binds the adjacent layers together.

Figure 4.2 Crystal structure of the common clay minerals. Illite structure is similar to that of muscovite. The *b* crystallographic axis in horizontal and the *c* axis is vertical in these diagrams. *(Redrawn from Jackson, 1964, Figs. 2.3, 2.8, 2.9, 2.10, 2.14, and 2.15 in Chemistry of the Soil by F.E. Bear, ed., © 1964 by Litton Educational Publishing, Inc.; reprinted by permission of Van Nostrand Reinhold Co.)*

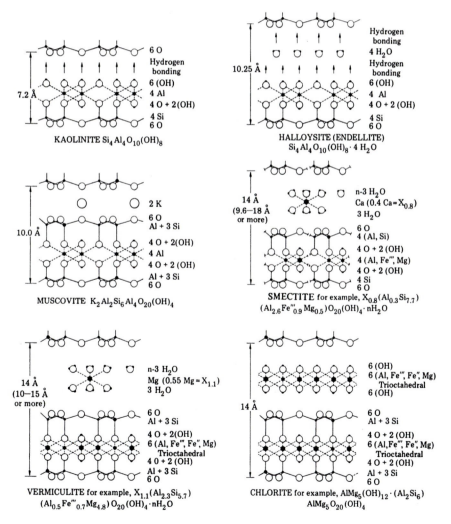

The smectite group of clay minerals makes up a wide variety of 2:1 layer clay minerals (Fig. 4.2). Isomorphous substitution is common, and this gives rise to many minerals that differ in chemical composition. Although Al^{3+} does substitute for Si^{4+} in the tetrahedral sheet, the more common substitution is Fe^{2+} and Mg^{2+} for Al^{3+} in the octahedral sheet. The resulting net negative charge is partly balanced by interlayer hydrated cations that bond adjacent layers. Because the negative charge is due mainly to substitution in the octahedral layer, and because that layer is some distance from the interlayer-bonding cation, bonding is relatively loose. Therefore, exchange of cations and water layers readily occurs. These minerals expand or contract as water layers are added or removed, and the amount of such change depends, in part, on the cation present.

Three forms of smectite occur in soils, based on their composition. Montmorillonite is the Mg variety of smectite, with both Al and Mg (Al>Mg) in the octahedral sheet. Beidellite is the Al variety, with Al in the octahedral sheet, and Al substituting for some Si in the tetrahedral sheet. Finally, nontronite is the Fe variety, with Fe in the octahedral sheet, and Al substituting for some Si in the tetrahedral sheet.

Illite is a 2:1 layer clay mineral with a composition close to that of muscovite (Fig. 4.2), but with less K for an approximate formula of $K_{0.8}$ Al_2 $(Si_{3.2}$ $Al_{0.8})$ O_{10} $(OH)_2$ (Eberl, 1984) . Some illites seem to be micas intermixed with layers of smectite, vermiculite, and chlorite. Although the basic structure is similar to that of montmorillonite, most of the isomorphous substitution is Al^{3+} for Si^{4+} in the tetrahedral sheet. The interlayer bonding cation is K^+, and because it is close to the site of the negative charge, bonds are especially strong, and interlayer expansion does not occur.

Vermiculite is also a 2:1 layer clay mineral (Fig. 4.2). Like smectite, both dioctahedral and trioctahedral types are recognized, and like illite, the major isomorphous substitution is in the tetrahedral layer. This results in a fairly tight bond between the adjacent layers and the interlayer hydrated cations, and thus interlayer expansion is limited.

Chlorite is the last 2:1 clay mineral common to soils (Fig. 4.2). It consists of alternating 2:1 layers and octahedral layers. The 2:1 layers are commonly trioctrahedral, and substitutions are mainly Al^{3+} for Si^{4+} in the tetrahedral sheet, which results in a negative charge. The octahedral interlayer contains Mg^{2+}, Fe^{2+}, and Fe^{3+} in addition to Al^{3+}, which results in a positive charge. Bonding between the two oppositely charged layers is strong and is enhanced by hydrogen bonding; expansion does not take place.

Mixed-layered clay minerals result from the interstratification of 2:1 layer clay minerals and octahedral sheets. Only rarely is a 1:1 layer mineral involved. Common interlayering is possible because many of the clay-mineral species are structurally similar, and because some expand readily in water or on slight weathering. Common ones are illite/smectite, chlorite/smectite, chlorite/vermiculite, and mica/vermiculite; kaolinite/smectite seems to be less common (Moore and Reynolds, 1989). Mixing varies from a regular repetition of the components in the direction of the c-axis to a wholly random arrangement. Such mixed-layered clays are quite common, and their detailed identification is difficult (see Moore and Reynolds, 1989). There has been some discussion about whether illite is a discrete mineral or is a mixed layered mineral.

Two clay minerals have a fibrous morphology in contrast to the platy morphology of most of the above clay minerals; the one exception to the latter is tubular halloysite. Although the fibrous minerals have a 2:1 structure, the lateral extension of the layer is limited in the b dimension but not the a dimension, thus producing linear chains and a fibrous morphology (Singer, 1989a, Fig. 17.1). The two minerals of this group are palygorskite and sepiolite, with the former having near equal proportions of Al and Mg in the octahedral sheet, and the latter having mostly Mg. There is some substitution of Al for Si in the tetrahedral sheet.

The chemical composition of the various clay minerals is quite variable (Newman and Brown, 1987). Representative analyses are not easy because of the problems of obtaining pure samples, substitution is common with many of them, and the common occurrence of mixed-layered minerals, the detection of which is not easy. Minerals or means of groups of minerals for which analyses are complete are given in Table 4.2. It was difficult to always select representative minerals, so I selected those in soils or the result of weathering where I could, and tried for a global representation. Water contents are included in the data set; recall that H_2O^+ indicates weight loss due to crystal lattice OHs, and H_2O^- values have little use when discussing mineral chemistry and structure (Chapter 3). Brief comments follow. Kaolinite is a relatively pure aluminosilicate, and thus is commonly used in

Table 4.2 Chemical Composition of Selected Clay Minerals

Oxide	Kaolinite	Illite	Smectite	Vermiculite	Al-Chlorite	Mg-Chlorite	Palygorskite	Sepiolite
SiO_2	45.44	49.78	59.49	43.21	39.85	30.89	55.86	52.5
Al_2O_3	38.52	26.35	21.93	17.21	33.52	18.55	10.54	0.6
Fe_2O_3	0.8	4.3	3.77	8.03	4.56	1.69	3.23	2.9
FeO		0.61	0.2	0.6		1.42		0.7
MnO				0.21				
MgO	0.08	2.75	3.55	25.71	4.58	33.74	9.2	21.31
CaO	0.08	0.32	1.18	5.1	0.15		1.56	0.47
K_2O	0.14	7.02	0.34	0.32	1.61		0.05	
Na_2O	0.66	0.25	0.82	0.05			0.68	
TiO_2	0.16	0.42	0.25	0.06	1.03	0.05	0.47	
H_2O^-	0.6	1.48				1.08	8.71	12.06
H_2O^+	13.6	7.12	8.38		13.18	12.97	9.13	9.21
Total	100.08			100.5		100.38	99.58	
Molar SiO_2/Al_2O_3	2.02	3.22	4.61	4.26	2.02	2.83	9.04	148.31

From Weaver and Pollard (1973): Kaolinite from a weathered pegmatite, Virginia (Table LIX). Mean of 24 illites from several countries (Table III). Mean of 101 montmorillonites–beidellites (Table XXV). Al-chlorite from a soil, B.C., Canada (Table XLII). Palygorskite from Attapulgus, Georgia (Table LII). Sepiolite from Madagasgar (Table LVI). Mg-chlorite from Foster (1962), sample no. 50. Ca-saturated vermiculite from Australia (Newman and Brown, 1987, Table 1.18).

writing weathering reactions (Chapter 3). The K content in illite is high because K is the interlayer ion, and fits tightly into the space outlined by the silica tetrahedra (Figs. 4.1 and 4.2). The smectite data combine montmorillonite with beidellite, so one does not see the chemical contrast between the two. Ideal montmorillonite has Mg substituting for Al at one of four octahedral sites, and ideal beidellite has Al substituting for Si at one of eight tetrahedral sites. Newman and Brown (1987) present chemical data for both; SiO_2 percentage is about the same for both, but Al_2O_3 is close to 20% for montmorillonite and 30% for beidellite, and MgO is in the 3–5% range for montmorillonite and 2–3% for beidellite. Vermiculite has the highest Fe_2O_3 content, and one of the higher MgO contents. High Al_2O_3 vs high MgO separates the chlorites into the two common species. The fibrous clays (attapulgite and sepiolite) are marked by the lowest Al_2O_3 contents, and both are relatively high in MgO. MgO also makes up a dominant part of the structure in vermiculite and Mg-chlorite.

Because silica tetrahedrals and Al octahedrals are essential components of almost all the minerals, and

H_2O is included in the analyses, we can most readily compare the minerals on the molar SiO_2/Al_2O_3 ratio (Table 4.2). Low ratios (near 2) are characteristic of kaolinite, halloysite, and Al-chlorite. In fact the ideal Al-chlorite can be thought of as a kaolinite with a different stacking arrangement (both have alternating Si-tetrahedra and Al-octahedra in the direction of the c axis, but in chlorite the 2:1 layer that is similar to the other 2:1 layers has an octahedral sheet between them, rather than hydrated ions). The 2:1 layer clays (illite, smectite, vermiculite) have ratios in the 3–5 range; Al-chlorite mimics the kaolinite value, and Mg-chlorite is close to 3. The low Al_2O_3 contents of the fibrous clays result in the highest ratios, especially sepiolite.

The SiO_2/Al_2O_3 ratios can be used to indicate the amount of SiO_2 leaching that has to take place to weather either a rock or mineral to the various clay mineral end products. Recall that in many environments Al_2O_3, once released from the parent mineral, is relatively insoluble and remains near the site to potentially link with SiO_2 and form crystalline clay minerals. For example, in Table 3.4 the ratio changes from 5.8 (rock) to 2.13 (saprolite sample 4). Provided saprolite sample 4 is composed entirely of crystalline products produced during weathering, the products could be kaolinite, halloysite, or Al-chlorite. The authors report that halloysite is common (Hendricks and others, 1967), so the chemical data would suggest that only crystalline products are present (halloysite), and that little of the original rock remains. Some molar SiO_2/Al_2O_3 ratios for common rock-forming minerals are as follows:

Muscovite—2.0
Biotite—4.6
K-feldspar—5.6
Na-plagioclase—6.0
Ca-plagioclase—2.0
Hornblende—9.4
Augite—29.3
Hypersthene—222.5
Olivine—340

This means that Ca-plagioclase or muscovite can convert to kaolinite with little loss of SiO_2, as can biotite to vermiculite. Other reactions involve increasingly greater losses of SiO_2, such as Na-plagioclase, and hornblende and augite to montmorillonite. Hypersthene and olivine have such high ratios that an aluminosilicate weathering product is unlikely; any clay mineral that forms from them probably would have to scavenge Al_2O_3 from some adjacent weathering mineral higher in Al_2O_3 content. Finally, palygorskite and sepiolite have such high ratios that only small amounts of Al_2O_3, but high amounts of MgO are needed; these require special conditions for formation.

Various quantitative methods using X-ray diffraction data have been employed to determine the content of clay minerals in soils and sediments (Schultz, 1964, 1978; Ruhe and Olson, 1979). Unfortunately, many commonly used methods do not give comparable results (Pierce and Siegel, 1969). The problem is that the X-ray diffraction pattern of the various clay minerals is a function of several factors (e.g., crystallinity, composition, and sample preparation) in addition to abundance. Attainment of results within 1% seems unlikely, although sometimes these are reported. The latest word seems to be from Moore and Reynolds (1989), who give these broad estimates of accuracy: ±10% for major constituents, and ±20% for constituents that might be present in less than 20%. A common alternate procedure is to report results in peak heights or peak areas, or in qualitative terms such as low, moderate, and high, accompanied by sketches

of representative diffractograms. Other methods can also be combined with X-ray diffraction to estimate clay–mineral abundances (Amonette and Zelazny, 1994).

■ Crystalline and Noncrystalline Compounds of Aluminum and Iron

Products that form in soils can be crystalline, such as the clay minerals mentioned above, or noncrystalline. In the latter materials, the atomic order is local and not repetitive, sometimes called short-range order (Wada, 1989). The term "amorphous" may not be appropriate always because some of these compounds do have form. There is no sharp dividing line between the two. Imogolite is termed paracrystalline because of the tubes indicate long-range order in one direction. Identification of the noncrystalline materials is difficult, in contrast to the relative ease of identifying clay minerals by X-ray diffraction methods.

Allophane and imogolite are the common noncrystalline aluminosilicate materials (Wada, 1989). Approximate compositions are 1–2SiO_2 Al_2O_3 2.5–3H_2O for allophane and Al_2O_3 SiO_2 2.5H_2O for imogolite. Electron micrographs show that imogolite occurs as smooth and curved threads with an outer diameter of about 2 μm, and allophane as hollow aggregates up to 5 μm in diameter that may be spherules or polyhedrons. Imogolite threads in cross section are depicted as a rolled kaolinite structure, but with planes of OHs on both the exterior and interior of the tube (Parfitt, 1980, Fig. 2). The two materials have the following oxide compositions:

Imogolite: 30–40% SiO_2, ~40% Al_2O_3, 15–20% H_2O^+
Allophane: ~30% SiO_2, ~47% Al_2O_3, 20–24% H_2O^+

Molar SiO_2/Al_2O_3 ratios are 1.3–2.0 for imogolite and ~1.1 for allophane.

Crystalline aluminum minerals are present in some environments. Common ones are gibbsite [$Al(OH)_3$] and boehmite (AlOOH). If these form from common aluminosilicate rock-forming minerals, intense leaching of SiO_2 is required for their formation.

Several iron minerals may form during weathering (Table 4.3); of the common minerals, only magnetite does not form in soil environments (Schwertmann and Taylor, 1989; Schwertmann, 1993). Goethite is the most common iron mineral in most well-drained soils with yellowish brown color. In contrast, lepidocrocite forms the orange mottles and bands in noncalcareous, clay-rich, soil formed in anaerobic conditions (presence of Fe^{2+}) in the aquic soil-moisture regime. Hematite is the second most important iron mineral in oxidizing conditions, and seems to be the pigment responsible for the red color of old soils and of soils formed under relatively high surface temperatures (deserts, subtropics, tropics). Maghemite is common in soils of the tropics and subtropics that apparently have burned. Finally, ferrihydrite is most likely the Fe material present in Bs horizons of Spodosols and produces the bright colors of some soils formed in the aquic soil-moisture regime.

Identification of the noncrystalline aluminosilicate and iron materials and of the iron minerals is difficult. Part of the problem is that the materials occur in low amounts, and the usual X-ray diffraction does not always yield good results. More complicated tests usually are needed (Wada, 1989; Schwertmann and Taylor, 1989; Schulze, 1994). One can dissolve some of these materials using various reagents to get an estimate of the amounts in the soils. Parfitt and Childs (1988) provide a useful table on the most effective reagents for this purpose (Table 4.4); these methods are sometimes termed extractive chemistry. The idea is to dissolve the material of interest (imogolite, ferrihydrite, etc.), which puts into solution the dominant ion of interest (Al, Fe, Si), and analyze for the latter. For example, dithionite–citrate–bicarbonate treatment extracts Fe from most of the pedogenic iron minerals (termed Fe_d), but the oxalate extract (termed Fe_o), is

Table 4.4 Effectiveness of Reagents for Dissolving Soil Components[a]

Soil component	Pyrophosphate	Oxalate	Dithionite-citrate
Al humus	Good	Good	Good
Ferrihydrite	Poor	Good	Good
Goethite	Poor	None	Good
Hematite	None	None	Good–mod
Lepidocrocite	None	Poor–mod	Good
Maghemite	None	Poor–mod	Good
Magnetite	None	Poor–mod	Mod
Allophane	Poor	Good	Mod
Imogolite	Poor	Good	Mod
Gibbsite	None	Poor	None

Taken from Parfitt and Childs (1988, Table 11).

[a]Poor, <10%; moderate (mod), 10–80%; good, >80%.

Table 4.3 Characteristics of Iron Minerals Common in Soils

Mineral	Formula	Color
Goethite	α-FeOOH	7.5 YR–2.5Y
Lepidocrocite	γ-FeOOH	5YR–7.5YR, value ≥ 6
Hematite	α-Fe_2O_3	7.5R–5YR
Maghemite	γ-Fe_2O_3	2.5YR–5YR
Ferrihydrite	$Fe_5HO_8\cdot4H_2O$ or $Fe_5(O_4H_3)_3$	5YR–7.5YR, value ≤ 6

Taken from Schwertmann and Taylor (1989, Table 8.1) and Schwertmann (1993).

selective mainly of ferrihydrite. This latter is such a good test that Fe_o content multiplied by 1.7 is a useful estimate of ferrihydrite (Parfitt and Childs, 1988). The difference, $Fe_d - Fe_o$, is useful in estimating the Fe sequestered in goethite and hematite. Al and Si of the oxalate extract (Al_o and Si_o, respectively), combined with Al of the pyrophosphate extract (Al_p), can provide an estimate of the allophane and imogolite in the soil. Pyrophosphate extract, analyzing for Fe (Fe_p) is a poor indicator of any iron mineral; this was not always thought to be the case as the original Soil Taxonomy (Soil Survey Staff, 1975) based part of the Spodic horizon definition on Fe_p and Al_p contents. Because magnetite dissolution takes place during dithionite–citrate–bicarbonate extraction, it is removed with a magnet prior to analysis for the pedogenic iron minerals (Walker, 1983). Because of their rather complex compositions, these iron and aluminun materials are sometimes called oxyhydroxides. To call attention to their origin in soils, they also are called pedogenic Fe and Al.

Several Fe ratios are commonly used. Fe_o/Fe_d is a measure of the proportion of the total pedogenic Fe that is amorphous plus ferrihydrite. This is sometimes called the Fe activity ratio (Blume and Schwertmann, 1969). Fe_2O_3(dithionite)/Fe_2O_3(total) is a measure of the proportion of the total Fe that is sequestered in the soil. Note that iron can be expressed both as an element or an oxide, and the conversion is (elemental Fe)(1.43) = Fe_2O_3.

There is a rapid test for the presence of noncrystalline Al-oxyhydroxides (e.g., allophane and imogolite) in soils (Fieldes and Perrott, 1966). Because of the abundant exposed OHs, the soil is treated with an NaF solution, the Fs displace the OHs, and the pH rises rapidly (read in the laboratory with a pH meter). Significant amounts of these materials are present if the pH rises above 9.4 within 2 minutes. Neall and Paintin (1986) used a 5-minute reaction time and got good results. This procedure might also indicate the presence of ferrihydrite. NaF pH and Fe_o both have similar buildup curves with time in the Southern Alps of New Zealand (Rodbell, 1990), suggesting a correlation, but the influence of Al oxyhydroxides could also be a factor. Because these materials accumulate in the B horizons and are leached from the A and E horizons, NaF pH remains low in the latter two horizons over 10 ka, in contrast to the increase in NaF pH with time in the B horizons (Fig. 4.3).

■ Cation-Exchange Capacity of Inorganic and Organic Colloidal Particles

Colloidal material in soils carries an electrical charge. Although both negative- and positive-charge sites exist, the origin of the negative sites is the focus here.

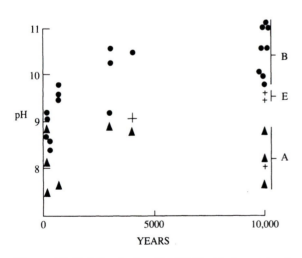

Figure 4.3 Variation in the pH of NaF with time and soil horizon, Southern Alps, New Zealand (*unpublished data of author*). Soils are in Birkeland (1984).

The negative charge is approximated by the cation-exchange capacity (CEC) of the material, and most of the cations attracted to these sites are exchangeable. The CEC varies in amount and origin with the soil material. In humus, the charge originates from the dissociation of H^+ from carboxyl and phenolic groups at the particle surface and within the particle. In clay particles, charge originates at the clay edge or along the surfaces parallel to the *a–b* crystallographic plane (Dixon and Weed, 1989). Charges along the clay edge originate from unsatisfied (broken) bonds at the edge of the particle, say, between Si—O or Al—OH, or from the dissociation of H^+ from OH^-. Charge originating in this manner is common in the 1:1 layer clays and depends on particle size, because the number of exposed edges increases as particle size decreases. Ionic substitution in the clay lattice (low charge ion for a large charge ion) results in a negative charge for the particle, mainly along the interlayer surfaces. This is the main origin of charge in the 2:1 clays, although some charge originates at the clay edge.

The CEC is expressed in milliequivalents (meq)* per unit dry weight of the material. The CEC can be calculated from the clay mineral formula. If we refer back to the formula for montmorillonite (early in this chapter), we can calculate the formula weight. This is the summation of the subscripts multiplied by the atomic weights (e.g., 1.67×26.97 for Al and 0.33×24.32 for Mg); in this case the formula weight is 367 g. Because of substitution in the crystal lattice, there is a charge deficit of 0.33 eq/367 g, and in this particular case the balancing, interlayer cation is Na. Values must be recalculated to 100 g, so the above is 0.09 eq/100 g, or 90 meq/100 g. SI units are the same numerical value, but are given in centimoles per kilogram (cmol/kg).

CEC varies with clay mineral species and organic matter content (Table 4.5). The highest values are for organic matter. The variation with clay mineral is due to a combination of ionic substitution and its extent, the degree of hydration, and the number of exchange

---

*One milliequivalent is defined as 1 mg of H^+ or the amount of any other cation that will displace it. If, for example, the CEC is 1 meq/100 g oven-dry soil, 1 mg of H^+ is adsorbed. If Ca^{2+} displaces the H^+, the amount of Ca^{2+} has to be quivalent to 1 meq of H^+; this amount can be calculated: Ca has an atomic weight of 40, compared to 1 for H, with a positive valency of 2; one Ca^{2+} ion, therefore, is equivalent to two H^+ ions because the latter ion has only one charge. The amount of Ca^{2+} required to displace 1 meq of H^+ is 40/2, or 20 mg, the weight of 1 meq of Ca^{2+}.

sites at the edges of particles. Because organic matter can have such a high CEC, the presence of a small amount of organic matter can greatly affect the CEC of the soil. In contrast, nonclay minerals and rock fragments have a negligible effect on the CEC; zeolites, because of their structure, are an exception. One can obtain a rough estimate of the CEC of the clay fraction, and hence of the possible clay minerals present, by knowing the CEC of the soil and the amount of clay present (in percent), and by sampling deep enough to avoid sizable amounts of organic matter. For example, if a soil has a CEC of 20 meq/100 g soil and is 20% clay, the CEC for the clay fraction would be approximately 20 meq/20 g clay, or 100 meq/100 g clay. The clay mineral could be smectite or vermiculite (Table 4.5).

If organic matter is present in the latter soils, the estimation of the clay mineral(s) present must account for that, as organic matter has a great influence on CEC. The CEC per 100 g clay is calculated as above. In a similar fashion, calculate the grams of organic matter or carbon per 100 g clay. (If a soil has 5% or-

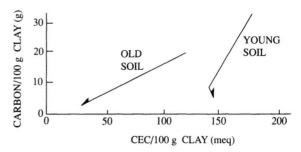

Figure 4.4 Diagram of change in CEC and mass of carbon per 100 g of clay, for soils in northern California. *(Courtesy of Dr. Esther Perry, University of California, Berkeley.)*

ganic carbon and 20% clay, this converts to 25 g organic carbon/100 g clay). Plotting one against the other (Fig. 4.4) one can follow the trend of CEC/100 g clay as the amount of carbon decreases. In this case, at 0 g carbon, the young soil has a CEC of about 150, whereas the old soil has one of <20. Provided both soils have pure clay species (see Table 4.5), this could suggest a conversion of smectite to kaolinite over time. If more than one mineral is present in both soils, their weighted sum would have to equal the indicated values (e.g., a CEC of 100 could indicate equal contributions from both vermiculite and chlorite).

■ Ion Mobilities

The behavior of ions, once released by weathering to the soil solution, varies; some ions participate in reactions involving mineral synthesis, some are adsorbed to colloid surfaces, and some are removed with the downward-moving water. Only when specific ions in the soil solution occur in the right proportions can mineral synthesis take place. Here we look into the factors governing the mobility of the more common ions.

Ionic potential, a concept of the geochemist, V.M. Goldschmidt, provides broad guidelines as to the fate of ions released by weathering (Chesworth, 1990, 1991). This is done by calculating the ratio of the charge in valency units to the ionic radius in Angstrom units (Å)(called the z/r ratio). Ionic potential, therefore, is a measure of the intensity of the positive charge. A plot of radius vs charge delineates several fields. Ions with a ratio <2 remain in solution as simple cations (some become exchangeable cations), al-

Table 4.5 Representative Cation-Exchange Capacities for Various Materials

Material	Approximate Cation-Exchange Capacity (meq/100 g dry weight)
Organic matter	150–500
Kaolinite	1–10
Halloysite	5–10
Hydrated halloysite	40–50
Illite	10–30
Chlorite	10–40
Smectite	80–150
Vermiculite	100–200+
Palygorskite	5–30
Sepiolite	20–45
Allophane	20–50
Hydrous oxides of aluminum and iron	4
Feldspars	1–2
Quartz	1–2
Basalt	1–3
Zeolites	230–620

Taken from Carroll (1959), Grim (1968), Weaver and Pollard (1973), Dixon and Weed (1989), and Norrish and Pickering (1983).

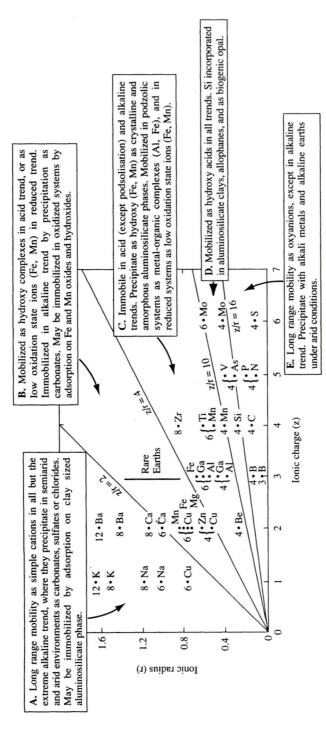

Figure 4.5 Mobility and immobility characteristics of various elements in terms of ionic potential. Numbers beside each point are the coordination number. Acid trend refers to acidic, oxidizing conditions, alkaline trend to alkaline, oxidizing conditions, and reducing trend to reducing conditions. (*Redrawn from Chesworth, 1992, Fig. 2.4, Weathering systems, p. 19–40. In Martini and Chesworth, eds., Weathering, soils & paleosols, © 1992, with kind permission from Elsevier Science, The Netherlands.*)

though in the aridic moisture regime they can precipitate out as carbonates or chlorides (Fig. 4.5). Ions with a ratio of 2–4 are mobilized either as hydroxy complexes (Mg), or as cations in the reduced state (Fe, Mn). Ions with ratios of 4–10 can precipitate as oxyhydroxy materials (Al, Fe, Mn) in both acid and alkaline solutions; however, both Al and Fe are soluble either under very acid conditions (Fig. 4.6) or if they form metal–organic (chelating) complexes. Note that Al is also highly soluble under alkaline conditions (Fig. 4.6), but it is not leached from the system because those pH values are not compatible with a leaching environment. Si is the only common element with a ratio between 10 and 16, and it is mobilized as a hydroxy acid (H_4SiO_4), the form shown in most weathering equations. Si is soluble over all pH values, and solubility is greatest in alkaline conditions. Ratios greater than 16 indicate mobility as oxyanions, but under alkaline conditions of the desert they precipitate out (C as carbonate under various conditions, and N as nitrate under hyperaridic conditions such as that of the desert of southern Peru and northern Chile; see Fig. 8.32).

The above behavior of various ions can be partly explained as a result of (1) the bond strength between cations and anions (Chesworth, 1991, Table 1.4) and (2) the attraction of the particular ion in solution for the O ion of the H_2O molecule. If the attraction of the ion for O^{2-} is weak, the ion remains in solution surrounded by water molecules (e.g., Na^+). If, however, the attraction of the ion for the O^{2-} is comparable to that of the H^+ for the O^{2-}, one H^+ from the water molecule is expelled, and the ion is precipitated as a hydroxide (e.g., $FeOOH$).

From the above, several generalities can be made with regard to the fate of ions in solution. Iron and aluminum can remain close to the site of weathering and partake in synthesis reactions under oxidizing conditions over the pH range of most normal soils (Fig. 4.6). Silicon seems to be available to combine with aluminum to form clay minerals if conditions favor them. Some ions will become fixed into the crystal lattices of the clay minerals. A familiar example is the fixation of potassium between layers of illite where it is not readily exchangeable. Another example is the introduction of magnesium into the octahedral layers of montmorillonite. Ion exchange also is an important factor. Although all cations potentially are attracted to the mineral surfaces, certain ions can replace others. The usual replacement series is $M^{2+} > M^+$, where M refers to any cation. Within cation groups of similar

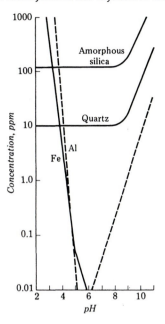

Figure 4.6 Relation between pH and solubility of Al, Fe, amorphous Si, and quartz. *(Data for amorphous Si and quartz from Krauskopf, 1967, Fig. 6.3, and that for Fe and Al from Black, 1967, Figs. 1 and 2, © 1967, John Wiley & Sons.)*

valency, the replacement series is dictated by the size of the hydrated ion. Smaller size ions are held more tightly because the site of the positive charge of hydrated ions of small size is closer to the negative site on the clay mineral or organic compound than is an ion with a larger hydrated radius. A general overall replacement series of the common ions in the average soil is $Al^{3+} > Ca^{2+} > Mg^{2+} > K^+ > Na^+$ (Singer and Munns, 1987). Replacement rank may be modified by a number of factors (Wiklander, 1964). One obvious one is that a high concentration of any cation in the soil solution will bring about the exchange of that cation for the preexisting cations on the exchange sites. For example, with highly acid solutions, H^+ can replace the other cations, and in alkaline calcic horizons Ca^{2+} can replace the other cations.

■ Genesis of Noncrystalline and Crystalline Products

The genesis of minerals in the soil is complex due to the variety of weathering environments, the possibility of unlike microenvironments in close proximity to one another, and the fact that some minerals are derived by precipitation from solution, whereas others are derived from solid-state alteration of preexisting phyllosilicates. To further complicate matters, some minerals are of eolian origin and may have no relationship to the chemistry of the soil. It is not surprising, therefore, that most soil samples are characterized by several clay-mineral species. Further complications to understanding the origins of all the clay minerals are (1) reaction rates are very slow at surface conditions, (2) duplication of natural conditions in the laboratory is difficult, and (3) some minerals may not be part of the current mineral–water equilibrium, but rather may have formed in the past under different conditions or even have been derived from the parent material. However, some general statements can be made regarding the chemical conditions favoring the formation of certain minerals.

An unusual environment is that in which volcanic glass weathers in a humid environment to produce the noncrystalline materials (Al and Fe oxyhydroxides) common in Andisols. Dahlgren and others (1993) mention the following important factors. Volcanic glass is one of the least stable materials in a weathering environment. Rapid weathering releases elements faster than crystalline minerals can form, oversaturates the soil solutions, which, combined with rapid precipitation, favors the formation of the noncrystalline, metastable solid phases. Continued weathering can deplete the glass fraction, leach more Si, and lead to a transformation to more stable, crystalline forms (e.g., halloysite and gibbsite). Another factor of possible importance is that layer-lattice clay minerals are absent in the parent materials and soils; if there, these could have served as templates for the formation of clay minerals from solution.

Formation of secondary iron-bearing minerals in the soil depends partly on pH–Eh conditions (covered in all geochemistry texts; e.g., Langmuir, 1997, Chapter 11). Eh is a measure of the ability of the environment to bring about either oxidation or reduction, important also to the behavior of Mn in solutions (see redox conditions in Chapter 5, also Fig. 5.13). Under oxidizing conditions, the Fe^{2+} released from primary minerals during weathering is oxidized to Fe^{3+}. Because ferric compounds are insoluble in oxygenated waters over the normal pH range for soils (Figs. 4.5 and 4.6), precipitation takes place close to the source of iron. In contrast, if conditions are reducing (low

Eh), Fe^{2+} is the stable aqueous species, and under appropriate conditions, it can migrate far from the site of release by weathering before precipitation takes place. Reducing conditions usually are associated with the aquic soil-moisture regime.

The formation of the iron minerals is complicated and discussed and diagrammed by Schwertmann and Taylor (1989, their Figs. 8.8 and 8.9). As with the clay minerals, one first has to establish which of these are inherited, and their contents (e.g., Fe_d or Fe_o%). Under oxidizing conditions with relatively slow release of Fe by weathering and low soluble, organic matter content (common to many surface environments), goethite is usually formed. If, however, much soluble organic matter is present (inhibits crystallization), conditions are acid, and Fe release is relatively rapid, ferrihydrite probably will form. Ferrihydrite seems to be a necessary precursor for the formation of hematite, especially in tropical environments. Adsorbed organic matter will retard this transformation, which takes place by dehydration and structural rearrangement. Goethite does not convert to hematite, nor does hematite convert to goethite, strictly by dehydration or hydration. In the first case, goethite must dissolve, ferrihydrite must form, and the latter then converts to hematite, a transformation enhanced by high temperatures, near-neutral pH, and well-aerated conditions. Hematite can convert to goethite, but only by dissolution and reprecipitation; a change to cool-moist climatic conditions could initiate such a conversion. Under reducing conditions with low partial pressure of CO_2, Fe^{2+} Fe^{3+} hydroxy compounds can form, and subsequent transformation to other minerals is complex. Under oxidizing conditions, these can convert to lepidocrocite. Most of the maghemite in soils seems to come from the conversion of various iron oxides during fires, thus their common occurrence in the topsoils of the subtropics and tropics (Schwertmann, 1993).

Clay minerals occur in soils by way of three pathways (Eberl, 1984). One pathway is inheritance from the parent material. A second pathway is transformation, in which the clay mineral has retained some of its inherited structure during chemical weathering. Two forms are recognized: (1) ion exchange in which cations attracted to the negative charges on the clay particles are exchanged for those in the surrounding solution and (2) layer transformation, in which the octahedral, tetrahedral, and fixed interlayer cations are altered to varying degrees. The third pathway is neoformation, in which the clay precipitates either from solution or by some reaction involving amorphous material, such as allophane.

Inheritance is suggested if both the parent material and soil have layer-lattice clay minerals in common. If the parent material is igneous or metamorphic rock lacking in mica or pure limestone, there would be no inheritance. Almost all other parent materials, especially surficial materials (till, river alluvium, loess, etc.), could contain clay minerals. Inherited clay minerals persist in the soils either if they are in equilibrium with the soil environment or if the soil is young and, reaction rates being slow, transformation has not been complete. Illite is commonly pointed to as an inherited clay mineral, but as will be shown later, this is not always the case. Derivation unaltered from atmospheric dust is here included as inheritance. Such minerals are detected if weathering has not altered them. A famous example is the mica present in soils of the Hawaiian Islands, odd because the rocks are low in K. Workers tried to explain how pedogenic mica could be produced, until numerical dating showed that the micas were much older than the islands. The obvious conclusion is that they have an eolian origin from some distant source (Dymond and others, 1974). Most Quaternary soils will have some inherited clay minerals unless there has been intense alteration over sufficiently long periods of time to eradicate them (e.g., Singer, 1988). Marine sedimentary rocks represent inheritance over many rock cycles. Many of the clay minerals in ocean-floor sediments are inherited (Simonson, 1995), and these move through the rock cycle and become rocks, which later become the parent material for a soil; some of the minerals might trace their heritage to an ocean-floor environment one or more rock cycles back in time.

The second pathway is transformation, which takes place essentially in the solid state. Various processes have been proposed for transformation (Jackson, 1964; Barshad, 1964; Allen and Fanning, 1983). Changes from one mineral to another almost always involve the interlayer areas, because that is where ions can be exchanged, hydroxy ions introduced, or a silica sheet removed. Reactions usually occur preferentially along a layer or are initiated at the crystal edge and proceed inward.

The third pathway is neoformation, that is the clay mineral forms by precipitation in a soil solution. This depends on many factors, including the concentration of the common framework elements (Si and Al), the kind and concentration of cations present, and the soil pH; all of these correlate with the amount of leaching (Pedro and others, 1969; Drever, 1982; Dixon and Weed, 1989; Velde, 1992; Langmuir, 1997). Both iron and aluminum released by weathering precipitate as oxides and hydrous oxides over the normal soil pH range and so generally remain within the soil for possible reaction to form clay minerals. As mentioned earlier, many of the other common cations in the soil have relative mobilities much greater than iron and aluminum, and they can be leached from the soil environment. Because silicon and aluminum form the basic framework for many clay minerals, their presence in the correct proportions is essential. Davis (1964) has shown that silicon is common in most natural waters, so the amount of silicon available for reaction depends less on its rate of release during weathering and more on the rate of leaching from the soil environment. The same can be said for many other cations, but their presence probably depends to a large extent on rock type. One unequivocal example of neoformation is in New Mexico, where palygorskite fibers radiate perpendicularly into pores (Monger and Daugherty, 1991).

The SiO_2/Al_2O_3 molar ratio of both primary and secondary minerals indicates the amount of silica leaching required to form most secondary products. As pointed out earlier in this chapter (Table 4.2), the ratio for some primary $\Rightarrow$ secondary mineral pairs is the same, but for most reactions a loss of silica is required. For the fibrous clays, however, there has to be a gain in silica relative to alumina.

A hypothetical transect from high to low leaching environments helps one visualize the requirements for neoformation of the common clay minerals (see various chapters in Dixon and Weed, 1989). If the soil solution has very low silica content (e.g., high rate of leaching in a tropical environment), gibbsite and boehmite would form. Those leaching conditions that would keep silicon content low would also result in low cation concentrations. With increasing amounts of silicon, aluminosilicates form. Relatively high leaching, a low cation content, and a pH less than 7 favor the formation of kaolinite and halloysite. Halloysite seems to form primarily from volcanic materials and, in time, could convert to kaolinite. The conditions necessary to form the 2:1 clay minerals are quite varied. Montmorillonite forms in near neutral to alkaline pH and relatively high concentrations of Ca^{2+}, Mg^{2+}, and Na^+. In contrast, biedellite seems to form under conditions of acid pH and high leaching, such as in the E horizons of some Spodosols (Gjems, 1960; McKeague and others, 1978; Ross, 1980). Illite forms under these same conditions when the concentration of K is high. Although we sometimes arbitrarily ascribe illite as an inherited mineral or by subtle transformation, Singer (1989b) makes a convincing case for its transformation from smectite in hot aridic environments. If the solution is high in Na_2CO_3 and $NaHCO_3$, extremely alkaline conditions prevail and zeolites form (Baldar and Whittig, 1968; Hay, 1963,1964). Palygorskite and sepiolite form in soils characterized by an alkaline pH (8.9), impeded drainage (relatively high water table or water movement impeded at a textural contact), abundant $(Ca,Mg)CO_3$ to supply the Mg^{2+}, and a low Al^{3+} content. These conditions are met in fairly old carbonate-enriched horizons (Bk and K) (Bachman and Machette, 1977). If the minerals from which they form have normal contents of Al_2O_3, there has to be a sink for the Al [note the high solubility of Al under these conditions (Fig. 4.6)]. Smectite formation could take up the Al. Vermiculite formation is favored by a high concentration of Mg^{2+} and a pH less than 7. Chlorite with a gibbsite-layer interlayer would probably form under these same or more acid conditions; if, however, conditions are more alkaline and Mg^{2+} is abundant, the interlayer probably would be brucite-like.

Phase diagrams have been constructed to help quantify the relationship between the ionic concentration (geochemists prefer activity, which is close to concentration) (Drever, 1982) of the solution and the mineral(s) in a particular soil (Helgeson and others, 1969; Drever, 1982; Dixon and Weed, 1989; Velde, 1992; Langmuir, 1997). Data on free energies of formation and on the chemistry of the species are used to construct such diagrams. Many diagrams are available; Fig. 4.7 is a good example as it is three dimensional, it includes primary and secondary minerals, including zeolites, which can be pedogenic under unusual conditions. The diagrams should be considered only approximations because the soil is such a complex system, with microsite influences, eolian contributions, and slow reaction rates. At any rate, the diagrams are useful in that they show the approximate stability fields for clay minerals precipitated from solutions of varying chemical composition. They also indicate the stability relationships of one or more minerals; minerals adjacent to one another in the diagram can be in equilibrium (e.g., kaolinite and montmoril-

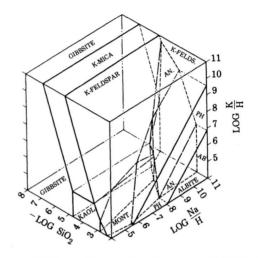

Figure 4.7 Phase diagram for the system K_2O–Na_2O–Al_2O_3–SiO_2–H_2O at 25°C and 1 atmosphere. KAOL, kaolinite; MONT, montmorillonite; PH, phillipsite; AN, analcite; AB, albite. *(Redrawn from Hess, 1966, Fig. 1.)*

lonite), but those not adjacent to one another are not in equilibrium (e.g., gibbsite and montmorillonite). Furthermore, the use of such diagrams helps to predict possible successive alteration products with gradual variation in ion-activity ratios, and activity of H_4SiO_4. In any study involving equilibrium diagrams, it is important to compare the mineralogical data with the water data from the same horizon; river water collected several miles away may not always be representative.

A usual extension of this mineral-equilibria work is to plot the water data on the same diagram to determine if the minerals are in equilibrium with the solution (Tardy, 1971; White, 1995; diagrams in the texts mentioned above). If the water-chemistry data plot in a particular mineral field and the X-ray data confirm the presence of that particular mineral, the mineral probably is in equilibrium with the solution. This is the standard interpretation. However, equilibrium conditions may not exist for samples in which the water and X-ray mineralogical data do not coincide. As an example, plots have been made for perennial and ephemeral springs in granitic terrain of the Sierra Nevada (Fig. 4.8). K-mica and montmorillonite are present in some samples, but the water data all plot in the kaolinite field. It is suggested that this means that the water composition may later shift to activity values at which all three minerals are in equilibrium,

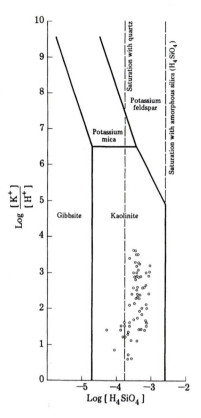

Figure 4.8 Phase diagram for the system K_2O–Al_2O_3–SiO_2–H_2O at 25°C and 1 atmosphere. Circles are data for water from springs and seeps in a granitic terrain in the Sierra Nevada. *(Redrawn from Feth and others, 1964, Fig. 13.)*

or perhaps the minerals present will in time convert to more stable phases (Feth and others, 1964). Garrels and Mackenzie (1971) plot data from many rock types and environments, and most fall within the kaolinite field. It is suggested that kaolinite is the stable end product in the chemical weathering of silicate minerals. This is not always borne out by field studies, however; perhaps more time is needed to reach equilibrium, or some of the data used to construct the diagrams are not too accurate. Finally, in a broad study covering several major environments and rock types, Tardy (1971) found good correspondence between clay mineralogy and water chemistry in only some cases.

A few examples can be given of these water–mineral equilibrium data for soils. In any such application, it is important to know which, if any, miner-

als were inherited from the parent material, and the direction in which mineral alteration schemes might be proceeding. Cleaves and others (1974), for example, report that because montmorillonite and chlorite are present in the soils they studied and stream-water data plot in the stability fields for both minerals, the latter could have been synthesized in the present pedologic environment. Pavich (1986) used a similar argument to verify the stability of kaolinite as a by-product of saprolite formation in the Virginia piedmont. It took a while for workers to accept a pedogenic origin for palygorskite, and the stability field diagram of Singer and Norrish (1974) for pH, Mg^{2+} and $Si(OH)_4$ helped make the case more convincing. Analogous data can be used to help prove a nonpedogenic origin for some clay minerals. For example, mica is present in humid soils formed from mica-free basalt in Hawaii, but because the soil-solution data do not plot in the mica field of the stability diagram, a contemporary pedogenic origin is not supported (Swindale and Uehara, 1966). In fact, as mentioned earlier, we now know that later studies showed the micas to be of eolian origin and to be much older than the island (Dymond and others, 1974).

Within any one soil or soil horizon there is usually a variety of clay minerals (Jackson and others, 1948; Allen and Fanning, 1983, especially Table 6.3; Allen and Hajek, 1989, Fig. 11.15). Various hypotheses have been advanced to explain this diversity. One explanation is that the clay minerals in the assemblage are not in equilibrium with present conditions and are being altered to other more stable clay minerals. Some of the clay minerals, for example, could have been inherited from the parent material, or be of eolian origin. Another possible explanation is that some of the clay minerals could have formed in the past under different environmental conditions, and reaction rates being so slow at surface conditions, these clay minerals are metastable in the present environment. If one considers the diverse microenvironmental conditions within a soil, perhaps diversity of clay minerals should be expected. For example, ionic concentration might vary from place to place due to the variation in charge of the particles. Furthermore, the primary minerals weather at slightly different rates, releasing a variety of substances into the soil. Thus, there will be times when the ionic species do not have a uniform distribution. There could even be slight variations in pH that could be important to any ongoing reactions involving mineral synthesis. Such variation could be caused by proximity to a root, as they commonly have adsorbed H^+, or root CO_2 respiration and the local concentration of H_2CO_3. Data on abrasion pH of various primary minerals (Table 3.2), as well as their composition, also might lead one to suspect that similar local variations in the soil exist during mineral weathering and synthesis.

The above examples do not exhaust the list of possible chemical conditions at the microscopic level. They are mentioned to point out that a soil may have a variety of clay minerals due to a variety of local conditions. If this is the case, one may have to look carefully at the use of clay-mineral stability diagrams to determine whether a clay mineral or a clay-mineral assemblage is stable or not. The soil-solution chemical data needed to test for equilibrium on any diagram represent only the average conditions within the soil. On a microscopic level, the average may be difficult to find, controlled as it is by the rate of ion release by weathering, the chemical conditions at the site, and the rate of leaching (Tardy and others, 1973).

There is much conjecture regarding the manner in which clay minerals form from a solution (Jackson,

1964; Siffert, 1967; Dixon and Weed, 1989; Velde, 1992), because the reactions are very difficult to reproduce under laboratory conditions that even approach those in the field. One way in which crystallization might proceed is by the precipitation of colloidal SiO_2 and $Al(OH)_3$. One problem here is the orientation of the combining constituents as tetrahedral and octahedral sheets. One way in which this could take place is by adsorption on both nonclay and clay-mineral surfaces. Orientation into sheets may be facilitated with substrates with an atomic lattice structure similar to that of the forming clay minerals. Because clay-mineral surfaces carry a negative charge, one might visualize colloidal $Al(OH)_3$ first being attracted to the mineral surface followed by colloidal SiO_2 to build up a particular clay mineral. With the similarities in clay-mineral structures and dissimilarities in microenvironments, mixed-layer clay minerals could be intermediate products whether the clays form by transformation or neoformation.

Alteration among the 2:1 clay minerals probably takes place quite readily because of their similar structures and interlayer-bonding mechanisms (Fig. 4.9).

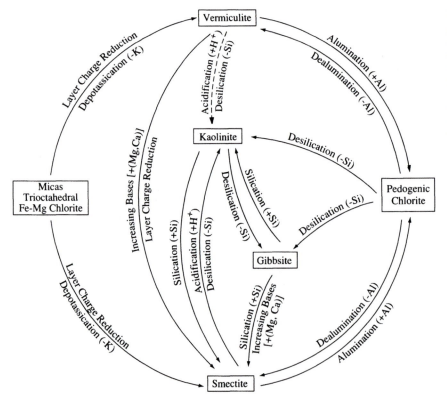

Figure 4.9 Transformation paths among some clay minerals. *(Redrawn from Allen and Fanning, 1983, Fig. 6.3, Composition and soil genesis, p. 141–192. In Wilding et al., eds., Pedogenesis and soil taxonomy, 1. Concepts and interactions, © 1983, with kind permission of Elsevier Science, The Netherlands.)*

In general these involve changes in layer charge due to substitution within the layers, and the addition or subtraction of the base cations (K^+, Ca^{2+} and Mg^{2+}), and/or the elements making up the bulk of most layers (Si and Al). In most weathering environments, Si is lost, so the reaction should proceed in that direction. However, if Si is gained (Fig. 4.9), Allen and Fanning (1983) believe that is mainly in areas of high water table influence.

Ion exchange is the simplest reaction to envisage (Fig. 4.10A). By this process, ions from the soil solution replace interlayer ions, and there may be some replacement within the crystal lattice. The resulting product is mostly a function of the host mineral and the replacing ion. For example, replacement of interlayer K^+ with Mg^{2+} could alter muscovite or illite to montmorillonite or biotite to vermiculite. If, however, Al-hydroxy groups are present, they could replace the interlayer ions, because they too carry a positive charge, and form chlorite (Fig. 4.10B). In a similar manner, if the soil solution is rich in Mg^{2+}, brucite layers could form and become interlayered with 2:1 clays to form chlorite. Both trioctahedral montmoril-

lonite and vermiculite could be altered to chlorite by these mechanisms. Such an alteration might proceed more rapidly for montmorillonite than for vermiculite, because the former is characterized by an expandable lattice. This expansion would allow for more rapid interchange of interlayer ion and hydroxy ion, or hydroxide sheet.

The alteration from 2:1 to 1:1 clay minerals involves a greater amount of structural reorganization. Two mechanisms proposed for this transformation are (1) gibbsite interlayering to form a chlorite–ormation, and (2) the removal of silica tetrahedral sheets from montmorillonite layers (Fig. 4.10C and D).

It is also known that gibbsite and kaolinite can form from one another. The formation of gibbsite from kaolinite involves the removal of a silicon sheet, perhaps in the manner shown in Fig. 4.10D. The reverse reaction, the formation of kaolinite from gibbsite, is thought to involve partial dehydration of the gibbsite structure, the entry of silicon-enriched solutions between gibbsite layers, with the oxygen of the silicon tetrahedra occupying the lattice positions vacated by hydroxide ions during dehydration (Tamura and Jackson, 1953).

Figure 4.10 Several clay-mineral alteration schemes. (A) Alteration of illite to montmorillonite by ion exchange along the edges of the layers. (B) Alteration of vermiculite to chlorite by replacement of interlayer ions by Al-hydroxy groups to form a gibbsite layer. *(Redrawn and modified from Jackson, 1964, Fig. 2.12 in Chemistry of the Soil by F.E. Bear, ed., ©1964 by Litton Educational Publishing, Inc.; reprinted by permission of Van Nostrand Reinhold Company.)* (C) Alteration of smectite to chlorite to kaolinite. *(Redrawn and modified from Brindley and Gillery, 1954, Fig. 1; see also Glenn and others, 1960.)* The Si tetrahedra in the middle kaolinite layer must invert so that the interlinked bases of the tetrahedra all face upward. (D) Alteration of smectite to kaolinite by stripping Si sheets from smectite. *(Redrawn and modified from Altschuler and others, 1963, ©1963, by the American Association for the Advancement of Science.)* Kaolinite growth can take place laterally from a newly formed kaolinite layer, and additional layers can become oriented on these layers. Hydroxyls occupy the O sites in the octahedral layers left vacant by Si sheet removal.

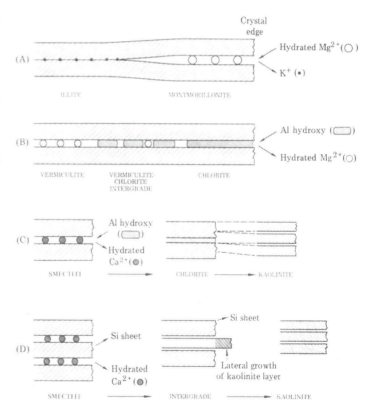

Some clay mineral transformations are reversible and others may not be. Reactions between the 2:1 clay minerals and between kaolinite and gibbsite seem to be reversible (Fig. 4.9). The reaction 2:1 to 1:1 clay, however, may not be readily reversible in the solid state under surface conditions. The reverse reaction would involve addition of a silicon sheet between each kaolinite layer, and this might be difficult because the hydrogen bond between the kaolinite layers is fairly strong, the layers are held closely together, and there are no interlayer cations. Perhaps the reverse reaction proceeds by solution of the original kaolinite followed by precipitation of the constituents as montmorillonite.

Bachman and Machette (1977) present data on transformations involving the fibrous clay minerals. Their studies suggest that montmorillonite and mixed-layer illite–montmorillonite transform to palygorskite. This transformation is indicated by the usual interpretation of X-ray diffraction data, that is, the intensities of the peaks of the former minerals decrease as that of palygorskite increases systematically in a soil profile. Singer (1989a) states that this transformation is not likely, however, because of the energy requirements for the breaking of the bonds. A further transformation of palygorskite $\Rightarrow$ sepiolite takes place only in fairly old soils (middle Pleistocene or older). The source of the Mg^{2+} might be partly from the carbonate phase as with age the Ca:Mg ratio in this phase increases in the older soils with greater carbonate contents. This occurrence of sepiolite in old soils concurs with the review by Singer (1989a).

To summarize some of the above, clay minerals can form both by transformation and by precipitation from solution. However, because mica is so common in rocks, transformation may be the main process responsible for most soil clay minerals. Mica, or other clay minerals, can also be introduced to the soil by eolian processes, and this will further confound our ability to decipher the origin of specific clay minerals. Even if the original mineral is no longer present, that does not rule out transformation as a main process; it does make it more difficult to prove, however. Finally, Colman (1982) challenges the assumption that well-crystallized clay minerals form in abundance in soils from primary minerals and rock particles in some soils about 100 ky or older. He bases this on the fact that weathering rinds of basaltic and andesitic clasts contain few crystalline clay minerals, whereas the clay fraction of the Bt horizon in which the clasts occur has abundant crystalline clay minerals. His study

should force workers to look more carefully for the origin of soil clay minerals and not always to accept weathering of nonclay minerals as the primary source of such minerals. In this particular case we want to be assured that conditions for clay formation are met in the rind, before comparing them with that in the adjacent soil.

■ Clay-Mineral Distribution with Depth in Soil Profiles

Clay minerals vary in amount with depth in some soil profiles, but they remain relatively constant throughout in others. Such relationships seem to be closely associated with the leaching conditions within the soil. Three leaching conditions can be examined. With extensive leaching, many cations and silica may be carried to great depth. This can result in relatively uniform chemical conditions with depth in the soil and therefore a rather uniform clay-mineral distribution. The same uniformity involving different clay minerals would be expected in arid regions, because the little leaching that takes place may not change the chemical conditions enough to favor differences in clay mineralogy with depth. Between these two extremes, however, differences in leaching with depth might produce enough variation in chemistry to favor the formation of different clay minerals. Put another way, the ions and colloids released by weathering in the higher levels of the profile become reactants in mineral synthesis at depth.

The most common vertical variation is one in which the amount of silica and bases in the weathering product increases with depth. Thus, gibbsite at the surface grading downward to kaolinite or kaolinite at the surface grading downward to montmorillonite are common trends. Although it is possible that one or the other mineral may no longer be in equilibrium with the present environmental conditions, this need not be the case. Both minerals in the vertical sequence can be in equilibrium with the environmental conditions, and because the latter changes with depth, this change is reflected in the clay mineralogy. If conditions change, for example, due to erosional lowering of the surface with time, the montmorillonite at depth eventually may be in that part of the profile most conducive to kaolinite formation, and it may alter to kaolinite (Fig. 4.11). The variation of clay mineral with depth does not by itself constitute proof that one mineral is

Figure 4.11 Vertical distribution of clay minerals as a function of leaching conditions within the soil. (A) Leaching conditions favor the formation of kaolinite and montmorillonite; each forms at different depths. (B) Rapid lowering of the surface by erosion results in montmorillonite residing in the soil environment of high leaching. (C) With time, the montmorillonite in the surface becomes desilicated and alters to kaolinite. Rates of ground-surface lowering, mineral alteration, and amount of leaching control the resulting clay mineral(s). With slow lowering, alteration may proceed at the same rate, and the clay-mineral distribution always appears as in (A). If lowering far exceeds the rate of clay-mineral alteration, a distribution such as (B) occurs temporarily.

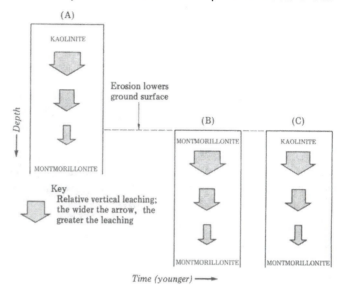

Figure 4.12 Approximate clay-mineral distributions in surface and C horizons of soils in North and Central America. *(Reprinted from Gradusov, 1974, Figs. 1–4. A tentative study of clay mineral distribution in soils of the world. Geoderma 12, 49–55, © 1974 with kind permission from Elsevier Science, The Netherlands.)*

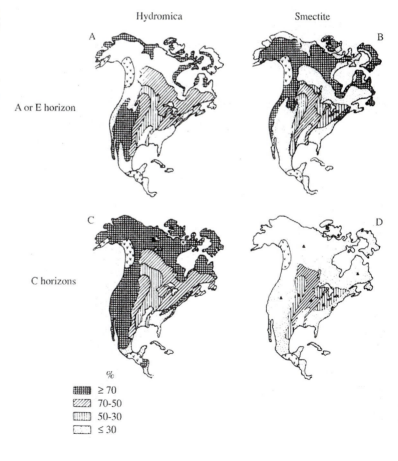

changing over to another. In this case, the lowering of the ground surface so changed the conditions within the soil that the original clays were no longer stable; they therefore equilibrated with the environment.

■ Continental and Global Clay Formation

Ideally, there should be a rather consistent clay-mineral assemblage with climate (see Chapter 10), and thus with position of soils on a continent. However, this is not always the case. Allen and Fanning (1983, Table 6.3) and Allen and Hajek (1989) compiled data between Soil Taxonomy and mineralogy classes for the United States. Not surprisingly, most suborders have a mixed mineralogy as the major class. No doubt this is because many factors control the clay mineralogy at that classification level. Only three orders have an association with one clay mineral: allophane is associated with Andisols, montmorillonite with Vertisols, and oxides and kaolinite with Oxisols. All other orders encompass such a wide variation in the soil-forming factors that a wide variety of clay minerals is present.

Several workers have shown global distributions of clay-mineral processes. Pedro (1976, and Fig. 1.4 in Duchaufour, 1982) partitions the Earth into several types of weathering. Neoformation is dominant in humid climates within about 35° of the Equator. North and south of that is a large area of transformation. In the deserts it takes place under nonacid conditions. In contrast, in grasslands and forests it takes place under acid conditions; this is shown only for the northern hemisphere, and it extends to the forest–tundra boundary. No weathering occurs in hyperarid areas (west coasts of central South America and southern Africa, continental areas of Sahara-Sinai, southern Arabian Penninsula, and central Asia). Gradusov (1977) has produced a global map of some clay minerals, and North and Central America [within Pedro's (1976, in Duchaufour, 1982) transformation and solution areas] are shown here (Fig. 4.12). A broad interpretation of the maps is that C horizons are high (most ≥ 50%) in hydromica (~illite), and a small area (central United States) is high in smectite. In the A horizons, however, the hydromica content remains the same in the desert and coasts of the northernmost areas, but smectite is prominent along the west coast, and across a large part of the area north of the treeline, as well as Alaska. This could be interpreted as transformation of hydromica ⇒ smectite in the wetter areas [solution areas in the scheme of Pedro (1976, in Duchaufour, 1982)].

Chapter 5

Processes Responsible for the Development of Soil Profiles

Because many processes act together to form any one soil profile, it is difficult to discuss soil formation as a function of a specific process. The formation of a soil profile is viewed by Simonson (1978) as the combined effect of additions to the ground surface, transformations within the soil, vertical transfers (up or down) within the soil, and removals from the soil (Fig. 5.1; see definitions of processes in Buol and others, 1997). Many of these processes are discussed by Nahon (1991), who incorporates micromorphology into their discussion.

For any one soil, the relative importance of these processes varies, and the result is the variety of profiles seen in any landscape. The main additions to most soils are organic matter from the surface vegetation and their contained elements, ions and solid particles introduced with rainfall, and particles carried by the wind. Transformations include the multitude of organic compounds that form during organic matter decomposition, the weathering of primary minerals, the formation of secondary minerals and other products, the formation of redoximorphic features, and changes in bulk density. Transfers generally involve the movement of ions and solids with the moving soil water, the eluviation of substances from one level and their illuviation into a lower level. Soluble substances (e.g., Fe, Al, Si, the bases, anions such as the common one, HCO_3^-, organic compounds) move with the percolating water unless changing chemical conditions cause

them to precipitate out of solution. In places where capillary rise of water is important, ions can be transferred upward and precipitated as salts high in the soil profile. Ions can move upward through plants and can be returned to the surface with litterfall. For solid particles, clay translocation is most common, but silt can be translocated in sandy soils. Soil-dwelling fauna can actively move solid particles in any direction. Examples are alpine gophers moving soil into overlying snow, to form pseudo-eskers as they are gently let down to the ground when the snow melts. Paton and others (1995) review bioturbation in soils, conclude that animals move more material than do plants, and that the global ranking is earthworms > ants = vertebrates > termites. They point out that the importance of some animals will be specific to various environments. Filled in borrows, called krotovinas, are common in many soils. Finally, when water moves through the profile, it removes substances still in solution; these substances then become part of the dissolved constituents of the groundwater or surface waters. Surface erosion should also be included in removals. Because transformations have been discussed elsewhere in this book, the emphasis here will be on other processes that form the various horizons diagnostic of soil profiles.

It should be pointed out that it is difficult to determine soil processes because few actual measurements can or have been made. Consider, for example, clay

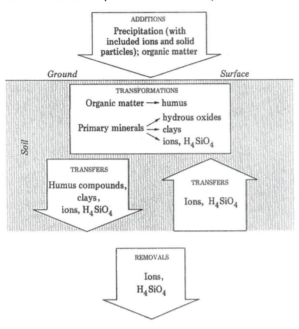

Figure 5.1 A flow chart of major processes in soil-profile development. Surface erosion should be added, as erosion rates have to be less than soil-formation rates for profiles to form. *(Redrafted and modified from Simonson, 1978, Fig. 3.)*

translocation from the A to the B horizon. This process probably is so slow that measurements taken over a year or two may not be representative. In addition, laboratory simulations may or may not be good approximations of what happens in nature. In lieu of these approaches, we commonly infer the processes from combined field and laboratory data. Process interpretation is confounded because, although process is commonly tied to particular climates, the Quaternary record is clear that climates have varied during the formation of many soils; hence, what might be considered a contemporary process actually could have been active at some time in the past. One should not always expect variation in process or its rate with climatic change, for it might be that neither of these changed markedly in the past, even though climatic change seems fairly well documented.

In this chapter, both laboratory chemical methods and trends with depth will be mentioned, as they are used to help identify various processes. The reader should become familiar with the expected trends, for any deviations from such trends for seemingly uniform profiles might indicate subtle parent material layering, eolian additions at the surface, presence of buried soils, climatic change, or more complicated processes.

A brief look at the composition of the Earth's crust gives one an appreciation of the marked changes brought on the variety of soil processes. The Earth's crust consists of these weight percentages of the common elements: O (47), Si (27), Al (8), Fe (5), Mg (2), Ca (4), and three each of Na and K. Soil-forming processes act on this average material to bring about dramatic chemical changes within the top several meters of the Earth's surface. For example, podzolic processes can produce an E horizon with the composition of a quartz sand; lateritic processes can produce a Bo horizon with over 90% $Al_2O_3 + Fe_2O_3$; and calcification processes can produce a K horizon with a total chemical composition that mimics some limestones.

■ Relative Importance of Processes by Soil Order

A listing of the relative importance of various processes can be used to show the degree to which several processes, working in concert, produce the various soil orders. Smeck and others (1983) discuss soil development in terms of entropy, to which the interested reader is referred, and also rank (by length of an arrow) each major process as it contributes to the soils of a particular order. I have converted the arrow length to a relative ranking of highly important, moderately important, and of little importance (Table 5.1). Entisols, because they are so little developed, have been impacted by few processes. For the same reason, the weakly developed Inceptisols have been influenced by many processes, but none to a great extent. Two orders form under one major process, physical mixing for Vertisols, and organic matter accumulation for Histosols. Mollisols form under the moderate influence of four of the six listed processes. Eluvial-illuvial processes are highly important to the formation of both Spodosols (Fe, Al, and organic compounds) and Alfisols (clay), and four other processes are moderately important in soil formation in both orders. Eluvial–illuvial processes are also highly important to Aridisols, but the importance of the other factors is generally ranked low. The orders with the highest number of highly important processes (3) are the Ultisols and Oxisols; both rank high in the formation of secondary minerals and leaching, and whereas eluvial–illuvial is ranked high in the Ultisols, weathering of the primary minerals is ranked high in the Oxisols. Obviously such a ranking is difficult to make as the soil-forming factors at each specific site can help determine the ranking for that soil.

■ Formation of the A Horizon

The organic matter content of a soil is a function of gains in organic matter to the soil surface and upper layers of the soil and losses that accompany decomposition. Ideally, a newly formed surface has no organic matter. Once vegetation is established, an A horizon begins to form. Early in the formation of a soil, the gains exceed the losses, organic matter gradually accumulates, and the A or A and O horizons thicken. With time a steady-state condition is reached, and the gains equal the losses. Thereafter, even though the system is a dynamic one with continuing gains and losses, the amount of organic matter in the soil and its distribution with depth remain essentially constant. The length of time needed to reach the steady-state condition and the amount of organic matter in the soil at that time vary with the soil-forming factors (Chapters 8 and 10).

The decomposition of organic matter to form humus is complex and best presented in soil texts (e.g., Singer and Munns, 1987). What is involved is the carbon and nitrogen cycles, and the general trend is a loss of carbon and sequestering of nitrogen. These result in a narrowing of the C/N ratio from the high ratios of plants to the low ratios of soil fauna, and the ultimate ratio for the soil is a reflection of the environment (Fig. 10.9).

Table 5.1 Ranking of the Importance of Various Processes in the Genesis of Soils of the Soil Orders[a]

Process	Entisol	Inceptisol	Vertisol	Histosol	Aridisol	Mollisol	Spodosol	Alfisol	Ultisol	Oxisol
Physical mixing	l	l	h	o	l	m	o	o	o	o
Mineral weathering	l	l	l	o	l	m	m	m	m	h
Formation secondary minerals	o	l	o	o	m	l	m	m	h	h
Leaching	o	l	o	o	o	l	m	m	h	h
Eluvial– illuvial	o	l	o	o	h	m	h	h	h	l
Organic matter accumulation	o	m	m	h	o	m	m	m	l	l

Modified from Smeck and others (1983, Table 3.2).

[a]Explanation of letters in ranking: o, none to slight; l, little; m, moderate; h, high.

Because of the dynamic nature of the A horizon its age is difficult to define. However, in some stratigraphic studies, workers have tried to differentiate the A horizon from underlying horizons by age. Thus, there might be a description of a soil in which a modern A horizon overlies a Sangamon (O-isotope stage 5; see Chapter 8) B horizon. Such differentiation is possible if the modern A horizon has formed from a younger parent material that was deposited on the Sangamon B. If this is not the case, such age differentiation is difficult to justify. For example, the primary mineral fractions of both the A and B horizons are the same age in a soil formed from one parent material in a stable landscape. Clays and other weathering products are younger than the primary minerals, and the organic constituents are youngest of all. The organic constituents, however, vary from vegetative matter just deposited and in the initial stages of decomposition, to humus that may have resided in the soil several thousand years. In this sense, then, all A horizons at the land surface are modern or have a considerable modern component. If it is postulated that the A horizon is modern because the older A horizon was removed at some time in the past by erosion, evidence for the erosion that removed the A horizon may be preserved in the stratigraphic record as an unconformity.

■ Translocation of Iron and Aluminum (Podzolization)

Many soils show evidence of podzolization, a process that involves a pronounced downward translocation of iron, aluminum, and organic matter to form an Si-enriched eluvial E horizon overlying an illuvial spodic or B horizon enriched in some combination of Al, Fe, and organic matter (Fig. 1.3). In many podzolized soils, the processes for such translocation are not readily apparent because the pH, although acid, can be in the range in which Fe^{3+} and Al^{3+} are essentially insoluble (Fig. 4.6). Furthermore, many of these soils are characterized by an oxidizing environment, which means that iron cannot move as the more soluble ionic species Fe^{2+}. Favorable conditions are a Si-rich, sandy parent material, good internal drainage, and a high latitude or altitude cool-climate forest that gives way to the tundra. Although forest is the dominant vegetation association, Spodosols are present in many alpine areas around the world (Burns, 1990), and they are lo-

cally present in the tropics, mainly as Aquods (Buol, 1990).

A fair amount of work has gone into the study of the processes by which iron and aluminum move in Spodosols or Spodosol-like soils (Stobbe and Wright, 1959; Ponomareva, 1969; Peterson 1976; Duchaufour, 1982; McKeague and others, 1983; Buurman, 1984; Kimble and Yeck, 1990; Gustafsson and others, 1995; Jersak and others, 1995). The key to understanding their origin is in soil chemistry. Total soil chemistry generally does not provide many clues as to process. Two analyses presented in the original Soil Taxonomy (Soil Survey Staff, 1960) illustrate the point (Table 5.2). The soil from Washington has average values for the major oxides, and no unambiguous podzolic trends in either Al or Fe oxides; trends like these could be inherited from the parent material or even be due to the formation of an argillic horizon. In contrast, because the soil from Georgia is so silica rich, the amounts of Al_2O_3 and Fe_2O_3 in the presumed spodic horizon are so high above background that a pedogenic trend is strongly suggested.

A more common approach in studying podzolization is to chemically extract pedogenic forms of Al and Fe from the soil; historically, three chemical extractions of Al and Fe were done (Fig. 5.2; discussed in Chapter 4). This is because the materials of interest form thin films on the grains, and we want a chemical characterization of such films. In this way, data are obtained on the overprinting material and the latter is not diluted by the chemistry of the grains making up the soil, be they original or pedogenic. For many years the materials were extracted using three treatments: dithionite-citrate (d), oxalate (o), and pyrophosphate (p). The idea was that Fe_p was the organic-bound Fe, Fe_o was the amorphous and some organic-bound Fe, and Fe_d was all pedogenic Fe (crystalline + amorphous + organic-bound). Characteristic trends were depletion in the E horizon, and accumulation in the B horizon, as well as a secondary increase in organic matter in the B horizon (Fig. 5.2). Different countries did not always use the same extract and the same amounts in their classification scheme. Later, an important piece of work by Parfitt and Childs (1988) showed that the oxalate extract was the most specific to materials involved in podzolization.

The Fe and Al extracts can be performed on any soils in any environment to quantify Fe and Al relationships, but they are most commonly used in char-

Table 5.2 Data on Spodosolic Profiles[a]

Basal Depth (cm)	Clay (%)	Organic Carbon (%)	C/N	SiO$_2$ (%)	Al$_2$O$_3$ (%)	Fe$_2$O$_3$ (%)	Fe$_2$O$_{3d}$ (%)	SiO$_2$/ R$_2$O$_3$ (%)	SiO$_2$/ R$_2$O$_3$, clay (%)
3	3.5	1.7	18	62.1	15.7	4.6	0.4	5.7	6.2
11	4.1	2.7	27	54.6	19.7	5.5	1.2	4	1.6
33	5.1	2.3	20	56.7	20.1	5.6	1.6	4.1	1.1
58	4.8	1.7	18	59	19.2	5.5	1.8	4.4	1.1
91	2.6	0.7	18	64.1	15.5	5.9	0.6	5.6	
122	2.3	0.4	13	66.9	15.5	5.8	0.4	5.9	1.7
8	0.5	1.1	29	97	0.2	0.1	<0.1	621	
41	0.4	0		99.8	0.1	0.1	<0.1	1040	42.3
43	2.7	1.1	30	96.6	0.4	0.1	0.1	357	
46	4.6	2.5	69	92.3	2.1	0.3	0.1	68	2.6
51	4.2	1.1	38	94.6	1.6	0.2	<0.1	93	2.7
69	1.7	0.1		97.6	0.6	0.2	<0.1	226	3.3
99	0.7	0					<0.1		
112	4.7	0.1						0.2	

From Soil Survey Staff (1960).

Data are for <2-mm fraction, except where indicated. Upper profile is from the state of Washington, lower one from Coastal Plain of Georgia. Illuvial horizon for the upper profile is the interval 3–58 cm; for lower profile 43–51 cm.

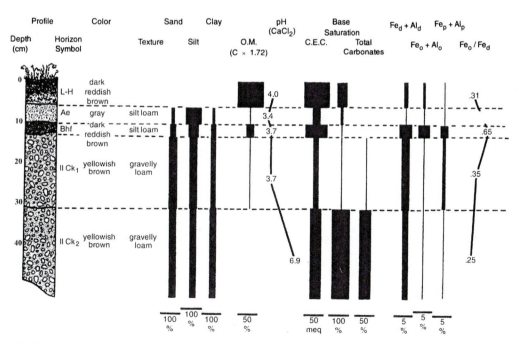

Figure 5.2 Physical and chemical characteristics of a podzolic soil in the Canadian Rocky Mountains. Horizon nomenclature is the Canadian system, and conversions to the system used here are L–H, O; Ae, E; Bhf, Bhs; and Ck1 and 2, C. Because the parent material is high is carbonate, the bases have to be depleted (leached) before podzolization can begin. *(Reprinted from King and Brewster, 1976, Fig. 2.)*

acterizing defining Spodosols of Soil Taxonomy and Podzols of other classifications. For example, with increasing degree of podzolization, amounts of all extracts increase and display eluvial–illuvial relations down the profile (Table 5.3). In addition, the ratio Fe_o to Fe_d, called the Fe activity ratio (Blume and Schwertmann, 1969), is commonly used to indicate the proportion that the extractable Fe, which is probably ferrihydrite, makes up of the total pedogenic Fe. This ratio increases with degree of podzolization (Fig. 5.2; Table 5.3), but it is also used in soil studies on pathways of Fe mineral genesis that do not involve Spodosols (McFadden, 1988; note that he reports the extracts in the oxide form, rather than as elemental, so a conversion will be necessary for a comparison with elemental Fe; divide Fe_2O_3 value by 1.43). In regard to soil classification using these extracts, the definitions are rather detailed and specific to each country.

Because of the association between spodic horizons and organic matter, Daly (1982) proposed a simple method for differentiating podzolized soils from other soils by measuring the optical density of the oxalate extracts of the soils. It is assumed that the optical density is due mainly to extracted fulvic acids and that appreciable amounts of the latter indicate podzolization as a major process. He suggests calculating the optical density for both the B and A horizons; if the B:A ratio is greater than 1, podzolization has occurred.

There are two main hypotheses invoked to explain the podzolic mineralogical and chemical trends. One involves an organic (chelate) and the other an inorganic connection. Both have been used to explain the mobilization and subsequent arrest of Fe and Al over short vertical distances.

The Chelate Connection

One hypothesis is that Fe and Al probably move in the soil as soluble metalloorganic chelating complexes (see Fig. 3.10). Fulvic acids are thought to be the common chelating agents for a number of soils (Kononova, 1961; Wright and Schnitzer, 1963; Schnitzer, 1969). Berner and Berner (1996, p. 154) have written a re-

Table 5.3 Iron and Aluminum Extracts of Some Canadian Soils

Approximate Classification	Horizon	Organic Matter (%)	Organic Carbon (%)	Dithionite Fe (%)	Dithionite Al (%)	Oxalate Fe (%)	Oxalate Al (%)	Pyrophosphate Fe (%)	Pyrophosphate Al (%)	Fe Activity Ratio
Mollisol[a]	A	14.99	—	1.28	0.34	0.38	0.34	—	—	0.30
	Bw1	2.83	—	1.51	0.34	0.39	0.36	—	—	0.24
	Bw2	1.61	—	1.14	0.26	0.32	0.29	—	—	0.28
	C	0.19	—	0.64	0.11	0.14	0.07	—	—	0.22
Spodosol[b]	E	1.88	—	0.18	0.16	0.05	0.15	—	—	0.28
	Bs1	7.26	—	3.26	1.40	2.60	1.38	—	—	0.80
	Bs2	2.13	—	1.22	0.79	0.56	1.01	—	—	0.46
	C	0.22	—	0.56	0.38	0.18	0.18	—	—	0.32
Spodosol[c]	E	—	1.4	0.1	—	—	—	0.02	0.01	—
	Bhs	—	12.0	3.3	—	—	—	2.6	1.6	—
	Bs	—	2.8	1.5	—	—	—	0.6	1.1	—
	Cx	—	0.2	0.6	—	—	—	0.03	0.2	—
Spodosol[d]	O	—	23.04	0.70	0.10	0.22	0.08	0.05	0.06	0.31
	E	—	2.15	0.34	0.10	0.22	0.10	0.06	0.10	0.65
	Bs	—	3.98	2.35	1.50	1.34	2.50	0.27	0.07	0.57
	2C1	—	0.27	2.30	0.23	0.70	0.25	0.16	0.10	0.30
	2C2	—	1.10	2.30	0.10	0.54	0.04	0.06	0.10	0.24

[a] Clarkaway soil (McKeague and Day, 1966).
[b] Holmesville soil (McKeague et al., 1978).
[c] Laurentide soil (McKeague et al., 1978).
[d] Profile 2 (King and Brewster, 1976).

action to show this, using oxalic acid as the chelating compound:

$$2H_2C_2O_4 + 4H_2O + NaAlSi_3O_8 \Rightarrow$$
oxalic acid   albite

$$Al(C_2O_4)^+ + Na^+ + C_2O_4^{2-} + 3H_4SiO_4$$
Al oxalate   oxalate ion silica (aq)

Here the Al is carried downward as a chelate, but eventually bacterial decompositon causes oxidation of the oxalate to form HCO_3^- and CO_2 and the precipitation of an Al oxyhydroxide or kaolinite. They point out that because of this there would be no memory of the oxalate in the soil or related stream waters.

The process of translocation envisaged begins with the production of fulvic acids in the O or A horizon. A stable fulvic acid chelate is formed with Al^{3+} and Fe^{3+} and because these chelates are water soluble, they move downward with the percolating soil water. Precipitation of these complexes can be brought about by a number of conditions at some depth in the soil profile: (1) In some cases quite small changes in ionic content can bring about precipitation (Wright and Schnitzer, 1963). (2) Another way to precipitate Fe or Al or both is by increasing the ratio of chelated ion to fulvic acid, for at relatively high ratios precipitation occurs (Lee, 1980). (3) Petersen (1976) presents a theory that the complexes that move have variable ratios of organic matter to metal. These complexes move in the descending water, but those with high organic matter to metal ratios precipitate first, whereas those with lower ratios can move to greater depths before encountering conditions for precipitation. Thus, a spodic horizon can be relatively enriched in organic matter near its upper boundary, a common attribute of Spodosols (Table 5.3). (4) Another way to precipitate aluminum and iron from the chelating complex at depth is through decomposition of the chelating complex by microorganisms at depth, leaving the metal ions free to precipitate as oxides or hydroxides, as shown in the above equation. (5) A process for deposition that does not involve chemical interaction is deposition during partial dessication, as might be expected to happen during dry seasons (Flach and others, 1980).

Although conditions for chelate formation and movement of iron and aluminum are best for the Spodosols, movement can also take place in other soils with E horizons. In general, the degree of podzolization declines in a transect going from boreal forest to grassland (Kononova, 1961), that is in a transect of

Spodosols $\Rightarrow$ Alfisols $\Rightarrow$ Mollisols. In one such transect, the humic acids:fulvic acids ratio increases from about 0.5 to 1.5–2.0. Furthermore, the humic acids in the Spodosols are similar to the fulvic acids in that they are dispersed and mobile. Calcium–humic acid complexes that are both relatively stable and immobile form in Mollisols and thus cannot translocate iron and aluminum.

Parent material controls the formation of Spodosols to some extent, because they are most common in sandy materials. Petersen (1976) suggests that as Ca^{2+}, Al^{3+}, and Fe^{3+} are released from the parent minerals the Ca^{2+} can be quickly leached from the soil and so does not interfere with and cause the metalloorganic complexes to precipitate. In addition, the relatively slow release of Al^{3+} and Fe^{3+} keeps the metal:chelate ratio low enough so that the complex is soluble and moves to greater depths. In contrast, clayey parent material might release the three ions at such a rapid rate that the metalloorganic complexes do not remain mobile.

The Inorganic Pathway

The competing hypothesis is that podzolization occurs without the interaction of chelating complexes (Farmer and others, 1980; Farmer and Fraser, 1982; Anderson and others, 1982; Childs and others, 1983; Wang and others, 1986). The suggestion is that Al and Si can form a positively charged, soluble hydroxy-aluminum silicate complex, called protoimogolite, that is stable at low pHs (<5). This protoimogolite could represent the form in which Al is eluviated. Mixed Al_2O_3–Fe_2O_3 sols appear to be stable, with and without silica, and also can be eluviated. Illuviation also would show as relative enrichment of Si_o and Al_o. These inorganic sols precipitate to form the Bs horizons. Eventually, the Al–Si complex converts to imogolite if organic matter contents are not too high, and the iron to ferrihydrite. Some Fe might move due to low pH. Soluble organic matter complexes can form and move down and, because they carry a negative charge, are precipitated on the imogolite and ferrihydrite, which can have a positive charge (Parfitt, 1980, Fig. 1). Thus, most of the Al and Fe move as inorganic complexes or as ions and precipitate as materials of low crystallinity; most of the organic matter is precipitated later. This can result in organic-matter enrichment in the upper part of the Bs horizon.

The debate continues as to which origin is dominant in podzolization. Chesworth and Macias-Vasquez (1985) and Macias and Chesworth (1992), for example, use arguments from a series of pe–pH diagrams to argue for the importance of organometallic complexes. Ugolini and others (1990) use chelation to explain the distribution of Al and Fe in the E and Bhs horizons, but in situ weathering and release of both elements to explain their presence in the underlying Bs horizon. Perhaps all these hypotheses are viable and local environment dictates which is dominant at any one place.

A good way to test the many hypotheses for podzolization, and for many other pedogenic processes, is to analyze samples of water moving through the soil under field conditions. Ugolini and co-workers (1982; Dawson and others, 1978; Singer and others, 1978) did this in studies of podzolization and the environments in which the process can be detected. One study is on a Spodosol-like soil in western Washington (Table 5.4). The soil data meet most, but not all, of the requirements for a Spodosol. Organic data on the soil indicate a buildup of fulvic acids in the B horizons, as well as a higher proportion of pyrophosphate-extractable C; these apparently move downward from the O horizon and are responsible for the movement of Fe and Al from the E horizon. Soil water analyses are used to show that these processes are ongoing. Since Si solubility is not affected by organic matter complexes, Si is nearly uniform in amount with depth and is being leached from the system. In contrast, Fe and Al probably are affected by the complexes, and although not much Fe moves below the 2Bhs horizon, Al moves to a greater depth, where it is precipitated.

■ Translocation of Clay-Sized Particles

The distribution of clay-sized particles in many moderately to strongly developed soils is marked by relatively low contents in the A and C horizons, with the maximum amount in the B horizon, generally in the upper part. Several processes may account for this distribution (Hubble and others, 1983): (1) The constituents of clay are derived by weathering higher in the profile and they move downward in solution with the percolating water and precipitate as clay minerals in the B horizon. (2) The clays have formed in place from mineral weathering in the B horizon. (3) The clays have moved as particles in suspension in the downward-percolating water to accumulate in the B horizon because of flocculation or constrictions in the pores through which the water moves, or because the base of the B horizon marks the lower limit of most water movement (McKeague and St. Arnaud, 1969). No doubt, in most soils, clays in the B horizon form by all three processes, but the relative importance of

Table 5.4 Data for a Spodosol-like Soil in Western Washington

Horizon	Depth (cm)	pH	Organic C	Cp (1%) <2 mm	$Fe_d + Al_d$	$Fe_p + Al_p$	Humic: Fulvic Acids	Fe	Al	Si
					Soil Analysis[a]			Soil Water Analysis (mg/ml)[b]		
01	5–3	3.9	56.0	15.4	—	—	0.36	—	—	—
02	3–0	3.6	53.6	15.7	—	—	0.33	0.04	0.75	3.37
E	0–11	4.3	2.4	0.8	0.3	0.2	0.25	0.03	0.68	4.36
2Bhs	11–31	4.7	6.9	5.4	2.3	2.6	0.10	<0.01	0.35	3.71
3B	31–53	4.8	9.7	6.4	3.9	3.9	0.12	—	—	—
4B	53–100	4.6	18.3	8.4	7.4	4.5	0.13	0.02	0.66	3.61

Data from Singer et al. (1978, Tables 1, 2, and 3).

[a]p is the pyrophosphate extract, d the dithionite-citrate extract.

[b]Precipitation values are <0.01 Fe, 0.03 Al, and 0.09 Si, and amounts for the precipitation intercepted by the vegetation are 0.02 Fe, 0.06 Al, and 0.09 Si. Data for soil horizons generally are for the base of the horizon.

Table 5.5 Clay Data on a Degraded Bt1 Horizon and the Underlying Bt2 Horizon for a Soil Formed from Till, New York

	Bt1 Horizon (41 cm)		Bt2 Horizon (68 cm)	
	Clay (%)	Coarse: Fine Clay	Clay (%)	Coarse: Fine Clay
Ped surface	9.3	2.1	75.2	0.4
Ped interior	20.5	1.1	23.2	1.2

Taken from Bullock and others (1974, Table 1).

each may vary from soil to soil. It should be possible to differentiate clay particles translocated from those formed in place, but I know of no criteria that can be used to identify clay precipitated from downward-moving solutions. A nonpedogenic cause of the clay-content contrast of the A versus the B horizon is that the horizons represent different deposits, such as sandy sheetwash deposit overlying clayey colluvium; subsequent pedogenesis, acting on both deposits, could so mute the contact that it might seem as if the soil formed from a uniform parent material.

Field and Laboratory Data

In the field, clay films can be used to suggest B-horizon clay origin by translocation. However, Federoff (1997) warns that in places it might be difficult to be sure the films are indeed illuvial features without thin-section verification. Clay films coating grains and clasts or lining ped surfaces or voids generally are taken as evidence for clay-particle translocation (Fig. 1.14; also, Soil Survey Staff, 1975, Plates 3A and 3B). Such translocation can be verified by comparing detailed chemical analyses of the horizon with those of the film; the two usually differ because the film has been emplaced by downward movement more recently, and therefore the film has properties that more closely resemble those of the A than of the B horizon (Buol and Hole, 1959). As clay content increases, especially if the clay minerals are of the 2:1 swelling variety, shrink–swell cycles may orient the ped surface films enough that they have a slight shine in the field. In this case, it may be difficult to separate illuvial clay from stress-oriented clay as the former could be giving way to the latter with time.

Ratios of clay fractions, such as fine clay:total clay, can be used to indicate clay translocation using the particle-size laboratory data (Nettleton and others,

1975; Gile and Grossman, 1979, Table 7.2; Walker and Hutka, 1979). The hypothesis is that the finer clay sizes are more mobile and so will move from the A to the B horizon, or downward within the B horizon, as long as the pores will allow it. An excellent example of this is a soil in New York in which the Bt1 horizon shows signs of being degraded to an E horizon. Here the percentages clay in the ped interior of both the Bt1 and Bt2 horizons are comparable, but the data on total clay and on the coarse clay:fine clay ratio demonstrate removal of the fine clay fraction from the ped surface in the Bt1 horizon and its deposition as nearly pure films on ped surfaces of the Bt2 horizon (Table 5.5). The match here was good because of the selective close-spaced sampling. In most cases, however, the interpretation may not be as clear as in the above example, because some ratios can be explained by parent material layering and others by the fine clay being newly formed clay minerals.

Clay bands or lamellae provide good field evidence for clay translocation (Dijkerman and others, 1967; Gile, 1979; Miles and Franzmeier, 1981; Berg, 1984; Bond, 1986; Schaetzl, 1992). Such bands usually consist of several thin layers with relatively high amounts of clay (ca. 10%) and pedogenic Fe that have an abrupt contact with interband layers of similar or greater thickness, but much less clay (ca. 1%) and Fe. The banded material is in the B-horizon position and is common in sandy parent material, such as eolian sand. The bands increase in number and thickness with time, and can merge into a normal Bt horizon (Fig. 5.3). Although geological layering is a possible origin for the banded materials, they commonly join and split in an anastomosing fashion, quite unlike a geologic depositional pattern. Hence, the field and laboratory relations support pedologic translocation. The clay can be eolian, inherited, or formed within the soil. The problem in explaining the specific origin is to arrest clay move-

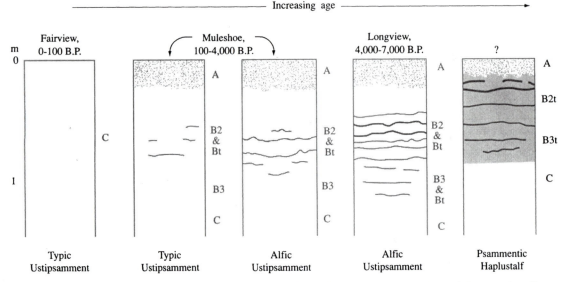

Figure 5.3 Development of clay lamellae with time for dune sands of different ages, Texas. In the field some lamallae anastomose more than depicted here. To the right and with time they merge into a Bt horizon in which a few lamellae still can be recognized. *(Redrawn from Gile, 1979, Fig. 3.)*

ment into a series of thin bands. Several processes that could be responsible for the deposition of bands include (1) depth of water penetration for particular storm events or seasons, (2) subtle changes in grain size to retard downward movement of water, and (3) flocculation of clay at particular levels, say by $CaCO_3$, by high levels of Ca^{2+} or Mg^{2+}, or by Al and Fe released during weathering. Once the process begins and the initial clay bands form, these become the sites for future clay deposition via a sieving action, and band formation is accentuated.

Thin-Section and Scanning Electron Microscope (SEM) Analysis

Thin-section analysis (Bullock and Murphy, 1983; Stoops and Eswaran, 1986) is a powerful tool for separating in situ clay from illuvial clay. Unfortunately the method is not used as much as it should be, perhaps because soil studies are so laboratory intensive. Irrespective of that, T.R. Walker (oral communication, 1997; Walker and others, 1978), who has worked mainly with sedimentary rocks, notes that in situ-formed clay occupies the pitted margins of grains that were originally smooth; the same criteria should work with soils. The clay can be optically oriented or have a random arrange-

ment, and the clay–grain boundary can be sharp or diffuse. In one case where kaolinite replaced quartz, SEM examination clearly shows that the embayment has the same hexagonal shape as the kaolinite, thus proving the replacement hypothesis. The latter could not be seen in thin section, however. In time, the original grains converting to clay might become so weathered that the grain $\Rightarrow$ clay connection becomes lost. Hopefully, if preservation is sufficient, one might be able to detect that the clays are psuedomorphous after the original grain. The above connection might be best elucidated using a chronosequence of soils.

Illuvial clay has been identified in a multitude of thin-section studies (Bullock and Thompson, 1985; Kemp, 1985). As with field occurrences, the clay forms coats on grains, in voids, and along ped surfaces. Sandy or loamy soils offer the best substrates for illuvial clay features. In many cases the clay is rather pure, it can have a laminated microstratigraphy, and it commonly has a color different from the matrix. The contact of the clay with the associated feature is sharp, and the clay platelets are oriented so that their 001 axes is parallel to the associated surface. Orientation on grains is related to soil drying, whereas that in pores and on ped surfaces probably is due to deposition as the waters are drawn to the relatively dry ped interiors. Some workers assign various kinds of films to different times in Quaternary climatic cycles (Federoff, 1997).

One problem with using clay films as the only basis for demonstrating clay illuviation is that they can be destroyed by a variety of process, as reviewed by Kemp (1985). This destruction could be as soon as they form, or later. In many places, the lose or disruption of the films is greatest higher in the profile. A common process is loss during shrink–swell cycles within the soil. Nettleton and others (1969) report clay-film destruction due to shrinkage and swelling in soils formed in the desert and Mediterranean climates of the southwestern United States. They found, for example, that soils with a low shrink–swell potential and a lower than 40% clay content can retain clay films, and that soils with a higher shrink–swell potential and a higher than 40% clay content cannot retain films. Thus, many soils may have translocated clay, but the thin-section evidence for it is destroyed as the films become incorporated into the B-horizon matrix. Thin-sections analyses have the potential to identify such incorporated oriented masses. In areas that have undergone dramatic climatic change (warm inter-glacial–cold glacial), films formed during the warmer part of the cycle are disrupted by frost action during the colder part of the cycle (Van Vliet-Lanoë, 1985). Gile and Grossman (1968) and Gile and others (1981) note that clay films in aridic soils are most stable on sand and pebble surfaces, but they can be destroyed during the accumulation of $CaCO_3$. Yet another process that destroys films is soil mixing by roots and fauna.

Many kinds of data have been used to assess the relationship between the amount of clay formed in place from that introduced by translocation (Oertal, 1968). One common analysis involves the calculation of the volume of clay films in the soil, by horizon, to determine if the horizon of maximum clay content coincides with the horizon exhibiting the greatest amount of clay films. In many cases the match is not good because the films are commonly best expressed (and preserved?) below the zone of maximum clay content (Brewer, 1964; Buol and Hole, 1961). The amount of illuviated clay, as evidenced by the percentage of clay films in thin section, commonly falls short of the total B-horizon clay (Fig. 5.4A), suggesting either that the latter may not have been derived entirely by illuviation, or if it was, the thin-section evidence for such an origin has been subsequently eradicated. In other

cases, however, the match between the argillic B horizon and maximum expression of clay films is good (Gile and Grossman, 1968), and in one aridic soil virtually all of the clay in the profile is accounted for by

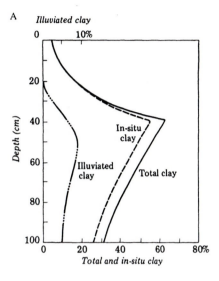

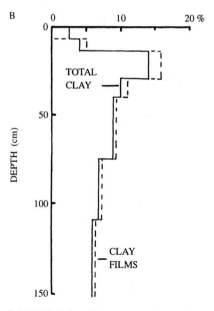

Figure 5.4 (A) Relationship among total clay, illuviated clay (recognized by oriented clay films), and by difference, clay formed in situ. In situ clay represents that formed by weathering and/or that originally in the parent material. *(Redrawn from Brewer, 1968, Fig. 3.)* (B) Relationship between contents of oriented clay films and total clay for a Haplargid in Nevada. *(Data from Nettleton and others, 1975, Table 7.)*

clay films measured in thin sections (Nettleton and others, 1975) (Fig. 5.4B).

Finally, the SEM might be used to help identify translocation of soil clays, as Walker and others (1978) have demonstrated for young sediments (Fig. 5.5). Chemically precipitated clay minerals form a crystalline texture on the host grain, and the morphology of the crystalline clay varies with mineral species. In contrast, mechanically translocated clays are recognized by platelets aligned parallel to the grain surfaces on which they are deposited and by platelet deposits that have the overall shape of menisci between grains, with an orientation more or less parallel to the former meniscus surface. For sediments, and perhaps for

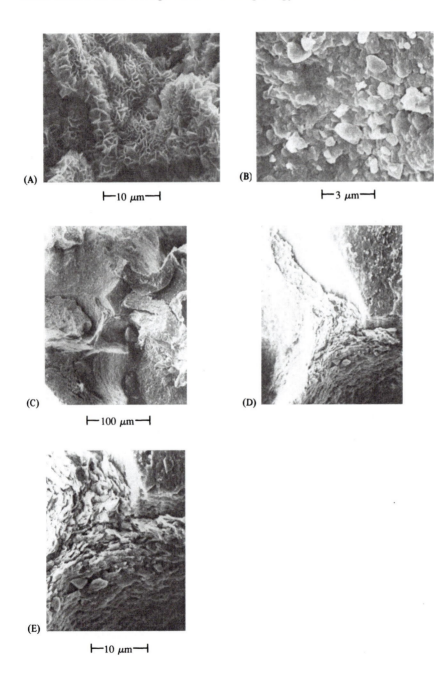

(A) ⊢—10 μm—⊣

(B) ⊢—3 μm—⊣

(C) ⊢—100 μm—⊣

(D)

(E) ⊢—10 μm—⊣

Figure 5.5 Scanning electron microscope (SEM) photomicrographs comparing the morphology of chemically precipited clay with that of translocated clay. (A) Chemically precipitated smectite; (B) translocated clay oriented parallel to the grain surface on which it was deposited (more or less parallel to the page); (C) translocated clay platelets that form bridges between grains; the bridges mimic the form of the water minisci; (D) and (E) are progressively closer views of the bridge in (C). *(Courtesy of T.R. Walker.)*

Figure 5.6 Comparison between original (nonclay) mineral loss due to weathering and clay formation from constituents released during weathering, by the index-mineral method. Nonclay mineral loses and clay formation are estimated to be restricted to the upper 61 cm; this thickness would depend on the choice of parent material. *(Data from Barshad, 1964, Table 1.13 in Chemistry of the Soil by F.E. Bear, ed., © 1964 by Litton Educational Publishing, Inc., Reprinted by permission of Van Nostrand Reinhold Company.)*

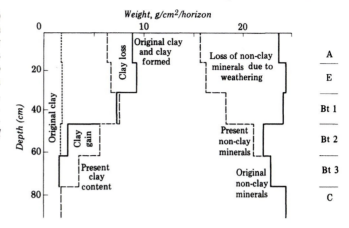

soils, they indicate that SEM photomicrographs could point more clearly to a translocation origin than does thin-section evidence.

Mass-Balance Studies

Quantitative mineral-analysis techniques have been developed by various workers to determine weight gains and losses of both clay and nonclay (or primary) minerals within the profile; these techniques can be used to assess clay formation in place from a clay distribution due to translocation (Barshad, 1964; Brewer, 1964; Marshall, 1977). First, one has to prove parent material uniformity (Chapter 7), and because few soils have a uniform parent material, this limits the usefulness of these techniques. All changes within the soil are based on comparisons with the unweathered parent material, which might be either rock or unconsolidated deposits at depth. A resistant immobile primary mineral, such as zircon or quartz in the sand fraction, is used as an index mineral. If one assumes that the amount of weathering of the index mineral is negligible, the ratios of all other primary minerals to the index mineral can be calculated. These ratios will decrease from the parent material upward into the soil as the less-resistant primary minerals weather, and they should be lowest where the weathering has been most intense. There should be some relationship between the loss in primary minerals and the formation of clay-size particles. Once the relationship is established, one can estimate the amount of clay formed per horizon and the amount of clay now present per horizon; the difference is the amount of clay that has been gained or lost by translocation (Fig. 5.6). In this particular case the reconstruction indicated that 22.9 g of nonclay minerals was lost while 20.9 g of clay was formed. This indicates that in going from parent material ⇒ soil, 91% of the original parent mass is retained.

Brewer (1964) raised some objections to the above approach and integrated both mineralogical and thin-section data to determine the formation and migration of clay-size material in an Ultisol in Australia (Brewer, 1955). Knowing the approximate amounts of the rock-forming minerals that have weathered, as well as the clay mineralogy, he calculated the maximum amount of clay that could form due to weathering in place; this amount of clay will vary with the chemical composition of the clay fraction. Calculations were made for both kaolinite and for "complex" clay minerals (a variety of 2:1 minerals considered end products of weathering and found to be present on X-ray analysis). A comparison of either the weight of kaolinite or that of "complex" clay mineral that could form from weathering in place with the actual clay present showed that much of the material released by weathering was removed from the soil and that only a small portion remained to form the clays (in contrast with the 91% retention of mass, above). By these data alone, it is not necessary to hypothesize clay illuviation to explain present clay distribution. Thin-section study revealed a lack of oriented clay and, thus, no evidence of clay illuviation. There is the possibility, however, that the thin-section evidence for clay movement has been destroyed, because the B horizons contain 30–43% clay.

In all of these studies, the ultimate origin of the clays is tantamount to the success of the calculations. For example, it is assumed that the clay comes from the weathering of the nonclay, and the C-horizon clay is used as the starting point. But, if the parent material is igneous rock, it has no clay, and so the amount of pedogenic clay should reflect this. In places where the soils form from a mixture of sedimentary rocks (e.g., a till with a mixture of shale, sandstone, and granite clasts), the main supplier of clay could be the physical breakdown of shale particles, and the rate of supply could be fast. Finally, as pointed out in Chapters 7 and 8, clay in aridic soils can come from eolian dust, and this would have no relation to the weathering of the nonclay minerals; this seems to be the case for circum-Mediterranean soils too (Yaalon, 1997).

Summary

The examples given here demonstrate how difficult it is to document and quantify evidence that translocation of clay-size particles has occurred in soils. Oriented clays in thin section generally are accepted as evidence for such movement, as long as stress orientation can be excluded. However, subsequent pedogenesis can destroy this evidence. Furthermore, for in situ weathering, one should expect to find some of the weathered mineral forms shown in Fig. 8.7 and in the photomicrographs of Walker and others (1978). Clay and nonclay mineral relationships also can be used, but it is extremely difficult to calculate the gains and losses accurately, and identifying parent materials is a continuing problem. However, the above studies, before being considered solved, should be examined in the field again for evidence of erosion or atmospheric additions, etc.

Much of the above can be combined to examine the origin of duplex or textural contrast soils in Australia, soils characterized by an abrupt increase of clay from the A to the Bt horizon (Chittleborough and Oades, 1979, 1980a, 1980b; reviewed in Paton and others, 1995). Clay content is ca. 20% in the A horizon, increases to >80% in the Bt horizon, with much of the clay in the finer fraction, and decreases to 35–40% at greater depth. Mineral and resistant oxide data are sufficiently uniform to suggest one parent material. If the clay is due to in situ weathering in a leaching environment, this should show up in nonresistant:resistant oxide ratios, yet the $CaO:TiO_2$ ratio of both sand and silt remains relatively constant with depth. Illuviation was checked for via thin sections, but there are no clay films; perhaps with these high clay contents, films are destroyed. Several different calculations of clay gains and losses show (1) little profile mass loss due to clay formation, and (2) a mis-match between clay illuviation into the Bt horizon and clay eluviation from the A horizon of about 2:1 on the basis of grams/100 g parent material. The conclusion is that translocation of clay inherited from parent material is the most likely reason for the extreme contrast in particle size distribution with depth.

The latter study illustrates the difficulty in obtaining definitive results on the origins of clay distribution in profiles, in spite of intensive and diverse laboratory work. In soils such as these, the field setting is so important to the interpretation. For example, one could look again at the mineralogy and chemistry, for it is possible for them to indicate parent material uniformity in a texturally layered parent material. Furthermore, weathering can go on, yet if the soil is not leached, the $CaO:TiO_2$ ratio may not change. Finally, surface erosion could help account for the A–B loss–gain mismatch, as could translocated eolian clays. A later study of Australian duplex soils suggests that several processes were operative—initially, there was translocation of inherited parent material clay, and this was followed by weathering in a soil-moisture regime characterized by strong seasonal fluctuations (Walker and Chittleborough, 1986). Still later detailed studies indicate that atmospheric dust contributes to the genesis of textural contrast soils (Chartres and Walker, 1988; Chartres and others, 1988; Walker and others, 1988).

Dispersion and Flocculation

Clay migration requires that the clay be dispersed so that it can remain in suspension and be translocated by water moving slowly through pores or cracks in the soil. Dispersion is favored by several factors, among them a low electrolyte content in the soil solution and the absence of positively charged colloids, such as iron and aluminum hydroxides (Olphen, 1963). Flocculation is induced by high electrolyte content in the soil solution and the presence of positively charged colloids; clays under these conditions cannot migrate.

The dispersion or flocculation of clay-size particles also depends on the thickness of the ion layer that satisfies the negative charge of the particle, and this thickness can vary with the ion present. Ions attracted to particle surfaces are distributed so that their concentration is highest close to the surface, and concentration diminishes away from the surface (Fig. 5.7A). When two particles, each with positive ions attracted to their surfaces, move toward one another, the initial reaction is one of repulsion (Fig. 5.7B). Thus, the clays

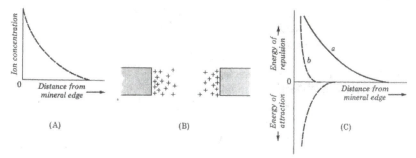

Figure 5.7 Conditions for clay dispersion and flocculation. (A) Change in concentration of ions with distance from a charged surface. The total positive ion charge satisfies the negative clay-particle charge. (B) Distribution of positive ions attracted to two clay surfaces. If the two clays are brought close together, and the ion layers are thick enough, flocculation cannot take place and the system remains dispersed. (C) Energy relations for repulsive and attractive forces between two clay particles with positive surface ions. In plot a, the ion layer is thick so flocculation does not take place. In plot b, the ion layer is thin because of either high electrolyte content or higher valency ions; clays can come into close contact because of short-range van der Waals forces of attraction between particles, and flocculation occurs. *(Redrawn and modified from Olphen, 1963, Figs. 9 and 12, © 1963, John Wiley & Sons.)*

are dispersed. If, however, the clay particles can move closer together, van der Waals force takes over, and attraction and flocculation occur. The particles also can come close together if the thickness of the ionic layer is reduced (Fig. 5.7C). This reduction can be brought about in two ways: by increasing the electrolyte concentration of the soil solution or by replacing the ion layer with ions of higher valency.

These flocculation–dispersion effects can be seen in soils (McKeague and others, 1969). Calcium clays would have a thin ion layer; they commonly are flocculated and so commonly do not migrate. Soil parent materials high in $CaCO_3$, for example, show little evidence of clay migration; migration can take place only after the carbonate has been leached from the soil and some of the Ca^{2+} in the exchange complex is replaced (e.g., Schaetzl, 1996; Timpson and others, 1996). In a study in Pennsylvania a base (Ca + Mg) saturation of 40% was sufficient to flocculate clays (reported in Ciolkosz and others, 1989). Thus, a common interpretation of a Btk horizon is that clay movement preceded carbonate accumulation. In contrast, clays with appreciable amounts of Na^+ in the exchange complex remain dispersed because the ionic layer is thick (reviewed in Naidu and others, 1995). The pedogenic translocation and subsequent accumulation of Na-clays produce a natric horizon. In some desert basin deposits of the western United States, natric horizons can form in about one-half the time it takes argillic horizons to form (Alexander and Nettleton, 1977; Peterson, 1980). Coastal regions are characterized by high inputs of atmospheric Na^+ (Yaalon, 1963), and transects in Australia show the rapid decrease in salt content inland (Isbell and others, 1983). The result is that the rate of formation of Bt horizons can be accelerated close to the coast (Shlemon and Hamilton, 1978). Some of these effects can be seasonal, with Na^+ concentration during the dry season, and dilution during the wet season. Finally, at still higher concentrations of Na^+, flocculation results because the high electrolyte content in the soil solution compresses the adsorbed ion layer.

Some experiments have been performed to determine the factors responsible for clay migration (Brewer and Haldane, 1957; Hallsworth, 1963). Besides verifying, to some extent, the cation relationships discussed above, the experiments indicate an upper limit of 20–40% clay, depending on the clay mineral, above which migration in the pore spaces virtually ceases. At these clay contents, the pore spaces may be small enough to limit movement, or the swelling of clays on hydration may close off some pores. Clay migration at greater clay contents would have to take place along soil–structure discontinuities (prismatic structure), and, as mentioned earlier, soils with such high clay contents tend to lose the clay-film evidence for migration quite rapidly.

Conditions for clay migration can vary with depth in the profile, or even seasonally. Quite commonly, the electrolyte content will increase with depth because the upper parts of the profile have more water moving through them. Thus, clays could be dispersed near the surface, but be flocculated at depth. In soils with a high clay content, high shrink–swell potential, and distinct dry and wet periods, clay movement might take place only at the onset of the wet season, when an open prismatic soil structure with wide cracks reaches the surface and thus provides avenues for rapid clay translocation.

Still another way in which clays can be translocated in a soil is by suspension originally in rainwater. Holliday (1988) mentions this as a mechanism to form argillic horizons within about 450 years at Lubbock, Texas. In that area it is not uncommon for a dust storm to precede a thunderstorm, and the subsequent rain comes through the dust, producing a mud rain. One appealing aspect of this process is that it does not require that clays already present in the soil be brought into suspension before translocation by slowly moving soil water. A similar mechanism might also be responsible for the movement of clays in calcareous materials. It was argued above that such movement would not be expected, but it has been demonstrated in an artificial field experiment (Goss and others, 1973).

■ Translocation of Silt

Some coarse-grained soils display an accumulation in silt with depth. An example of this is a soil formed in ca. 20-ka till in the eastern Canadian Arctic, in which the silt-enriched horizon underlies the Bw horizon (Table 5.6). Commonly, the silt-enriched horizons also have silt caps on the clasts, leading Forman and Miller (1984) to devise the silt-cap-morphology stages (Appendix 1). Similar silt accumulations occur in soils of the Peruvian hyperarid desert (Noller, 1993). Laboratory work demonstrates that water can transport silt vertically in coarse-grained soils, provided the voids are of the proper size (Wright and Foss, 1968). To invoke a pedogenic origin for such a particle-size distribution, other possible origins, such as parent-material layering, should be discounted. Because many of these soils are in cold climates, one must consider frost sorting as a possible mechanism to redistribute inherited silt (Bockheim, 1979; Washburn, 1980; Van Vliet-Lanoë, 1995; Frenot and others, 1995), as well as silt movement during seasonal melting. That is not the only mechanism, however, for soils at high latitudes and altitudes or in other areas. Some sites are on well-drained moraine crests; arguments can be made that these soils are relatively dry and frost

Table 5.6 Data for a Soil with a Silt-enriched Baffin Island, Eastern Canadian Arctic

			Particle-Size Distribution in <2 mm Fraction (%)		
Horizon	Depth (cm)	Color	Sand	Silt	Clay
A	0–1	10YR 4/2.5	82.3	14.6	3.1
Bw	1–12	10YR 4/4.5	91.6	5.4	3.0
B*l*	12–28	2.5Y 5.5/4	66.4	29.3	4.3
Cox1	28–106	5Y 7/4	90.1	7.3	2.6
Cox2	106–116+	5Y 7/3	89.4	7.8	2.8

Data from Birkeland (1978, Table 5).

stirring is minimal. Locke (1986) analyzed the particle-size distribution of both matrix and silt caps in Baffin Island soils, and found that there is a surplus of silt in both materials; silt is greatest in the caps, at over 15 times the amount in the presumed parent material. The depth distribution is a classical one for distribution by vertically moving soil water, the hypothesis he favors. Finally, because there is a silt surplus, it must be slowly added to the soil surface over time, and then moved into the soil as fast as it is added. This helps explain the lack of loess in the area, as the meager amount of eolian silt becomes sequestered in the soils.

There are several ways of determining whether the silt was inherited from the parent material and later redistributed within the profile, or was introduced from an extraneous source. (1) See if there is a silt surplus, as Locke (1986) did in his mass-balance approach. (2) Determine if the silt can come from weathering. Soils studied on Spitsbergen are very coarse grained (>85% gravel, 1% silt + clay), in contrast to the maximum silt accumulation of 73% (Forman and Miller, 1984). As >80% of the clasts are carbonate rocks, and those near the surface show evidence of dissolution, an obvious silt source is the insoluble residue left from dissolution. (3) Determine the silt mineralogy, for in one high-altitude study, this proved an eolian silt source from nearby basins (Dahms, 1993). (4) Check for evidence for movement and accumulation using thin-section methodology (Frenot and others, 1995). (5) Finally, the oxygen-isotope composition of quartz grains also could help identify materials in the soil suspected of being of eolian origin from a distant source. Drees and others (1989) review the literature on this and point out that different geological environments in which quartz forms are characterized by different oxygen-isotope compositions.

Fragipan and Its Origin

Fragipan is a slowly permeable subsurface soil horizon that is brittle and rigid when moist (hard to very hard when dry), yet yields suddenly in a brittle fashion when pressure is applied to a sample (Smeck and Ciolkosz, 1989). The style of breakage is similar to that of a dry graham cracker. Such behavior would make one suspect a cementing material, but if one exists it must be weak, for the material slakes in water. Bulk density is high, usually greater than 1.6 g/cm^3. Thickness of fragipan ranges from 15 to 200 cm, with the top at about 33–100 cm depth. Fragipan is most common in loamy materials (till, loess, lake, alluvial and colluvial deposits), in climates characterized by water moving through the soil at some time of the year, in warm or cold temperatures, and with tree vegetation. A very coarse prismatic structure is common, as are bleaching and mottling along the prism walls, and clay films can be present. In glacial terrain, a dense till can be mistaken for fragipan.

Fragipan is found in four soil orders: Alfisols, Inceptisols, Spodosols, and Ultisols, and is most common in the United States and New Zealand. In wetter soils, it is found beneath an eluvial horizon, but not necessarily immediately below it. It may underlie a spodic, cambic, or argillic horizon, and the fragipan is in the B or C horizon. Commonly, an E horizon is at the top of the fragipan, formed perhaps by lateral flowage of water because vertical water movement is impeded in the fragipan. The horizon sequence gives rise to bisequal soils, that is soils with two vertical and separate eluvial-illuvial sequences. The lower E horizon in a bisequal soil is denoted by a prime. Thus, a bisequal soil with fragipan can have a A/E/Bt/E′/Btx horizon sequence, with x designating fragipan (Table 1.1).

The origin of fragipan is obscure (Smeck and Ciolkosz, 1989). There seems to be no chemical or mineralogical association, other than the fact that fragipans are low in organic matter and usually noncalcareous. Several origins have been invoked: (1) Initial drying out of a wet soil mass, called physical ripening, can result in the dense fabric and prismatic

structure of a fragipan; (2) silicate clay minerals can bind the grains together; or (3) silica-rich amorphous materials can be the binding agent. If origins (2) or (3) are involved, apparently they are rather thin and difficult to detect. It could be that depending on local circumstances, all three hypotheses play a role in fragipan formation.

Morphology and Origin of Lateritic (Oxisolic) Profiles

The origin of Oxisolic or lateritic profiles has been reviewed by several workers, from which this review is taken (Sivarajasingham and others, 1962; McFarlane, 1976, 1983, 1987; Paramananthan and Eswaran, 1980; Eswaran and Tavernier, 1980; Division of Soils, CSIRO, 1983; Goudie, 1983; Buol and others, 1997; Tardy, 1992; papers in Schwarz and Germann, 1994). Rather than be restricted to the definition of the Oxisol, this review will include the origin of a broader group of profiles collectively known as lateritic, thus the discussion will take into account the origin of the altered materials beneath the oxic horizon. Here we are concerned with the lateritic profiles that are thick (measured in meters), old (measured in My), highly weathered material, characterized by (1) massive enrichment of Fe, Al, or both, and associated oxide, hydroxide, and oxyhydroxide minerals, (2) massive depletion of silica that was combined with other oxides in the common rock-forming minerals exclusive of quartz (in fact, few original minerals are left), and (3) extensive depletion of bases; in addition, (4) they can contain quartz and various layer-lattice clay minerals and (5) part of the profile may harden irreversibly on drying, or be capable of so hardening. If Fe is the cement, the material is called ferricrete, a form of duricrust (others are calcrete, cemented with $CaCO_3$; silcrete, cemented with silica; Goudie, 1973). It is generally conceded that alteration took place on low-relief stable landscapes under a hot, humid tropical climate, with or without a dry season. Because of these conditions, lateritic soils are found on stable parts of continents, characterized by little tectonic influence and no Pleistocene glaciation (e.g., Brazil, central Africa, parts of Australia). Probably more books have been written about these materials than any other soil, perhaps because their origin has stirred the imagination of soil and earth scientists, and various ores (Fe, Al, Ni, Co, Cr; even Au) are associated with them.

Valeton (1994), for example, shows that major lateritic bauxites have formed only during times of global wet and warm periods (other conditions being favorable), and there have been no more than five such periods between the Precambrian and the Quaternary. In addition, many of the areas with lateritic profiles are in areas of growing and/or high population, so there are concerns about the agricultural productivity of these soils. Many hypotheses have been put forward to explain lateritic soils, perhaps because they have been difficult to study (old landscapes and present-day processes may not be those under which the soils formed), but also because many processes may be involved, depending on the local environmental setting. Lateritic profiles are present in a variety of field settings. First, they occur on all common rock types. Second, they commonly are associated with essentially flat plateau remnants, formed during subsequent incision (see Fig. 10.20), although such topographic inversion is challenged in places (Conacher, 1991). Without this incision, workers would not have been able to work out the details of the profiles. Not uncommonly, lateritic profiles occur at several levels in the landscape, leading workers to suggest several possible origins for the levels: (1) surfaces are of different ages, left behind during downcutting; (2) the original surface was a rolling one of one general age, and with subsequennt dissection parts of it are now at different levels; and (3) lateral geochemical differentiation on a sloping plain produced different lateritic profiles that, upon dissection, appear to form a stepped sequence, attributed by some to age. Third, the profiles are deep, with reported thicknesses of 10–150 m. Fourth, there is a wide range of old ages for them—10 to 100 My—and global climates have varied greatly over this time span. These attributes, combined with the dense vegetation in places, make it difficult to be certain of the genetic pathway for their formation.

The Lateritic Profile

The common lateritic profile with Fe accumulation is red ferricrete or an oxic horizon overlying saprolite, as described by Tardy (1992; Fig. 5.8). Clay content commonly exceeds 50%, and the aluminosilicate layer-lattice clay minerals would be kaolinite in the main part of the profile, if well drained, with smectite at greater depth. In places the parent rock is recognizable at depth. Overlying the rock, generally with a

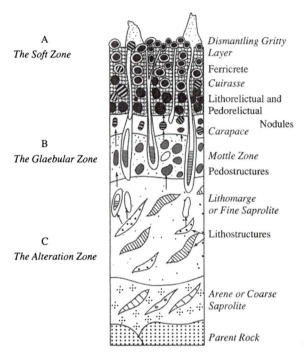

Figure 5.8 Schematic lateritic profile with ferricrete, showing the three zones, from top to bottom: (A) soft zone (termite mounds at the land surface), (B) glaebular zone, and (C) alteration zone. In the glaebular zone hematite nodules are set in a kaolinitic matrix, and in the soft zone the nodules are dismantled and secondary goethite developed. *(Redrawn from Tardy, 1992, Fig. 15.2 in Martini and Chesworth, eds., Weathering Soils & Paleosols © 1992, with kind permission from Elsevier Science, The Netherlands).*

diffuse contact along an irregular weathering front, is saprolite, characterized by in situ weathering with little change in volume but a decrease in bulk density relative to the parent rock. Coarse saprolite at depth gives way upward to fine saprolite. Ideally, the former has a greater proportion of less-weathered rock masses, and kaolinite and gibbsite under good drainage conditions and smectite under poor drainage. In contrast, the latter have a greater degree of weathering, with quartz the main survivor of the parent minerals, and kaolinite, hematite, and goethite the main secondary minerals. Within the saprolite, Fe and Al are liberated by weathering, form secondary products, and although much of it remains close to the sites of release, there can be some local areas of leaching and accumulation. The saprolite zone is sometimes called the pallid zone (Fig. 10.20), because of its pale color.

Above the saprolite zone is the glaebular zone with a yellow soil matrix, subdivided into a lower mottled subzone and an intermediate nodular subzone, and an upper oxic horizon or ferricrete. The mottled zone is made up of bleached areas of quartz, the only parent mineral survivor in the more weathered profiles, and kaolinite, from which Fe has been removed. If leaching is extensive, macrovoids or tubules form, and at a later stage these can be partly filled with kaolinite, by precipitation from solution or physical translocation; these are called pedorelictual features. The Fe is transported to sites where it accumulates as mottles. Two common sites of Fe deposition are association (1) with the pedorelictual features and (2) with masses in which the parent rock structure still is visible (lithorelictual mottle). The nodular subzone is characterized by hard masses cemented with Fe. Lithorelict nodules are more common in schists and amphiboles, whereas pedorelict nodules dominate in rocks high in quartz (sandstone, granitic rocks). The uppermost subzone is red and can qualify as an oxic horizon; if it is ferricrete it consists of a hematite-cemented mass of pedorelictual or lithorelictual nodules, some of which are pisolitic (near spherical with concentric bands), and nodule density is at its greatest for the profile (Fig. 5.9). If clay minerals are in the matrix, they would be kaolinite. Tubes or channels are present, and the walls can be lined with goethite and the channel partly filled with kaolinite.

The top of the ideal laterite profile with ferricrete is the soft zone. It is loose, sandy to pebbly, and contains all the materials of the underlying ferricrete. If Fe, Al, and Si are leached from it and precipitate at the base of the soft zone, they form a reconstituted ferricrete.

The tubes and voids of the ferricrete, structure of the Bo horizon, and extensive isovolume alteration in the saprolite result in a good internal drainage throughout the thick profile.

Other lateritic profiles have a near-surface or surface horizon rich in aluminum, with the main minerals gibbsite and boehmite; these are termed lateritic bauxite (Bárdossy and Aleva, 1990; Tardy, 1992; Valeton and Wilke, 1993). Bauxite is a general term for materials in which Al minerals predominate. As with the laterite with ferricrete, there is a soft zone at the surface, pisolitic structures are common, and saprolite is present at depth; the latter is light colored and can be mottled, and, being a porous material, internal drainage is good. Valeton and Wilke (1993) give

a general profile in which the Bo horizon is subdivided into an upper ferricrete and a lower bauxite (can be termed alucrete if cemented) (Fig. 5.10). The underlying saprolite is of two kinds: one is kaolinite rich, and the other consists of kaolinitic saprolite overlying smectitic saprolite. Two general forms of profiles seem to occur, according to Tardy (1992). One has a gibbsite-enriched zone above the saprolite, and a hematite-enriched zone at the surface, as in Fig. 5.10. The other has ferricrete consisting of hematite, goethite, and kaolinite above the saprolite, a layer of gibbsite, hematite, and goethite above that, and a layer of boehmite and hematite at the surface. The first of the above forms is found in humid tropical climates, and the second in tropical climates with a dry season. Many of the lateritic bauxites illustrated in Bárdossy and Aleva (1990) are red, no doubt reflecting the presence of secondary Fe minerals, but those that are low in Fe minerals are almost white.

Chemistry and Mineralogy of Lateritic Profiles

The most developed horizons of the lateritic profiles are highly enriched in Fe and Al. The chemistry of ferricretes changes with time, and it is presumed that the chemistry of bauxitic laterites does too. Young ferricretes on either granite or basic rock carry a legacy of the rock chemistry and mineralogy. In time however, as pointed out by Chesworth (1973; see also Fig. 3.12), chemistries of ferricrete over both kinds of rock become similar (Table 5.7). All bases are drastically reduced to less than 0.2%, silica is reduced to 25–30%

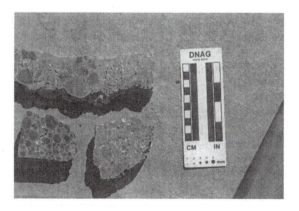

Figure 5.9 Polished sections of ferricrete. *(Samples courtesy of H.M. Churchward.)*

HORIZON · MINERALS

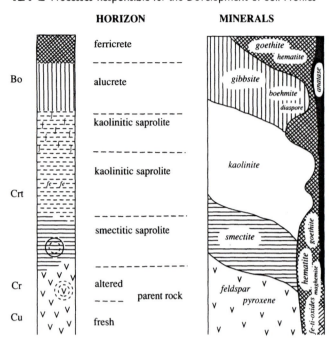

Figure 5.10 Schematic profile of bauxite-bearing laterite and saprolite over basalt in India. *(Redrawn from Valeton and Wilke, 1993, Fig. 14).*

of the original value, and Fe_2O_3 values are 5 to over 25 times greater than parent material values. Molar SiO_2/Al_2O_3 ratios are close to 2, in line with the statement by Tardy (1992) that the ratio does not exceed the value required for kaolinite. Any excess silica would be sequestered in quartz. Hematite, geothite, and kaolinite make up close to one-third each of the minerals, and quartz is <5%. Many trace elements display enrichment, but Sr and Ba are depleted.

Kronberg and others (1982) compare various bauxites with the Earth's crust. The one with the highest Al_2O_3 (60% or a four times enrichment) is extensively leached of bases (all <<0.1%), SiO_2 (at 3%) is about 6% of the crustal value, Fe_2O_3 (at 3%) is 40% of the crust, TiO_2 (at 0.7%) is 70% of the crust, and the molar SiO_2/Al_2O_3 value is 0.1. Bárdossy and Aleva (1990, Table 5.4) present bauxite data from around the world, and most have these values: Al_2O_3, 40–52%; SiO_2, 4–8%; Fe_2O_3, 10–25%; and TiO_2, 1–4%. Vertical plots through thick profiles demonstrate overall trends of progressively greater Al_2O_3 from the parent rock toward the surface, extensive leaching of SiO_2, and Fe_2O_3 and TiO_2 both fluctuating and increasing. Of the trace elements, Cr, Ni, Sn, and Zr are enriched, and Co and Zn are depleted. The mineralogy is dominated by 40–70% gibbsite, with boehmite the other common Al species.

Lateritic profiles are intimately involved with tropical climate agriculture. Given the above oxide and hydrous oxide mineralogies, the soils have exceeding low CECs (ca. 4 meq/100 g; see Table 4.5). Those soils with values below 1.5 meq/100 g of clay are denoted by the prefix "acr" at the great group level of the Oxisols (Soil Survey Staff, 1994). This is about as low in CEC and exchangeable bases that a soil can get. Because plant nutrition depends on a supply of bases, usually from weathering of parent minerals now long gone, these soils can recycle the exchangeable bases only from soil to plant and back again. If this cycling is disturbed, bases are leached and the soil is rendered devoid of nutrients (Fyfe and Kronberg, 1980). This has major ramifications to agriculture in the tropics (Kronberg and Fyfe, 1989). Some bases come from distant sources via atmospheric transport of solids (later).

Genesis

Lateritic profiles develop on stable landscapes over millions of years, where climatic conditions are such as to create intense weathering conditions. Leaching of almost all elements within the soil is rapid, due to the porosity created either by in situ weathering, the large voids between nodules, or tubules. The ground water table is depicted as being at a depth shallow enough to influence the profile, or is within the zone of the maximum weathering; the water table can fluctuate seasonally. The water table can influence the behavior of the accumulating elements: Al and Fe. Many workers mention that reducing conditions can be obtained below the water table, and this provides appropriate conditions for separating the two elements. Fe is mobilized as Fe^{2+} under reducing conditions, but Al is not. Thus, under a groundwater influence, Fe can be separated laterally from Al, in laterally flowing groundwaters. Valeton and Wilke (1993) depict

Table 5.7 Average Chemical and Mineralogical Composition of Granitic (PRA) and Basic (PRB) Rocks, Moderately Evolved Ferricrete on Granite (FMA) and Basic (FMB) Rocks, and Highly Evolved Ferricrete (Averaged) for Both Rocks (MEN)

	PRA	PRB	FMA	FMB	MEN
SiO_2 (%)	70.75	55.76	30.81	25.81	16.8
Al_2O_3 (%)	14.56	15.12	14.87	18.7	14.07
Fe_2O_3 (%)	2.01	9.17	42.67	42.9	57.18
MgO (%)	0.68	4.95	0.1	0.11	0.06
CaO (%)	2.39	7.6	0.21	0.21	0.2
Na_2O (%)	4.88	3.19	0.07	0.09	0.08
K_2O (%)	2.14	0.64	0.11	0.2	0.09
Mn_3O_4 (%)	0.03	0.17	0.2	0.15	0.12
P_2O_5 (%)	0.11	0.16	0.27	0.23	0.28
TiO_2 (%)	0.27	0.67	0.83	0.75	0.84
H_2O (%)	1.27	3.17	9.45	10.79	9.94
Sum (%)	99.17	100.66	99.45	99.8	99.47
Trace elements (ppm)					
Sr	449	317	14	20	10
Ba	656	314	206	101	59
Ni	9	55	54	49	70
Co	8	35	44	61	32
Cr	11	187	358	485	451
Zn	45	123	57	67	122
Zr	127	75	214	210	139
Selected minerals (%)					
Quartz	28	8	17	9	4
Kaolinite			31	41	30
Goethite			32	33	35
Hematite			17	16	31
Gibbsite			2	2	0

From Tardy and Roquin (1992, Table 16.1).

this with a series of diagrams (Fig. 5.11). With a low water table (several 10s of meters down), Al and Fe are enriched in the Bo horizon, but they can both migrate downward to the water table, as well as laterally at the water table. With a shallow water table one ends up with a Bo horizon in which Al and Fe are enriched above the water table and Al below it. Some of the Fe enrichment is upward from the water table (see also McFarlane, 1976, Fig. 17). Finally with a water table at the surface, Fe is removed and Al remains as the residue.

Because of the antiquity of the system, the origins of the profiles have been elusive. Three origins are generally considered (McFarlane, 1976, 1983, 1987; Goudie, 1983; Bárdossy and Aleva, 1990; Tardy, 1992; Tardy and Roquin, 1992; Valeton and Wilke, 1993; Schellmann, 1994). Here only general concepts are given, and the original papers should be consulted for the details, including thermodynamic considerations regarding the stability of the resulting minerals. One origin is that the profiles have formed in place from the presumed parent material at depth. The loss of the majority of the elements is expected, and the enrichment of Al and Fe is as an in situ residue. The latter two elements can continue to accumulate with the slow lowering of the land surface. One has to search for the mineralogical or elemental clues that connect the weathered mass to the parent materials, and gold could even be one (Tardy, 1992, Fig. 16.8). Schellmann (1994) suggests that the preference of Al over Fe concentration is related to a greater degree of leaching for the former, combined with such low levels of dissolved silica that Al hydroxide minerals form in preference to kaolinite. Such intense leaching can even dissolve quartz. The second origin is the lateral movement of the element of interest in laterally moving ground water, followed by its subsequent precipitation and eventual concentration. This could be the way some ferricrete forms (McFarlane, 1976, Fig. 18; see also Schwarz, 1994). In this case the parent material and its weathering products stay in place, and Fe is added from outside the system laterally, giving an Fe concentration far greater than one could envisage for derivation from the presumed thickness of overlying rock. I may have seen this along the banks of the Tambopata River of southeastern Peru. The river banks are several meters high, consist of sandy alluvium, and abundant Fe is being precipitated from ground water; it oozes from the banks at river level. The third origin is the mechanical transport of previ-

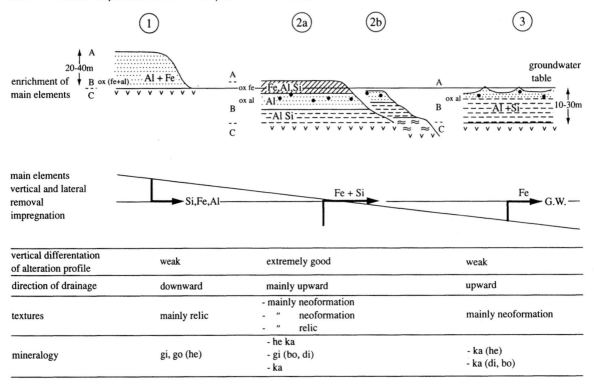

Figure 5.11 Schematic diagrams depicting three main types of bauxite formation. (1) Formation above the water table produces a red-brown Bo horizon without separation of Al and Fe, and mainly goethite and gibbsite. Saprolite is reduced or missing. (2) Formation near the water table produces a surface red, hematitic layer, overlying a whitish, gibbsitic bauxite layer, overlying thick saprolite with kaolinite or smectite. (3) Formation below the water table with good drainage produces kaolinitic saprolite with Al minerals, but no Fe minerals. *(Redrawn from Valeton and Wilkes, 1993, Fig. 43.)*

ously weathered material, including lateritic materials, followed by subsequent pedogenesis, including cementation, or solution and reprecipitation of exiting materials. Cementing materials, such as Fe, can move to the site vertically or laterally, or be derived from the solution and reprecipitation of transported materials. In such deeply weathered profiles, this latter origin may be difficult to differentiate from the other origins.

Stone lines may help in recognizing the depositional origin of lateritic soils. Stone lines are thin, planar concentrations of resistant clasts. They can have many origins (reviewed by Johnson and Watson-Stegner, 1990; Johnson and Balek, 1991), even concentration by termites. In the context of lateritic profiles they could occur as a concentration of vein quartz clasts within a lateritic mass (Ruhe, 1959; Wells and others, 1990; see particularly good photographs in the

latter article). One origin suggests that as the landscape lowers, geomorphic processes move and concentrate resistate clasts into a thin surface mantle, which subsequently can be buried by materials of varying thicknesses (Parizek and Woodruff, 1957); the buried concentrate is the stone line. With further weathering, the stone line could appear at any level within the lateritic profile, and serve as a hint as to the erosional–depositional origin of part of the profile. In Madagascar, the geometry of the stone line marks the positions of ancient hills and valleys in the old landscape (Wells and others, 1990). This latter work supports other work showing the inversion of segments of these old landscapes.

In addition, the materials from which the lateritic profiles form can be derived from distant sources (transoceanic) via eolian transport. Muhs and others (1990) mention this as a possible origin for bauxite parent materials for Jamaica and Haiti, and Brimhall and others (1988) argue for an eolian contribution to the lateritic profiles of western Australia. From a plant nutrition point of view, dust from the arid Sahara crosses the Atlantic Ocean and contributes unweathered minerals to the rainforests of the Amazon. On weathering, the minerals release needed nutrients to the soils. In fact, given certain assumptions, data indicate that the forest is maintained because of the nutrient supply from atmospheric input, as there is thought to be little nutrient input from mineral weathering, as few primary minerals are left (Jordan, 1982). In the example from western Australian, the dust was thought to be already weathered, so perhaps it was not as good a supplier of nutrients.

Clays form in lateritic profiles, but they do not seem to migrate very far, even though most oxic horizons are highly permeable and can translocate large quantities of rainfall. Perhaps the mutual attraction of the negatively charged clay and the positively charged sesquioxides to give stable aggregates prevents much clay translocation; this seems to be one reason why dispersion of laboratory samples is so difficult. Some clay moves and partly fills the tubes in the glaebular zone, however (Fig. 5.8).

The soft surface zone of lateritic profiles has several origins (Tardy, 1992). One origin is that some are sufficiently thick, measured in meters, to be considered a distinctly separate deposit, much younger than the lateritic profile. Fluvial quartz-rich sands in Australia may have this origin (Asumadu and others, 1991). Second, some result from the mechanical and chemical disaggregation of the underlying ferricrete. This is termed dismantling. And third, termites can bring saprolitic materials from depth and deposit it on the surface, where it remains or is redistributed. Termite mounds are common in places (Fig. 5.8), they can bring materials to the surface from as deep as 13 m, and some workers attribute some of the vesicular and tubular structures to termite activity.

Some oxic horizons are hard due to exposure to drying conditions (Alexander and Cady, 1962). Plinthite, as used in Soil Taxonomy, is a soft sesquioxide-rich material capable of hardening irreversibly on drying, and Daniels and others (1978) give clues for its recognition in the field. Sufficient iron and dehydration seem to be necessary for hardening to occur, concomitant with increased crystallinity and continuity of the crystalline phase of already existing iron compounds. This results in the formation of goethite and hematite in a rigid network of crystals that cements the material together. Dehydration necessary for the hardening may come about by a natural change in vegetation from forest to savanna, by the clearing of forests for agricultural and other purposes, or by exposure following erosion.

Climate and Plate Tectonics

Climate is a factor in determining whether bauxite or ferricrete is the main lateritic material. Tardy and Roquin (1992) give these general parameters: bauxite under hot and humid equatorial climate (MAT, 25°C; MAP, 1.8 m), and ferricrete under seasonally contrasted tropical climate (MAT, 30°C; MAP, 1.3–1.7 m).

In a series of diagrams Tardy and Roquin (1992) review the plate tectonics connection of lateritic profiles of South America and Africa (their Figs. 16.13–16.15). First, however, bauxitic profiles are depicted as forming along the equator, whereas ferricrete profiles form a symmetrical belt both north and south of the bauxitic zone in the seasonally contrast climates. At 140 Ma, the plate tectonic position for both continents puts the bauxitic zone in northernmost Colombia and Venezuela in South America, and angling across northern Africa from Mauritania to Libya. The lateritic belts move southward in time, as the continents drift to the north and rotate into their present positions. While the belts sweep southward, and depending on specific locales, the bauxitic laterites and ferricretes are both forming and being destroyed as a function of the position of the equator versus that of the continents.

■ Origin of CaCO₃-Rich Horizons, and Those with More Soluble Salts

Soils in semiarid and arid regions commonly have carbonate-rich horizons at some depth below the surface, or if the climate is dry enough or the surface erosion intensive enough, these horizons may extend to the surface. These are called calcic soils. Although several origins have been presented for some of these horizons (Goudie, 1973; Reeves, 1976; Goudie, 1983; Wright and Tucker, 1991; Dixon, 1994), our concern here is with CaCO₃-rich horizons demonstrated to be of pedogenic origin (Bachman and Machette, 1977;

Machette, 1985; Harden and others, 1991). Some of these horizons are the caliche and calcrete of present and past geologic literature. Pedologists call these accumulations Bk horizons, and because there are recognized and defined stages in the buildup of $CaCO_3$-bearing horizons (Gile and others, 1965, 1966; Machette, 1985) (see Appendix 1), this terminology, plus use of the K horizon, seems preferable to the general term caliche or calcrete. Both calcium and magnesium carbonates are present in these soils, with the former dominant.

The origin of carbonate horizons involves carbonate–bicarbonate equilibria, as discussed in many geochemistry texts (Drever, 1988; Krauskopf and Bird, 1995; see also McFadden and others, 1991, and Wright and Tucker, 1991) and shown by the following reactions:

$$CO_2 + H_2O$$
$$\quad\;\; g \qquad l$$

$$CaCO_3 + H_2CO_3 \Rightarrow Ca^{2+} + 2HCO_3{}^-$$
$$\quad c \qquad\; aq \qquad aq \qquad\quad aq$$

An increase in CO_2 content in the soil air or a decrease in pH will drive the reaction to the right; carbonate will dissolve and move as Ca^{2+} and $HCO_3{}^-$ with the soil water. Dissolution is also favored by increasing the amount of water moving through the soil, as long as the water is not already saturated with respect to $CaCO_3$. Precipitation of carbonate occurs under conditions that drive the reaction to the left, that is, a lowering of CO_2 pressure, a rise in pH, an increase in ion concentration to the point where saturation is reached and precipitation takes place, or evapotranspiration of the soil moisture.

All the above conditions are found in calcic soils. Carbon dioxide partial pressures in soil air are 10 to more than 100 times that in the atmosphere (Brook and others, 1983); this decreases the pH, which, in turn, increases $CaCO_3$ solubility (Arkley, 1963). The partial pressure of CO_2 is high as a result of CO_2 produced by root and microorganism respiration and organic matter decomposition. Thus, one would expect the highest CO_2 partial pressure to be associated with the A horizon, with values diminishing down to the base of the zone of roots. The amount of water leaching through the soil also is greater near the surface than at depth, so as the water moves vertically through

the soil, the Ca^{2+} and $HCO_3{}^-$ content might increase to the point of saturation, after which further dissolution of $CaCO_3$ is not possible.

Combining the effects of high CO_2 partial pressure and downward-percolating water, we might visualize the formation of a $CaCO_3$-rich horizon as follows. In the upper parts of the soil, Ca^{2+} may already be present or may be derived by weathering of calcium-bearing minerals. Due to plant growth and biological activity, CO_2 partial pressure is high and forms $HCO_3{}^-$ on contact with water. Water leaching through the profile carries the Ca^{2+} and $HCO_3{}^-$ downward in the profile. Precipitation as a $CaCO_3$-rich horizon would take place by a combination of decreasing CO_2 partial pressure below the zone of rooting and major biological activity, and the progressive increase in concentration with depth in Ca^{2+} and $HCO_3{}^-$ in the soil solution as (1) less water percolates downward and (2) water is lost by evapotranspiration (Schlesinger, 1985). The position of the $CaCO_3$-bearing horizon is, therefore, related to the depth of leaching, which, in turn, is related to the climate (Fig. 10.4). The carbon- and oxygen-isotope composition of pedogenic carbonate has been used to differentiate between precipitation by CO_2 loss and that by evapotranspiration (Salomons and Mook, 1976; Salomons and others, 1978). Models for the formation of carbonate horizons, incorporating much of the above as well as the dust flux, have been presented by Mayer and others (1988), McFadden and others (1991), and McDonald and others (1996). The latter work links the carbonate distribution with the yearly variation in rainfall (i.e., wet versus dry years).

Temperature also affects $CaCO_3$ equilibria. Because CO_2 is less soluble in warm water than in cold water, $CaCO_3$ solubility decreases with rising temperature (Arkley, 1963). This temperature effect may not be too great in one profile, but it is important in comparing the depths to the tops of Bk or K horizons between regions of different temperature.

Finally, evidence is accumulating for a biogenic origin of at least some of the carbonate precipitated in Bk and K horizons (Monger and others, 1991b; Wright and Tucker, 1991; Amit and Harrison, 1995). The main criterion is microfabrics, mainly seen in SEM photomicrographs, that appear to have been precipitated by soil microorganisms. One process envisaged is that excess Ca^{2+}, secreted by microorganisms, concentrates on their external surfaces. This subsequently combines with CO_2 respired by the microorganism.

Stages of Accumulation

Several stages in the buildup of carbonate horizons are recognized (Appendix 1, Fig. A1.5 and Table A1.5), and the morphology is controlled by the presence or absence of gravel (Gile and others, 1981; Machette, 1985). Monger and others (1991a) provide thin-section data on the stage progression. In gravelly material, the first stage is the appearance of carbonate coatings on the undersides of gravel clasts; in non-gravelly material, the first stage is the occurrence of thin filaments. The undersides of clasts probably are favored sites initially for deposition because downward-moving water would tend to collect there. In contrast, the filaments reflect $CaCO_3$ precipitated on roots and the walls of pores (Monger and others, 1991a). With time, in both parent materials, the horizon is increasingly impregnated by carbonate deposition on solids until the voids become constricted and eventually plugged and water percolation through the horizon is greatly restricted. At this point, water tends to collect periodically over the plugged horizon; the resulting solution and reprecipitation produce the laminated part of the upper K horizon (stage IV). At this point in the development the horizon builds upward and so is younger in that direction. During the buildup of a K horizon, the volume of pedogenic carbonate eventually exceeds the original volume of the pores. In most cases, this is attributed to the carbonate, upon crystallization, forcing the silicate grains and gravel apart (Watts, 1978; Machette, 1985), resulting in volumetric expansions of 400–700%. However, there also are instances of carbonate replacing silicate minerals (Reheis, 1988; Wang and others, 1994). In their study, Monger and others (1991a) report that many primary grains have ragged edges due to solution, and if effecting Ca-bearing minerals, this could provide Ca^{2+} to the system. The buildup of morphological stages of $CaCO_3$ accumulation is more rapid in gravelly alluvium than in nongravelly material because the former has less surface area per unit volume, as well as less pore space (Flach and others, 1969; Gile and others, 1981). Finally, progressively higher stages are usually attained in the older soils of a chronosequence; in any one soil profile, however, the highest stage attained is usually near the top of the carbonate-enriched horizon, and this changes to progressively lower stages with depth. Note that the parent-material influence on carbonate morphology is diminished by stage III.

A different origin of calcic and petrocalcic horizons is advocated for carbonate rock parent material (Rabenhorst and others, 1991). Whereas the scheme presented above, of translocation and accumulation, is repeated many times across the western United States in noncalcareous gravelley alluvium, the scheme for formation in carbonate rock involves the alteration and rearrangement of inherited carbonate. First, however, they separate features of the rock from those of pedogenic origin, and in places this can be verified by the different isotopic signatures of rock versus pedogenically altered rock. Five stages of morphological development were recognized, with the fifth being a petrocalcic horizon with a laminar top. Translocation of $CaCO_3$ probably has taken place, but the distances involved may not have been great. Some calcic horizons have even lost some $CaCO_3$, but still classify as calcic because of the alteration recognized. Differences within the parent rock, such as degree of induration, influence the morphological stage to the point that some Km horizons are a replacement feature of strongly indurated limestone. Finally, they suggest caution in attributing all laminar tops to inorganic processes, as some may be more of a biogenic origin (e.g., algal mats formed at the land surface).

Several morphological features are present in indurated K horizons of stages IV, V, and VI (Fig. 5.12; Bachman and Machette, 1977; Machette, 1985). One is ooids, which are sand-sized round particles in which a nucleus (could be a sand grain) is surrounded by one or more concentric layers of fine-grained calcite grains (micrite). There could be several origins of ooids, and those studied by Hay and Reeder (1978; for others, see Hay and Wiggins, 1980) seem to result from carbonate replacing clay coats on sand grains. Another common feature is pisoliths, which are subangular to spherical bodies, 0.5 to more than 10 cm across, surrounded by many thin layers of micrite. The nucleus of pisoliths can be anything from rock fragments to pieces of broken and rotated indurated K horizon material, some of which display internal laminar layers and pisoliths from previous pedogenesis. Both of these features usually are set into massive carbonate cement. Machette (1985) points out that collectively some of these features, plus the laminar structure, have led some investigators to suggest a lacustrine origin for some calcretes.

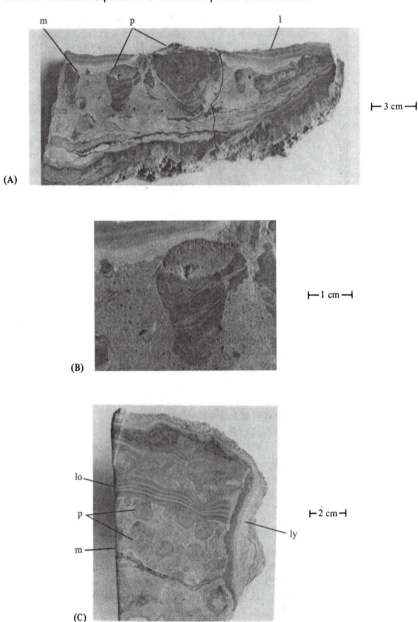

Figure 5.12 Polished sections of a Km horizon with stage VI morphology, Mormon Mesa, Nevada. *(Samples courtesy of M.N. Machette.)* (A) Sample is almost entirely CaCO₃ with pisoliths (p) set in a massive matrix (m), capped with laminar carbonate (l). (B) Closer view of pisolith in (A). (C) Sample with complex carbonate relations in which old pisoliths (p) are set in a massive groundmass (m); still younger are two sets of laminar carbonate deposition (lo is older, ly is younger).

Chemistry and Mineralogy

The chemistry and mineralogy of K horizons have interested workers because some of the best developed ones approach the composition of limestone (Table 5.8); in these, CaCO₃ dominates the analyses, and in the best developed K horizons approaches 90%. Some of the MgO is in the form of dolomite (Hutton and Dixon, 1981), most of which probably is primary, but in one instance is pedogenic (Hay and Reeder, 1978). The rest of the MgO can be in original silicate grains, as exchangeable ions of the clay minerals, or in pedogenic clay minerals. In this last group are the pedogenic Mg-enriched fibrous clay minerals. Bachman and Machette (1977) point out that in K horizons of stages IV to VI with high carbonate contents, the Ca:Mg molar ratio of the carbonate fraction increases relative to that of the parent material and that these increases coincide with the clay-mineral changes to the more Mg-rich clays:

Increasing total pedogenic carbonate $\Rightarrow$
    Montmorillonite $\Rightarrow$ Illite-Montmorillonite $\Rightarrow$
    Palygorskite $\Rightarrow$ Sepiolite

The Ca:Mg ratio for carbonate in the parent material can be 10–20 and increase to 100–160 in the best developed K horizons. The Mg is moving from the carbonate to the clay mineral fraction. The rest of the oxides in the total chemical analysis probably reside in silicate grains or in engulfed B-horizon material, but some of the SiO_2 could be pedogenic opal, chalcedony, or silica gel.

Source of the Carbonate

The $CaCO_3$ of the carbonate horizons may come from several sources (Reeves, 1976; Dan and others, 1981; Gile and others, 1981; Machette, 1985; Reheis and others, 1995). One is from the parent material. If the parent material is limestone, the source is obvious (Rabenhorst and others, 1991). For other parent materials, Ca^{2+} released by weathering could combine with HCO_3^- deeper in the profile, and, if this is the

origin, there should be a close relationship between the CaO content of the parent material, the amount of weathering in the upper part of the soil to release Ca^{2+}, and the amount of $CaCO_3$ in the Bk or K horizon. For most K horizons in the western United States, derivation of the carbonate by parent material weathering is highly unlikely. Gardner (1972) calculated that 37–90 m of material would have to be weathered to release the amount of Ca present in the K horizon at Mormon Mesa, Nevada. Not only does the lack of mineral weathering negate such an origin, but geomorphologically, it is highly unlikely. For example, tens of meters of weathered material would have to be removed from the mesa and the carbonate continually concentrated in a horizon of several meters thickness, while preserving a well-formed terrace with a surface that has a consistent relationship with the position of the shallow K horizon. This sequence of events seems impossible.

To address problems such as the above, an external source of Ca^{2+} seems most likely. The first detailed study in the United States was in the Las Cruces region of New Mexico, and indicated that the atmosphere is an important source for both Ca^{2+} and $CaCO_3$ (Ruhe, 1967; Gile and others, 1981). Dust-trap data at Las Cruces suggest that the dust contains <5% carbonate and that the latter is added to the soil at a yearly rate of about $0.2–0.4 \times 10^{-4}$ g/cm². More data on this have been obtained from a large part of the western

Table 5.8 Comparison of Chemical Analyses for Calcretes, K Horizons, and Limestones

| Oxide | Average of 300 Calcretes | K Horizons | | Average of 345 Marine Limestones |
		Pleistocene Age, Tanzania	Pliocene Age, New Mexico	
CaO	42.6	44.4	45.4	42.6
SiO₂	12.3	8.5	13.2	5.2
MgO	3.1	1.1	0.4	7.9
Al₂O₃	2.1	2.5	1.4	0.8
Fe₂O₃	2.0	2.7	0.5	0.5
K₂O	—	1.0	0.2	0.3
Na₂O	—	0.6	0.1	0.1
CO₂	—	36.0	36.0	41.6
Total CaCO₃	79.3	81.8[a]	81.8[a]	94.5[a]

Taken from Goudie (1973), Hay and Reeder (1978), Aristarian (1970), and Clark (1924).
[a]Based on CO₂ content and includes MgCO₃.

United States (Chapter 8). Another source is the amount of Ca^{2+} in yearly precipitation (maps such as those of Junge and Werby, 1958, and NADP/NTN, 1984). At Las Cruces the Ca^{2+} from this source is enough to produce perhaps three times the amount of carbonate as can be produced by dust alone, giving a total annual carbonate production of about 2×10^{-4} g/cm^2. This can be an estimate only because, although the Ca^{2+} in precipitation probably is carried into the soil, the carbonate in dust could be carried to other sites before precipitation events dissolve it and move it into the soil. A similar atmospheric origin for most pedogenic carbonate is favored for semiarid to arid regions in the western United States (Machette, 1985; references in Chapter 8) and other parts of the world (Yaalon and Ganor, 1973). A locally important source for the Ca^{2+} in carbonate is atmospheric gypsum, for on dissolution, Ca^{2+} is released (Lattman and Lauffenburger, 1974).

Isotopic analysis can be used to isolate carbonate origins of soils that have at least some parent-material carbonate. The problem is to determine the fraction of the carbonate that has gone through dissolution and reprecipitation and thus can be termed pedogenic. Carbon-isotope composition can help solve this problem, since it has been shown that the isotopic composition of the parent-material carbon can be much different from that which undergoes pedogenic alteration, as well as that of the CO_2 in the soil atmosphere. If the carbonate equilibria equation is recalled, a pedogenic composition will probably be between the two isotopic values, and from this it is possible to calculate the percentage of the total carbonate that is pedogenic (Salomons and Mook, 1976; Salomons and others, 1978; Magaritz and Amiel, 1980).

Accumulation of $CaCO_3$ by capillary rise from a perched high water table may explain some $CaCO_3$ occurrences (Malde, 1955), but such occurrences are not thought to be widespread. This is because capillary rise in coarse alluvium is nil, and impermeable beds necessary to perch the water table are uncommon. Also, in many areas, streams have downcut following deposition of alluvium, which has lowered the water table far below the level that could possibly be reached by the capillary rise of water with its dissolved carbonate.

Finally, it is important to be able to differentiate between pedogenic carbonate and groundwater carbonate when working with buried soils because it is obvious that the latter may be of little pedologic or geomorphic importance. Pedogenic carbonate should have features such as those of the six morphological stages with the correct depth distribution. In addition, in most cases there should be overlying soil horizons, and their development ideally should bear some relationship to the carbonate stage attained (see descriptions in Gile and Grossman, 1979). For soils with high amounts of carbonate (K horizons), pedogenic carbonate is mainly micrite and has dry colors of 7.5YR 7/4 to 10YR 8/3, and the original silicate grains have been forced apart. In contrast, groundwater carbonate is coarser grained, fills only the original pore spaces, and because of overburden pressure has not forced the original grains apart. Isotopic analysis can be used to help differentiate groundwater carbonate from pedogenic carbonate, as was done at Yucca Mountain, Nevada (Chapter 11).

Silica Accumulation

Pedogenic silica occurs in some soils, commonly in arid and semiarid climates (Drees and others, 1989). Morphological stages of development are recognized, more or less parallel to those for carbonate (Harden and others, 1991). In parts of the world, especially Australia, there is material so well cemented by silica and indurated that it is termed silcrete (Milnes and others, 1991). Many origins have been proposed, some pedologic and some geologic, and many are associated with ancient landscapes. Precious opal can be associated with these materials.

More Soluble Salts

In areas of aridity greater than that required for carbonate accumulation (precipitation < 25–30 cm/year), more soluble salts may accumulate at depth in well-drained soils above the influence of the water table (Watson, 1979, 1983, 1985; Dan and Yaalon, 1982; Nettleton and others, 1982; Szabolcs, 1989; Eswaran and Zi-Tong, 1991; Harden and others, 1991; Chen, 1997). They also occur in dry polar regions (Tedrow, 1977; Bockheim, 1980; Campbell and Claridge, 1987). Common salts are gypsum ($CaSO_4 \cdot 2H_2O$) and halite (NaCl) (Table 5.9). There are many similarities in morphology of these horizons versus those of carbonate accumulation (Harden and others, 1991), as well as similarities in origin. Some petrogypsic horizons can contain over 90% gypsum, and at such high

Table 5.9 Analysis of a Gypsiorthid Formed in Gravel in the Central Sinai with about 50-mm Annual Precipitation

Horizon	Depth (cm)	Electrical Conductivity (mmho/cm)	Cations[a] (meq/100 g soil)				Anions[a]			Gypsum (meq/100 g soil)	pH
			Ca	Mg	K	Na	HCO₃	SO₄	Cl		
A1	0–2	9.4	0.8	0.38	0.051	2.5	0.093	1.31	2.3	0.3	7.8
A2	2–9	34.9	8.0	4.47	0.028	5.8	0.079	6.44	11.8	8.1	7.4
Bz1	9–19	46.0	14.2	3.65	0.024	7.6	0.07	8.0-5	17.4	7.7	7.4
Bz2	19–35	86.3	23.3	7.23	0.028	43.1	0.09	23.22	50.4	139.0	7.3
Byz	35–47	182.6	18.8	5.34	0.018	142.4	0.097	32.00	134.5	103.0	7.2
Cy1	74–75	24.6	6.9	3.40	0.013	11.9	0.13	10.46	11.6	404.0	7.8
Cy2	75–100	48.1	7.6	3.88	0.012	21.3	0.10	9.60	23.1	293.0	7.8
Cy3	100–110	44.0	7.8	3.52	0.012	19.1	0.10	8.12	22.2	331.0	7.7
C	110–120	51.8	7.8	4.31	0.011	18.0	0.08	8.08	22.0	207.0	7.7

Data from Dan and Yaalon (1982, Table 4).

[a]Water extract from saturated paste.

contents the original silicate grains are forced aside during gypsum crystallization. These salts accumulate at depths reached by mean annual soil water, or by extreme rainfall events, and if both salt accumulations occur in one profile, the depth to the maximum amount of halite probably is deeper than that for gypsum because the former is more soluble. The above origin is for soil waters moving vertically downward from the surface. For some other gypsum accumulations, Watson (1985) has evidence for evaporative origins; one is for precipitation during evaporation of upward moving waters from a shallow water table, and the other is precipitation during evaporation of shallow ponds. The former could produce a subsurface accumulation, but the latter would produce a surface deposit.

The source of the salts is a problem. Some are derived from salts in the rock or the sediment from which the soils form. Another source for sulfur would be release during the weathering of pyrite. If the parent materials are sufficiently low in salts, however, pedogenic salts probably have an atmospheric origin, either as solid matter or dissolved in precipitation. Marine aerosols are commonly advocated for salts in soils adjacent to nearby oceans, even those of Antarctica (Campbell and Claridge 1987). In Australia, for example, the marine aerosol contribution to soils is highest within about 200 km of the coast (Isbell and others, 1983). Israel also has a marine aerosol contribution close to the ocean, but in the southern part of the country the ionic composition changes, reflecting a continental origin from dust (Singer, 1994).

A very uncommon salt, soda niter (NaNO₃), accumulated in soils of the Atacama Desert of northern Chile (Ericksen, 1981; see also the Peruvian desert transect in Chapter 8 and Figs. 8.31 and 8.32). The pedogenic accumulations occur at depth, presumably at the depth penetrated by soil moisture during the heaviest rains. It is known as caliche blanco, and is leached and reprecipitated at depth. This is one of the most soluble pedogenic salts known, and unusual conditions are called on to account for their accumulation. First, however, the source seems to be that which is normal to many environments: ocean spray carried inland, volcanic emissions, and dust. Conditions are as follows: (1) MAP less than 1 mm, and rainfall of 1 mm or more may occur only every 5–20 years; MAT is 18–20°C; (2) a long formation time, with some forming in about 10–15 Ma; and (3) a paucity of nitrate-utilizing plants and soil mircoorganisms. Over a very long time, with little increase in precipitation, pedogenic horizons of this composition (with associated halite) can accumulate.

It is not uncommon for the horizon of carbonate or gypsum accumulation or that of other salts to occur at the surface. This might be due to extreme aridity, to derivation from a former shallow water table, or to erosion of all overlying horizons. Knowing the present field relations, the geological history of the area, and the mean annual precipitation might favor one origin over another. For example, the precipitation in many parts of the western United States is such that Km horizons should be at some depth; when they are not, erosion of overlying horizons usually is suspected. Finally, the presence of the above soil accumulations can put limits on any scenario for climate change during the time of their accumulation.

■ Horizons Associated with High Water Table or Perched Water Table, and Redoximorphic Features

Some soil horizons are associated with high water contents due either to high regional or local water tables or to horizons within the soil that impede the downward movement of water. Many of these are termed hydromorphic soils, meaning that the morphology owes much of its origin to the high water content, many of which have an aquic soil-moisture regime. The processes that produce features formed under reducing conditions are termed gleization (Buol and others, 1997); the "g" horizon designation comes from this. Color and color patterns, and their position within the profile, can be used to indicate the state of the water-moisture regime (Duchaufour, 1982; Fanning and Fanning, 1989; Richardson and Daniels, 1993; Soil Survey Division Staff, 1993; Vepraskas, 1994). Uniform oxidized colors throughout the profile indicate well-drained conditions. If soil moisture content varies sufficiently between seasonally saturated (reducing) conditions followed by a season of low water content (oxidizing conditions), a mottled soil can be produced. Mottled color patterns are defined as repetitive color variation throughout the soil that cannot be attributed to compositional properties of the soil (Soil Survey Division Staff, 1993); an example is a repeating pattern of small masses of yellowish-brown to reddish Fe compounds and bluish-black Mn compounds set in a grayish groundmass. The alternating oxidizing and reducing conditions produce redoximorphic features, such as mottles and nodules involving Fe and Mn (Vepraskas, 1994). Gray, greenish-gray, or bluish-gray colors (chroma usually < 2) commonly denote reduced conditions, with the colors due to some combination of Fe loss, the color of the <2 mm material, and the compounds that form in such an environment. The lower case symbol "g" is the field designation for such horizons (e.g., Btg, Cg; Table 1.1), and some of these horizons are termed "G" by Duchaufour (1982).

Cutler (1983) has a classification that includes the kind of redoximorphic feature and its position within the profile. Soils are either mottled or reduced, and the latter feature is located at the top, middle, bottom, or throughout the profile. Such features at the top of the profile may require saturated conditions above an impermeable layer (e.g., a pedogenic feature such as an iron pan, or a parent material layer), whereas those that occur throughout the profile result from a water table that intersects the surface, at least for part of the year. A common topographic transect involving the above materials and patterns is as follows: a topographic high with well-drained conditions and oxidized soil colors changes downslope first to a soil with deep seasonally saturated conditions and bottom mottling, and eventually to the soil in the lowest part of the landscape that is saturated, conditions are reducing, and the soil has a uniform grayish color.

Redoximorphic features are divided into concentrations and depletions (Vepraskas, 1994). Redox concentrations consist of either soft masses of Fe/Mn accumulations, or hard Fe/Mn accumulations that occur as nodules or concretions (differ from nodules in having concentric layers). Concentrations can occur on ped faces, ped interiors, or line pores (possible root channel). Redox depletions are bodies in which Fe/Mn oxides have been removed (chroma ≤ 2, with value ≥ 4), or in which the latter plus clay has been removed. Depletions usually occur along ped faces or pores. The New Zealand soil horizon nomenclature might be the most detailed one for field descriptions of a large variety of mottling and reduced features encountered in hydromorphic soils (Clayden and Hewitt, 1989).

A general understanding of the formation of redoximorphic features can be obtained by examining an Eh-pH diagram (Fig. 5.13; Rowell, 1981). To the right of the dashed line, both Fe and Mn form precipitates. To the left of the dash-dot line, both Fe and Mn exist as +2 ions in solution. Between the two lines, Fe forms a precipitate and Mn^{2+} is in solution. Consider a soil at pH 6.5. Starting at the top of the diagram under oxidizing well-drained conditions, both Fe and Mn, when released from the primary grains by weathering, would form precipitates close to the site of release. As conditions become relatively reducing with saturated conditions (recall the aquic soil-moisture regime has little dissolved O_2), we move vertically down the diagram into the realm where only Mn^{2+} stays in solution. If subsequent seasonal conditions are more oxidizing (move up the diagram), Mn oxides could form dark-colored redox accumulations.

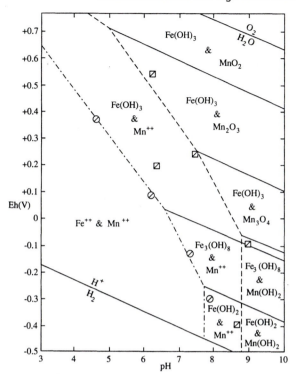

Figure 5.13 Eh-pH diagram for 25°C with approximate fields for the ionic and oxide phases of Fe and Mn. *(Redrawn from Collins and Buol, 1981. Effects of fluctuations in the Eh-pH environment on iron and/or manganese equilibria. Soil Sci. 110, 111–118, Fig. 1, © 1981, The Williams and Wilkins Co., Baltimore.)*

If, however, we continue to move down the diagram where conditions are more reducing (aquic moisture regime), both Fe and Mn are in solution. If the soil waters are moving, both ions will be removed from the site, leaving no Fe/Mn pigment and producing gray colors. If, in contrast, the soils are seasonally drained and oxidizing conditions are obtained, we move to the top of the diagram and both Fe and Mn precipitate as multicolored redox accumulations.

The above can be used to generalize on the origins of soil hydromorphic features using a useful diagram of Bouma (1983) (Fig. 5.14) and definitions of Anderson (1984) and Rowell (1981, Fig. 7.22). Under conditions depicted by type 1 (saturation < 1 day; short periods of near saturation), soil colors are uniform and chroma is high (>2). Type 2 soils have longer periods of near saturation, resulting in mobilization of Mn to form black accumulations on peds. Type 3 soil (saturation of a few days; near saturation of several months) moves into the Eh-pH realm, where both Mn and Fe are mobile, and they form accumulations during the oxidizing part of the year. Type 4 soils have continuous saturation for several months, producing chroma of <1 inside the peds, prominent Fe accumulations and lesser Mn accumulations, as well as depletions (termed neoalbans in figure), along ped faces and pores. Finally, type 5 (continuous saturation) is reduced, has few accumulations or deple-

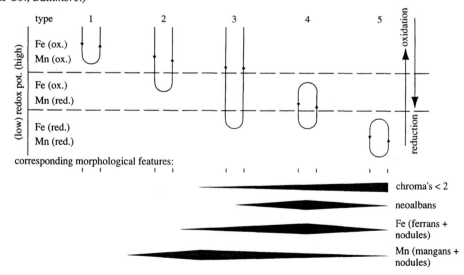

Figure 5.14 Schematic diagram of approximate soil-moisture redox conditions, and resulting redoximorphic features. *(Redrawn from Bouma, 1983, Fig. 9.8, Hydrology and soil genesis of soils with aquic moisture regimes, p. 253–281. In Wilding et al., eds., Pedogenesis and soil taxonomy, 1. Concepts and interactions, © 1983, with kind permission from Elsevier Science, The Netherlands.)*

tions, and is marked by uniform low chromas. Rowell (1981, Fig. 7.22) gives broad guidelines on the duration of saturation and shows that several of the above types can occur in the same profile, with the lower-numbered type overlying the higher-numbered type; the reason for this is that commonly the lower parts of profiles are saturated for longer periods of time relative to the upper parts. For specific interpretations and data, consult detailed studies, such as those of Evans and Franzmeier (1986), Mokma and Sprecher (1994a,b), and Stolt and others (1994).

Redoximorphic features can be imparted on a soil quite rapidly. Reider (1985 and personal comm., 1987) studied soils at an archaeological site that subsequently were under water periodically when a reservoir was constructed. Features are mostly mottles and formed in between <1 to ca. 3 decades.

A unique kind of redox feature occurs in acid sulfate soils, and various bacteria aid in the reactions (reviewed in Rowell, 1981; Duchaufour, 1982; Kittrick and others, 1982; Dent, 1986; Szabolcs, 1989; Fanning and Fanning, 1989; Fanning and others, 1993). These soils form in the strongly reducing environments adjacent to low-relief marshy coasts (marine, delta, estuary). Fe is supplied from the sediments, and SO_4^{2-} from the sea water. In the extremely reducing environment, they exist as Fe^{2+} and S^{2-}, and in places pyrite will form. With artificial drainage, such as along the coast of The Netherlands, or natural drainage, oxidizing conditions set in, sulfide is oxidized to sulfate forming sulfuric acid, soil pH values drop to 2 or less, and mottles form that contain the yellowish ferric sulfate mineral jarosite. These changes occur rapidly. If the lower part of the profile remains reducing, as many as 5 pH units might occur from the top to the bottom of the soil. Similar reactions can occur around bogs, or associated with minespoil containing pyrite.

Brinkman (1970) coined the term ferrolysis for a complex process that results in acid soils in which E horizons with poor structure form and grow downward at the expense of the underlying B horizon. A relatively impermeable barrier within the soil, such as a Bt horizon or fragipan, causes alternating oxidation and reduction in the overlying material, so the iron alternates between the Fe^{2+} and Fe^{3+} state. In the reducing mode, Fe^{2+} displaces the other exchangeable ions on the clay minerals, and the latter, as well as some Fe^{2+}, can be leached from the soil by laterally moving waters. In the oxidizing mode, however, ferric hydroxide forms as a precipitate and the soil be-

comes more acid. With progressively more acid conditions, some clay minerals are destroyed by acid weathering and others, because of the relatively high solubility and availability of aluminum at low pH, can be converted to Al-chlorite. The result is an acid soil with a prominent E horizon that has an abrupt and wavy contact with the underlying mottled Bt horizon. Although the impermeable barrier that sets up the process can have many origins, Pedro and others (1978) describe how the process can act on an originally uniform loamy parent material, and the Bt horizon that forms with time eventually sets up the ferrolysis conditions that result in the destruction of the upper parts of the B. The sequence is

Eutrocrept $\Rightarrow$ Hapludalf $\Rightarrow$
$$Fraglossudalf \Rightarrow Fragiaqualf$$

and was used as an example of B-horizon destruction and soil retrogression (Fig. 8.42). They also describe the similarities and differences between this evolutionary sequence and the one involving podzolization. Other soils may show a somewhat similar process. In Alaska (Ugolini and Mann, 1979) and New Zealand (Ross and others, 1977) iron-cemented B horizons, formed by podzolization processes, have so restricted soil drainage in the overlying material that iron is depleted to the point that an E horizon forms. In the Alaska example, continuing reducing conditions eventually solubilize the cementing materials, the B horizon disappears, and a bog results.

In areas of relatively high evapotranspiration, under the influence of a high local or regional water table, salts can accumulate to appreciable concentrations (Szabolcs, 1989). Common cations are Ca^{2+}, Na^+, Mg^{2+}, and K^+ and common anions are Cl^-, SO_4^{2-}, S^{2-}, and HCO_3^-. The position of the salt concentration can be either at the surface or at some depth, depending on the depth to the water table and the height of capillary rise. The latter will differ with the texture of the soil, being greater for finer-grained materials.

In geomorphological studies, it is important to differentiate salt accumulations of this origin from those mentioned earlier (also in Chapter 8) in well-drained soils. In the latter case, the salts commonly have an external origin, their position is very sensitive to mean annual precipitation as the transport mechanism is vertically moving soil water, and the more soluble ones occur at greater depth. With a capillary rise origin, however, the position of the salt accumulation would not bear much relationship to mean annual precipitation, the more soluble salts might be closest to the surface (Parakshin, 1982), and the rate of salt accumulation could be more rapid.

■ Phosphorus Transformations

In New Zealand, there has been a considerable amount of work on the forms of phosphorus that are significant in soil genesis (Walker and Syers, 1976). The total P in a soil is in four fractions, as follows:

P_{ca}	P of the original minerals, such as apatite
P_{oc}	Occluded P, or that fraction that occurs within coatings or concretions of oxides and hydrous oxides of iron and aluminum
P_{noc}	Nonoccluded P, or those P ions adsorbed at the surfaces of the above oxides and hydrous oxides and $CaCO_3$
P_o	That fraction bound with the organic matter

An earlier fractionation scheme that is probably just as useful for pedologic studies recognizes three fractions, with acid-extractable P (P_a) being approximately equal to $P_{ca} + P_{noc}$, P_f to P_{oc}, and the remainder reported as P_o (Walker, 1964). In areas of relatively high leaching, the original P_{ca} declines in total amount and

converts to the other forms (Fig. 5.15). As will be shown later, climate and the amount of leaching dictate the rate of transformation and the amounts in each form. Generally in a leached soil profile, P_{ca} and P_a are progressively depleted toward the surface, whereas P_o increases toward the surface in a fashion similar to organic matter trends. Compared to the above New Zealand example, depletion of P_{ca} is much less in the drier climate of the San Joaquin Valley, California (Meixner and Singer, 1985).

■ Radiocarbon Dating of Soil Horizons

Because soils contain both organic and inorganic carbon, radiocarbon dates can be obtained in an attempt to date the soil or some of its features. Such dates should not be accepted at face value, however, because the systems are very complex, and both new and old carbon can be introduced or exchanged in the soil. Graphs have been prepared to estimate the error due to contamination (Campbell and others, 1967a,b). A radiocarbon date of organic carbon from the A horizon of a surface soil includes a mixture of organic matter varying from that fraction being added daily to that synthesized and resynthesized over several hundreds or thousands of years, as reviewed by papers in Yaalon (1971), Matthews (1985), and Wang and others (1996). The dates that reflect this dynamic system are

Figure 5.15 Relative change in the amounts of forms of phosphorus (P) in soils with time. *(Redrafted from Walker and Syers, 1976, Fig. 1.)*

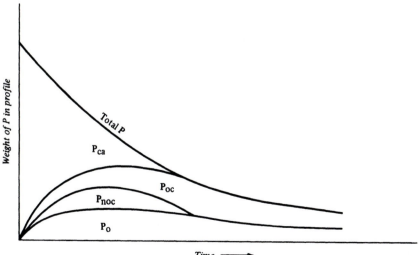

mean residence times (MRT). MRTs for western Canada range from about 0.2 to 2 ka (Paul, 1969), and values of 0.2–0.4 ka are reported for the central United States (Broecker and others, 1956). Dates for various fractions of the organic matter yield different values, a reflection of the relative stability of the fractions (Paul, 1969; Sharpenseel, 1971; Hammond and others, 1991; Martin and Johnson, 1995). The MRTs have little meaning with respect to the age of a surface soil, although they do give minimum ages.

MRTs commonly increase with depth in a variety of soils (Fig. 5.16; Sharpenseel, 1971; Matthews, 1985). Such dates for most surface soils have little meaning. They are important, however, in dating buried soils isolated from surface roots or vertically translocated organic materials. Dating the very top of the buried A horizon provides a limiting date on the time of deposition of the overlying deposits, and one could subtract the estimated MRT prior to burial as it is a built-in error (Benedict, 1966). The error will vary with the soil and environment and is difficult to estimate. For a minimum age on the deposits in which the soils formed, the deepest pedogenic organic matter could be dated, but this could be much younger than the deposit. Vertical mixing by organic or inorganic means can wreck havoc on any uniform depth trends, as can the introduction of old, wind-transported carbon. Finally, this scheme will not work with Spodosols because there is essentially no relationship of radio-

carbon age with depth, as the organic compounds are mobile and a ready contaminant (see also Hammond and others, 1991, although their data are not for soils).

Another problem is the manner of pretreating the samples prior to dating. Goh and Pullar (1977) and Goh and others (1977, 1978) used several chemical pretreatments for removing supposed contaminants from materials, and obtained dates that are older (by >10 ky) than those obtained with more traditional pretreatments. Older dates were even obtained for charcoal using their methods. The idea is that the oldest date is probably closest to the true age of the enclosing material. More recent work by Hammond and others (1991) came to similar conclusions. Martin and Johnson (1995) did not see a consistent age sequence with kind of pretreatment for late Quaternary buried soils.

A recent review by Wang and others (1996) reaffirms the above and presents a model to estimate MRT ages in soil chronosequences (Fig. 5.17). At best, the oldest radiocarbon ages are only about one-tenth the age of the soil. Curves of depth versus ^{14}C age of soil organic matter attain gross similarity, and therefore reach a steady state, in about 15 ky in a forest soil, 50 ky in a prairie soil, and >300 ky in a desert soil. This study again shows what we have known for a long time—MRT ages of soils have little value if the purpose is to estimate the age of a surface soil.

Radiocarbon dates on inorganic carbon of $CaCO_3$-enriched horizons have been used to try to date soils (Bowler and Polach, 1971; Williams and Polach, 1971). The problem is that $CaCO_3$ is readily soluble, and solution and reprecipitation can take place; every time this happens, new carbon is added to the system because of the CO_2 in the soil air. Moreover, the carbon in airfall carbonate may be of any age. An early application of this was by Gile and others (1970) on various soil materials with depth (Fig. 5.18). Although the $CaCO_3$ is mainly of airfall origin and enters the soil by solution and reprecipitation, paired dates on initial pedogenic $CaCO_3$ in young soils and on charcoal in the deposits from which the soils have formed indicate an initial age at time of deposition of less than 3 ky. The dates on the Btk horizon are consistent, because carbonate first coats pebble bottoms and later begins to accumulate in the fine-grained matrix. The soft upper laminae at the top of the K horizon probably undergoes occasional solution and reprecipitation as water is held at that impermeable interface, and thus it is younger than the B horizon pebble coatings. Hard

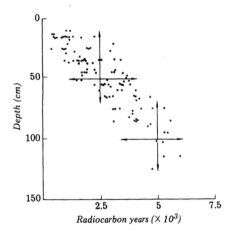

Figure 5.16 Radiocarbon dates on humus collected at various depths from 24 Mollisols. *(Redrafted from Scharpenseel, 1971, Fig. 2; published by Israel Universities Press, Jerusalem.)*

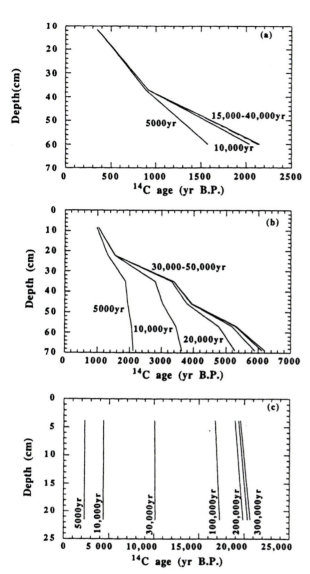

Figure 5.17 Calculated ^{14}C ages with depth for soils of different estimated ages (numbers along the curves) in different environments. Environments are (a) forest, (b) prairie, and (c) desert. *(Reprinted from Wang and others, 1996, Fig. 1.)*

of inorganic carbonate dates. One explanation of this discrepancy might be that roots in cracks in the K horizon introduced the material later; however, this problem requires further study. At greater depth in the K horizon, dates on pebble coatings are consistent with other dates, but that on the whole soil is much younger, again perhaps the result of solution and reprecipitation. The original stratigraphic work suggested that these dates are not all worthless because some workers thought that deposition ceased and soil formation began about 30 ka. A more recent interpretation of the radiocarbon dates on pedogenic carbonate from Las Cruces is that they are bare minimums (Gile and others, 1981), particularly those over 10 ky. In contrast, in the more arid parts of Australia, there is a fairly good match between dates on pedogenic carbonate and on the parent material deposits (Williams and Polach, 1971).

A challenging dating study was done at Kyle Canyon, Nevada, where the soils have formed in gravel with 90% or more carbonate lithologies (Amundson and others, 1989). ^{14}C ages for both inner and outer pedogenic carbonate rinds on clast bottoms were compared. Here the focus is on materials associated with surface 3, considered to be 5–15 ka (Reheis and others, 1992), taken at different elevations. Dates on the inner rinds range from 11 to 15 ka (one date of 4 ka is for associated organic C), and

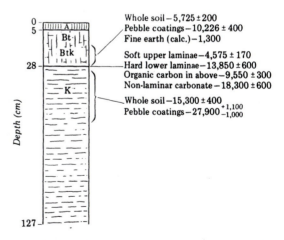

Figure 5.18 Radiocarbon dates on a soil near Las Cruces, New Mexico. This is the Terino soil, and laboratory data are presented in Fig. 1.2. Radiocarbon ages (Gile and others, 1970, p. 14A) are on inorganic carbon of the $CaCO_3$ fraction, with one date on organic carbon.

laminae in the upper K horizon is older, probably because percolating waters seldom have a chance to dissolve the more indurated carbonate. One important reversal in the dates is apparent: organic carbon sealed in the hard laminae is younger than that of the surrounding carbonate sealant. This reversal led Ruhe (1967) to recommend being very cautious in the use those on the outer rind, 9 to 30 ka. Problems with the dates involve contamination from both modern and dead (from limestone) C, as well as C from soil organic matter, solution/reprecipitation events with climate change, especially a problem if C from limestone contributes to the C of the pedogenic carbonate, and how closed the system remains after initial deposition (Sowers and others, 1988).

A recent study of ^{14}C dating of coats of pedogenic carbonate on gravel clasts resulted in a model for predicting soil ages by this means (Amundson and others, 1994). It was pointed out that the ^{14}C content of a sample will approximate the age of the sample only if these conditions are met: (1) ^{14}C content of the atmosphere is constant or its variation with time is known; (2) ^{14}C content of the sample is the same as that of the atmosphere, or the relationship between the two is known; and (3) there is no exchange of C with an external source, following pedogenic carbonate formation. One problem, for example, is the ^{14}C content of CO_2 contributed to the system by decaying soil organic matter, as it and the CO_2 from respiring plant roots are the main soil CO_2 sources. The model shows that the ages for the total carbonate increase with depth, that they are always younger than the soil age, and that the ^{14}C age of the rind and the soil age become closer as the rind that is sampled becomes thinner. For example, if the inner one-third of the rind is analyzed, the two ages are only 2 ky apart. Furthermore, climate change and the accompanying dissolution/precipitation will always be a problem, hence better success is predicted for Holocene soils than for late Pleistocene soils. This is a good start, and more field testing is needed.

U-series disequilibrium methods have been used to date pedogenic carbonate. A study in California demonstrates the use of the relationships between ^{230}Th, ^{234}U, and ^{238}U to date pedogenic carbonate (Ku and others, 1979). Samples were carefully taken of the innermost carbonate coating pebbles, and the dates are considered to be for the time the carbonate precipitated from solution shortly after the deposits were laid down. The dates obtained are encouraging, as they seem to match the estimated geomorphological ages for the deposits. The potential useful age range of the method is a few thousand to about 350 ky.

A later study by Slate and others (1991), however, demonstrated problems with the U-series method. Calcic soils formed on K–Ar dated basalt flows, and the U-series dates were obtained on the carbonate rinds on clasts within the soils. The basalts range in age from 140 ka to 1.1 Ma. Although the carbonate U-series ages increase systematically with age of flow, they are only 20% of the K–Ar ages, and so provide a bare minimum age for the soils. Reasons given for the discrepancy in ages are the time required to produce a clast and coat it, contamination by additions of U, and solution and reprecipitation of the carbonate rinds.

A study at Kyle Canyon compares the ^{14}C and U-series methods of dating inner carbonate rinds (Sowers and others, 1988). Surface 3, mentioned above as being about 5–15 ka, gave U-series ages of 47 and 76 ka. The next older surface, surface 2, gives ages of 129 ka (U-series) and 34 ka (^{14}C), and Reheis and others (1992) give the best age estimate at 130 ka. Still older surface 1 has U-series ages of 88, 110, and >350, yet the best age estimate is 800 ka. Sowers and others (1988) leave us with these sobering thoughts (p. 142): "the results of rind dating raise as many questions as they answer" and "we make no conclusions at this time as to the true ages of these surfaces."

My goal here is not to discredit the dating methods, but to show the difficulty of getting meaningful numerical ages from soil materials. The system is a very complex one, as various materials are being both introduced and removed, and climate change and changing soil moisture pose additional problems. Finding the best materials in the right field context is a vexing problem. Hence, in many Quaternary studies using soils as a dating tool, greater priority is often given to using the estimated rate that a soil constituent is introduced into the soil to estimate ages, than is some of these numerical dating methods. We can point to the carbonate influx method of dating as an example of success in this regard (Fig. 8.27); other examples are given in Bull (1991).

Chapter 6

Factors of Soil Formation

Since the early work of Dokuchaev in Russia and Hilgard in the United States, pedologists have been trying to describe the main factors that define the soil system and to determine mathematically the relationship between soil properties and these factors. Jenny (1941, 1946, 1958, 1961, 1980) has made many important contributions to this facet of pedology. This work has been shown to be so fundamental to pedology that a "factors" symposium was held on the 50th anniversary of the 1941 book and published a few years later (Amundson and others, 1994). The work of many people served as the background from which Jenny derived the factors; these are well reviewed by Tandarich and Sprecher (1994). Because of the success, simplicity, and practicality of of this approach, this edition of the book retains the same factorial framework, modified by the progressive/regressive pathways of Johnson and Watson-Stegner (1987).

■ Factors and the Fundamental Equation

Five factors are usually used to define the state of the soil system. They are climate, organisms, topography, parent material, and time. Other factors may be important locally. The factors theoretically are independent variables, in that field sites can be found in which the factors can be considered to vary independently. Although the factors can be dependent variables in some field sites, their real value in a quantitative factorial treatment is as independent variables.

Factors and processes should be clearly distinguished. In previous chapters, we have been concentrating on the processes that are operative in soils. These processes form the soil. The factors, in contrast, define the state of the soil system. If the combination of factors that describe a soil system are known, the soil properties might be predicted. A change in a factor would change the soil. However, at present, the factors cannot be defined this precisely, and may never be. In spite of this, some valid qualitative and sometimes quantitative predictions can be made.

With the recognition of the factors, equations were formulated to establish the dependence of certain soil properties on the factors. The most widely quoted is the fundamental equation of Jenny (1941, 1980)

$$S \text{ or } s = f(cl, o, r, p, t, \ldots)$$

where S denotes the soil, s any soil property, cl the climatic factor, o the biotic factor, r the topographic factor, p the parent material, and t the time factor. These are the main factors. The dots after t represent unspecified factors that might be important locally or even regionally, e.g., airfall dust or salts that are important to the formation of Aridisols. In this equation, S and s are the dependent variables; the factors are the independent variables. More elaborate forms of this equation have been proposed, but none has been solved, for if all factors are allowed to vary, it would be difficult to sort out the effect of each factor on the soil property studied.

Jenny overcame the dilemma of solving the equation by solving an equation for one factor at a time. To do this, one factor is allowed to vary while the others are held constant. He established the following functions:

$s = f(\underline{cl}, o, r, p, t, \ldots)$ climofunction
$s = f(\underline{o}, cl, r, p, t, \ldots)$ biofunction
$s = f(\underline{r}, cl, o, p, t, \ldots)$ topofunction
$s = f(\underline{p}, cl, o, r, t, \ldots)$ lithofunction
$s = f(\underline{t}, cl, o, r, p, \ldots)$ chronofunction

To solve each function, the first factor listed (underlined) is allowed to vary while the others remain constant: the dependency of one (or more) soil property on a single factor is therefore determined by appropriate statistical methods. This can be extended to include more factors, and eventually it might be possible to rank them on their relative importance to that soil property or properties.

There are two ways in which a factor can be considered constant: (1) if the range in the state factor is quite small and (2) if variation in the state factor is large, yet has a negligible effect on the soil property. To illustrate this latter case, let us say that the functional relationship between a soil property and factor a has been determined (Fig. 6.1). In the same sampling area, factor b also varies, so its relationship to the soil property can also be established. The plot indicates a close dependence of the property on both factors between 0 and x on the horizontal axis, and a simple functional relationship with one factor cannot be established. Between x and y, however, ds/dFb approaches zero. Because variation in factor b has little effect on the soil property, it can be considered a constant. Variation in s, therefore, between x and y is ascribed to variation in factor a in this simplified model.

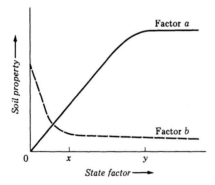

Figure 6.1 Hypothetical variation in one soil property with the variation in two state factors. The vertical scale is not necessarily the same for both plots. On the horizontal axis, x and y are arbitrary values for the state factors.

Beyond y, both factors can be considered constants because the slopes of the functions are close to zero. To establish the dependence of s on factor a between 0 and x, sites would have to be chosen in which variation in factor b is small.

The relative importance of the factors varies from soil to soil. In the early days of soil science, however, rock type was generally considered to be most important factor. Following the early Russian work, especially that of Dokuchaev, the effect of climate was considered to be most important, and many soil-classification schemes were based on climate and processes thought to be related to climate. Although climate may be the most important factor in the gross, worldwide distribution of soils, the other four factors are equally important in describing soil variation in a landscape.

In a study of soils in the northeastern United States, Ciolkosz and others (1989, Table 2) present a ranking of the importance of the soil-forming factors in determining the occurrence of various soil orders. Parent material has a strong influence on many of the orders. Others factors having a strong influence are time on the Ultisols, topography on the Histosols, and organisms on the Spodosols. Of those with a moderate influence, they list climate on Ultisols, Spodosols, and Histosols; topography on Entisols and Inceptisols; and time on Inceptisols, Alfisols, and Spodosols.

Jenny (1946) was one of the first to describe the influence of the factors on the soil mapping units, soil series, and types. In fact, mapping can be viewed as a field solution of the fundamental equation. Mapping involves drawing contacts between contrasting soils. In the field, coring and mapping the soils, one has to wonder why the soils differ. The differences in soil may be due to differences in parent material, topographic position, slope steepness, redistribution of moisture, vegetation, age of the associated landscape, etc. Because all of these can be seen or visualized in the field, in time a correlation of factors with mapping boundaries develops. Thereafter, the mapping proceeds at a more rapid pace and one can predict the location of contacts better.

One problem in solving the functions set down by Jenny is to find appropriate field sites at which the factors operate as independent variables. Crocker (1952) attributes a part of Jenny's success in solving the functional equations to the definitions given the state factors, which ensure that they could be independent of each other. The definition of the climatic factor will illustrate the problem. One can speak of the regional climate or the climate at the ground surface immediately above the soil. The two climates are measurably different. The regional climate is measured above the canopy of vegetation, so it does not depend on any ecosystem property. The climate at the ground surface, however, depends on the vegetation at the site or on the orientation of slope with respect to the position of the sun. In this example, the regional climate is the independent variable, whereas the soil climate is a dependent variable. Thus in solving for the functions, the regional climate must be used.

The definitions of the factors by Jenny are as follows:

Climate (*cl*): The regional climate, commonly given as mean annual precipitation and mean annual temperature. In some studies (i.e., McFadden and Tinsley, 1985) the leaching index is used, as it is a simple calculation that combines the effects of both precipitation and temperature (Table 10.1). Yaalon (1975) introduces another climatic parameter, the moisture available for leaching (*L*). *L* is the part of the precipitation that is in excess of actual evapotranspiration and that satisfies the water-holding capacity of the soil. The soil-moisture regime of Soil Taxonomy could also be used as the climatic factor. Because precipitation and temperature are not interdependent, they can be treated as separate functions.

Organism or biotic factor (*o*): Because vegetation is most important, this factor is the summation of plant disseminules reaching the soil site or the potential vegetation; it is approximated by a list of species growing in the surrounding region that could gain access to the site under appropriate conditions.

Topography (*r*): Included are the shape and slope of the landscape related to the soil, the direction the slope faces, and the effects of a high water table, the latter being commonly related to topographic position. Geomorphic slope processes that tend to result in thin soils upslope and cumulative soils downslope should be included. We find it useful to classify the slope according to various parameters given by Ruhe (1975; see Appendix 1). This quickly indicates whether the slope is susceptible to erosion or deposition, relative dryness as moisture flows from the site, or relatively greater moisture as upslope moisture is focused on the site.

Parent material or initial state of the system (*p*): Included are materials, both weathered and unweathered, from which the soil formed. If the parent material is an unweathered granite, we know that the initial condition is 0% clay, 0% organic matter, near 0 CEC, etc. If the parent rock is weathered, it would have properties of saprolite (Cremeens and others, 1994). Parent material could also be a soil when one wishes to study the effect of climatic change on a preexisting soil. For example, in determining the influence of Holocene climate on the distribution of carbonate in a Pleistocene soil, the initial state of the system is the Pleistocene soil (McFadden and Tinsley, 1985). The initial state of soils formed on colluvium is even more difficult to determine because the properties of colluvial deposits are a function of the initial rock that was weathered, soil properties formed during differing times of slope stability, and mixing and layering that take place during downslope transportation (i.e., Ciolkosz and others, 1979; Graham and others, 1990; Graham and Buol, 1990).

Time (*t*): Elapsed time since deposition of material, the exposure of the material at the surface, or formation of the slope to which the soil relates; if a study is being made of the effect of climatic change on a preexisting soil, time is the time since the change.

Soils are difficult materials on which to obtain a numerical date for their time of initiation. Where one can date the deposit from which the soil has formed, the date is the maximum age for the soil. Several processes (widespread erosion, uprooting of trees, etc.) can reset the time that soil formation was reinitiated. Easterbrook (1993) has a useful summary of the pertinent dating techniques for obtaining numerical ages.

These definitions raise several problems, discussed by Jenny (1941, 1980), Crocker (1952), and Yaalon (1975). One is that the biotic and parent-material factors cannot be quantified; for the most part, they can be described only qualitatively. An example of what can be done, however, is to determine the function for each of several rock or vegetation types and then compare one function with the other. As another example, if one were comparing soil formation of different volcanic rocks, $\%SiO_2$ could serve as a quantitative measure of the different parent materials, and then quantitative statements made on various aspects of soil development. For example, conduct a chronosequence study of each rock type and then compare the chronosequences, as was done by Bockheim (1980). Jenny has been criticized for including time as one of the factors, because, by itself, time does nothing to a soil. Its importance, however, lies in the fact that most soil-forming processes are so slow that their effect on the soil is markedly time dependent (e.g., Harden and Taylor, 1983). The one definition that probably has been most controversial, however, is that of the biotic factor, because it is defined and used in two different ways. When deriving functions for the other factors, the biotic factor is taken as the potential vegetation or species pool. The actual vegetation at the site is a function of the same set of factors as is the soil; both vegetation and soil develop concurrently and can and do influence each other (see Major, 1951). Therefore vegetation cannot be taken as an independent variable. Biofunctions can be derived, however, for areas in which vegetation is the variable and all other factors are constant and thus do not influence the vegetation. Instances of this are quite rare, but a commonly studied example is the prairie–forest transition. Some workers might prefer a bioclimatic factor to include both present climate and vegetation. This will be done when discussing soil transects across climatic gradients (Chapter 8).

Not all workers are in agreement with the independence and dependence of the factors. Chesworth (1973) argued that time is the only independent factor. In addition, he feels that because the effects of parent material are nullified with time, parent material cannot be considered independent. The argument is that, with long intervals of time, soils formed from different parent materials converge toward common chemical properties. This is true, but probably only for some very old soils. Surely for most Quaternary soils parent material has a significant effect on soil properties, and it is the one factor strongly influencing the formation of the most soils orders in the northeastern United States (Ciolkosz and others, 1989).

■ Arguments for and Against the Use of Factors, and Monogenetic Versus Polygenetic Soils

Several criticisms have been made of the use of factors in the study of soils, some of which are discussed by Crocker (1952) and Bunting (1965). One is that the general equation has never been solved. By this I mean that we cannot apply quantitative data on the factors to predict the resulting soil or soil properties adequately. This is true now, and it may never be otherwise. Moreover, many soils have formed under conditions in which several factors have varied, and functions for these soils may never be derived. Individual functions for other soils have been derived, however, and they are very useful in pedology.

Another problem is that many soils can be classed as either monogenetic or polygenetic (definitions modified from those of the working group on definitions in paleopedology, compiled by John Catt in the Paleopedology Newsletter of October 1995). Monogenetic soils have formed during a time in which variations in the factors, mainly bioclimate, have been too small to produce detectably different assemblages of soil features in different parts of that time span. Although these reasonably constant conditions are certainly preferred in factoral analysis, such soils are not widespread. Examples of monogenetic soils are those formed during parts of the Holocene in many areas, and perhaps those formed in tropical environments for longer periods of time. In contrast, polygenetic soils have formed during two or more times in which the factors were sufficiently different to produce detectably different assemblages of soil features. Thus, the present-day factors may not adequately define the state of these particular soils. These are the more common soils, even though one cannot always isolate which parts of the soil or which soil properties are not in tune with the present factors. A common example is those soils in the desert that have two depths of carbonate accumulation, a deep one formed during the wetter Pleistocene climate, and a shallower one formed during the drier Holocene, and the latter is superimposed on an older Bt horizon (McFadden and Tinsley, 1985). With the glacial–interglacial cycles of the Quaternary, and the corresponding extremes in climate (e.g., Peterson and others, 1979), many soils will have been impacted by events associated with climate change. Johnson and others (1990) make the case that all soils are polygenetic, but for some workers their definition might be too restrictive.

A problem with using the above classification is that the features have to be detectable, and this is as it should be. In general, however, finer-grained soils (e.g., silty loam) preserve the pedological record of climate change better than do coarser-grained soils (gravelley sandy loam). It is feasible that soils composed of these contrasting materials undergo the same climatic change, yet only the finer-grained soil preserves a clear record of it. Although the coarser-grained soil is theoretically polygenetic, in this case there might not be any obvious detectable features to prove it.

There are two ways to surmount the problems brought out by monogenetic vs polygenetic soils in factoral analysis. One would be to study only those soil properties that form rapidly and reach a steady state in a rather short period of time, that is less than the duration of the Holocene. This would limit the kinds of properties that could be studied by the use of factors. Many argillic horizons, for example, take longer than that to form. The other solution is to allow for climatic change and to assume that, at most sampled sites in a particular region, the differences between the climates of the past and the present were relatively constant, that is, all sites experienced a similar variation in climate (Harradine and Jenny, 1958). This seems reasonable as long as soils of similar age are being studied, because then all these soils may have gone through similar climatic cycles. Furthermore, as pointed out by Jenny (1941), in solving for climofunctions one main interest is in gradients rather than in absolute values of precipitation and temperature. Past climatic gradients may have been similar to present ones in some, but certainly not all, regions. It might be difficult to solve some chronofunctions, however, because older soils probably have undergone more changes in climate than have younger soils. Perhaps these problems are not insurmountable when one considers the problems of locating and collecting truly representative samples in the field. Furthermore, as more data become available on past climates and their duration, perhaps these can be taken into account in quantitative studies. Examples of this will be presented in later chapters.

Another problem cited in the use of factors is that one learns a lot about the factors, but little about the soil (Bunting, 1965). However, Barshad (1964) believes that this approach is important in studying soil-forming processes, and the rates at which they proceed, and that it is valuable in predicting the soil properties that might be encountered at a particular site. As an example, we can consider the origin of a clay-enriched horizon near the surface. It could be an argillic horizon or a depositional layer. A knowledge of the variation in clay chronofunctions with other factors, and the factors at the locality studied, would help

set upper and lower limits on the probable amount of pedogenic clay. If further study indicates that more clay is pedogenic than first was thought probable, other factors, such as time, would have to be reassessed. Another example might be the determination of the origin of an E horizon in a soil. It could result from podzolization; if it did, it usually would be indicated by a combination of factors. Or, the E could be a carryover from a past vegetation and/or climatic factor. It might be found, however, that the E horizon is not related to podzolization, but rather to either a layered parent material or a strongly differentiated soil with a perched water table that produced the E horizon by removal of Fe^{2+} in laterally draining soil water. Although more examples could be cited, the point is made. The use of factors, along with geological data and data on soil processes, provides the tools necessary to evaluate the origin of the soil in the field partly because it forces one to keep an open mind to alternate hypotheses.

The derived functions are increasing at a measureable rate, and although one may question how quantitative the use of the factors is, the approach is basic to geomorphological research. Derived functions can be used in a qualitative sense, for surely they indicate the trends that one can expect throughout a region. Such qualitative expressions of the variables give rise to what Jenny (1980) calls sequences rather than functions. One can study chronosequences or climosequences, and these are not without value in research. If the trends are not those that were predicted, perhaps the factors for that site will have to be redefined.

A final justification for the continued use of the Jenny factors is that most published works in pedology start with a section discussing the factors, and compared to some models presented in the next section, one beauty of this approach is its simplicity (see also Yaalon, 1983). Holliday (1992), for example, has found this approach useful in his work on the use of soils in the interpretation of archeological sites, and Retallack (1994) notes its usefulness in paleosol studies.

■ Other Models of Pedogenesis

There have been other models of pedogenesis put forward, and the excellent reviews by Smeck and others (1983) and Johnson and Watson-Stegner (1987) will not be repeated here. One model commonly mentioned is the energy model of Runge (1973), which states that soil is a function of O (organic matter production), W (water available for leaching), and time. I do not know of examples in which this equation has been solved. Yaalon (1975) discusses the problems in using this model.

Another model is the evolutional model of pedogenesis of Johnson and Watson-Stegner (1987), which states that

$$S = f(P, R)$$

where S is the soil, P represents progressive pedogenesis, and R regressive pedogenesis. Long tables are included to give examples of progressive and regressive pedogenesis. Basically, the progressive pathway includes factors and processes that promote horizonation, soil deepening with time, and the assimilation of any materials added to the surface. Examples are the Bt horizon, which becomes more clay and Fe enriched with time, as well as thicker. In contrast, the regressive pathway includes factors and processes that promote a more simplified profile with time. Examples are (1) erosion of the earlier, more developed profile (A/Bt/C) and its replacement by a simpler profile (A/Bw/C), (2) mixing during treethrow of a previously strongly layered profile to produce one less strongly layered, and (3) the buildup and subsequent depletion of the Bt horizon during ferrolysis (Brinkman, 1970; Pedro and others, 1978, see also Fig. 8.42). Both pathways can act together, but one usually dominates.

As in the Runge model, I do not know if this latter equation has been solved, in the same sense that the factoral equation has. Examples in Yaalon (1975) and Bockheim (1980) demonstrate reasonable mathematical relationships between soil properties and various factors, with time being the most common one. Phillips (1993) has a good discussion of the above evolutional model of pedogenesis as it relates to the development index value of soils in the eastern United States coastal plain soils, but no equations are solved to explain the relationships.

■ The Approach Used Here

In the following chapters each factor will be discussed separately and combined with the evolutional model. As the factors are discussed, both the progressive and regressive pathways will be presented where appropriate. This distinction is important in geomorphic studies because if soils are used to estimate the age of a landscape, and the more recent soil pathway is regressive, the age of the landscape could be underestimated. Furthermore, although each factor is taken up as if it is the variable and the other factors are constant, in nature these sites are hard to locate. In spite of this the focus is on one factor at a time, and the other factors will be discussed where they play a major role in explaining a particular feature or features.

■ Steady State

Soils often are described as being mature or in equilibrium with their environment (see discussion in Smeck and others, 1983). Lavkulich (1969) argues that equilibrium is not a good term in the sense that reactions at dynamic equilibrium go in both directions with no apparent change in the system. This condition is probably uncommon in the soil because many reactions, such as primary mineral $\Rightarrow$ clay, are not reversible in the soil environment. He believes steady state is a better term to describe soil conditions. At the steady state, energy is being applied to the system, and reactions are going on, but properties are either not changing or their rate of change is too slow to be measured. This is similar to the concept of a soil-forming factor being considered constant even though it varies. That is, for a change in the factor, the soil property does not change.

The time factor is a familiar example of soil steady state. In a plot of soil property as a function of time, the curve is usually steep during the initial stages of soil development (Fig. 6.2). After some time the soil property shows little change, even though reactions are still going on. We can say that the soil property has reached a steady state with the environmental factors. If we consider a soil at one locality, each soil property commonly will require a different amount of time to attain the steady-state condition (Fig. 6.2). The soil profile could be said to be in a steady state when its diagnostic properties are each in a steady state. Examples of curves that flatten with time are those for various weathering phenomena (Colman, 1981), those for soil properties (Bockheim, 1980), and those for soil-profile development indices (Harden, 1982; Harden and Taylor, 1983).

There is some doubt as to whether soils reach the steady state. Early in the use of soils to estimate the ages of deposits and landscapes (Richmond, 1962;

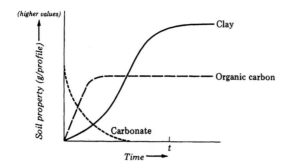

Figure 6.2 Hypothetical variation in several soil properties with time. The parent material contained carbonate, which has been leached from the soil. Organic carbon reaches the steady-state condition before clay, and the soil profile might be considered at the steady state at time *t*.

Morrison, 1964), it was noticed that the older soils all had morphological similarities. It seemed as if they had reached a steady state; hence, they could not be used to discriminate between landforms that, from topographic arguments, were of different ages. Once quantitative data were obtained on properties of some old soils (ca. 1 My), however, it was shown that for some the curves do not flatten, but continue to increase at a slow rate (Markewich and others, 1987). In this context it is important to differentiate between a morphological steady state and a steady state of various soil properties. In a sequence of soils there could be a morphological steady state, yet individual properties (e.g., pedogenic Fe) might continue to change. Tonkin and Basher (1990; their Fig. 5) illustrate this concept for young soils. For the first 2 ky of soil evolution there are characteristic morphological changes related to age. However, for the next 4 ky of soil evolution the profile form remains unchanged, a state they call persistent. During this time in which the profile form is not sensitive to age, one should quantify other soil properties to see if they continue to change with time, if the goal is to use soils to estimate deposit or landform age. To sum, in discussing the soil steady state, one has to carefully define the criteria on which it is based as well as the age span of interest.

Chapter 7

Influence of Parent Material on Weathering and Soil Formation

Parent material influences many soil properties to varying degrees. Its influence is greatest in drier regions and in the initial stages of soil development. In wetter regions, and with time, the influence of parent material is less clear. Here we will deal with the effects of minerals, rocks, and unconsolidated deposits on weathering and soils. We also will explore soil development along an extensive loess transect, as this involves parent materials of different texture and in environments of varying sedimention rate.

■ Mineral Stability

Minerals vary in their resistance to weathering; some weather quite rapidly (10^3 years), whereas others weather so slowly (10^5 to 10^6 years) that they persist through several sedimentary cycles (reviewed by Nahon, 1991). The release of elements at various rates is important to the kinds of pedogenic products that form in soils.

Minerals can be ranked by their resistance to weathering in several ways. One is to study the soil or weathered material and compare the minerals there with minerals in the unweathered material on the basis of their respective depletion, evidence for alteration to clay, or etching of individual grains (Fig. 7.1). Goldich (1938) established the mineral stability series for the common rock-forming minerals in this way. A second

way is to assess the occurrence of various minerals in sedimentary rocks of varying ages; older rocks should have relatively greater amounts of the more resistant minerals (Pettijohn, 1941; Pettijohn and others, 1987). Brewer (1976) compared several rankings and showed that all rankings are not consistent; perhaps there are mineralogical reasons for this (Raeside, 1959). A rather complete list of minerals ranked according to relative stability has been compiled by Allen and Hajek (1989), and in general those between zircon and quartz (Table 7.1) have a high resistance to weathering and can be used as stable components in mineral-depletion weathering studies (Marshall, 1977). However, even the most resistant minerals may show weathering features in some environments (Tejan-Kella and others, 1991). The more easily depleted minerals would be augite and those below it. In the secondary mineral list (Table 7.1), calcite and those below it can be primary minerals and are among the most rapidly weathering. Drever and Clow (1995, Table 5) tried to quantify the differences in weathering between the various minerals.

Several factors account for the stability of the common rock-forming minerals. Some are related to the mineral itself, and some are related to the weathering environment. Only those related to the mineral will be considered here.

Mineral structure plays a major role in the resistance of a mineral to weathering. For some minerals

Figure 7.1 Photomicrographs of thin sections showing mineral weathering. (A) Two etched pyroxene grains in weathered ash. Left: Hypersthene occupies a cavity that retains the original shape of the crystal. Right: Augite is surrounded by halloysite subsequently deposited in the cavity formed during grain weathering. Such sharp cockscomb terminations are characteristic of weathered pyroxene and hornblende grains in many areas. Lack of clay formation in place might be related to insufficient Al in the parent grain. *(Reprinted from Hay, 1959, pl.2,D, J. Geol. © 1959, The University of Chicago.)* (B) Halloysite replacing interior of a zoned plagioclase. The core of this grain probably was more calcic than the rim, thus the selective replacement. *(Reprinted from Hay, 1959, pl. 3,A, J. Geology, © 1959, The University of Chicago.)*

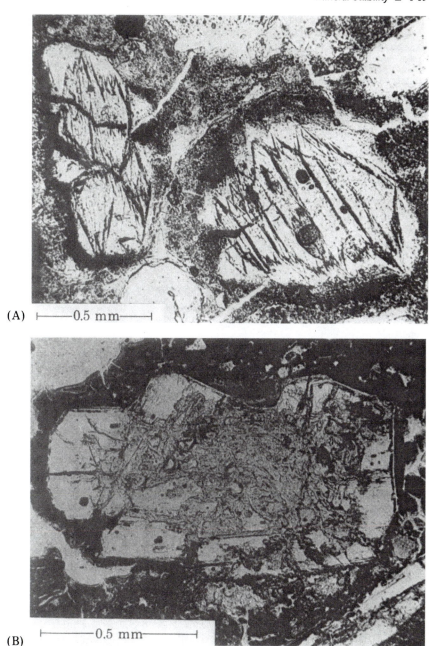

(A) ├────0.5 mm────┤

(B) ├────0.5 mm────┤

the mineral-stability ranking parallels a ranking based on mineral structures: that is, the progressive increase in sharing of oxygens between adjacent silica tetrahedra correlates with increased resistance to weathering. This is especially true for the sequence olivine–pyroxene–hornblende–biotite–quartz. Most of the cation–oxygen bonds in minerals are intermediate between ionic and covalent, and Curtis (1976a,b) points out that the ionic bonds are the weak link in the weathering of the minerals. Hence, the more ionic the sum

Table 7.1 Relative Resistance of Minerals to Weathering, From More Resistant (Top) to Least Resistant (Bottom).

Primary Minerals (Mainly Sand and Silt Size)	Secondary Minerals (Mainly Clay Size)
Zircon	Anatase
Rutile	Gibbsite
Tourmaline	Hematite, goethite
Ilmenite	Kaolinite
Garnet	Hydroxy-interlayered
Quartz	Vermiculite (pedogenic (chlorite)
Epidote	Smectite
Sphene	Vermiculite
Muscovite	Illite
Microcline	Halloysite
Orthoclase	Sepiolite, palygorskite
Na-plagioclase	Allophane, imogolite
Ca-plagioclase	Calcite
Hornblende	Gypsum, pyrite
Chlorite	Halite
Augite	More soluble salts
Biotite	
Serpentine	
Volcanic glass[a]	
Apatite	
Olivine[b]	

From Allen and Hajek (1989, Table 5.1).

[a]Although not a mineral, this material is listed for comparative purposes.
[b]Hay (1959) found that the Fe-rich variety weathers faster than the Mg-rich one.

of the bonds, the more susceptible the mineral is to weathering. For the common cations, the approximate percent ionic bond ranking is

$$Si = Ti > Fe^{3+} > Al > Fe^{2+}$$
$$= Mg > Ca > Na > K$$

Therefore, mineral resistance to weathering should be reflected in the structure, composition, and relative importance of ionic bonds in the mineral.

Mineral breakdown during weathering probably is initiated at the site of the weakest bond, and the greater the number of sites exposed to weathering solutions, the more rapid the weathering should be (Barshad, 1964). Olivine weathers rapidly because the separate tetrahedra are linked only by the fairly weak Mg– or Fe–oxygen bonds. The other minerals in the stability series contain tetrahedra linked together by the stronger Si—O and Al—O bonds and so are more resistant to weathering. In these minerals weathering is most effective if the weathering solutions have access to sites of the weaker cation–oxygen bonds. Such sites link the chains of the inosilicates together (pyroxenes and amphiboles), and the sheets of the phyllosilicates together (micas), or offset the charge deficiency brought about by the substitution of Al^{3+} for Si^{4+} in the tectosilicates (feldspar group).

Other factors seem to explain the relative resistance of minerals classified in the same structural groups (Barshad, 1964). The plagioclases, for example, show a ranking from the more stable sodium-plagioclases to the least stable calcium-plagioclases. This sequence is due in large part to the substitution of Al^{3+} for Si^{4+} in the tetrahedral position as the plagioclase becomes more calcic; thus, stronger Si—O bonds are replaced by weaker Al—O bonds. Biotite and muscovite, although both phyllosilicates, differ in their weathering. Biotite is more susceptible to weathering, probably because the Fe^{2+} is readily oxidized to Fe^{3+}. This change in valency disrupts the electrical neutrality of the mineral, which, in turn, causes other cations to leave and thus weaken the crystal lattice. The tightness of packing of the oxygens might be a factor with some minerals. In both of two mineral pairs, sodium-plagioclase and orthoclase, and olivine and zircon, the latter minerals are packed more tightly and are more resistant to weathering. This is particularly important in the olivine–zircon pair, because it includes one of the least and one of the most resistant minerals.

■ Rock Chemical Composition

Rocks vary in their chemical composition because of their mineralogy. Compositions of a few rocks are given in Table 7.2 as an aid in understanding the kind of controls each has on the direction of pedogenesis. The igneous rocks go from more felsic on the left to more basic on the right. One major control on SiO_2 content is quartz content, which varies from a high of 30% in granite to none in basalt and dunite. Feldspar content influences K_2O and CaO distribution, with high K-feldspar content (ca. 70%) in the felsic rocks and high Ca-plagioclase (ca. 60%) in the basic rocks. Finally, the MgO and Fe oxides increase in the more mafic rocks because of the increase in mafic minerals (hornblende, pyroxenes, olivine). The ultrabasic rock dunite (mainly olivine) is included, as it is an extreme composition, and, as such, the direction of pedogenesis can be much different than in adjacent rocks of more average composition.

Average sedimentary rock compositions are quite variable (Table 7.2). Sandstone compositions reflect the source area rocks, and their high SiO_2 content is due to high quartz content. The cementing agent will vary, and common ones are carbonate and silica. Shale composition is dominated by that of the included clay minerals, with illite being common to many. Finally, limestone composition is dominantly that of CaO in $CaCO_3$ (CO_2 content usually not given). The few included mineral grains explain the rest of the chemistry.

Molar SiO_2/Al_2O_3 ratios (Table 7.2) give an idea of the changes that would have to take place during soil formation to produce the characteristic clay minerals from rocks of different composition (compare with Table 4.2).

■ Susceptibility of Rocks to Chemical Weathering

Rocks weather and erode at different rates, as can be seen in the variations in topographic relief that ac-company variations in rock type. For rocks from widely spaced localities, however, factors other than rock type might influence the weathering variations observed. Nahon (1991) reviews data for several kinds of rocks in a variety of climatic settings. To compare rocks of differing lithology under similar conditions of weathering, however, it is best to study a sedimentary deposit, such as bouldery till or outwash, which includes a variety of rock types. Rocks from the same depth below the surface should have weathered under similar conditions.

Goldich's stability series for minerals (Table 7.1) can be used to predict igneous rock stability in the weathering environment. Rocks with a high content of more weatherable minerals should weather more rapidly than rocks with a high content of minerals resistant to weathering. To make a valid comparison, however, the rocks should be similar in crystallinity and grain size. For igneous rocks, therefore, resistance to weathering should be

$$\text{rhyolite} > \text{granite} > \text{basalt} > \text{gabbro}$$

Clay production should follow these trends, and it seems to (Barshad, 1958).

Grain size has an effect on the rate of weathering, for it is observed that coarser-grained igneous rocks commonly weather more rapidly than finer-grained rocks (Smith, 1962). This is readily seen in many tills in the Cordilleran Region. In the Sierra Nevada, for example, till of probable O-isotope stage 6 age has the

Table 7.2 Chemical Composition of Selected Igneous and Sedimentary Rocks

Oxide	Granite	Granodiorite	Diorite	Basalt	Ultrabasic rock	Average shale	Average sandstone	Average limestone
SiO_2	71.3	66.1	57.5	49.2	38.3	63.3	78	5.2
TiO_2	0.3	0.5	1	1.8	0.1	0.8		
Al_2O_3	14.3	15.7	16.7	15.7	1.8	17.2	7.2	0.8
Fe_2O_3	1.2	1.4	2.5	3.8	3.6	0.8	1.7	
FeO	1.6	2.7	4.9	7.1	9.4	5.5	1.5	0.5
MgO	0.7	1.7	3.7	6.7	37.9	3	1.2	8
CaO	1.8	3.8	6.6	9.5	1	3.5	3.2	43
Na_2O	3.7	3.8	3.5	2.9	0.2	1.5	1.2	0.1
K_2O	4.1	2.7	1.8	1.1	0.1	3.6	1.3	0.3
Molar SiO_2/Al_2O_3	8.5	7.3	6	5.5	32	6.5	18.6	9

Data from Garrels and Mackenzie (1971, Table 9.1), Best (1982, Appendix D), and Boggs (1991, Table 7.7).

following variation in weathered clasts: coarse-grained granitic stones are weathered to grus, coarse-grained porphyritic andesitic pebbles and cobbles have thick weathering rinds or are weathered to the core, and fine-grained volcanic rocks of intermediate to basic composition have weathering rinds up to 4 mm thick (Fig. 8.2). Moreover, it is a common observation in large boulders that mafic inclusions, generally of finer grain size than the enclosing granitic rock, stand in relief above the latter due to weathering (Burke and Birkeland, 1979). In the high Peruvian Andes, Rodbell (1993) describes unusual weathering-post features that protrude above the boulder surface, but he could not find a reason for their presence.

Weathering rind studies help to quantify some of these relations. Weathering rinds are rather uniformly thick zones of chemical alteration around the periphery of clasts; most thicknesses are measured in millimeters. A measure of chemical alteration is the thickness of the rind. Because rinds become thicker with time, rind data have been collected to estimate the ages of glacial deposits in various parts of the western United States (Table 7.3). The differences in rate of rind thickness development probably are due to climate and specific rock type. In contrast to the thin

rinds on volcanic rocks, those on granitic clasts at the ground surface are much thicker. Few data are available for rind development on sedimentary clasts, but those on surface graywacke clasts in New Zealand are some of the most thoroughly studied and reach about 6 mm thickness in about 10 ky (Whitehouse, 1986; see also Fig. 8.4).

Rocks with high amounts of glass weather more rapidly than rocks with low glass contents (Lowe, 1986), and rocks of finer grain size weather more slowly than rocks of coarser grain size. One reason for the latter may be that the intergranular surface area increases with a decrease in grain size; hence more energy probably would be required to disintegrate the finer-grained rock.

There are several reasons for the above differences in weathering with rock type. One is biotite content. High biotite-bearing rocks commonly weather more rapidly than low-biotite bearing rocks, as seen in many igneous and metamorphic rock outcrops. The reason for this behavior was presented earlier (Chapter 3). Biotite, one of the first minerals to weather in granitic rocks, forms alteration products that can occupy a greater volume than did the original biotite (Nettleton and others, 1970); the result is mineral expansion, with numerous localized points of stress within the rock that eventually shatter the rock and form grus. This mechanism loosens the intergranular contacts between minerals and probably explains, in part, why the zone between fresh and weathered granitic rock usually is gradational over about 0.5 m or more. Basalt is an extreme case of rock lacking biotite. Weathering proceeds inward, grain by grain, and the boundary between fresh and chemically weathered rock can be quite sharp (<1 mm) (Cady, 1960; Porter, 1975, Fig. 2; Colman and Pierce, 1981). The point to be made is that the chemical alteration necessary to disaggregate a biotite-bearing granitic rock is not comparable to that necessary to weather basalt to a similar depth. Much of granitic rock weathering is mechanical shattering induced by slight chemical weathering, whereas basalt rock weathering is mainly chemical and proceeds slowly inward. This, along with the variation in susceptibility of minerals to weathering, explains the common observation that soils formed from granitic rock are higher in sand and lower in clay than are soils formed from basalt, other factors being equal.

The weathering of sedimentary rock differs from that of igneous rock, but there is no absolute rule to determine which rocks are most susceptible to weath-

Table 7.3 Thicknesses of Weathering Rinds (mm) of Clasts in the Western United States

Estimated Age (years)			
Basalts[a]			
	West Yellowstone, Montana	McCall, Idaho	Yakima Valley, Washington
10,000–20,000	0.10 ± 0.07	0.25 ± 0.14	0.25 ± 0.04
140,000	0.78 ± 0.19	1.61 ± 0.41	1.05 ± 0.17
Coarse-Grained Andesites[a]			
	Truckee, California	Lassen Pk, California	
10,000–20,000	0.18 ± 0.13	0.18 ± 0.12	
140,000	0.97 ± 0.23	0.82 ± 0.25	
Granitic Rocks			
	Mt. Sopris, Colorado[b]	Central Sierra Nevada, California[c]	
10,000–20,000	6–45	1.7	
140,000		2.9	

[a]Data for clasts in soils from Colman and Pierce (1981).

[b]Data for surface clasts from Birkeland (1973).

[c]Data for surface clasts from Burke and Birkeland (1979).

ering. In some mountain ranges the evidence favors more rapid weathering of sedimentary rocks (Hembree and Rainwater, 1961), but this can be reversed in other areas with other rock types. Sedimentary rocks can have any combination of nonclay and clay minerals, and the weathering of these rocks could proceed as predicted by their mineralogy (Table 7.1) if the cementing agent is kept constant. Rocks with a considerable clay-size component may break down more rapidly than rocks low in clays because clays by themselves do not bind minerals tightly together and they might expand and contract with variation in moisture content. The cementing agent can have a marked effect on the weathering of sedimentary rocks (Fig. 7.2). In stream terrace deposits in the Colorado Piedmont that are >100 ka old, for example, some clasts of silica-cemented arkoses in soils are intact, whereas adjacent granitic clasts of seemingly similar mineralogy are weathered to grus. Either the silica cement keeps weathering solutions from reaching the biotites, or biotites were weathered prior to deposition, or the cement is strong enough to withstand stresses set up by weathering biotites. In the same soils, clasts of silica-cemented quartz sandstones also are intact, but nearby clasts of shale are weathered.

Limestones present a special case in the weathering of sedimentary rocks. The rock weathers by solution of the readily dissolved carbonate minerals, and the rate of weathering is determined by carbonate equilibria. Thus, in some desert environments, the surface of a limestone clast is more weathered than the subsurface of the same clast (Fig. 3.6).

Till deposits in Illinois contain a wide variety of rock types and thereby provide field sites for relative weathering rates. The data of Willman and others (1966) suggest the following rock-stability ranking, from more to less stable:

quartzite, chert > granite, basalt > sandstone,
    siltstone > dolomite, limestone

Although there are local variations in the ranking due to variations in specific lithologies, the overall trend seems reasonable.

Finally, the time factor exerts some control on the influence of rock composition on soil composition. Early in the evolution of a soil, rock composition exerts a much greater control on soil composition than it does late in the evolution of a soil (Chesworth, 1973; see Fig. 3.12). In time, soils formed from a variety of rock composition tend to converge to a similar composition, especially in humid climates. Thus, the parent-material signal is muted with time.

Figure 7.2 Variation in boulder weathering as a function of rock type, Eagle River terrace gravel, Colorado. Some granitic clasts have weathered to grus (g), whereas other granitic clasts (light tone) and sedimentary clasts (dark tone) are unweathered.

■ Influence of Parent Material on the Formation of Clay Minerals

Parent material, whether mineral or rock, exerts some control on the clay minerals that form because weathering releases constituents essential to the formation of the various clay minerals. The specific ions, their concentration, and the molar $SiO_2:Al_2O_3$ ratio are all important aspects of the parent material (Table 7.2). However, once weathering releases the elements, the micro- and macroenvironments within the soil determine whether certain ions or other constituents are selectively removed or remain behind (Fig. 4.5), and this then determines which clay mineral forms. To keep factors constant, samples should be collected from a certain depth within the soil and comparisons should be made between the weathering products and the parent minerals or rocks.

Even the clay mineral that forms may not be a stable end product because it too can subsequently change to other products with changing conditions within the soil. There is a large literature on clay-mineral formation relative to the parent mineral or rock, and almost any product is possible, probably because of the large variety of environments that are possible. These will not be reviewed here because many of the relationships could involve parent material masked by other environmental factors. A review of some of this work is given in Loughnan (1969).

Rocks or minerals of varying composition weathering in the same environment can produce different clay minerals. Barnhisel and Rich (1967), for example, sampled boulders of different lithologies within the same weathered zone and analyzed the clays that formed. Granites and gneisses, rocks low in bases, produced kaolinite predominately, whereas gabbro, a rock high in bases, produced mostly smectite. These same extremes in clay-mineral formation can be seen at the mineral level. In Hawaii, Bates (1962) reports plagioclase altering to halloysite, whereas adjacent olivine grains alter to montmorillonite. Tardy and others (1973) added to the latter by finding that the specific alteration product can also vary with climate (Table 7.4), specifically at sufficiently high rainfalls climate overrides the parent-material influence. As a generalization, the parent mineral will greatly influence the mineralogy of the weathering by-product in those soils, or places within soils, characterized by low leaching, and the influence of the parent rock or min-

Table 7.4 Variation in Clay-Mineral Alteration Product from Various Primary Minerals, as a Function of Climate

Climatic Regime	Primary Mineral	Clay Mineral(s) Formed
Arid	Biotite	Smectite
	Plagioclase	Kaolinite
Temperature	Biotite	Vermiculite
	Plagioclase	Smectite, kaolinite
Humid tropical	Biotite	Kaolinite
	K-feldspar	Kaolinite
	Plagioclase	Kaolinite
Very humid tropical	Biotite	Kaolinite
	K-feldspar	Kaolinite, gibbsite
	Plagioclase	Gibbsite

Taken from Tardy and others (1973).

eral will gradually diminish with greater leaching under higher rainfall.

Velde (1992) finds agreement with the above at the profile level along the Atlantic region of western France, under a leaching regime and temperate climate. Granite weathering produced a kaolinite–illite–smectite–vermiculite assemblage. Weathering of a gabbro with 10% $FeO + Fe_2O_3$ produced Fe-beidellite, Fe- and Mg-smectites, and vermiculite. In contrast, kaolinite, vermiculite, Mg-smectite, and beidellite are associated with an Al-amphibolite rock (23% Al_2O_3). Finally, ultrabasic rocks with 35% MgO produced Fe- and Mg-smectites, oxides, and talc (rather pure Mg-rich 2:1 layer clay mineral not common in soils).

Parent-material influences also can be studied by sampling over large regions. Barshad (1966) compiled data on the uppermost 15 cm of soils from a large part of California. This ensured representation of a large range in climate and in parent materials. Kaolinite, halloysite, montmorillonite, illite, vermiculite, and gibbsite are the major minerals present. The main variation in clay minerals is a function of precipitation (Fig. 10.15), but parent material affected some clay-precipitation relationships. The parent materials form two main groups: mafic and felsic igneous rocks. The influence of parent material was recorded in two ways. Illite is associated with felsic rocks only, probably because of their higher mica content as well as the availability of potassium. Montmorillonite, although abundant at low precipitation with both rock types, persists as a prominent clay mineral at higher precipitation in the mafic parent materials than in the felsic parent materials, probably because of the greater content of bases in the mafic rocks.

The eastern side of the Sierra Nevada offers another sampling region suitable for studying parent-material effects on clay mineralogy because tills and stream gravels with a variety of rock types are abundant (Birkeland, 1969; Birkeland and Janda, 1971). Again, the major effects probably are climatic, although parent-material influences are recognized in the drier areas. In soils formed from a parent material that is partly andesitic, montmorillonite can be present at an MAP of 48 cm (Fig. 10.16). In contrast, soils formed from materials of predominantly granitic lithology, at even lower rainfalls, contain illite and kaolinite, and montmorillonite only at depth. Again, the parent material helps determine the kind of clay that forms through availability and kind of bases.

For some sediments, the parent-material clays have to be assessed prior to the soil study. R.V. Ruhe has done much work in the Mississippi River basin (Chapter 10), but before a climate-clay mineral study can be taken across that huge region, it is necessary to know the parent material clay mineral distribution. In working with soils on Peoria loess (O-isotope stage 2), for example, Ruhe (1984) recognized four clay-mineral provinces, dictated, no doubt, by (1) the geological terrain over which the glaciers advanced, (2) ice-flow directions, and (3) floodplains from which the loess was derived. To sum, each of these different inherited clay-mineral suites is the parent material for subsequent clay-mineral transformations.

■ Influence of Original Texture of Unconsolidated Sediments on Soil Formation

Much of the research undertaken by geomorphologists involves unconsolidated deposits of Quaternary age; hence, a brief treatment of the effect of textural variation of this material on soil formation seems in order. Although we try to keep lithology and mineralogy of parent material constant (see parent-material uniformity section), this is not always possible in the field. When all other factors are equal, however, the texture of the parent material has a great influence on the course of soil formation.

Texture influences the rate and depth of leaching, and this is related to many soil properties. Textural influence can be such that soils that generally occur only in different climatic regions occur side by side, and yet each is stable for the prevailing site conditions. Figure 7.3 depicts some possible field conditions. The depth of leaching is governed by the texture, being greater in gravel (clasts occupy space and contribute little to soil formation unless highly weathered), less in nongravelly sand, and least in finer-textured material. For the same duration of soil formation, an argillic horizon could form in the finer-textured material but not in the sandy material, because it could only be a function of translocation of original clay-size material. If sufficient time has passed for clays to form via weathering, it probably would proceed more rapidly in the finer-textured material (Miles and Franzmeier, 1981) because the weathering rate of nonclay minerals increases with greater surface area per unit volume due to the availability of more water for weathering

Figure 7.3 Hypothetical depth of leaching related to the texture of the parent material.

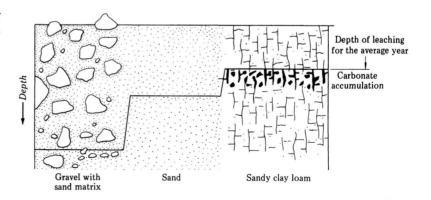

Depth of leaching for the average year

Carbonate accumulation

Gravel with sand matrix   Sand   Sandy clay loam

over a longer period of time each year. Clay formation by precipitation from solutions is also enhanced by increased surface area, because surfaces promote retention of clay-forming constituents. In contrast, constituents released by weathering in the more gravelly materials may be leached from the soil before they have the opportunity to react to form clays. This observation may help explain the presence of well-expressed argillic horizons in O-isotope stage 2 tills and loesses in the midcontinent and the general lack or poor expression of such horizons in gravelly tills and outwashes of the same age in the Sierra Nevada and other parts of the Cordilleran Region (Fig. 8.15). In addition, there might be more release of Ca^{2+} by weathering in the finer-textured material, which, in combination with less leaching, might lead to formation of a Bk horizon. Greater leaching in the coarser grained materials, especially during the wetter years, might keep the Ca^{2+} content too low to form a Bk horizon. However, if the soil-moisture regime is such that carbonate precipitates in both gravelley and nongravelly materials, these material differences dictate the morphology of the precipitate (Appendix 1, Fig. A1.5). The clay minerals that form also follow these trends. Those in the finer-textured material might have a higher ratio of 2:1 layer clay to 1:1 layer clay than those in the more permeable material. These relationships can be seen in soil samples spaced closely together; in Hawaii, kaolinite occurs in freely drained parts of the soil, whereas montmorillonite forms in the same soil in local areas of slightly restricted drainage beneath stones (Sherman and Uehara, 1956). Finally, organic matter contents should be higher in finer-textured materials, a relationship that has been demonstrated in California (Harradine and Jenny, 1958).

■ Examples of Extreme Rock-Type Control on Soil Formation

The composition of some rock types controls the direction of soil formation to an unusual extent. Here we will look at the control of parent material composition on processes and resulting pedogenic materials.

It is well known that podzolization takes place best in acid, sandy materials poor in bases. Cline (1953) demonstrated that podzolized soils in New York formed in sandy, mainly crystalline parent materials. In contrast, soils formed from nearby till with high carbonate content do not exhibit podzolization char-

acteristics. The difference in podzolization trend probably results from the fact that the high base content in the latter keeps pH high and prevents translocation of Fe and Al. Podzolization trends in southern Alaska soils are also controlled by the parent-material composition (Alexander and others, 1994). Soil formed from ultramafic rock has been affected by podzolization but has no E horizon. Nearby till with less ultramafic rock content is podzolized to the extent that the soil is a Spodosol. The rate of release of bases from the contrasting parent materials is thought to control the extent of podzolization. In the Coastal Plain of the eastern United States texture controlled the pedogenic pathway such that sandy parent materials evolved into Spodosols, whereas clayey parent materials followed an argillic pathway (Chapter 8).

As another example, it is commonly known that Bk horizons form in dry moisture regimes. However, at the margins of areas with these regimes, the Ca content of the parent material could determine whether such horizons form (Jenny, 1941). If, however, the Ca is from an atmospheric source, either in ionic form or as $CaCO_3$, as shown in the previous section, soil texture might dictate whether a Bk horizon forms.

Soils Formed on Limestone

Soils formed in limestone are an example of seeming bedrock control on soils, for a fine-grained soil with or without carbonate commonly overlies the limestone bedrock with a sharp contact. Commonly, these soils are red and have chemical analyses much different from the parent rock (Ruhe and others, 1961). Muhs (in preparation) reviewed the literature on the origin of these soils and lists these primary origins: (1) residual origin, that is the insoluble residue of the parent limestone accumulates as the limestone fraction dissolves; (2) fluvial or colluvial deposition of fine-grained materials from higher surrounding terrain onto a topographically low carbonate landscape; (3) weathering of volcanic ash that has fallen on limestone surfaces; and (4) weathering of fine-grained eolian dust that has distant source areas 1000s of kilometers away. Detailed field work or laboratory analyses may be required to determine the correct origin. In areas where the degree of weathering is weak, or the deposits are young, the soil may retain properties that make it easy to determine its origin. However, many of these soils are in areas of rapid tropical weathering, which quickly obscures many of the properties needed to determine origin and source. Also, if weathering is weak the soil carries a legacy of the parent-material properties, whereas with intense weathering there is little influence of the parent-material on the soil.

The above four origins can be differentiated in ideal circumstances. For the residual origin one should do a mass balance and calculate the thickness of rock needed to produce the thickness of soil. Check to see if the amount seems reasonable. It would seem that removal of a large amount of rock would result in karst topography. Furthermore, the properties of the insoluble residue should match those of the soil. For the fluvial–colluvial origin, there should be topographically high terrain from which to derive the soil parent materials, again with the correct properties. Such an origin is advocated for red soils in the karst of Indiana (Olson and others, 1980). The ash and eolian origins require mineralogical or chemical signatures that differ from those in the insoluble limestone residue, and these should match those of nearby volcanos or deserts (Chapter 8).

Several studies have shown the importance of eolian materials in interpreting silicate soils on limestone. Macleod (1980) reported on the red, clay-rich soils that overlie limestone in Greece. Most of the soils are <0.5 m thick, dominantly silt and clay, with kaolinite the main clay mineral, and >4% pedogenic Fe_2O_3. Using various reasonable assumptions, it was calculated that 130 m of limestone would have to be dissolved to form 40 cm of soil. This would require landscape stability for several million years, unlikely in that area characterized by Quaternary tectonics and climate change. An eolian origin seems more likely for the soil as it is known that sirocco winds transport reddish dust northward from the Sahara (e.g., Rapp and Nihlén, 1991). When the northward-moving dust is washed out by rain, the result has been called "blood rains." Some storms have carried dust as far north as southern England. In another study, Mizota and others (1988) used the contrast in $\delta^{18}O$ values of quartz grains in both Greek soils and underlying limestone to support the eolian origin of such soils. Yaalon (1997) has dust as a component of most soils of the Mediterranean region. After the dust is deposited pedogenesis takes over and forms red clay-enriched soils.

Many carbonate islands of the Pacific Ocean have thin silicate soils overlying uplifted Quaternary coral reef deposits. One studied by the author (unpublished data) is Rota Island, one of the southern Mariana Islands. Most of the soils are <1 m thick, and data for one (Table 7.5) show the common properties: neutral to slightly acid pH, >90% clay, >70% total Al_2O_3 + Fe_2O_3, and gibbsite is the more common clay mineral. Making various assumptions, it can be shown that to form 0.5 m of soil by dissolution of the nearly pure limestone (0.03% insolubles) would require the weathering of nearly the entire 500-m-high island! This, of course, is not possible. Another observation against a dissolution origin is that it should result in a karst landscape, yet karst is rare. This soil is so altered in the tropical environment that it is difficult to pinpoint the source of the dust. Obvious sources are dust from Asia, known to be transported across the northern Pacific (Hovan and others, 1989), and volcanic ash. Rare-earth-element analysis may help to resolve the problem of source (Chapter 8).

Table 7.5 Laboratory Data for Soil on Rota Island

Horizon	Base Horizon (cm)	pH, CaCl$_2$	Carbonate (%)	Sand (% < 2 mm)	Silt (% < 2 mm)	Clay (% < 2 mm)
A	11	6.2	1.7	0.9	8.4	90.7
Bw[a]	18	6.8	3.2	0.5	6.7	92.8

Horizon	Chemical Composition (%)					
	SiO$_2$	Al$_2$O$_3$	Fe$_2$O$_3$	CaO	Other bases	TiO$_2$
A	4.9	34.4	38.9	4.1	0.9	2.8
Bw[a]	5.1	36.4	36.7	3.8	1.1	2.7

[a]Soil rests abruptly on Quaternary carbonate reef.

Although the red soils of Caribbean and western Atlantic Ocean carbonate islands were thought to be of residual origin by early workers, work by Muhs and others (1987, 1990) points to an airborne origin. The residual origin was based mainly on the similarity in clay minerals between the soil and the insoluble residue of the underlying rock. The problem was to distinguish eolian dust from the Sahara, as it is known to cross the ocean, from volcanic ash derived from nearby volcano sources. Intense weathering masks many of the obvious properties that could be used to pinpoint the source. The strategy was to estimate the masses of various immobile elements in the soils, calculate profile ratios of the elements, and compare these with characteristic ratios for Sahara dust and volcanic ashes. Ratios examined were Al_2O_3/TiO_2, Ti/Y, Ti/Zr, and Ti/Th. Most of the soil ratios plot closer to the envelope curve for Sahara dust than those for the ashes (Fig. 7.4), implying that the soils are mostly the weathering product of Sahara dust.

There are two important implications of the above study. One is that because there are predictable morphological and geochemical trends of the soils from younger to older landscapes, the dust seems to have had a similar trajectory for about 0.9 Ma. This is important to long-term climatic reconstructions. The second is that bauxite on Haiti has immobile element ratios close to those of the soils and Sahara dust, implying that they also might be of eolian origin. Brimhall and others (1988) also present evidence for an eolian dust origin for bauxites in Western Australia. Thus, economic geologists think about long-term eolian depositional and preservation patterns as they seek to understand the genesis of some ore deposits. A complication is that the dust composition can vary from unweathered (ground up rock) to weathered, depending on the conditions at the source area.

The conclusion on many soil-carbonate rock studies is that there is minimal influence of the latter to the former, because the two materials commonly are so different. Bioturbation, however, can bring carbonate fragments upward into the soil, adding easily released bases to the latter.

Soils Formed from Volcanic Ash Deposits

Volcanic airfall deposits (ash and pumice) probably control soil formation more than any other parent ma-

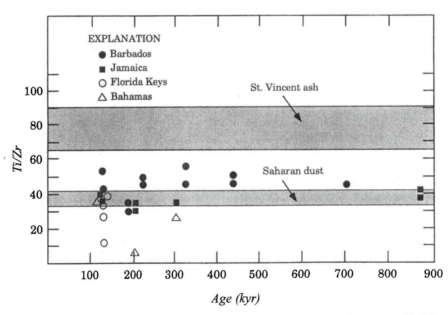

Figure 7.4 Ratios of Ti/Y in soils as a function of age, compared with data for St. Vincent ash (Caribbean volcano) and dust of the Sahara Desert. Shaded areas represent the mean values ± 1 standard deviation. *(Redrawn by D.R. Muhs from Muhs and others, 1990, Fig. 9.)*

terial, particularly if soil formation takes place in a humid environment (Flach and others, 1980; Leamy and others, 1980; Lowe, 1986; Shoji and others, 1993). Indeed, these soils have such unique properties that a new order was created—Andisols (Soil Survey Staff, 1994). It was the study of such soils, half of which are located in the tropics, that helped with the characterization of short-range-order minerals and noncrystalline colloidal materials (e.g., imogolite, allophane, ferrihydrite, and Al/Fe humus complexes) (Wada, 1989).

Andisols commonly have A/B/C profiles, with Bw horizons more common than Bs horizons. Because volcanic ash deposition is episodic and buried profiles are common, there is a problem in differentiating light-colored ash layers from albic horizons and buried A horizons from Bhs horizons; chemical tests can help with the latter (Ito and others, 1991). One horizon is common to many Andisols; some surface epipedons are so dark that melanic epipedon was added to Soil Taxonomy. These are dark horizons (moist value and chroma ≤ 2), with $\geq 6\%$ organic carbon (mean weight), and something called a melanic index of ≥ 1.70 (Soil Survey Staff, 1994). Andisols must have $<25\%$ organic carbon and andic soil properties, and for the more highly weathered soils these are (<2-mm fraction)

A. $Al_o\% + 1/2Fe_o\% \geq 2\%$, and
B. bulk density of ≤ 0.9 g/cm^3, and
C. phosphate retention of 85% or more.

The rapid alteration of volcanic glass in a humid environment produces a clay-size assemblage dominated by allophane, imogolite, opaline silica, and ferrihydrite. Many of these materials are extracted by the oxalate procedure (Table 4.4). Particle-size analysis is difficult with these materials because they are hard to disperse. Clay minerals can form and, depending on the environment, these can be gibbsite, halloysite or smectite. A very porous structure forms in these allophanic materials with high content of organic carbon, and results in bulk densities ≤ 0.9 g/cm^3. The phosphate test is important to agriculture. Phosphorus is present in unaltered volcanic ash, either in glass or apatite. When freed by weathering it is not always available to plants, but instead is tied up in or sorbed on Al and Fe compounds.

It has long been known that these soils in humid regions are dominated by ECDAM—exchange capacity dominated by amorphous material. A simple test for these materials is the NaF pH test (Fieldes and Perrott, 1966). The sample is placed in an NaF solution, F replaces the OH of the amorphous and low crystallinity materials, and if there is sufficient materials present, the pH increases to 9.2 in 2 minutes (see Fig. 4.3). Neall and Paintin (1986) investigated the method further, and used it to indicate the degree of weathering and production of ECDAM materials with time. Less than 1 ky is needed to form such materials in the western United States.

■ Test for Uniformity of Parent Material

In any soil study, the identity of the parent material from which the soil formed (at time = 0, referring to initiation of pedogenesis) should be established beyond a reasonable doubt. In some places this is no easy task, particularly if the soil is quite weathered, and overlies seemingly unaltered rock. A case in point would be a soil formed from weathered colluvium resting on granite. The problem here would be to identify the state of the colluvial materials following deposition at the particular site and prior to pedogenesis. These problems have been addressed in the Appalachian mountains of the eastern United States (Hoover and Ciolkosz, 1988; Graham and others, 1990). If parent material cannot be identified in the field, several laboratory tests for parent-material uniformity can be made. All the field tests should be exhausted before resorting to laboratory tests.

The main reason why parent-material uniformity is important is that it provides clues as to the original particle-size distribution of the soil in question. The depth distributions, particularly of silt and clay, provide good clues as to the soil-forming processes and factors operating at the site, and thus help ensure that pedogenesis, rather than some parent-material layering that can mimic pedogenesis, has taken place.

In the field, the least weathered material, the C or R horizon, is carefully examined to determine if the overlying soil could have formed from that material. For example, if the C or R horizon is granitic bedrock and the soil is colluvial, the soil should contain only granitic clasts and minerals; rounded clasts should be absent unless a round shape can be shown to have resulted from weathering. All soil horizons should have a spatial relation to both overlying and underlying horizons that is pedogenic in character, including cumulic properties if on a slope. Be mindful of the criteria for buried soils (see criteria in Chapters 1 and 11), so they are not confused with parent material layers. Gravel content, particle-size distribution, mineralogy, and chemistry all can be used to infer whether the parent material was uniform or layered when soil formation was initiated. Furthermore, the materials used to infer such uniformity should be those most resistant to weathering. One final warning—the contacts between parent materials may be obvious and occur over a short distance in the field, or they may be more subtle, occur over a greater distance, and require laboratory analysis for identification (Schumacher and others, 1988).

The gravel fraction in surficial materials (stream, glacial, colluvial deposits, etc.), because of its size, is the most resistant of all the particle sizes. Uniformity is suggested by a more or less constant gravel content with depth. If there are fewer clasts toward the surface, as a result of weathering, these should be progressively greater clast alteration toward the surface. This could be shown by more intense grussification of granitic clasts, or thicker weathering rinds on basaltic clasts. However, some depositional environments are commonly distinguished by finer materials overlying coarser materials. In river environments, fine-grained overbank alluvium can overlie high-energy mainstream gravelly alluvium. If the sites are stream terraces, and they are near valley walls, there could be fine-grained side-canyon alluvial-fan material overlying the main valley gravel deposit; the position of the soil (on versus buried by the fan deposit) indicates the soil-geomorphic history of the site. In Alpine areas worldwide, it is common for gravel content to decrease upward, and in many places this is due to younger eolian deposits diluting the gravel content (Fig. 7.5), as long as other geomorphic processes can be ruled out. In these latter three cases, any clasts that were originally present in the overlying materials at time = 0 should have weathering characteristics equal to that expected for the weathering and soil history of the site. As an example, in many eastern Sierra Nevada sites on moraines the youngest deposits with grussified granites are also the youngest with argillic horizons (Burke and Birkeland, 1979), and Berry (1994) has shown a good correlation between soil profile development and clast weathering. Still older soils have redder, more clay-rich argillic horizons and grussified granites that are more oxidized and more clay enriched (Birkeland and others, 1980). Many western U.S. sites display similar clast weathering-soil relations (McCalpin, 1982; Colman and Pierce, 1986). Finally, if the weathering and even loss of gravel toward the surface is due to weathering, upward change in weathering character should be gradational rather than sharp.

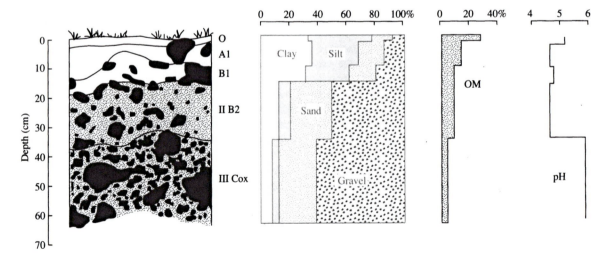

Figure 7.5 Soil formed on a 10-ka moraine in the Colorado Front Range. The lowest B horizon could be a Bt horizon. Roman numerals denote parent material layers (arabic numerals are now used). Particle size includes all fractions, and organic matter (OM) is based on the <2-mm fraction. *(Redrawn from Benedict, 1981, Fig. 24.)*

The above gravel distribution analysis does not provide a unique answer to the parent-material question. Gravel content that varies with depth at time = 0 indicates that the initial <2-mm fractions also may (not has to) have varied with depth, to the extent that the clay distribution could be misinterpreted as a pedogenic feature. However, in fluvial or glacial deposits gravel content may vary with depth, yet particle-size distribution remains constant.

Some unique parent materials in the right environment can produce layering of the gravel fraction. Consider a rather homogeneous gravel, poorly sorted with high smectite clay content at time = 0, perhaps deposited by a debris flow. If a Vertisol forms and the clasts are larger than the cracks, fines fall into the bottom of the cracks during the dry season, and soil swelling during the wet season moves the clasts toward the surface, producing a surface gravel layer, a stone-free layer at depth, and perhaps a deeper gravel layer (process depicted in Fig. 7.6). In contrast, animal activity in a sandy gravel can move fines upward and thereby concentrate the original gravel at depth producing sandy layer/high-gravel-content layer/ original-gravel-content layer (Johnson and Stegner-Watson, 1990, Fig. 17). In both of these cases the gravel layering can be considered part of the overall pedogenic process, not inherited.

Whether or not the gravel content is uniform, the next test for parent-material uniformity is to examine the <2-mm fraction. Field texture is the first clue, but laboratory data should be obtained. For easy viewing, a cumulative curve can be used (see Folk, 1974, for their construction). An instructive study on using such curves to identify parent materials is that of Farrand (1975), who worked on deposits in caves associated with archeology in France. In analyzing the data set, keep in mind that with mechanical weathering any particle fraction can weather in only one direction—to the next smaller size fraction. Furthermore, as usual, the coarser fraction is more resistant than a finer fraction. A soil formed from a 140-ka till can be used to demonstrate the steps (Fig. 7.7A). When total particle-size data are plotted, the curves are grossly similar, but depart from one another in the silt and clay ranges. Silt content is higher in the E horizon than in the other two horizons, shown by the steeper slope of the plotted line through the silt range. In the clay range, the A and E horizons are similar, and different from the Bt, no doubt due to pedogenic clay. One can also make a similar plot of the sand fraction (the most

STAGE 1
Time zero in Pleistocene

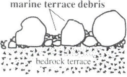

subaerially exposed
marine terrace debris

bedrock terrace

STAGE 2
2:1 lattice clay accumulates by *in situ* weathering and eolian-slopewash accretion

STAGE 3
Shrink-swell action; small-size fraction falls to bottom of cracks; upward translocation of terrace debris

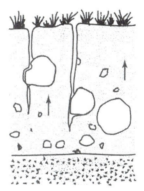

STAGE 4
Well-developed Vertisol with cobble-boulder surface

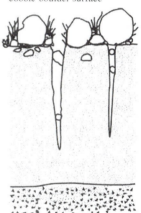

Figure 7.6 Model depicting the upward movement of clasts in material high in 2:1 clay minerals, and a climate characterized by wet and dry seasons (Vertisols). In this case the clay is deposited after the gravel, but the same sequence should happen if the original deposit has clasts in a clay matrix. *(Redrawn from Wood and Johnson, 1978, Fig. 9.22, © 1978, by permission of Academic Press Inc., Orlando, Florida.)*

resistant of the <2-mm fraction), after normalizing the fraction to 100% (divide the percentage of each sand fraction by the percentage of sand in the <2-mm fraction). In this case (Fig. 7.7B) similarity in the curves suggests uniformity in parent material at time = 0, thus supporting the interpretation that the B horizon is argillic.

Particle-size distribution can help identify the contribution of eolian fines to the soil surface. Pye (1987)

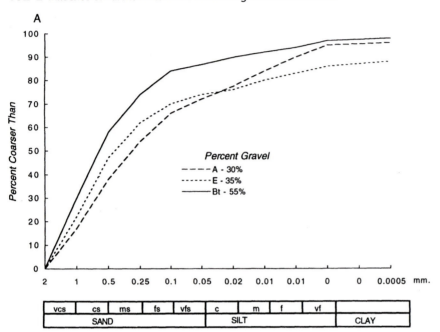

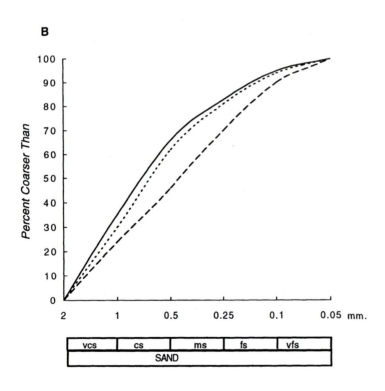

Figure 7.7 Cumulative curves for a soil on a 140-ka moraine in the Colorado Front Range. (A) Curves for the <2-mm fraction. (B) Curves for the sand fraction, normalized to 100%. *(Data from Madole and Shroba, 1979, Table 4.)*

and Simonson (1995) review global dust storms, and atmospheric dust transport and contributions to soils (see also Chapters 5 and 8). These leave little doubt that many soils carry the imprint of a dust contribution. In many localities, even those far removed from eolian source areas, parent-material layering might be due to eolian deposition. Loess forms a thin surface layer high in silt and clay in soils in many diverse environments (Bouma and others, 1969; Yaalon and Ganor, 1973; Muhs, 1983; McFadden and others, 1986; Litaor, 1987; Reheis and Kihl, 1995; Reheis and others, 1995; Yaalon, 1997) and if it is not recognized as such, one might err in describing the soil-forming processes at a particular site. In the western United States, for example, loess is an important component of the A and some B horizons of many gravelly mountain soils (Shroba and Birkeland, 1983) (Fig. 7.8). Many of the soils have a sand component, and in places this also is eolian, picked up from local sources by fierce winter winds. Animal burrowing also will introduce sand into the loess mantle, but this time from the subsurface (Litaor and others, 1996), and freeze–thaw processes can do the same. We therefore call these mantles that are dominantly of loess origin, mixed loess. An additional problem is that of the timing of mixed loess deposition—shortly after or long after the deposition of the underlying deposit, and continuous or episodic through time.

Fluvial gravel terraces commonly have a fine-grained surface layer. The origin of this could be fluvial (overbank deposit), or eolian. If the latter, it may have either a thickness or textural relation to a source floodplain (Reheis, 1980). The unambiguous field test for origin is that loess mantles more than the top of the terrace—it extends down the scarp and onto the nonfluvial landscape.

Mineral characterization is a sure test for parent material uniformity as long as the source area and site of interest have different mineralogies. Marshall (1977, Fig. 32) shows that mineral content varies with grainsize, so one should stick with one grainsize in such studies. Uniform silty soils are present along the western side of the Teton Range in Idaho, in spite of variation in bedrock type (Boulding and Boulding, 1981). Although texture suggested a loess origin, examination of the A-horizon heavy minerals showed hornblende and hypersthene contents at least three times higher than in the C horizons, and the high amounts match those in extensive loess deposits of the Snake River Plain, located upwind of the study area.

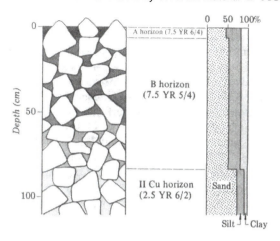

Figure 7.8 Soil formed on a 20-ka rock-glacier deposit, Mt. Sopris, western Colorado. The upper 83 cm is mixed loess, and the underlying material (IICu, old parent material designation) is rock-glacier debris. The latter contains large grains of quartz and feldspar, much too large to break down to the sizes in the overlying materials in the time available. Hence, much of the upper mantle is best attributed mainly to eolian processes. *(Redrawn from Birkeland, 1973, Fig. 10.)*

Dahms (1993) did a similar study in the nearby Wind River Range, Wyoming, but concentrated on soils formed from till of several ages. Volcanic minerals in the coarse silt fraction of the A and B horizons differ from those in the granitic- and metamorphic-composition tills and suggest a mixed loess origin. Silt production via mechanical weathering in the surface of alpine soils is also possible (Munn and Spackman, 1990; Dahms, 1992; Munn, 1992), and would be difficult to argue against if the mineralogical argument was not so compelling. In an isolated alpine area in eastern California, this kind of contamination has been shown to account for as much as one-half of the soil material (Marchand, 1970).

Eolian source areas do not have to be near the area of interest, but can be distant, and transport distances global in scale (reviewed by Simonson, 1995). A classical example of this is quartz and mica in the soils of the Hawaiian Islands, virtually impossible via weathering because the basaltic rocks of the islands do not contain them (Jackson and others, 1971; Dymond and others, 1974). In addition, K–Ar dating by the latter workers shows that the soils are older than the islands, another unlikely relationship.

Once the more obvious ways to ascertain uniformity have shown that the material is probably uniform, another mineralogical check can be made. If weathering has occurred within a specific sand or silt fraction, minerals that are relatively resistant to weathering should show a relative increase in abundance from the parent material toward the surface, whereas more weatherable minerals should show a relative decrease in the same direction (Marshall, 1977). According to Brewer (1976), plots such as these should change gradually and uniformly with depth; sharp inflections or reversals in trends might indicate inherited layering. Ratios of resistant mineral to nonresistant mineral also should show the same trends (Fig. 8.5).

A final mineralogical test for uniformity with depth is to determine the ratio of two resistant minerals (Table 7.1); if the ratio remains nearly constant with depth and matches that of the parent material, uniformity is strongly suggested (Barshad, 1964; Marshall, 1977). This may not be true in every case, however, because sedimentary layers of different textures could have similar ratios of resistant mineral to resistant mineral in the same size fraction.

In places where the suspected eolian surface material and the parent material are mineralogically similar or have some minerals in common, eolian additions might be detected by O-isotope ratios ($\delta^{18}O$ value) of quartz grains. Drees and others (1989) have shown that the ratios vary with rock type and, in general, the higher the temperature of formation the lower the δ value. A change in δ values downprofile could signify parent-material layering. Indeed, this is a good method for tracking quartz grains to their sources, and led to the realization that Asia was a likely source for quartz grains spread across the floor of the northern Pacific Ocean and into the soils of the Hawaiian Islands (Rex and others, 1969; Jackson and others, 1971) and northern Japan (Mizota and others, 1992).

Total chemical analyses are sometimes used to depict chemical trends in soils and paleosols, but here also one should first try to address parent-material uniformity. Ratios of element oxides that reside in resistant minerals are usually used. Common oxide ratios are Y_2O_3/ZrO_2 and ZrO_2/TiO_2 (Marsan and others, 1988; Reheis, 1990). These should remain constant with depth in a uniform parent material.

Combining eolian influx and total chemical analysis brings up a problem in identifying pedogenic trends in surface soils and especially in old paleosols. Mc-Fadden (1991) has modelled what happens to the chemical analysis if atmospheric dust is physically mixed with the upper part of an alluvial-fan deposit in an arid region. Depending on the chemistry of the components, the mixed chemistry trends with depth can mimic that of older soils and even paleosols, even though no pedogenesis has taken place. This example should make workers cautious when interpreting total chemical trends in paleosols (Chapter 11).

Magnetic susceptibility (MS) is a property of unconsolidated materials (see loess in Chapter 10), and in some instances the materials responsible for MS are pedogenic. Fine and others (1992) suggest that depth trends in MS that are pedogenic can be recognized, and that deviations from that can be due to parent-material changes.

The Okarito soil (Aquod) of the west coast of New Zealand demonstrates many of the problems workers face in separating parent-material(s) influences from pedogenic trends (Ross and others, 1977; Mew and Lee, 1981, 1988; Robertson and Mew, 1982; Mew and others, 1983; Thomas and Lee, 1983; Birkeland, 1994). The soil is characterized by a silty surface layer (A and E horizons) overlying a sandy gravel deposit (Bh, Bs and C horizons) (Fig. 7.9). The soils are acid (pH 3.6–5.2, increasing with depth), MAP is ≥ 2 m, and MAT is ca.11. Acid leaching over perhaps 100 ky could explain the textural and mineralogical contrast, but this would require a sharp weathering front, intense enough weathering to eradicate most stones from the (now) silty surface layer, alteration of much sand to silt, to produce a product high in %SiO_2. All of this is possible. However, because the cumulative curves of the A and E horizons are similar to many loesses, workers began to think of it as originally loess. Although loess is not a common sediment in such a high-rainfall area, periods of dry conditions, coupled with winds sweeping across large unvegetated floodplains, could deliver loess, albiet slowly, to Okarito soil sites. An eolian origin was proven when volcanic ash fragments were discovered in the silty layer. The ash is about 20 ka, and much younger than the gravel. A final test was $\delta^{18}O$ analysis conducted on quartz grains in an Okarito soil located farther south of the above sites (Mokma and others, 1972). The δ values are uniform with depth in the profile and the same values are present in both younger soils in the chronosequence and in dust from Australia. This is a case where δ values are of little help. Of course, if the bulk of the eolian sediment is derived from the adjacent coastal shelf

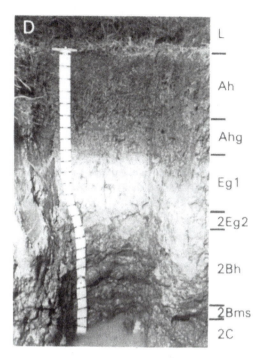

Figure 7.9 Okarito soil, an Aquod common along the west coast of the South Island, New Zealand. *(Reprinted with permission from Mew and Lee, 1981: Investigation of the properties and genesis of West Coast wetland soils, South Island, New Zealand. New Zealand J. Sci. 24, 1–24, Fig. 4A.)*

during a low stand of sea level, the wind is just recycling the same quartz grains back onto the land.

In all of these manipulations, one has to ask if the test is unambiguous with regard to parent-material uniformity, and with regard to the soil property of interest, and does it even matter. For example, say we are interested in clay production and it comes mainly from the weathering of feldspar. Many of the above tests could indicate layering of one sort or another (but not in clay), yet the feldspar content could have been nearly constant. In this case the parent material (feldspar) for the property of interest (clay) could be considered constant. In another case we might be interested in pedogenic iron buildup in time. Perhaps there are sufficient iron-bearing minerals weathering that in spite of possible parent-material layering, pedogenic Fe overprints the layering. A final example is carbonate buildup in deserts. If the carbonate is of atmospheric origin, it will overprint any parent-material layers in which it precipitates. Identifying unambiguous overprinting materials might be the key to pedogenesis in originally layered parent materials.

■ Cumulative Soil Profiles

Some soils receive influxes of parent material while soil formation is going on; that is, soil formation and deposition are concomitant at the same site. Nikiforoff (1949) named these soils cumulative (Fig. 7.10). In such soils, the A horizon builds up with the accumulating parent material, and the material in the former A horizon can eventually become the B horizon. In contrast, other soils gradually lose material through surface erosion, so that the A horizon eventually forms from the former B horizon (Fig. 7.10). These soils were called noncumulative by Nikiforoff (1949). Both

Figure 7.10 Cumulative and noncumulative soil profiles. Material of the noncumulative soil helps dictate the properties of the parent material of the cumulative soil. If erosional removal of material at the noncumulative site is matched by soil-profile development, the profile does not change with time. If erosion is rapid, however, bedrock can be exposed, and its eroded materials are the parent material of the downslope soils. At the cumulative site, horizons are commonly thickened, due to processes at the site and to materials deposited from upslope. Furthermore, subtle sedimentary layering might be discernible at the cumulative site. See Chapter 1 for meaning of arrows.

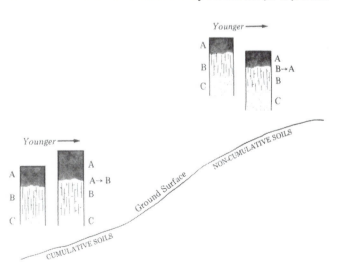

kinds of soils are common along many slopes. Some examples of cumulative profiles will be given here, and others are given in Chapter 9, where lateral soil variations with slope position are discussed.

Because cumulative soils have parent material continuously added to their surfaces, their features are partly sedimentologic and partly pedogenic. In a soil study, therefore, it is important that sedimentologic features are not ascribed to pedogenesis, thus the importance of the tests for parent-material uniformity. Some topographic positions are favorably situated for cumulative profile formation. They are especially common in colluvial and fan deposits at the base of hillslopes (Chapter 9). In such environments cumulative profiles are recognized by properties not consistent with those of soils in the surrounding area. Overthickened A horizons are common, due either to deposition of organic-matter-rich material from upslope, to organic-matter accumulation at the site while sediment is accumulating, or to a combination of both processes. Burial has to be young enough so that the relatively high organic material content at depth, where losses exceed gains, is still present. Cumulative profiles also may contain more clay to a greater depth than adjacent noncumulative soils. This occurs because clay content in the cumulative soil can be a function of both clay-enriched materials derived from upslope as well as clay formation and/or translocation at the site in question.

I wonder if cumulic profiles have been inadvertantly included in Soil Taxonomy. The criteria for both Paleborolls and Paleboralfs require that the top of the argillic horizon be at least at 60 cm depth. The material would be either an A horizon or A + E horizons. This thickness of A or A + E horizons seems extreme for many stable sites of both Alfisols and Mollisols, and I wonder if instead of indicating an age-related property (pale-), these are overthickened A horizons more related to hillslope position.

Alluvial fans offer an opportunity to study the genesis of cumulative soils. One fan studied in southern California will illustrate this (Yerkes and Wentworth, 1965, and unpublished work of the writer). The fan is at the mouth of a very small watershed underlain with sedimentary rocks, some of which are calcareous. The cumulative origin of the soil profile on the fan is suggested by the distribution of organic carbon with depth in the profile (Fig. 7.11). Organic carbon in the cumulative soil is higher in content, and the depth to a level that is less than 1% organic carbon is greater than it is in a nearby stable, noncumulative soil profile. Radiocarbon dates suggest that deposition commenced about 10 ka, and this agrees with stratigraphic and radiocarbon data from nearby areas. Textural variation with depth is rather uniform, implying that deposition has been rapid and continuous enough for sedimentation to mask any B-horizon development.

Carbonate in the fan deposit does not help identify it as a cumulative profile, however (Fig. 7.11). The soil carbonate originates from the weathering of rock fragments and calcareous sediments transported from the drainage basin. The radiocarbon date on the carbonate, although subject to error, is a minimum for sediment in that part of the profile, and it is consis-

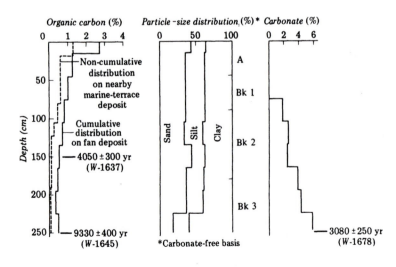

Figure 7.11 Data for a cumulative soil formed from an alluvial-fan deposit. The fan deposit overlies a coastal marine terrace deposit at the mouth of Corral Canyon, west of Santa Monica, California. Radiocarbon dates are from a nearby soil with similar morphology; two dates are on organic carbon, one on carbonate. Soil data from unpublished work of the author; radiocarbon dates by M. Rubin, U.S. Geological Survey.

Figure 7.12 Schematic diagram showing relations between dust accretion rate (A) and weathering rate (B) along a climatic gradient from a desert dust source area downwind to areas of progressively greater precipitation and vegetation cover. *(Reprinted and modified from Pye, 1987, Fig. 9.16, Eolian dust and dust deposits, © 1987, by permission of Academic Press Ltd., London.)*

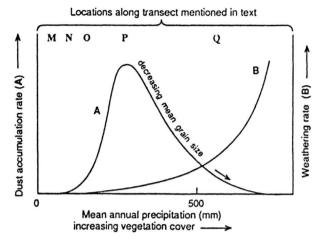

tent with the other dates because carbonate can undergo solution and reprecipitation (Chapter 5). Carbonate translocation may have gone on during sediment deposition, or the bulk of it may have been translocated at a later time. If carbonate translocation went on during deposition, thick fan deposits should have overthickened Bk horizons that extend to the base of the deposit. A deep artificial exposure in colluvium adjacent to the fan in similar parent material disclosed that the Bk horizon does not extend throughout the deposit, but has a base, below which is colluvium with a relatively low carbonate content. This can be taken to mean that carbonate did not accumulate with the buildup of sediment on the surface or that its absence at depth is explained by other factors, which may be paleoclimatic or lithologic. That some other factor is involved is supported by rough calculations on rates of sedimentation and carbonate translocation; these suggest that translocation can keep pace with sedimentation under the present climatic conditions.

Cumulative soils are also common along many river floodplains (Jenny, 1962). In the latter environment, Skully and Arnold (1981) showed that soils within the zone of flooding, however infrequent, are characterized by various numbers of buried A (and sometimes B) horizons. Once the river downcuts and isolates the landform from future flooding, the cumulative process is cut off and the profile slowly converts to the noncumulative A/B/C form. In fact, in a transect away from a river, a shift from soils with many buried A horizons to one with no buried A horizons might mark the edge of the present floodplain.

■ The Loess Transect: Cumulic Soil Formation on Parent Materials with Different Textures and Different Depositional Rates

Transects across loess sheets illustrate soil formation in a cumulic regime, as well as a controlled experiment on the influence of texture on soil development. A useful diagram depicting this is that of Pye (1987) (Fig. 7.12, with letters denoting localities discussed in the text). In this case the loess has a desert source (loc. M), perhaps dry river beds, playas, or salt-influenced weathering. Adjacent to the sources (loc. N), the atmospheric dust influx is low and the dust forms an Av horizon associated with the desert pavement and some is translocated into the underlying gravelley alluvium to eventually form a desert pavement and Av horizon over a stone-free B horizon (McFadden and others, 1987). In other places with higher dust accretion rates (loc. O), the soil is partly cumulic as soil formation continues as material accumulates at the surface, and some soil features resulting from past, alternating soil-moisture regimes can overlap (Chapter 10). At still higher accretion rates (loc. P), pedogenesis cannot keep pace, and soils form only during episodes of nondeposition or slow deposition dictated by climate change (soil-loess sequences of the loess plateau of China, Chapter 11). Finally, at great distances from the source toward the distal edge of the loess sheet (loc. Q), the loess is clay rich, either because of downwind sorting or weathering concomitant with deposition under the humid weathering regime or some combination of the two (examples in Chapter 8). Any vestage of the episodic nature of loess deposition there probably would be masked by pedogenesis.

The above model probably was derived partly from work on loess in the midcontinent. Smith (1942) studied loess in Illinois and suggested an influence of parent material on the soil pattern. Many properties of the parent-material loess are related to distance from the source of the loess, which was usually a glacial-age river floodplain. The loess is commonly thicker and coarser grained near the source and becomes thinner and finer grained downwind. The soils developed on the loess show a downwind increase in clay content, accompanied by the gradual loss of $CaCO_3$. Smith reasoned that both loess deposition and soil formation were intimately involved in producing the soil pattern. Close to the source rivers, the loess was deposited too rapidly for soil formation to keep up, and mostly unweathered calcareous loess was deposited. At localities farther from the rivers, however, soil formation could keep up or even exceed the rate of deposition, so that carbonates, if originally present, would have been leached as rapidly as the material

was being deposited. Moreover, weathering could have gone on for longer periods of time on loess materials away from the river where the depositional rate was less. Thus, cumulative soil profiles could be found at some distance from, but not close to, the rivers. The effect of parent material, therefore, is twofold. A downwind decrease in particle size results in finer-textured parent material and soils, and a downwind decrease in rate of deposition results in cumulative soils in which carbonate leaching and other soil-forming processes can go on concomitant with deposition; thus, in the absence of short intervals of soil formation between episodes of thick loess deposition, soil formation of longer duration is suggested for the downwind localities. Hutton (1951) explains the loess-soils distribution in Iowa in a somewhat similar manner. In Alaska, variation in Spodosol soil development with distance from the loess source is attributed to variations in the rate of loess deposition (Rieger and Juve, 1961).

Detailed work by Ruhe (1969a,b, 1973) in the midcontinent demonstrates the complexity of the loess landscape. In particular, in northeastern Kansas, the decrease in parent-material particle size away from the loess source is not the sole cause of regional variation

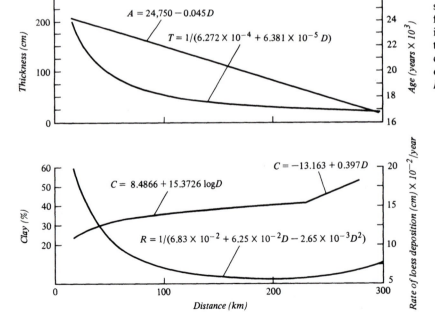

Figure 7.13 Data for O-isotope stage 2 loess in Iowa. D, distance from source; A, age of A horizon in buried basal soil in loess; T, thickness of loess; C, maximum clay content in surface soils; R, rate of loess deposition. *(Redrawn from Ruhe 1969b, Fig. 7.)*

in B-horizon clay contents. He showed that the age of the base of the loess in Iowa is younger away from the source area (assumes minimal contamination of radiocarbon samples), and therefore it no longer can be assumed that the more distant soils in a loess landscape have weathered longer (Fig. 7.13). Moreover, both the surface and buried soils have more strongly expressed Bt horizons with distance from the source area, and the relative increase in development of this horizon takes place over a shorter distance for the Sangamon Soil (probably formed mostly during O-isotope stage 5) than it does for the surface soil (Fig. 7.14). Although part of the Bt-horizon development trend could be a function of clay formation and translocation, part of that for the surface soils is due to greater moisture in the progressively thinner parent loess sheet away from the source because of a perched water table on the underlying buried soil. Greater moisture enhances clay formation. The explanation of the soil landscape as a result of this study involves many factors, both geological and pedologic, and parent material is only one of many factors.

An interesting transect demonstrating the influence of loess on pedogenesis is from the Sinai Desert northward into Israel (Dan and Yaalon, 1966; Yaalon and Ganor, 1973, 1979; Yaalon and Dan 1974; Dan and others, 1981; Gerson and others, 1985; Dan, 1990). Fine-grained material is deposited by infrequent floods in the arroyos of the Sinai. Vegetation is scarce,

and winds transport the dust northward into Israel, mainly from Sinai, but also from North Africa. The rate of dust accumulation and the mean size of the dust vary systematically northward. In parts of the more extreme desert, deposition is slow and the material is recognized as a fine-grained surface layer overlying gravelly materials, such as the ubiquitous alluvial-fan deposits. Where rainfall is sufficient, part of the loess is translocated to form part of the Av and B horizons of the soil. Still farther north, in the higher rainfall of the Negev (MAP 30), the loess is thickest and most widespread, and cumulative soils are formed entirely in loess. North of the Negev in the vicinity of Jerusalem (MAP 50), the rate of deposition declines and the loess is thin and more clay rich. Along the Mediterranean Sea, however, there are sand dunes of various Quaternary ages in which the airborne clay has infiltrated to produce a soil-development sequence. Soils formed in young sand dune deposits have Bw horizons, but in time enough eolian clay has infiltrated to produce prominent Bt horizons whose clay content increases with relative age. In other nearby places the clay-rich dust produced a deposit in which Vertisols have formed. Here parent material in the form of dust accretion plays an important role in explaining the regional soils, but other factors are also important. These include rainfall, which influences both the location of maximum loess deposition (rainfall removed dust from the atmosphere) and the infil-

Figure 7.14 Clay data for <20-ka-old surface soils formed in O-isotope stage 2 loess, and buried Sangamon Soils formed in O-isotope stage 6 loess in Iowa. The changes in the younger soils take place in about 260 km from the source area, and those in the older soils between about 18 and 33 km from the source area. (*Redrawn from Ruhe, 1969b, Fig. 3.*)

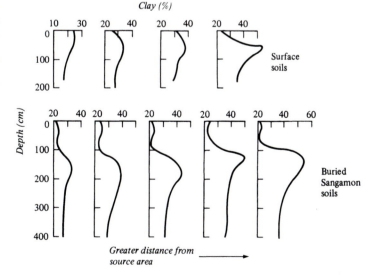

Chapter 8

Weathering and Soil Development with Time

Rock and mineral weathering and the development of prominent soil features are time-dependent. The time necessary to produce various weathering and soil features varies, however; those soil properties associated with organic matter buildup develop rapidly, but those associated with the weathering of the primary minerals develop rather slowly. In this chapter, data are presented on the rates of development of various weathering and soil features, soil orders, and clay-mineral alteration products in various environments. Although the soil data are soil chronosequences, together they form a climochronosequence.

■ The Time Scale

To compare soil data a time scale must be adopted. The dating of the deposits and soils discussed in this chapter, and in others, will vary in accuracy. Quaternary dating methods are reviewed by Colman and others (1987). Some deposits are dated directly by numerical-age methods, others are bracketed by numerical ages, and still others by relative age methods (lichenometry, progressive landscape modification, etc.), or by correlation with dated deposits or events elsewhere (paleomagnetic direction, marine O-isotope record, etc.). Although there is no consensus on the absolute ages for the deposits to be discussed, I will use those of the authors matched to either one of the following age schemes (approximate age estimations from Shackleton and Opdyke, 1973; Rich-

mond and Fullerton, 1985; Bradley, 1985, Table 6.2; Martinson and others, 1987):

Geochronologic unit	Age lower boundary (ka)	O-isotope stage
Holocene	10	
	12	1
	24	2
	59	3
	74	4
Late Pleistocene	130	5
	190	6
	244	7
	279	8
Middle Pleistocene	788	
Early Pleistocene	1,650	

These ages are reasonable estimations, acceptable to many workers. In most places, these ages and O-isotope stages will be used to plot data for the figures, as well as discuss data in the text. The alternative, using local geological names, would be confusing to readers not familiar with the local successions; these latter can be found in the references cited.

In the above chart and throughout the book, ka (thousands of years before the present) is used. These will usually be rounded to the nearest 1 or 10 ka. Time intervals will be given in thousands of years, as ky. For older deposits and soils, Ma (millions of years before the present) and My (millions of years) are used.

■ Rate of Rock and Mineral Weathering

As discussed in Chapter 7, rocks and minerals weather at different rates. Some quantitative data on the rate of weathering are presented here. Some data on initial rates of weathering come from the study of tombstones and other cultural features, but the majority of the data on longer durations of weathering for which the time factor is reasonably well known come from the study of Quaternary unconsolidated deposits. Much of the latter are taken from a review by Birkeland and Noller (1998), and this should be consulted for references.

Tombstones or other manmade structures are good for indicating the rate at which weathering can proceed above the ground surface. Data cover a variety of rock types in a variety of climates, and they are commonly referred to in most physical geology textbooks, as well as in other books on weathering (Ollier, 1969; Winkler, 1975). Rocks that weather quite rapidly, such as some limestones, can lose their tombstone inscriptions in as little as 100 years in a humid climate. With more resistant rock, such as some sandstones and igneous rocks, it may take several centuries before the tombstone shows distinct signs of weathering (Rahn, 1971). Several centuries are sufficient for weathering to be visible on almost any rock type in a humid climate. Arid climate weathering proceeds at much slower rates (Barton, 1916). Recent studies by Meierding (1981, 1993) on the weathering of marble indicate a methodology that could be used in future quantitative studies of tombstone weathering, and how this is used to show variation in rate with climate.

Studies of tills and glacial outwash deposits are ideally suited to the determination of weathering of different lithologies as a function of time because most factors, except paleoclimate, can be kept reasonably constant. Because quantitative weathering studies usually are made in conjunction with stratigraphic studies, these are data from which fairly sound conclusions can be drawn. Most of these methods were pioneered by Blackwelder (1931) and have been continued by other workers. Tills with deep exposures are best for these comparisons because, in places where only surface boulders are present, it is not possible to be certain that changes in lithologic composition are due solely to weathering subsequent to deposition. Some variation in the percentage of fresh boulders, and in boulder and lithologic frequencies at the sur-face, for example, could be a function of the weathered nature of the terrain over which the glaciers advanced. If the terrain had been highly weathered, only the more resistant lithologies might have been picked up, transported, and deposited as boulders. If the terrain had not been weathered, boulders of all lithologies might be plentiful. Little is known about the degree of weathering of landscapes between glacial advances. In a study in Idaho, however, the concentration of thorium-bearing minerals as placer deposits in outwash seems to be correlated with the duration of the interglaciation immediately prior to the glacial advance that produced the outwash (Schmidt and Mackin, 1970).

In weathering studies of tills, the amounts of subaerial versus subsurface weathering can be compared with time (Fig. 8.1). Many tills in the Cordilleran Region display the general weathering succession shown in Fig. 8.2. Weathering is recognized by surface roughness due to grain relief, oxidation, and development of weathering pits. Weathering of 20-ka tills has taken place on the exposed surfaces of all granitic stones so that they show signs of oxidation and have some relief due to weathering. Weathering rinds are present on surface clasts, and maximum thicknesses vary with rock type and climate. The same rock types in till at depth are less weathered or not weathered, and some retain glacial abrasion and striations. Hence, for that period of time subaerial weathering exceeds subsurface weathering. In contrast, 140-ka tills contain granitic stones at the surface with weathering relief, greater degree of oxidation, prominent pits, and somewhat thicker rinds; they are partly or wholly weathered to grus at depth. Therefore, for this period of time, weathering of granitic clasts at depth is equal to or exceeds that at the surface. For better comparison in future studies a clast disintegration index, which combines the percentage of clasts weathered to grus and the largest size so weathered, can be calculated (Berry, 1994). Although the data overlap, Berry (1994) calculates index values in the Sierra Nevada of 0.2 for 20-ka till, 0.4 for 140-ka till, and 0.75 for still older till.

Under extremely arid conditions, salt weathering can help produce rapid weathering of both surface and subsurface clasts (Campbell and Claridge, 1987; Amit and others, 1993). Surface weathering of some clasts is accelerated in these salt environments; southern Israel exhibits some of the most rapid I have observed (e.g., see Bull, 1991, Chapter 3); clasts on Holocene

(A)   (B)

Figure 8.1 Variation in stone weathering in the Rocky Mountains with age of till and position with respect to surface. (A) Rotated boulder on 20-ka till; the part exposed to surface weathering (left) is quite weathered, whereas the part beneath the surface (right) is virtually unweathered. (B) Rotated boulder of 140-ka till; weathering is extensive both on the portion of the boulder that was at the surface (right) and the portion that was beneath the surface (left).

surfaces are highly weathered. One can expect a weathering gradient along some large alluvial fan surfaces, with the clasts on the distal parts of the fan surface, close to a playa salt source, more weathered than clasts farther up the fan.

Weathering rinds on volcanic clasts show a progressive increase in thickness with age of till (Porter,

1975). Volcanic clasts are good for such studies because the inner boundary of the rind is sharp and easily measured. In contrast, the inner rind for granitic clasts commonly is diffuse enough so that it might be difficult to obtain consistent results. If growth rate is slow, it may be difficult to measure the thickness of thin rinds on young volcanic clasts. The most exten-

Figure 8.2 Variation in subaerial and subsurface weathering as a function of time, Sierra Nevada, California. *(Data from Burke and Birkeland, 1979; Birkeland and others, 1980; Colman and Pierce, 1981.)*

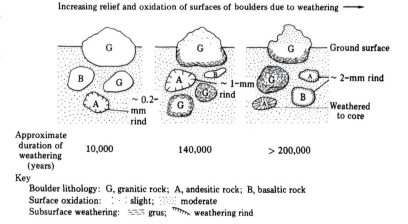

sive study is for the western United States by Colman and Pierce (1981, 1992) in which clasts were collected from soil B horizons to avoid the problem of surface spalling due to occasional fire. Logarithmic curves fit the data best, and show a gross similarity in shape for basalts and andesites (Fig. 8.3). Many studies have shown that logarithmic curves best fit the data for a variety of rocks, and that the growth rate of rind thickness for basalts and andesites is relatively slow when compared with other rock types in other environments (Fig. 8.4). Note also that some rocks (those with a steep slope of rind thickness vs time) are better for dating Holocene deposits (graywacke in New Zealand), whereas others with a much flatter slope might be better for dating pre-Holocene deposits (basalt in the western United States).

Rinds offer a means of estimating ages of soils in places where other methods may not work well. Material for adequate absolute dating of parent materials seldom is available in many Quaternary terrestrial environments, and there are problems with contamination and resolution with almost all methods. Rinds collected at the soil surface or within the soil offer a satisfactory means for dating soils, provided the relationship between rind thickness and age is fairly well known, and one can live with a rather large ± in numerical age. The methology is explained by Colman and Pierce (1981), as well as how to estimate maximum and minimum ages for a landform of unknown age.

To summarize, initial weathering of granitic clasts is characterized by rates of subaerial weathering that exceed those of subsurface weathering; with time, however, subsurface rates are greater than subaerial rates. This change in rate with position relative to the surface takes place sometime between 20 and 140 ka. Thus, in old tills, most granitic stones at the surface have been at the surface for a long time; they probably are not lag gravel from depth. This difference in weathering with position relative to the surface is basic to an understanding of the origin of topographic features in granitic terrain (Wahrhaftig, 1965), and to the persistence of tors once the rock of the tors reaches the surface environment. Finally, it is important to remember that in the time necessary to alter a sound granitic boulder to grus in a soil, dense volcanic rocks may show only thin weathering rinds.

The rate of subsurface weathering of granitic stones east of the Rocky Mountains is somewhat similar to that reported above. Granitic clasts in outwash in Ohio show progressively greater alteration with age, but only in possible pre-O-isotope-stage 6 outwash are they thoroughly decomposed (Lessig, 1961).

Surface-weathering studies have be used to estimate ages of tills, so they also indicate the amount of granitic-rock breakdown with time. One commonly measured feature is the depth of near-circular weathering pits. Along the base of the eastern escarpment of the Sierra Nevada, these reach a maximum of 75 mm on 20-ka boulders and 370 mm on 140-ka boulders (Burke and Birkeland, 1979; see Fig. 3.4). Another measure is the number of boulders classed as weathered. One definition of a weathered granitic clast is that, on over one-half of the exposed surface, weath-

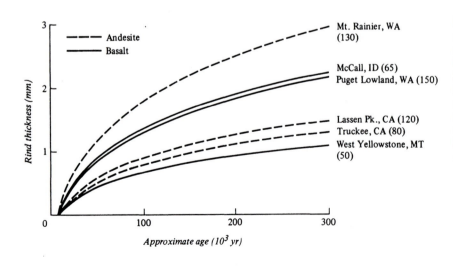

Figure 8.3 Curves for weathering-rind thickness of andesite and basalt versus time for tills in the western United States. Numbers in parentheses are MAP in centimeters. The curves for the Puget Lowland and Mt. Rainier are the more rapid rates of the two calculated. *(Redrawn from Colman and Pierce, 1981, Fig. 19.)*

Figure 8.4 Comparison of trend lines for rind thickness versus age of surface by different workers for different areas. Rock types are granite (g), graywacke (gr), dolerite (d), andesite (a), and basalt (b). *(Redrawn from Watanabe, 1990, Fig. 4.)*

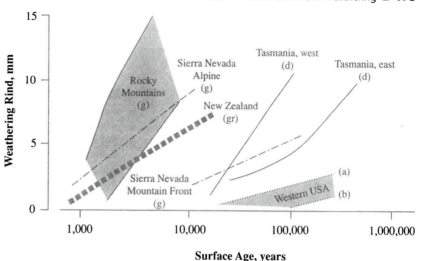

ering penetrates to the depth of the average grain diameter, thus producing surface roughness (Birman, 1964). These surfaces are characterized by loose minerals or minerals that can be readily loosened with a light hammer blow. It should be noted that those weathered boulders that have spalled subsequent to deposition can be considered fresh; the only requirement is the condition of the mineral grains on the present surface. A third measurement is the number of surface boulders per unit area; this is termed surface-boulder frequency. Data collected from widely separated areas give somewhat comparable results (Table 8.1); the scatter in the data could be real or be partly due to different definitions used by the various workers. It can be concluded that at least 100 ky is necessary for most stones to obtain weathered features, and much more than that to decompose all surface granitic stones. In many places, however, surface granitic stones persist for much longer periods of time, perhaps 1 My or more.

Benedict (1985) has improved on the methodology of counting weathered clasts. Rather that trying to decide if the clast is either weathered or fresh, he measures the percentage of weathered surface for each boulder, then constructs curves for each deposit of cumulative percentage vs classes of the percentage of surface area that is weathered. For an alpine area in Colorado, the resulting plots distinguish four age groups of deposits within the last 12 ky.

Surface-boulder weathering features develop at relatively rapid rates above the tree line in high moun-

tains. One plot for the Sierra Nevada indicates that early Holocene deposits in the alpine environment have weathering characteristics that are similar to those of deposits about 10 times older in the mountain-front environment (Fig. 3.4).

In the high mountains of Peru, an unusual weathering form on seemingly uniform granitic rocks is the formation of a protuberance of rock (or knob), called a weathering post (Rodbell, 1993); young posts mimic the forms carved on rock surfaces by Incas at Machu Pichu. In time, the height of the posts increases and this was used as one measure of estimated age. The oldest large boulders not only have >50-cm-high

Table 8.1 Weathering Data for Tills in Various Mountain Ranges, Western United States

Age (ka)	Fresh Granitic Boulders (%)	Surface-Boulder Frequency[a]	Area
20	88–95	96–180	Colorado
	78–88	38–53	California[b]
	50–90	180–300	California[b]
140	40–66	62–95	Colorado
	18–38	33–66	California[b]
	5–20	60–115	California[b]

From Sharp and Birman (1963), Miller (1979), and Burke and Birkeland (1979).

[a]Number of boulders in 186 m².

[b]Data from the same set of moraines.

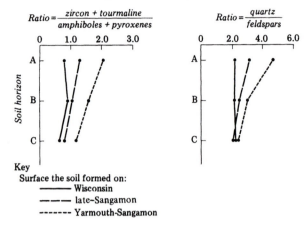

Figure 8.5 Average ratios of resistant to nonresistant minerals for soils developed on surfaces of different age from Kansan Till in Iowa. Wisconsin would be ca. 20 ka, late-Sangamon ca. 100 ka, and Yarmouth–Sangamon ca. >100 ka. *(Redrawn from Ruhe, 1956, Figs. 5 and 6, © 1956, The Williams & Wilkins Co., Baltimore, and Ruhe, 1969, Table 3.1, reprinted by permission from Quaternary Landscapes in Iowa, © 1969 by the Iowa State University Press, Ames.)*

minerals are depleted. Ruhe (1956, 1967) has done this for an area in Iowa in which erosion surfaces of varying ages are cut on Kansan (>0.5 Ma) till (Fig. 8.5). Soils that formed over a longer period of time have higher ratios of resistant to nonresistant minerals. The duration of weathering is difficult to estimate because the landscape has undergone burial by loess, and, in some places, it has been stripped by erosion. Nevertheless, Ruhe (1956) estimates that the soil related to the Wisconsin surface may have weathered for 7 ky, the soil related to the late-Sangamon surface no less than 13 ky and probably much longer, and the soil related to the Yarmouth–Sangamon surface 100 ky or more. The change with time is considerable, and 10 ky or more is necessary to record detectable variations in the ratio. The ratio plots also show that the surface horizons are the most strongly weathered and that weathering extends into the C horizons of the older soils. Brophy (1959) analyzed Sangamon soils (ca. 100 ka) in Illinois in a similar way and described the influence of parent-material texture on weathering in addition to that of age (Fig. 8.6). Outwash had lost 90% of the original hornblende compared to a 60%

posts but a flanking rock minipediment at ground level that is 1 or more meters wide. Such forms are informally called hat rocks, as they mimic hats with brims.

There are few data on surface weathering of nongranitic clasts, but several workers have published data on the ratios of granitic to nongranitic lithologies of surface stones. In most cases, the granitic stones are the more readily decomposed. For example, in one area in the eastern Sierra Nevada, the granitic to metamorphic clast ratio is 80:20 for 20-ka till and 30:70 for 140-ka till (Sharp and Birman, 1963). In northeastern Oregon, tills are composed of granitic, basaltic, and metamorphic clasts (Burke, 1979). Moraines of the 20-ka glaciation have 33–88% granitic boulders on the surface, whereas 140-ka moraines have a maximum of 4%; most of the latter localities, however, have no granitic clasts. Thus, resistant nongranitic boulders persist at the surface long after the granitic boulders have been reduced to grus. The example here from Oregon is one of the more rapid rates of granitic boulder eradication.

The rate of mineral weathering can be determined in two ways. In one, the uniformity of the parent material can be established, and then the ratio of resistant to nonresistant minerals for surfaces of different ages gives the rates at which the more weatherable

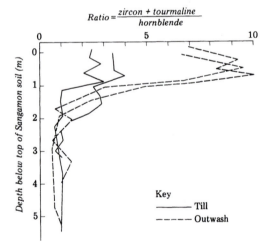

Figure 8.6 Relationship of resistant to nonresistant mineral ratios with depth and texture of the parent material. The ratio is for the indicated depth versus that for the three lowest samples. Clay contents in the assumed parent materials are 14–25% for the two tills and 3–10% for the two outwash samples. If the data of Ruhe for the oldest soils (Fig. 8.5) are recalculated on the same basis as the ratios here, the ratios for till in both areas are comparable. *(Redrawn from Brophy, 1959, Fig. 11.)*

Figure 8.7 Variation in etching of hypersthene grains in Sierra Nevada tills (Birkeland, 1964). Data from J.G. LaFleur (unpublished report 1972) for unetched (A) and slightly etched (B) grains in soils of 20-ka till, and for highly etched (C,D) grains in soils of >100-ka till.

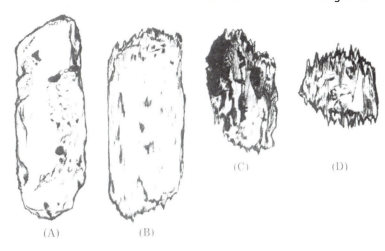

loss in till. The reason for the differences with texture is that outwash deposits are more permeable than tills, and, therefore, depletion in hornblende can proceed at a more rapid rate than it can in till. Guccione (1982) has conducted similar studies in Missouri with similar results, and cautions workers that reduction of grain size from one size fraction to another by weathering can confound the results of these kinds of studies.

Another way to study rates of mineral weathering is to examine individual mineral grains under a petrographic microscope for signs of weathering, such as etching (Locke, 1979, 1986; Hall and Martin, 1986) (Fig. 8.7). This method may be more sensitive than the ratio method in showing change, at least in the earlier stages of weathering, because minerals can show signs of weathering and yet not be depleted. Locke (1979) quantified etching studies by measuring the depth of etching of the cockscomb terminations and calculating the mean maximum etching depth. Measuring this for hornblende grains in cold, dry arctic soils, he found etching to be greatest at the surface and to diminish with depth, and minerals in older deposits are more deeply etched (Fig. 8.8).

Etching of quartz grains in B horizons takes place in old soils in the humid environment of the southeastern United States (Howard and others, 1995). The

Figure 8.8 Mean maximum etching depth with soil profile depth and age, Baffin Island. *(Redrawn by J.S. Noller from Locke, 1986, Fig. 7, © 1986, by permission of Academic Press Inc., Orlando, Florida.)*

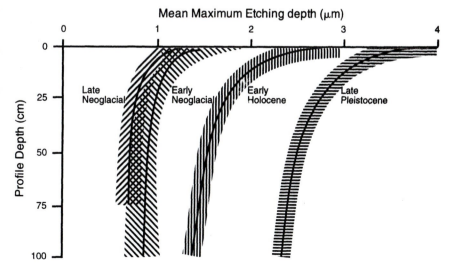

grains were studied using a scanning electron microscope. After 700 to 1600 ky of weathering, etching and pitting have been extensive enough that the grain shape is due mainly to chemical weathering. With more time the pits become deeper and interconnected, and by 13 Ma some pits extend through the grains.

Many of the above methods, as well as other reviewed in Birkeland and Noller (1998), have been used to estimate ages of or correlate deposits in a variety of environments (moraines, alluvial fans, etc.). A major hurdle has been to combine the data into a single value so that sites from various deposits can be compared. Berry (1994) suggests converting the measurements to z-scores:

$$z\text{-score}_i = \frac{x_i - x_m}{SD}$$

where x_i is the measurement of a given weathering parameter at a site (e.g., percentage weathered granitic boulders), x_m is the mean value of that parameter for all sites being compared (includes all ages), and SD is the standard deviation of the measurements around the mean. The z-scores for the different weathering parameters (e.g., percentage weathered, percentage pitted) are averaged to summarize weathering at each site, and then they are compared graphically with those at other sites. Berry (1994) demonstrates that deposits of 20 and 140 ka are readily distinguished by the z-score method.

■ Chronosequences and Chronofunctions in Different Climates

Many of the prominent properties of soil profiles require different amounts of time. Some, like organic matter content, form rather rapidly, whereas others, like high clay contents, take a much longer time. Perhaps surprisingly, many features of desert soils form relatively rapidly. These relations are of considerable interest to pedologists who work on soil-forming processes and to geomorphologists who might use the data to estimate ages of deposits or surfaces.

Classification of Chronosequences

Vreeken (1975) recognizes four kinds of chronosequences, based on the relationship between soils and deposits. Postincisive chronosequences are those in which there is a sequence of deposits of different age, and the soils related to each deposit have formed from the end of the time of deposition to the present. In contrast, a preincisive chronosequence is a sequence in which soil began to form on a particular deposit, but subsequent burial of the soil took place at different places on the surface at different times. A time-transgressive chronosequence without historical overlap is a vertical stacking of sediments and buried soils, so that the latter record times of nondeposition. Finally, a time-transgressive chronosequence with historical overlap is the most complicated sequence, and incorporates aspects of the other three. Each kind of chronosequence has its particular attributes for study of soil development with time. Vreeken argues that the time-transgressive sequence with historical overlap is the most valuable for soil chronosequence studies, but it could be argued that it is not always a very pure chronosequence because some properties might be as much related to postburial diagenesis or to slope position as they are to time.

Although Vreeken (1975) rightly notes the many problems with postincisive chronosequences, they are the most common (e.g., Stevens and Walker, 1970), and we have the most data on them, so most of the examples given here are of that kind. In a chronosequence we study soils related to landscapes of different ages, then make plots relating field or laboratory data to soil age (Fig. 8.9). In a study of this kind, the described sites must be carefully chosen to avoid subsequent erosion or deposition. In this example the data are plotted as either a linear function (solid line) or a logarithmic or power function (dashed line). If the ages of the older soils in the sequence are not known, one might use the method of Colman and Pierce (1981), who did this for rinds, to assign minimum and maximum ages to the soils. Alternatively, the linear function could have its origin in constant eolian influx with minimum subsequent weathering, whereas the logarithmic or power function is common to many weathering-related phenomena, rubification being a common one (Reheis and others, 1989).

Some workers are skeptical of the chronosequence approach. Daniels and others (1971) make the point well that older soils in a chronosequence may not have gone through the same history as younger soils; for example, climate could have changed or water table

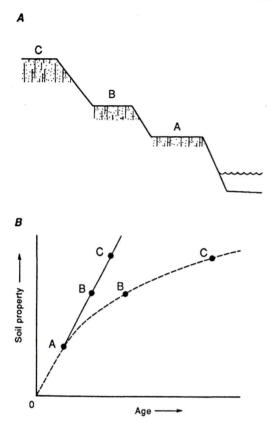

Figure 8.9 (A) Example of a postincisive soil chronosequence (soils A, B, and C) formed from river-terrace deposits of different ages. (B) Data for soils A, B, and C plotted against river-terrace age. Data could be PDI, g clay/cm²/profile, or other properties. A linear function (solid line) might define a minimum age relation or a property controlled by constant eolian influx, whereas a logarithmic or power function (dashed line) seems best for many soil properties versus age, and might define a maximum age relation (Colman and Pierce, 1981).

many environments, and that by the combined study of soils, surficial geology, and paleoclimate indicators, soil geomorphologists have made considerable progress in explaining soil formation in the context of Quaternary time and events (see supporting statement by Yaalon, 1983).

Landscape Evolution versus Soil Formation

The interval over which a chronosequence can be traced back in time is related to how long the landscape can survive. For example, consider a series of alluvial fans in a desert basin (Fig. 8.10). The younger soils are formed on relatively flat and stable surfaces, as the surfaces undergo change from their initial relatively rough bar-and-swale topography to a smooth desert pavement. Still older surfaces might be undergoing slight dissection, sufficient to erode the upper parts of the soils in places on the surface. Finally, the ballena stage can be reached in which the landscape is rolling, all original depositional surfaces are gone, and the soils are highly eroded. Obviously, in this case soil development will progress until erosion is more rapid than soil development; the apparent development then reverses as the soil thins, the regressive mode of Johnson and Watson-Stegner (1987). The time required to attain the ballena landform varies from place to place, as does the time to eradicate moraine or marine-terrace forms. Erosion of landscapes and soils in the western United States desert could be indicated by many surface clasts with attached carbonate rinds (clasts from the former Bk or K horizons).

Areas vary in the availability of sufficient landforms and deposits for chronosequence studies. The western United States and New Zealand are examples of areas with a wide variety of deposits and landforms (tills, alluvial fans, fluvial and marine terraces, sand dunes, loess, playa lakes, etc.) of many ages. Some could date to 1 Ma or more. In contrast, in the Great Lakes area, many deposits and landforms are present, but many are last glacial to postglacial in age. Finally, although in the southeastern United States the opportunity for chronosequence studies is mainly restricted to fluvial terraces, these make up a smaller proportion of the landscape than they do in the western United States, and some landforms and soils of great age (>10 my) are recognized and part of chronosequence studies.

history could differ. Later, Daniels and Hammer (1992) wrote that using soils to help date deposits has many weaknesses, and that in explaining soil formation, microenvironment and parent material are more important than duration of weathering. This sounds similar to the conclusion of Hunt and Sokoloff (1950) in their study of deep, red soils in the western United States. They concluded that time was the least important of all the factors that controlled formation of those soils, and opted for paleoclimate as the main factor. I argue that time is a major factor in soil formation in

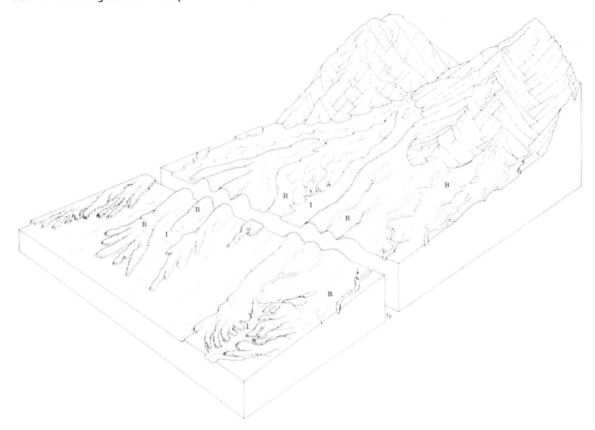

Figure 8.10 Diagram of the morphology of alluvial-fan river terraces, characteristic of those at the fronts of mountain ranges in the western United States. The youngest terraces (I) are connected to the drainages and little dissected. The next set of older surfaces are remnants (R) isolated from the parent stream, and have undergone slight dissection as shown by the drainage pattern on the surfaces. As erosion proceeds, all original topography is lost and eroded ridges (B, ballenas) are all that remain. *(Reprinted from Peterson, 1981, Fig. 5.)*

Chronofunctions

Many studies have been devoted to analyzing the shape of soil property vs age plots. Bockheim (1980a) produced a useful review using selected properties of chronosequences for 27 areas from the literature, and correlated properties with time using regression techniques and three models to produce chronofunctions:

$$Y = a + bX,$$
$$Y = a + (b \log X), \text{ and}$$
$$\log Y = a + (b \log X),$$

where Y is the soil property and X is time. Correlation coefficients were calculated, as were their statistical significance. The second model, $Y = a + (b \log X)$,

yielded the highest correlation coefficients for most soil properties, and 85% of these coefficients were statistically significant. This technique has been used by many subsequent workers. Switzer and others (1988) discuss statistical methods useful in making such plots.

Because the chronosequences varied in their parent materials and climate, Bockheim also was able to demonstrate the influence of these factors on soil properties. Although more data for more areas are needed, for the chronosequences he studied he showed that (1) rates of decrease in pH and in base saturation are similar, (2) the rate of increase in B-horizon clay content and in solum thickness is positively correlated with the clay content of the parent material, (3) rates of increases of solum thickness, depth of oxidation, salt content, and B-horizon clay content are positively correlated with mean annual temperature, and (4) the rate of change in the C:N ratio does not seem to correlate with either climate or parent material. To summarize, Bockheim's analysis is valid not only for the derivation of chronofunctions, but also for providing insight into the influence of the other factors.

The latest in-depth review of chronofunctions is that of Schaetzl and others (1994). They list some of the major shortcomings of such studies:

1. Mechanical problems, such as were a sufficiently large number of surfaces available for study, or were there enough numerical dates.
2. The site-specific nature of many studies. For example, do we know enough about the variation across a specific surface (such as parent material, or influx of eolian fines) to be assured that the site(s) picked for study are representative. How many sites should be studied per surface (e.g., J.B.J. Harrison and others, 1990). In defense of many of these studies, one has to appreciate the great time and effort that both field (especially hand-dug hole) and laboratory studies require. Most workers I know take great care in selecting sites.
3. The misapplication of soil chronofunctions for estimating ages of soils for which numerical dates are not available. I still feel this is a valuable exercise as long as one maintains a broad time frame.
4. The entire realm of how one goes about chronofunction curve fitting. They add hyperbolic and polynomial or nonlinear functions to the list of functions used by Bockheim (1980a). They make the point that the function used should not only fit the data, but also provide insight into pedological processes. They go on to say the use of selecting the highest r^2 value in helping to pick the correct curve is not always justified. For example, some pedologic systems are two phase, such as the threshold recognized by McFadden and Weldon (1987), or the leaching of carbonate prior to the translocation of clay (Muhs, 1984), but some regression models with a high r^2 value might obscure these relationships. Finally, they look in detail as to whether or not forcing the data set through 0 at time = 0 yr is warranted.

I take the view that we are not to the point where the chronofunctions are all that precise. Surely they would change if we doubled the number of surfaces in a particular area, double or triple the number of soil sites/surface, and were able to get numerical ages for all surfaces. Hence, in this book my intention is to present data to show broad, overall trends that few would argue with.

Thresholds

Not all soils develop over time at an even pace, as depicted by a logarithmic or power curve with a high r^2. Instead, the depiction of some properties might be a curve with a step or series of steps. Muhs (1984) suggests that some of these steps could signal the crossing of a threshold, much like the threshold concept of geomorphologists. He defines the pedologic threshold thusly (p. 100): "A pedologic threshold is a limit of soil morphologic stability that is exceeded either by intrinsic change of the soil morphology, chemistry, or mineralogy, or by the subtle but progressive change in one of the external soil-forming factors." Because extrinsic thresholds have to do with a change in factors other than time, the discussion here focuses on intrinsic thresholds.

Muhs (1984) presents several examples of intrinsic thresholds. One is clay movement in a carbonate-rich parent material. For much time early in the evolution of a soil, the clays are flocculated and do not migrate downward. However, once the carbonate is leached, an intrinsic threshold is crossed and clay migration can take place. A second is the influence of Na^+ on clay movement. Initially the exchangeable Na^+ content could be low and clay migration slow. With increasing Na^+ in the soil system, introduced from atmospheric sources, clays are dispersed and migrate rapidly. In time, however, the Na^+ content is sufficiently high, as is the clay content, that pores are plugged and clays are flocculated; the result is a decrease in clay migration. Both of these examples would have clay vs time curves with steps.

McFadden and Weldon (1987) report on a possible intrinsic threshold in southern California. There Fe_d builds up slowly in a linear fashion as silt and clay are added to the soil from atmospheric sources. In time the accumulation of the fines so alters the internal soil environment that the accumulation of pedogenic Fe_d and clay is greatly accelerated. We need to recognize such thresholds, and the key to this is reliable numerical ages for soils close to one another in age. In contrast, if soils are far apart in age, regression analysis to produce chronofunctions could produce smooth curves and obscure the evidence of thresholds.

Data Presentation

Soil data for chrono- and toposequences in this book will be portrayed several ways (Fig. 1.8) and, unfortunately, there is no uniformly accepted way. Generally most workers report color, clay and pedogenic Fe contents, and perhaps clay mineralogy, so these data will be used and compared throughout. In nonleaching environments, carbonate, gypsum, and more-soluble-salt accumulations are added to the discussion. In a few instances, data on parent materials or total chemistry are included. One could add a comparison based on the rate of soil deepening, but individual workers are not always in agreement as to what is pedogenic oxidation vs slightly oxidized parent material. Furthermore, it has been shown that the rate of deepening drastically decreases with time; for some California and Spanish soils the rate decreases from 0.1 mm/year at 10 ka to 0.001 mm/year at 1 Ma (Dorronsoro and Alonso, 1994). Hence, for very old chronosequences, comparing properties closer to the surface might be a better strategy.

The focus in the discussion to follow will be on the horizons of accumulation, such as pedogenic clay, Fe, carbonate, etc. Color hue (sometimes also chroma and value) is given as well as the contents of various properties, with the understanding that the parent material values are generally less. Where Bt is used, the horizon probably meets the requirements of the argillic horizon. As regards color, if the dry color is given it will be used. Furthermore, the color of the parent material is either the same hue as the horizon of interest, or a hue that is less red. For example, if the Bw is 10YR, the parent material hue is 10YR, 2.5Y, or even 5Y.

Climatic Gradient of the Soil Chronosequences

The climates under which the soils discussed here presently exist encompass the main ones of the world (Fig. 8.11). The range in mean annual precipitation (MAP) is 3 to >300 cm, and the range in mean annual temperature (MAT) is ca. ≤20 to +25°C (hereafter only the values will be given, and cm and °C assumed unless otherwise indicated). Although it would be better to use some combination of MAP and MAT, such as the leaching index of Arkley (1963) (Table 10.1), these index values have not been calculated. Instead, for each area the present values will be given.

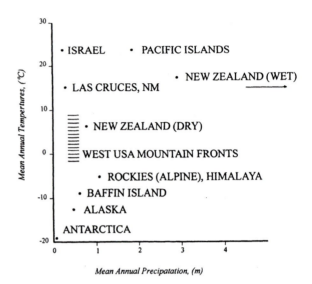

Figure 8.11 Climates for many of the chronosequences described in this chapter.

Paleoclimate will be considered only in a general way, when the author discusses it. For the terrestrial impact of Quaternary paleoclimate, consult review articles, such as those by Wright (1984) and Porter (1989).

This review will consider the changes in soil properties with time along a climatic gradient. Ugolini (1986) has done this for the average soil in the northern polar region, and Bockheim and Ugolini (1990) have repeated this for the southern polar region. They make the point well that many pedogenic processes change along such a climatic gradient, so we can also speak of a pedogenic gradient. Taking the southern polar region as an example (Fig. 8.12), the climatic gradient is MAP 36 and MAT −39 at the southernmost end, to MAP 67–144 and MAT 5–7 at the northern end. Key pedogenic processes vary along the gradient. The main ones in the cold, dry area are the formation of desert pavement at the surface, and salinization/alkalization (accumulation of soluble salts) and some rubification in the subsurface. In contrast, the main processes at the northern end of their study area are organic matter accumulation at the surface, rubification in the subsurface, and overall podzolization (formation of Spodosols). The review here is selective and will continue the climatic and pedogenic gradients of Bockheim and Ugolini into warmer and wetter areas, and add time as a major component.

Figure 8.12 Changes in pedogenic processes along a latitudinal transect in the southern circumpolar region. *(Reprinted from Bockheim and Ugolini, 1990, Fig. 3.)*

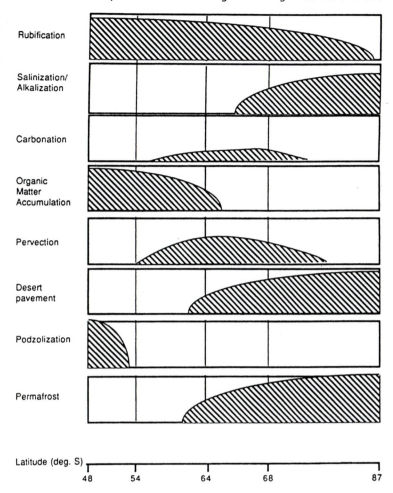

■ Chronosequences in Cold Polar Regions and High Mountains

Soil chronosequences form at increasingly more rapid rates in going from the the Antarctic cold desert to the northern polar desert (Baffin Island) to several alpine areas.

The slowest pedogenic rates in the world are found in interior Antarctica. Campbell and Claridge (1987, 1992) report that horizon development is weak, with 2.5Y–10YR hue in 0.5 Ma, distinct horizonation, close to the same colors, and some disintegration of clasts in the soil takes 1.5–2.1 Ma, and the combination of very distinct horizonation and colors as red as 2.5YR 3/6 takes 3.5 Ma or more. At the latter time Fe_d is usually about 1%, but in some soils it is 4–5%. Over the

same time span salt accumulation can increase from 0.01 to 10 g/cm^2.

The southernmost and coldest chronosequence (MAP 4; MAT −39) is that associated with moraines of the Beardmore Glacier (Bockheim, 1990; Bockheim and others, 1990). A horizons are not found in the unvegetated environment. Holocene soils show faint oxidation at most. The most rubified soil horizon (surface Bw) is 7.5YR, and that is achieved by 250 ka; Fe_d reaches a maximum of 0.4%. In contrast, soils formed on deposits of O-isotope stages 2 and 6 are 10YR, and Bw and Bz horizons are common only in the older soils. Taking the Transarctic Mountains as a whole, soils formed from sandstone and dolerite are sandy, translocation of fines is not detected (silt and clay contents are <5%), and soil features increase with time to reach the following values by 250 ka: oxidation of 20–30 cm with included clasts shattered by salt weathering, average electrical conductivity (EC; a measure of relative salt content) near 7 dS/m, 2–3 g/cm^2 of total salts to 70 cm depth, and the salt forms a strongly cemented pan (morphological stage III or IV of Harden and others, 1991; Appendix 1, Table A1.8).

Bockheim (1997) recognizes a variation in salt concentration with time inland from the Antarctic coast. Most of the salt is of atmospheric origin. The coastal soils are in the subxerous climate zone, there is enough moisture for ice-cemented permafrost to form, and Na and Cl are the main water-soluble ions. Inland sites are sufficiently dry that the permafrost is dry. The next climate zone inland is xerous, and the main enriched ions are Na and SO_4. The driest zone is ultraxerous, it is the farthest inland zone, and the main enriched ions are Na and NO_3. Comparing 10-ky-old soils, the ultraxerous soils have retained 91% of the salts introduced by precipitation, but the retention is only 35 and 15% for the xerous and subxerous soils, respectively. Salt minerals recognized are tachyhydrite, thenardite, soda nitre, halite, and gypsum. The salts accumulate at a linear rate over the last 250 ky:

$$y = 8.292x + 102.878 \quad \text{(xerous zone)}$$
$$y = 3.719x + 288.897 \quad \text{(ultraxerous zone)}$$

where y is the total salts (mg/cm^2) to 70 cm, and x is the age in ka. These equations show that the intermediate-climate soils accumulate salts about twice as rapidly as those in the driest climate, whereas Bockheim (1997) shows that the coastal soils have enough leaching that long-term accumulation is nil.

Salt accumulation can be used to help date deposits associated with the above soils. One Pliocene soil in the xerous zone has 10.1 g/cm^2 salt to 70 cm depth. Dividing by a conservative accumulation rate gives a soil-derived age of 2.8 My, well within the geologically determined age range of >2.2 and <3.5 my.

Soils formed from till on Baffin Island, Canada, are examples of development in a polar desert (MAP 30–34; MAT −8 to −11) (Birkeland, 1978; Bockheim, 1979; Evans and Cameron, 1979). Although most of the soil chronosequences were originally thought to extend back to ca. 100 ka, recent cosmogenic dating puts the ages of the older moraines as no more than ca. 20 ka (Marsella and others, 1996). The sparse lichen cover results in a thin A horizon with

6% or less organic carbon. The soils are sandy with clay contents of <5% and silt of <30%, with a silt maxima in the subsurface, suggesting at least some translocation. The tills are 2.5–5Y and oxidation extends slowly with depth; by 100 ka, 10YR oxidation typical of cambic horizons commonly extends to ca. 20 cm (Birkeland, 1978). The reddest color is 7.5YR 4/6 (Evans and Cameron, 1979). The average EC for one soil is 0.02 S/m, and Fe_d reaches a maximum concentration of 0.79% (Bockheim, 1979).

An unusual silt accumulation characterizes soil formation in western Spitzbergen (MAP >39, MAT −6; Forman and Miller, 1984). Plant cover is <10% in this extreme environment, and parent gravel clasts are mainly limestone (80–90%) and calcareous mudstone. It was this study that produced the silt-accumulation stages, as well as the subdivision of carbonate stages I and II for the gravelly facies (Appendix 1).

The latter soils all have similar A/B/k morphologies, with the main time-related changes being silt accumulation, 10YR 5/3 for all ages and no rubification trend, carbonate accumulation, and dissolution of near-surface clasts. Although even the youngest soils have stage 6 silt morphology, the basal depth of soil with stages 5 + 6 morphology increases with time. On a mass basis, about 2000 g/cm^2 of silt has accumulated to 130-cm depth, an amount of silt over an order of magnitude greater than that for clay in the chronosequences compared in Fig. 8.35. As the silt accumulated, it takes up more of the volume of the soil, so gravel percentage halves over time. Carbonate morphology reaches a maximum of stage IIb and these are found in soils as young as 12 ka. In general, more of the clasts have both thicker and more extensive carbonate coats with time. The carbonate clasts have dissolution morphologies (e.g., pits, quartz veins in relief), but there are no data on the thickness of a dissolution zone with time. The favored hypothesis is that the limestone dissolves, releasing both silt (insolubles in the limestone) and dissolved carbonate. Vertically moving waters translocate the silt to eventually form a B/ horizon and, at the depth of annual wetting, carbonate precipitates on clast bottoms. One could try a mass-balance approach with these soils to determine if the amount of silt accumulated matches that released by weathering.

The above suggests that the polar desert soils form faster than the cold desert soils. Bockheim's (1980b) review shows that if solum thickness is used as a measure of relative development rate, the difference in rate is at least one order of magnitude. In both deserts the origin of the salts seems to be marine aerosols, and the buildup in the cold deserts is at least one order of magnitude faster than in the polar deserts. The difference in rate could be related to the influx of salt, or perhaps polar desert soils periodically have some of the salts leached during high precipitation events, events that are unlikely in Antarctica.

Soil chronosequences for tills in various alpine areas in high mountains display varying rates of development. One for the Colorado Front Range has reasonable numerical age control (MAP 102, MAT -4; Birkeland and others, 1987). Alpine tundra vegetation ensures the development of A horizons that are 15 cm thick with 13% organic carbon on 10- to 12-ka moraines. A/Cox profiles 3 cm thick have formed on 0.1- to 0.4-ka tills. By 1–2 ka the soils are still A/2.5Y Cox profile form. The youngest deposits with a Bw horizon are 3–5 ka, and the latter have 10YR 5.5/3 color to 49 cm. A/Bt/Cox profiles are present on some 10- to 12-ka moraines, the maximum color-depth combination is 7.5YR 4/4 to 39 cm with 10–7.5YR hues to greater depth, the maximum clay content is 14%, and the maximum Fe_d is 1%. Harden (1990) points out that the field determination of rubification could be a better measure of soil age than laboratory determined Fe_d. The surface horizons of the older soils have a low volume of gravel and high contents of silt and clay, suggesting an admixture of eolian fines. Dixon (1986, 1991) concurs with this, based on clay mineralogy and total chemical analyses. Animal activity helps mix this material into the underlying sandy till (Burns, 1979; Litaor and others, 1996), so we call this mixed loess. Our experience in the Rocky Mountains has been that the development of these alpine soils is equal to, or greater than, that for soils twice as old formed on tills at mountain fronts in a drier, warmer environment under sagebrush (Shroba and Birkeland, 1983).

Soils on moraines of different ages in various alpine environments have been compared with each other and with the Baffin soils to determine relative development rates (Birkeland and others, 1989). The alpine areas are the Wind River Range, Wyoming, the Sierra Nevada, California, the Southern Alps, New Zealand, and the Khumbu Glacier area of Mt. Everest (climatic data in Fig. 8.13). The Wind River Range chronosequence is similar to the Colorado Front Range (above), and the Sierra Nevada chronosequence also has an eolian component but a less oxidized (10YR 5/3)

Bw horizon in 10 ka. In contrast, the older soils of the New Zealand and Mt. Everest chronosequences have formed in loess/till, profile forms in 10 ka are A/Bw/Cox, and the most oxidized B-horizon colors are 7.5YR 4/8 (New Zealand) and 5YR 3/4 (Mt. Everest).

These soils can be ranked in several ways: field properties, depth of oxidation denoted by a particular munsell color, or various laboratory properties. The way chosen was soil chemistry, as key properties vary in concert with the morphology (Birkeland and others, 1989). With age, the Sierra Nevada and Wind River data have some reversals when displayed graphically, whereas the New Zealand and Mt. Everest data have trends that progress in the predictable manner with age. By about 10 ky of soil formation, the New Zealand soils accumulate the most pedogenic Fe and Al and display the greatest depletion of P, and soils on Baffin Island have the least accumulation and depletion (Fig. 8.13). A good way to portray and compare profile trends is by the accumulation index for Fe_o and the depletion index for P_a (acid-extractable phosphorus) (Fig. 8.14). With age, the Fe_o accumulation (Fe_d has a similar trend; see Rodbell, 1990) matches that of P_a depletion; that is, high accumulation is matched by high depletion. The approximate ranking of of the areas on these indices is

Southern Alps $>$ Himalaya (Mt. Everest) $=$ Wind River Range $>$ Sierra Nevada $>$ Baffin Island.

Because Baffin Island soils form faster than those in Antarctica, an approximate trend for the latter is included.

Another way to rank some of the soils is by the half-life of mica. Mica is a primary phyllosilicate in many soils and, on weathering, converts sequentially to a variety of clay minerals (Fig. 8.38). A simple measure of weathering would be to follow the progressive reduction in the mica peak in an X-ray diffractogram, and determine the approximate age at which the peak is one-half that of the parent-material peak, here called the half life of mica. The half life of mica is about 5 ky in the New Zealand alpine environment (Table 8.2), about the same rate as for the Mt. Everest region (unpublished work of the author). The latter rate seems to be at least twice as fast as that in an alpine environment in the Colorado Front Range (Shroba and Birkeland, 1983), but only half as fast as the rate for

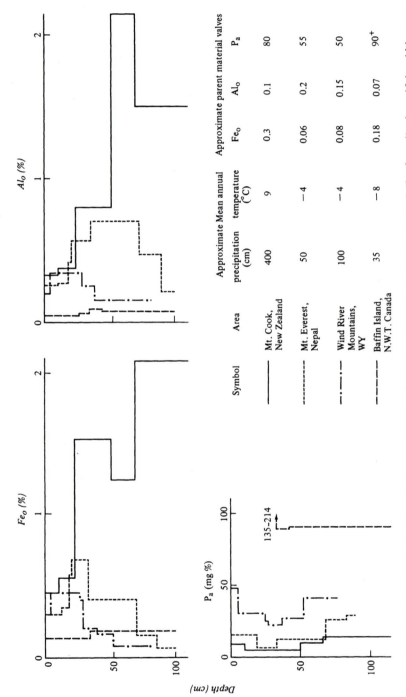

Figure 8.13 Vertical distribution of oxalate-extractable Fe and Al (Fe_o and Al_o) and acid-extractable P (P_a) for soils about 10-ka old in various alpine and arctic localities. (*Data from Birkeland and others, 1989.*)

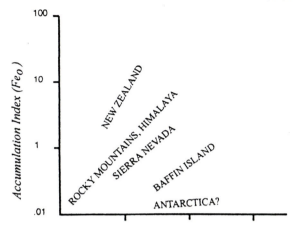

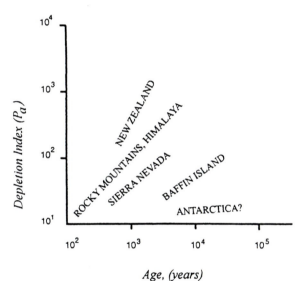

Figure 8.14 Generalized trends for accumulation and depletion values versus age for various alpine, arctic, and polar areas. Trend line for Antarctica is an approximation. *(Modified from Birkeland and others, 1989, Fig. 1.)*

an alpine area in the Wind River Range, Wyoming (Mahaney and Halvorson, 1986).

Longer mica half lives are reported for other areas. That for the humid environment of Missouri is 10 or several 10s ky (Guccione, 1985), that for sagebrush in the Wind River Range is about 100 ky (Mahaney and Halvorson, 1986), and that for the arctic tundra of Baffin Island is 100–200 ky (Isherwood,

1975). In one study of soils of Antarctica, mica not only does not alter, it increases with ages measured in My (Bockheim, 1982). This is caused by the physical breakdown from larger size fractions to clay size.

Degree of leaching and parent material seem responsible for many of the above trends in polar and alpine soils. Soils under the combination of relatively high precipitation and a significant surface loess layer (New Zealand and Mt. Everest) experience rapid alteration. The other soils in this comparitive study are sandy with only some having a slight loess influence; the low particle-size surface area results in slow alteration, and, within the latter group, the soils follow the precipitation gradient. Note that although some of the soils (Wyoming and Colorado) are frozen for about half the year, major leaching during snowmelt could be an important factor in explaining some of these trends.

Table 8.2 Clay Mineral Data for Soils in the Ben Ohau Range, New Zealand[a]

Age Estimated (years)	Horizon	Mica	Chlorite	Interlayered Mica-14 Å	Mica–Smectite
100	A	XXX[b]	XXX		
	B	XXX	XXX		
	C	XXX	XXX		
	Cu	XXX	XXX		
3,000	A	XX	XXX	XX	
	B	XX	XXX	XX	
	2C	XX	XXX	XX	
	2Cu	XX	XXX	X	
4,000	A	X	X	XX	XX
	E	X	X	XXX	XX
	B1	X	X	XXX	X
	B2	XX	XX	XXX	X
10,000	A			XXX	XX
	E		X	XXX	XX
	B		X	XXX	XX
	2B	XXX	XX	XX	X
	2C	XXX	XX	XXX	
	2Cu	XXX	XX	XX	

[a]Clay data courtesy of A.S. Campbell; age and soil data in Birkeland (1982, 1984).

[b]The number of Xs relates to the relative height of the peak, with XXX being highest, XX moderate, and X low; no X denotes absent.

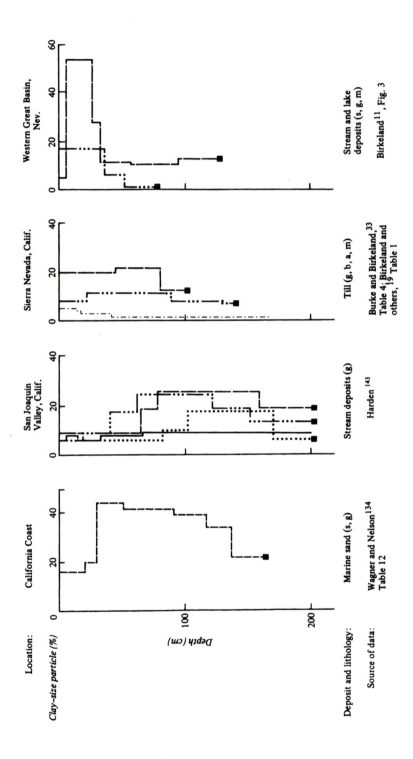

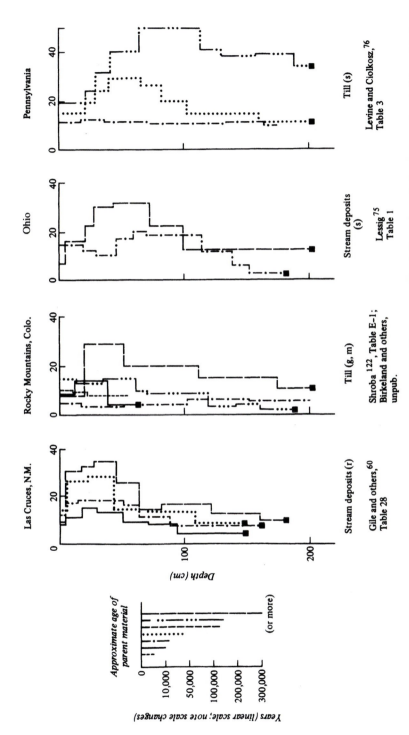

Figure 8.15 Variation in clay with depth for deposits of different ages in different areas. The indicated age is for the parent material, and that of the soil could be younger. Data sources are as follows: California coast, Wagner and Nelson (1961, Table 12); San Joaquin Valley, Harden (1987, Supplementary Table 2); Sierra Nevada, Burke and Birkeland (1979, Table 4) and Birkeland and others (1980, Table 1); Western Great Basin, Birkeland (1968, Fig. 3); Las Cruces, Gile and others (1981, Table 28); Rocky Mountains, Shroba (1977, Table E-1) and Shroba and Birkeland (1983, Figs. 8.3 and 8.6); Ohio, Lessig (1961, Table 1); Pennsylvania, Levine and Ciolkosz (1983, Table 3).

■ Noncalcic Soils in a Semiarid ⇒ Humid Transect

Key soil features vary with age along a climatic gradient from relative dry to wet. Here we will generally progress from selected places in the western United States to the eastern United States, then add an unusual chronosequence from the tropics.

Tills at Mountain Fronts: Western United States

Soils formed on moraine crests of tills of the major glaciations have been studied at numerous mountain fronts, from the Sierra Nevada on the west (MAP 40–45, MAT 7–8) to the Rocky Mountains on the east (MAP 40–90, MAT −1 to 6). Clay and Fe accumulation, as well as rubification, increase slowly across the region, perhaps because the soils are sandy and gravelly, and if a mixed-loess surface layer is present, it is thin (Fig. 8.15).

In general, soils on 20-ka tills in the Sierra Nevada (Burke and Birkeland, 1979; Shroba and Birkeland, 1983; Berry, 1994) have A/Cox and A/Bw/Cox profiles with a maximum oxidation of 10YR 5/4. The common clay curve with depth is one that displays a slight increase toward the surface, perhaps due to slight influx of eolian fines. Soils on 140-ka tills commonly are A/Bw/Cox profiles, but one has a Bt with 14% clay. Oxidation for the latter soil is similar to that of the younger soils (10YR 6/4), and the most oxidized of all the soils is 10YR 4.5/4. Fe_d commonly displays no trend with time. Berry (1994) measured pedogenic clay for three valleys and obtained these values (20-ka soil vs. 140-ka soil): 0.3 vs 1.0; 1.2 vs 1.4; 0.7 vs 3.4 g/cm^2. A still older soil (ca. 750 ka), supposedly on till (Birkeland and others, 1980), has a 145-cm-thick Bt, with maximum values of oxidation of 5YR 4/4, of Fe_d of 1.2%, and 40% clay.

Soils helped to solve an ongoing debate in the Sierra Nevada, that of the age of the Sherwin Till, the oldest diamicton that most workers agree is of glacial origin. At the till type locality a buried soil at the top of the till is buried by the 0.77-ka Bishop Tuff. Using the above rates of soil formation, the soil was considered to have formed for about 100 ky prior to burial by the tuff (Birkeland and others, 1980). Interestingly, cosmogenic dating of the same soil gave its age at 53–67 ky (Nishiizumi and others, 1989).

Soils on tills in the Rocky Mountains form at the same rate as those in the Sierra Nevada, or at a slightly more rapid rate (Shroba and Birkeland, 1983). Soils on 20-ka tills have A/Cox, A/Bw/Cox, or A/Bt/Cox profiles (Fig. 8.15). The Bt-horizons have an absolute clay increase (vs parent material) of 5%, and they are more common in sagebrush vegetation than in forests. For 140-ka soils, maximum oxidation is 10YR 5–6/3–4, and maximum clay is 15 to less than 20%, for an absolute clay increase of 5–13%. Note on the plot (Fig. 8.15) that some 10-ka soils have a clay maxima similar to that of 140-ka soils. In one study in sagebrush, Fe_d in the B horizon of the older soil is 0.6%, triple the value for the younger soil (Swanson, 1985). A soil on 400- to 500-ka till has a maximum of 28% clay, and oxidation of 7.5YR 6/6 (Shroba, 1977). Shroba (1984) gives these average clay-accumulation values (g/cm^2) for the different age soils: 0.4 for 10 ka, 0.6 for 20 ka, 4.5 for 140 ka, and 15.8 for 0.4–0.5 Ma. In contrast, Colman and Pierce (1986) get even higher rates for the McCall, Idaho area (MAP 65; MAT 4). The approximate amounts are 1 g in 20 ka, 9 g in 60 ka, and 15 g in 140 ka. There are many similarities between the Rocky Mountains and the Sierra Nevada, but Bt horizons seem to form faster in the former area.

Total chemical analysis data show trends not too different from the soil morphology and laboratory data (Taylor and Blum, 1995). Here are compared moraines younger than 11 ka in the alpine tundra (MAP 50–70, MAT −4), and moraines equal to and older than 21 ka in the mountain front sagebrush (MAP 23, MAT 2) (Table 8.3). For the 0- to 21-ka soils Ti enrichment is about 20%, the base cations (Ca, Na, K, Mg) are depleted from 10 to 20% in the younger soils (0.4 and 2 ka) and 10 to 40% in the older soils (11 and 21 ka), and, in general, the older the soil the deeper the alteration. However, the 11-ka soil in the alpine is remarkably similar to the soil twice as old at the mountain front. The trends continue for the 115-ka soil (1.5–2.0 times enrichment, 50% depletion) and the 297-ka soils. As with so many soil data, weathering rates are greatest with the youngest soils, and decrease with time. Although there are reversals, most of the data on element loss show a greater loss with age of soil; Ca shows the most consistent loss, whereas data for Mg is the most inconsistent. Before the above can be accepted, however, the influence of (1) eolian additions, (2) erosion, and (3) depth needed to obtain unaltered samples on the trends has to be evaluated (discussion by Dahms and others, 1997, with reply by Taylor and Blum, 1997).

An attempt has been made to use soils to identify the presence of O-isotope stage 4 till in the Wind River Range (Hall and Shroba, 1995). The range is the type locality for both the Pinedale (O-isotope stage 2) and Bull Lake (O-isotope stage 6) tills, and they hypothesized that the youngest Bull Lake moraine could have been laid down during O-isotope stage 4. Their argument is that data on soil properties for the moraine of interest (moraine V) are between those for the Bull Lake and Pinedale moraines, yet closer to the latter (Table 8.4), all in accord with a stage 4 age assignment. This does not agree with the cosmogenic isotope dating of one boulder on moraine V, however, which suggests a Bull Lake age (138 ka; Gosse and others, 1995).

Soils on Dune Sands

Soils formed in well-sorted dune sand sometimes progress along a pathway different from that of other soils. At the dry end is a chronosequence in northwestern Texas (MAP 44, MAT 14), in which some of the clay might have an atmospheric origin (Gile, 1979; see Fig. 5.3). Sand dunes less than 0.1 ka have no soil. Soils 0.1–4 ka are Ustipsamments (A/7.5–5YR B) with discontinuous to continuous, thin (up to 3-mm-thick), clay-enriched bands, beginning with a few bands and reaching a maximum of six. Such bands are called lamellae, and the horizons could be given

Table 8.4 Comparison of Soil Data for Moraines of Different Ages, Wind River Mountains, Wyoming[a]

	Estimated Age		
	140 ka	**Intermediate**	**20 ka**
A + B thickness (cm)	>125	58	38
B thickness (cm)	>78	45	28
PDI	39	28	25
Clay content (%)			
Profile maximum	19	11	10
Profile weighted mean	15	6	6
Accumulation index	1452	324	122
$CaCO_3$ content (%)			
Profile maximum	12	1	1
Profile weighted mean	7	0.5	1
Accumulation index	809	58	31
Number of profiles	3	3	6

From Hall and Shroba (1995, Table 5).

[a]All data are averages, except the maximum values which are for individual profiles.

the informal name, lamellar B. By 4–7 ka the bands are more continuous, thicker (up to 5 mm), and B-horizon hue is 5YR. Enough clay has accumulated between the bands by 10 ka to form an 5–2.5YR argillic horizon and a Haplustalf; faint bands can still be recognized, however. The bands form by translocation and are not a surficial deposit, as in many places the bands when viewed from the side have an anastomosing pattern.

In contrast to the above, rapid soil development (Btj/stage II Bk in 3–4 ka) in the Chaco dune field of New Mexico did not result in lamellae formation (Wells and others, 1990).

Such sequences are not restricted to dry climates. A somewhat similar development sequence has been described in Indiana (Miles and Franzmeier, 1981). Soils of 3.5 ka lack bands, 13-ka soils have deep bands and some have an argillic horizon, and 20-ka soils have argillic horizons (10–20% clay) and bands. The bands have two or three times more clay and up to two times more Fe_d, relative to the adjacent materials, and with time they become thicker, redder (7.5YR), and occur closer to the surface. Clay in this case is ascribed to feldspar weathering.

Table 8.3 Loss of Cations by Weathering in Soils Formed from Tills of Different Ages, Wind River Mountains, Wyoming (g/cm²/profile)

Estimated Age (ka)	Ca	Na	K	Mg	Sum
0.4	1.00	1.26	0.51	0.37	0.31
2	1.60	2.81	1.80	0.02	0.62
11	1.80	5.01	8.41	0.19	1.54
21	2.29	7.54	5.63	0.12	1.56
115	5.33	3.56	4.89	1.58	1.54
>297	17.27	9.77	8.88	0.78	3.67

From Taylor and Blum (1995, Table 1) with moles converted to grams.

River-Terrace Chronosequences: Western United States

The Great Valley of California has a chronosequence generally repeated from valley to valley for those rivers draining the Sierra Nevada (see Fig. 2.4). The most detailed work is by Harden (1987) for a chronosequence that ranges from 0.2 ka to 3 Ma (MAP 30, MAT 16). All Holocene soils are A/Cox profiles with 10YR hue. The youngest soil with a Bt horizon is 40 ka, the hue is still 10YR, and clay and Fe_d contents are 17 and 0.5%. By 130 ka the Bt is 7.5YR, clay is 25%, and Fe_d is 1.4%. Redness and clay contents of the Bt horizons continue to increase with time, being 5YR (25% clay, 0.8% Fe_d) by 330 ka, and 2.5YR (38% clay, 2.2% Fe_d) by 600 ka. Finally, after 3 My of soil development, the Bt horizon is over 7.5 m thick, has 10R hue, 63% clay, and 3.4% Fe_d. Over this time span profile mass of clay and Fe_d show significant increases (to about 200 and 7 g, respectively) as pH decreases from near 7 to 4–5.

Clay mineralogy and total chemistry change with time in the above chronsequence (Harden, 1988). There is an original mixed-clay mineral assemblage, and with time kaolinite increases in content as mica and vermiculite decrease. Using a percentage loss equation for the silt + clay fraction, by 3 Ma there has been about an 80% loss of Ca and Mg, 60% loss of Na, and 40% loss each of Fe and Al.

In contrast to the above, soils in areas of higher dust influx form more quickly. McFadden and colleagues (McFadden, 1982; McFadden and Hendricks, 1985; McFadden and Weldon, 1987; McFadden, 1988) studied chronosequences across southern California from aridic to xeric moisture regimes. If we compare noncalcic soils in the wetter part of the transect (MAP 40–78, MAT 16) relative to those in the Great Valley, Bt horizons form more rapidly. After about 10–15 ka, Bt horizons are 1 m thick with 7.5YR hue, 10% clay, and 1.4% Fe_d. Thicknesses of the Bt horizons increase to 5 m by 300–700 ka, with 2.5YR or 10R hue, 39% clay, and 2.9% Fe_d. Over this time span the original mixed-clay mineral assemblage converts to one of kaolinite and vermiculite, with the former dominant in one drainage.

Marine Terraces of the California Coast

Marine terraces are common along the California coast, so at any one location there is a chronosequence,

and this can be combined with climate in going from the dry south to the wet north. Too much data are available to summarize easily, so I will select particular ages for comparison.

At the dry end is San Clemente Island (MAP 17; MAT 16), with older terraces near 1 Ma (Muhs, 1982). The youngest soil (2.7 ka) has formed from an alluvial fan deposit and has a Bt horizon with enough Na^+ to qualify as a natric horizon; clay content is 36% and Fe_d is 0.4%. Pre-Holocene soils less than 200 ka are Alfisols or Mollisols with 10YR Bt horizons, whereas soils older than 200 ka are either Vertisols or soils with vertic properties. The oldest soil has a 77-cm-thick Bt that rests on bedrock, with maximum values for clay of 70%, Fe_d of 1%, and hue of 7.5YR; this is the only soil in the chronosequence that is not 10YR. Profile mass of clay shows a significant linear relationship with time (Fig. 8.16), but Fe_d does not. Atmospheric additions from mainland California have greatly influenced the textural properties of these soils (Muhs, 1983).

The above change from Alfisol ⇒ Vertisol is used as an example of the crossing of an intrinsic threshold (Muhs, 1984). In time, mica alters to smectite. For the younger soils the profile average smectite/mica ratio is near 0.3, and it increases gradually with time. As the ratio approaches 0.45 the pedogenetic pathway shifts for one dominated by clay illuviation (Alfisol) to one dominated by pedoturbation (Vertisol).

In central California, the grass-covered coastal terraces provide for a chronosequence that goes back ca. 600 ka (Aniku and Singer, 1990; J.W. Harden and others, written communication, 1996; MAP 72; MAT 14). A/Bt/BC or C is the common profile form for all ages (E horizon locally present), Bt horizons are between 100 and 200 cm thick, with 10YR hues dominant at 105 ka, 7.5YR at 370 and 490 ka, and 7.5–2.5YR at 600 ka. Over this time span A + B horizon clay contents increase from profile maxima of 17 to 57%, while Fe_d increases from 0.7 to 4.3%; goethite is the dominant pedogenic Fe mineral. The progressive drop in pH from 7 to 5 is matched by a decline in base saturation from 75–85% to 20–40%.

Marine terrace soils in Northern California form a 3.9- to 240-ka chronosequence also under grassland in a wetter climate (MAP > 100, MAT 12–14; Merritts and others, 1991). Parent materials are eolian sand and silt/marine sand and pebbly sand. During initial tectonic uplift eolian nearbeach sediments are deposited on the terrace, but this is markedly slowed in

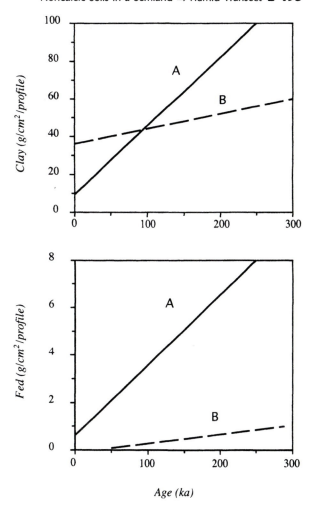

time when the surfaces are sufficiently uplifted above sea level. Nearby, eolian deposition is recognized by an increase in both silt and clay toward the surface (Burke and Carver, 1992), and surely this has implications toward assuming parent-material uniformity.

Morphology, clay, and Fe_d show significant change with time. Soils 3.9 ka are A/7.5YR Bs/C with generally <5% clay and 0.3–0.5% Fe_d in the A + B horizons. Soils 29 ka are A/10YR Bw/BC with generally <10% clay and 0.6–0.8% Fe_d. By 40 ka the soil has the same morphology as the latter, but clay has increased to the low 20%s, and Fe_d is 1.2–1.6%. The 124-ka soils have the same morphology as the latter, but now the clay is 18–43% and Fe_d is 1.1–1.9%. Finally, the 240-ka soil has a complex morphology,

A/Bw/10YR Bt/Bg/7.5YR Bs/BC, 34–46% clay and 1.4–4.6% Fe_d. Clay shows more reliable trends with time as redox conditions under a fluctuating water table cause various pedogenic Fe extracts to fluctuate. In situ weathering is considered to produce both the clay and the pedogenic Fe, hence some of the B horizons probably should be designated Bt.

Mass-balance analyses indicate trends in the various major elements with time (Merritts and others, 1992). Many of the major elements undergo losses over the time span, and these can be ranked: Si > Na > Ca > Mg, with some loses approaching 100% (e.g., Ca). In contrast, in this environment, Fe and Al contents remain near constant during pedogenesis; in the alteration of the original minerals, the original Fe converts to Fe oxyhydroxides, and the original Al to clay minerals.

Bockheim and others (1996) offer a comparison to the above with a somewhat similar chronosequence in southern Oregon (MAP 190, MAT 10). Soils up to 105 ka are Spodosols and those between 105 and 200 ka have a spodic/argillic horizon profile form, due to clay formation in situ as well as clay translocation. Soils >600 ka have argillic horizons masking spodic horizons as the early formed Spodosols convert to Ultisols. The oldest soil has a Bt horizon with 2.5YR hue, 39% clay, and 4.2% Fe_d. This sequence indicates what happens to Spodosols over large durations of time. As long as there is a parent mineral reserve of feldspars and other aluminosilicates, it seems logical to have the above sequence. In contrast, if weathering results in high quartz content in the silt and sand sizes, a Spodosol could persist for long times.

These coastal chronosequences can be compared on profile mass clay and Fe_d (Fig. 8.16). Both clay and Fe_d increase at fairly rapid rates in northern California; although not calculated or plotted, trend lines for central California and southern Oregon should be close to those for northern California. In contrast, the southern California clay and Fe_d increase at much lower rates and clay is present in high amounts, whereas Fe_d amounts are low.

These trends support the interpretations of the workers: in a humid weathering environment, clay and Fe_d should increase in concert, whereas in a dry environment with eolian influx, clay increases with time, but the lack of significant weathering might not release much Fe to recombine as pedogenic forms.

Chronosequences in Wetter Climates

Tills form parent materials for a soil chronosequence in Pennsylvania (Levine and Ciolkosz, 1983) (Fig. 8.16). Inceptisols with no accumulation of clay or Fe_d (mean = 0.7%) form in 15 ka, and all older soils are Ultisols. A prominent Bt has formed by 40 ka, and mean Fe_d has increased to 1.6%. The 140-ka soil has a Bt with over 40% clay and a mean Fe_d of 2.7%. Kaolinite shows a progressive increase in amount with time. Compared to the soils related to western U.S. tills, the younger soils are similar, but for the older soils those in Pensylvania exhibit much greater development. This was the study that provided the clay-accumulation index (Fig. 1.8), and values of the index

with time are 91–285 in 15 ka, 469–897 in 40 ka, and 1299–1387 in 140 ka.

River terraces provide the substrate for a soil chronosequence as old at 0.6–1.6 Ma in southwestern Washington (Dethier, 1987; MAP 125–150, MAT 11). Depth to unweathered material increases with age, being 0.3 m by 0.25 ka, 2.5 m by 65–170 ka, and >10 m by 0.6–1.6 Ma; this latter deposit has the unusual property of being so weathered that all clasts can be cut through with a shovel. The most oxidized hues are 10YR by 14 ka, 5YR by 24 ka, and 2.5YR, the maximum for the chronosequence, by 240–340 ka. The clay increase with time is 16% in 11 ka, 44% in 150 ka, and a maximum clay content of 61% for a 0.5-Ma soil. Bt-horizon Fe_d content increases from 1.5% at 11 ka to 2.2% at 150 ka and 4.1% in the oldest soil. Over this time span the subsurface CEC decreases from maxima of 24–20 meq/100 g as pH declines from 5.8 to 5.1. Kaolinite is the dominant clay mineral and two suites were identified; soils younger than 25 ka have kaolinite > smectite > vermiculite > illite = chlorite, whereas the three oldest soils have kaolinite > vermiculite >> smectite = chlorite = illite.

Total chemistry for the above soils gives the following mobility ranking:

$$Ca, Na, Mg > K > Si > Al > Fe, Ti, Zr$$

The first four elements are essentially depleted by 1 Ma, SiO_2 decreases over the time span of the chronosequence, and Al_2O_3 has only slight depletion because once released by weathering it goes to form clay minerals and pedogenic Al products. Fe_2O_3 shows little change with time, and the ratio Fe_d/total Fe_2O_3 increases from 0.1 to 0.35, suggesting conversion to pedogenic Fe oxyhydroxides, with some perhaps incorporated into clay-mineral lattices.

Soils in the Coastal Plain of the eastern United States follow two pathways, governed by the parent material (Markewich and Pavich, 1991). If the parent materials contain weatherable minerals the sequence is Entisol $\Rightarrow$ Inceptisol $\Rightarrow$ Alfisol $\Rightarrow$ Ultisol as clay builds up, and bases get leached from the system. If, on the other hand, the parent material is quartz sand, the pedogenic pathway is Entisol $\Rightarrow$ Inceptisol $\Rightarrow$ Spodosol. In this case, Spodosols are the end-product order. For contrast, Bockheim and others (1996) have Spodosol $\Rightarrow$ Ultisol in time in Oregon, with little of the above parent-material control.

Soils formed from marine and fluvial deposits in Maryland and Virginia (MAP 112, MAT 15) follow the first pathway, above, and display good chronosequence trends for the age span 30 ka to 1 Ma (Markewich and others, 1987) (Fig. 8.17). Although soils about 30 ka lack a Bt horizon, those between 60 and 125 ka have a 120-cm-thick Bt with 7.5YR hue, 22% clay, and 1.8% Fe_d. Values systematically increase with age, so that by 1 Ma the Bt is twice as thick, and has 2.5YR hue, 48% clay, and 3.5% Fe_d. This is an environment in which total $Fe_2O_3 + Al_2O_3$ increases as SiO_2 is leached, so the ratio of the former to the latter also systematically increases with time, and this ratio seems to be one of the more useful age indicators in the area.

In the inner Coastal Plain of Virginia (MAP 112, MAT 13) soil formation on alluvial deposits is thought to date to about 13 Ma (Howard and others, 1993) (Fig. 8.18). The 60- to 120-ka soil is a Hapludalf, similar to the same age soil above but with about 1.5 times more clay. Soils 0.7–1.6 Ma are Hapludults with 5YR Bts containing 42% clay. All older soils are Paleudults with E horizons, with a Bt hue and clay content of 2.5YR and 50%, respectively, by 3.5–5.3 Ma; soils 11 and 13 Ma have 10R Bts with 60% clay as well as plinthite, duripan, and ferricrete (Fe cementation). Mottling in the older soils, a feature common in other areas, is thought to reflect impeded drainage that results from the plugging of the horizon with high contents of clay. By about 2 Ma all original clay and nonclay minerals are depleted, and the generalized clay mineral abundances are kaolinite and Fe oxides > chloritized vermiculite and gibbsite.

Howard and others (1993) suggest that the above soil evolution involved two intrinsic thresholds (Fig. 8.18). The first took place at about 1 Ma when loss of most unstable minerals in the sand fraction was accompanied by formation of a stable clay-mineral suite of kaolinite and Fe- and Al-sesquioxides. The second threshold took place at about 10 Ma when pedogenic Fe increased markedly as heavy and opaque minerals declined, and kaolinite converted to gibbsite.

Soil formation in well-drained sandy soils on barrier islands in northern South Carolina (MAP 137, MAT 17) follow the Spodosol pathway (Markewich and others, 1986). The reason for this is the high quartz sand parent material with few weatherable minerals and little inherited clay. The result is a sequence of soils whose morphology changes little with time even though the age span is 200 ka to 3 Ma. More recent

work gives these regional values for Spodic horizon thicknesses (Markewich and Pavich, 1991): those <1 Ma are <3 m thick, those >2 Ma are >5 m, and one 2–3 Ma soil is 12 m thick. If age-dependent soil properties are being sought, the best parent material could be one with a high reserve of weatherable minerals.

The above Spodosol sequence is duplicated in the subtropical east coast of Australia near Brisbane (MAP 120–170, MAT 25). Quartz sand parent materials form six dune systems (Thompson, 1983; Thompson and Bowman, 1984). Both the E and Bh horizons progressively thicken with dune age, and the oldest soil has an E horizon 11 to >20 m thick. Dating is difficult, but one estimate for the oldest soil is 125 ka or older. Some of the thick B horizons are so colorful that they are tourist attractions.

Under appropriate conditions, Spodosols also form from parent materials high in weatherable minerals. Spodosolic properties are apparent in 230 years in acid soils (pH 4–5) on moraines in southern Norway (MAP 225, MAT 5; Mellor, 1985). In this time span a thin (25-cm) O/E/B profile has formed in which the Fe_d value of the B horizon is 1.5 times that of the E horizon and 2 times greater than that of the C horizon. Mica ⇒ hydrobiotite is the only clay alteration observed. Messer (1988) extends some of these data over a larger area.

Pedogenic Fe buildup is rapid in the heavily vegetated, wet and cool west coast of the South Island, New Zealand (Smith and Lee, 1984; MAP 640, MAT 11). The oldest soil is 1.3–3 ka, has an A/Eg/Bsg/Cg morphology, and is classified as an Aquod. Al_o and Fe_o reach maximum contents of 0.4 and 1.3% in the B horizon. pH decreases from about 6 to 4 for the time span of the chronosequence, while base saturation decreases from about 25 to 2%.

A river-terrace sequence that extends farther back in time is also present on the west coast of the South Island (MAP 190, MAT 11) and is characterized by Fe buildup followed by its depletion (Campbell, 1975; Ross and others, 1977; Dawson and others, 1991). After about 18 ka the soil has an A/10YR Bw/C profile with a maximum of 2% Fe_o in the Bw horizon. In marked contrast, the 130+ ka soil has an A/E/Bhs (with Fe pan) profile, the E horizon is gray (2.5–5Y or 5GY) and can be as much as 53 cm thick, and Fe_o is less than 0.1% in any horizon. If these soils indicate a genetic pathway, then pedogenic Fe accumulation is followed by its depletion. It is thought that a decreasing pH with time (6 to 4) and gleying are partly

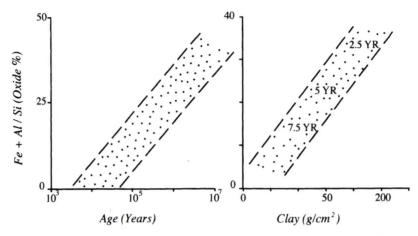

Figure 8.17 Relations between ratio of percentage Fe$_2$O$_3$ + percentage Al$_2$O$_3$/%SiO$_2$, age, hue of Bt horizon, and clay mass in the southeastern United States. *(Modified from Markewich and Pavich, 1991, Fig. 5.)*

responsible for the shift from Fe accumulation to depletion, with the formation of a thin iron pan at depth. This could be another example of the crossing of an intrinsic threshold, that is the gradual lowering of pH drastically alters the pedogenetic pathway. In some sequences there is a loss of clay with time, related to the drop in pH below 4.5.

Clay alteration and massive loss of elements take place during the development of the above soils (Dawson and others, 1991). Leaching and the lowering of pH favor a clay-mineral alteration scheme of mica ⇒ vermiculite ⇒ beidellite. By 12–14 ka, 20–30% of the Si has been lost and 16–20% of the Fe. These respective losses increase to 40 and 27% in 18 ka, 46 and 57% in >70 ka, and 92 and 97% in >130 ka.

Unusual Chronosequences of Tropical Carbonate Ocean Islands

Tropical areas present a problem for chronosequence studies in that finding landscapes of different ages is not always easy. A common dated substrate in the tropics is the uplifted coral reef deposits that were formed during the interglacials. Many of these deposits have been dated, and if not dated, one can estimate ages by assuming constant rates of uplift (Bloom, 1980). These studies differ from the studies above in that reef material at depth is not the parent material for the silicate and oxide minerals of the soils.

Muhs (in preparation) reviews pedogenesis on tropical carbonate islands. The ones of interest here have silicate soils overlying rather pure coral-reef limestone. Global dust is a major source of the silicate materials, and it is transported long distances from desert sources. Dust from the Sahara and Sahel regions of Africa contribute to the island soils of the western Atlantic Ocean and Caribbean Sea, and that from the Gobi desert of Mongolia, Takla Makan desert of China, and the Australian desert contribute to the Pacific Ocean island soils. Volcanic ash is also a potential source for silicate materials, as the oceans are dotted with active volcanoes. Muhs and others (1990) have shown how geochemical fingerprinting can be used to identify the contribution from either source (Fig. 7.4). On Bermuda, for example, many of the red soils have geochemical signatures more akin to Sahara dust than to other potential dust sources—Great Plains loess or Mississippi Valley loess (Herwitz and others, 1996)—a story consistent with that for several other western Atlantic and Caribbean islands.

In chronosequences that go back almost 1 Ma, Muhs and others (1990) and Muhs (in preparation) note several key trends with time for Barbados (MAP 110; MAT 26) and Jamaica (MAP 114–324; MAT 26). First, Bt horizons are common, and soil orders recognized are Vertisols, Mollisols, Alfisols, and Ultisols. With time, soils become more clay rich, they are redder (Barbados: 10YR for 125–320 ka, 7.5YR for 460

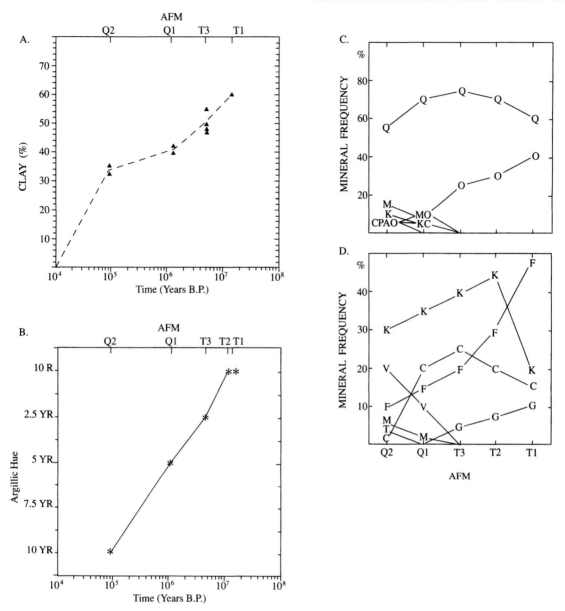

Figure 8.18 Relations between time and (A) percentage clay, (B) hue of the Bt horizon, and (C) silt and (D) clay mineralogy for soils in Virginia. In (C), Q = quartz, O = Fe oxides, M = mica, P = plagioclase feldspar, K = kyanite, A = amphibole, and C = undifferentiated clay minerals. In (D) K = kaolinite, C = chloritized vermiculite, F = Fe oxides, V = vermiculite, G = gibbsite, M = mica, and T = montmorillonite. Letter and number combinations on the horizontal axes are the alloformations, and their estimated ages are Q2 = 60–120 ka; Q1 = 0.7–1.6 Ma; T3 = 3.4–5.3 Ma; T2 = 10.8 Ma; and T1 = 13 Ma. *(Redrawn from Howard and others, 1993, Figs. 5, 7, and 8.)*

ka, and 5YR for 700 ka; Jamaica: 5YR for 125 ka and 10R for 870 ka), Fe_d average profile percentage for Barbados increases from about 0.5% at 125 ka to nearly 4% at 700 ka, and profile SiO_2/Al_2O_3 is reduced to about one-third the original value. Illite, common to Saharan dust, is not identified, presumably because it has been altered. On Barbados, the 125-ka soil has smectite and mixed-layer kaolinite–smectite, smectitie is gone by 460 ka, and by 700 ka only kaolinite remains. In contrast, on Jamaica kaolinite is common on the younger terraces and by 860 ka some has converted to boehmite.

Highly weathered soils are also present on some Pacific Ocean islands (unpublished work of author). Islands studied are Maré, east of New Caledonia (MAP 150, MAT 22), and Rota, north of Guam (MAP 250, MAT 25). Both have sequences of terraces composed of coral-reef limestone. Although both can receive volcanic dust, they can also receive dust from continental sources to the west, Maré from Australia and Rota from China and Mongolia. Dating is poor on the islands, but extrapolating back in time from dates on low terraces (B. J. Szabo, oral communication, 1996), the highest terraces (100 m on Maré and 500 m on Rota) could be close to 2 Ma.

The soils on both islands are similar in that thin (usually <0.5 m) A/Bo profiles, brown to red (7.5 to 2.5YR), are present on all terraces (see soil surveys by Latham and Mercky, 1983, and Young, 1989). Although they do not show much in the way of morphological or thickness trends with age, there are some chemical trends. Sand is <10% in all soils. On Maré clay contents reach 50–75% by 200 ka, and soils over 1 Ma are closer to 75%. Rota soils have 40–50% clay by 180 ka, and 80% or more in 340-ka and older soils. The resistate oxides (sum of total Fe_2O_3 and Al_2O_3) are 65–75% at 400 ka on Maré and increase to 80–90% in older soils; values for Rota are 65–75% in 180 ka, and 75–90% in older soils. SiO_2 displays massive depletion in the soils of both islands: the profile weighted mean molar SiO_2/Al_2O_3 value varies from 0.08 to 1.04 in <400 ka on Maré, and is less than 0.01 in all older soils, whereas Rota values are near 1 at 180 ka and <0.8 in older soils (Fig. 8.19A). As a reminder, kaolinite has a value of 2. Not surprisingly then, the common clay minerals are the Al oxides, boehmite and gibbsite.

In spite of all indications of depletion, the total bases/TiO_2 ratio decreases slowly but progressively with time on Maré (Fig. 8.19B). One explanation for

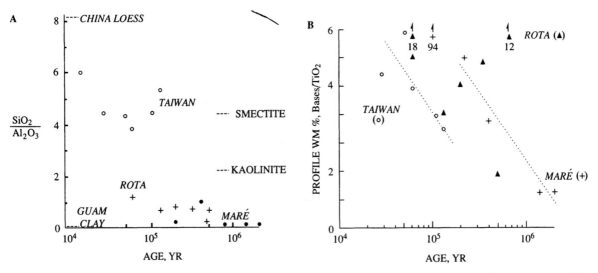

Figure 8.19 (A) Plot of profile weighted mean molar SiO_2/Al_2O_3 ratios with time for soils that overlie carbonate reef deposits on Taiwan (O), Rota (+), and Maré (●). For comparison, the red soil that overlies the Mariana Limestone on Guam is plotted without respect to its age. The same ratio is included for China loess, smectite, and kaolinite. (B) Plot of profile weighted mean sum of bases/TiO_2 ratios, and trend lines (except for Rota), with time for soils that overlie carbonate reef deposits on Taiwan (O), Rota (▲), and Maré (+). *(Unpublished data of author.)*

this anomalously slow depletion of bases (and random plots for Rota) is that treefall during storms, of trees rooted in shallow soils, recycles $CaCO_3$ upward from the coral into the overlying soil.

A similar study was done along the southern coast of Taiwan (unpublished work of author). There the silicate soils also rest on coral-reef limestone, and the silicate materials probably originate from eolian sources as well as from longshore drift of suspended sediment from rivers draining the interior of the island. The interior is underlain by a variety of sedimentary rocks. The silicate soils are 1 to ≥3 m thick, and red, thick Bt horizons are common.

The chemistry of the Taiwan soils differs from those on Rota and Maré, one reason being that the former are much thicker. Because weathering goes on throughout a greater thickness of silicate materials, the soils on Taiwan are not as intensely leached of silica; this is reflected in the molar SiO_2/Al_2O_3 ratios that vary from 6 (younger soil) to near 4 (older soils) (Fig. 8.19A). Furthermore, the soils of Taiwan, probably because their thickness isolates them from the underlying limestone, lose bases at a rate that is faster than that for the shallower soils of Maré (Fig. 8.19B).

Fe_d shows a progressive increase with time on Maré from <2 to 5% (200 ka), to a maximum of 10–13% by 1.4 Ma; on Rota, the maximum content of about 10% is reached in 300 ka. At the maximum, Fe_d makes up about 50–60% of the total Fe_2O_3 on Maré, if both values are expressed as the oxide, and 30–35% on Rota. These Fe_d values could be a minimum as only one extract was taken; some workers extract tropical soil samples more than once for Fe_d analysis.

The above silicate soils seem to come from external sources, rather than from the weathering of the underlying limestone. If, for example, they represent the insolubles left behind as limestone dissolves, one should see a karst topography, but one generally is not present (although the older studied surface on Taiwan has well-expressed karst). Furthermore, a rough calculation for Rota suggests that the entire 500-m-high island, composed of near pure limestone (0.03% insolubles), would have to dissolve to obtain the thickness of the present soils. Dust from nearby volcanoes or from the Asian mainland seem more plausible sources. Dust from both sources would have to be highly altered to produce the soils (Fig. 8.20), as would the original clay-mineral assemblages, which are illite, vermiculite, and chlorite for China loess (unpublished work of author), and illite, kaolinite, and chlo-

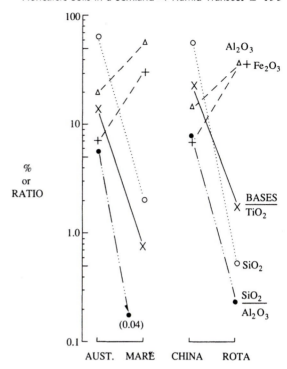

Figure 8.20 Plot of element oxide, nonmolar sum of bases/TiO_2, and molar SiO_2/Al_2O_3 ratios for a highly weathered soil horizon on both Rota and Maré compared to that for nearby continental atmospheric dust. Data for Australia are for dust that travelled to New Zealand (McTainsh, 1989), and for China are for loess. *(Unpublished data of author.)*

rite for Australian dust (McTainsh, 1989). Hence, one cannot identify source area by either major-element chemistry or clay mineralogy.

A promising way to trace these soils to their sources is by rare-earth-element (REE) analysis (data and analysis by E.E. Larson, personal communication, 1997). REEs are primarily sequestered in resistant minerals such as apatite, sphene, and zircon. On weathering, therefore, the REEs should be enriched in highly weathered soils that do not have a high quartz content. Here the soil REE data are normalized by dividing by the REE values in a chondrite. In an REE plot, the lightest REEs are listed on the left and the heaviest REEs on the right (Fig. 8.21). In many cases, the shapes of the plots can be identified with specific rock types (Wilson, 1989). Thus, it should be possible to detect if the soils are derived from continental crustal rocks (Asia) or oceanic ones (Pacific volcanos).

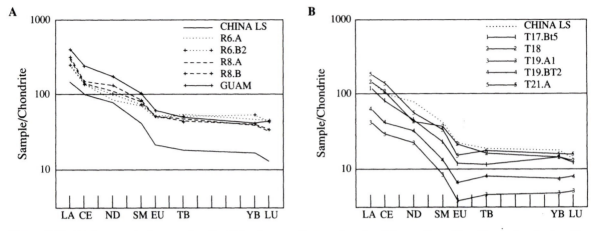

Figure 8.21 (A) Rare-earth-element plots for China loess, soils at various localities on Rota (R, with number the locality and letter the soil horizon), and red soil on Guam. (B) Rare-earth-element plots for China loess, shale on Taiwan (T18) and soils at various localities on Taiwan (T, number and letters same as in A). *(Diagrams courtesy of E.E. Larson.)*

Preliminary interpretation of REE plots for soils of this study show similarities with that of China loess (collected at the top of the loess section at Luochuan by the author). Soils on Rota have an REE plot that parallels that for China loess, but the ratios are greater (Fig. 8.21A). This suggests that the main source for the soils is atmospheric dust from China, and that intense weathering has enriched the REEs. The REE trend for soils on Taiwan has a different interpretation (Fig. 8.21B). Here REE data for China loess, soils on Taiwan, and a Tertiary shale on Taiwan are compared (recall that the sedimentary rocks in the interior are a potential silicate source for the soils). First, the plots for China loess and Taiwan shale are similar enough that one cannot distinguish one source from the other. Second, in contrast to Rota, the data plot lower than that for either China loess or shale. The reason for this could be (1) as intense weathering goes on, the quartz fraction of the weathered residue is enriched and (2) quartz might be contributed to the coast from the erosion of quartz-rich sedimentary rocks. Both of the latter would dilute the REE signal and result in a plot more or less parallel to the shale and loess, but with lower ratio values.

R.N. Lantz (written communication, 1997) used REE data to estimate the importance of volcanic ash in the soils of Rota. He modeled REE data for China loess, for ashes in cores taken from the West Mariana Ridge (may be similar to the average ash deposited on land), and for modern coral, in different proportions.

The best fit to the average Rota data was obtained with 70% China loess, 25% ash, and 5% coral. This appears to verify that atmospheric dust, which is the downwind equivalent of China loess, is the major contributor to the soils of Rota.

These chronosequences are what one might expect for the right-hand side of the Pye diagram (Fig. 7.12). Dust accumulates slowly on a limestone substrate as the island rises out of the sea. The high rainfall and temperature then combine to alter the dust at a rapid pace. Tsunamis have the potential to strip soils off low terraces, so a particular terrace might have to reach a certain height to avoid that. Another problem is that widespread treefall during intense storms could homogenize soils, introduce limestone clasts into the overlying soils, and the barren land could erode and thus foul any thickness vs age relationship. In spite of this, and even though the younger soils are highly altered, trends with time are discernible.

A somewhat similar highly altered, thin red (10R) Oxisol rests on pure carbonate rock in the Bahamas (Foos, 1991), and it may have a similar origin. Average total $Al_2O_3 + Fe_2O_3$ is 47%, average molar SiO_2/Al_2O_3 is 0.8, and boehmite is a common clay mineral. These soils are more weathered than those on islands farther south, and Muhs (in preparation) attributes this difference to a lower rate of dust influx, with weathering keeping ahead of dust additions.

The above examples of weathering are extreme, causing some people to look again at the origin of bauxites. Muhs (in preparation) argues for a distant dust plus weathering origin for Caribbean bauxites, as do Brimhall and others (1988) for some Australian bauxites.

■ Chronosequences Involving Transects from Noncalcic to Calcic Soils

Two workers have made wet-to-dry transects of soils formed on river-terrace deposits to determine the climates where particular morphologies and properties are best expressed.

McFadden (1988) studied a climochronosequence in southern California under three climatic regimes, arid, semiarid, and xeric, with increasing rainfall and decreasing temperature. The oldest soils are about 1 Ma. All soils have Bt horizons by mid-Holocene to latest Pleistocene. Buildup curves for both clay and Fe_d (here expressed as Fe_2O_3) flatten with time and

follow the climatic gradient, with accumulation in the wetter area being four times (clay) to about 15 times (Fe_d) that in the drier area (Fig. 8.22). A plot of clay vs pedogenic Fe_2O_3 shows three distinct slopes (Fig. 8.23). The lower slope (greater Fe_2O_3 per unit clay) indicates that Fe has been released by weathering, whereas the steeper slope is indicative of the relatively high clay influx as dust combined with a low rate of Fe release due to lower rates of chemical weathering in the more arid area. With fluctuating Quaternary climate, the dry area probably always was an area of carbonate accumulation and the wet area an area of carbonate leaching; the intermediate area may have experienced both carbonate leaching and accumulation as climate fluctuated.

The second traverse is in Montana, where the climate variation from relatively wet (MAP 64, MAT 6) to dry (MAP 37, MAT 8) is more subtle than in the above study (Reheis, 1987a). The oldest soils are 2 Ma. The drier soils have carbonate accumulations, but the wetter soils do not. All soils have Bt horizons by

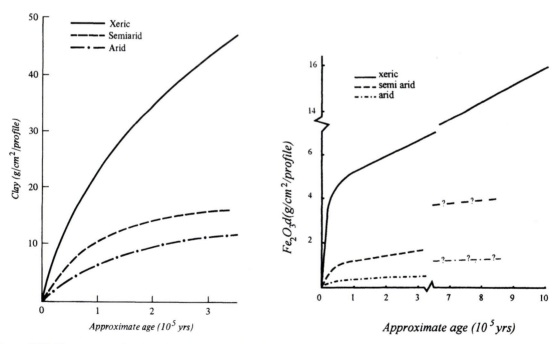

Figure 8.22 The amount of pedogenic clay (A) and Fe (B) as a function of time in three climatic regimes, southern California. The mean annual precipitations (cm) and temperatures (°C) for the three regimes are within these ranges, respectively: arid ≤12, <22; semiarid: 12–25, 15–22; and xeric: 25–75, 12–22. (*Redrawn from McFadden, 1982, Fig. 27, and McFadden, 1988, Fig. 17, from the Geological Society of America Special Paper 216, p. 153–177 © 1988, with permission from the Geological Society of America, Boulder, Colorado.*)

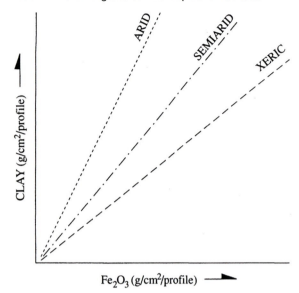

Figure 8.23 Generalized relation between masses of pedogenic clay and iron oxide in three climatic regimes, southern California. *(Redrawn from McFadden, 1988, Fig. 18, from the Geological Society of America Special Paper 216, p.1 53–177, ©1988, with permission from the Geological Society of America, Boulder, Colorado; for actual values, see latter.)*

20 ka. Color progression is slow: 7.5YR hue by 20 ka in the wetter area, and soils do not become much redder with time. In the dry area, all noncalcic horizons are 10YR. Maximum clay percentage in the wetter area is only 12%, vs 26% in the drier area. In addition, PDI values for the wetter soils are higher than those for the drier soils for much of this time span, and the higher values appear to be on the low side of the envelope of data presented by Harden and Taylor (1983). Erosion could limit the development of these soils.

Clay mineralogy and total chemistry vary with climate along the Montana transect (Reheis, 1990). In the wetter climate, vermiculite, mica–smectite, and smectite all increase with time; in contrast, in the drier soils, smectite increases with time, mica–smectite shows little change, and vermiculite is absent. Trends are best at depth. Total chemical analyses show that most major element oxides increase with time along the transect. Regression analysis of these increases vs time suggest a logarithmic relationship in the wetter area, and a linear relationship in the drier area. This is taken to mean that the eolian addition signal is altered by chemical weathering in the wetter area, hence the logarithmic relationship, but it is not in the dry area, hence the linear relationship.

One goal of many soil-transect studies along climatic gradients is to search for clues of climate change. Thus, do dry-climate soil signals persist into the wetter area and vice versa? Both of the above traverses address these issues.

■ Chronosequences in Aridic Regions with Gypsic and Calcic Soils

Salts accumulate in soils in areas of limited effective soil moisture. For example, in southern Israel carbonate accumulates in soils at less than about 50 cm MAP and salts more soluble than carbonate at less than about 25 cm MAP (Fig. 8.24). Below about 10 cm MAP both By and Bz horizons occur at shallow depths, and $CaCO_3$ accumulation is minor. The tops of the salt-accumulation horizons are used to estimate the depth of annual wetting, so the deeper they are, the greater the rainfall. Many of these soils are sufficiently dry, with enough fines, to have Av surface horizons. The order of presentation here will be from wet enough to produce carbonate accumulation, continuing with the above transects, to extremely dry.

Cold Desert versus Hot Desert Soils

The cold desert soils of Antarctica are excluded from this discussion as they were covered earlier. There are many similarities, as well as differences, between soil development in the cold deserts as compared with those in the hot desert (Claridge and Campbell, 1982; Singer, 1995). Similarities include a surface desert pavement underlain by a thin, light-colored A horizon, but the development of an Av horizon might be more common in a hot desert. Salts more soluble than carbonate are common in the cold desert, and common only in the driest of the hot deserts. The major differences are the common presence of these features in the hot deserts: (1) a horizon age sequence of Bw $\Rightarrow$ Bt, (2) Bk and K horizons at depth, (3) significant chemical alteration as shown by Fe_d buildup, rubification, and clay-mineral synthesis or transformation, and (4) soil morphology features that reflect a fluctuating climate. The main reason for the differences is that liquid water is extremely limited in the cold desert of Antarctica, whereas it is present in variable amounts in the hot deserts, and that, accompanied with the high temperatures, can bring about rapid alteration in the

Figure 8.24 General relation between kind of pedogenic salt, depth of the top of the salt-accumulation horizon, and precipitation in southern Israel. *(Redrawn and modified from Dan and Yaalon, 1982, Fig. 5.)*

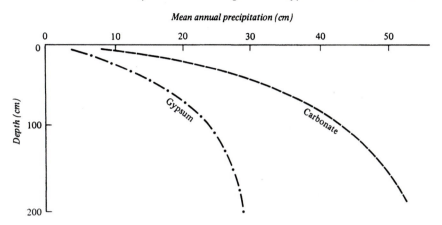

Chronosequences Involving Calcic Soils: Western United States

Much research on calcic soils has been undertaken in the Western United States (recent comparisons by Harden and others, 1991a, and Reheis and others, 1995). In general, pedogenic $CaCO_3$ accumulates in soils where MAP is less than about 50 cm, similar to the value for Israel (Fig. 8.24). The key work is that of Gile and Grossman (1979) and Gile and others (1981), who made many significant contributions to arid-region soil geomorphology. First, much of the material accumulating in the soils is of atmospheric origin, both solid and dissolved in rainfall. Hence, under appropriate conditions Bt horizons can form within the Holocene in gravelly materials, a rate considered too rapid if weathering processes are responsible for all of the clay. The red color of such horizons also appears in young soils, probably reflecting Fe released by weathering. Second, carbonate accumulation follows morphological changes in such a predictable manner with time that four stages were defined (Appendix 1, Fig. A1.5). These stages are a powerful tool in the correlation of deposits throughout the region. Finally, detailed maps are presented to demonstrate the relation of soils to landscape units; of importance

hot desert. The work below will demonstrate that soil development is rapid in the hot desert, and thus is a good tool for estimating the ages of a wide variety of events, such as climate change, depositional episodes, and fault movements. In contrast, soil development in Antarctica proceeds at a very slow pace.

is the fact that subtle variations in soils can be used to identify areas of erosion not always readily seen on surface examination alone. These maps are ideal models for understanding complex soil-geomorphic relations.

Bt horizons form rapidly in the Las Cruces area of New Mexico (MAP 21, MAT 16; Gile and others, 1981). Two pathways of buildup are recognized, depending on the gravel content. In low-gravel materials the initial stages of buildup are on ped faces. In contrast, in high-gravel materials the buildup is recognized as pebble coatings and filling of voids within the interpebble matrix. Because there is less available space for clays to occupy in gravelly materials (clasts take up a sizable proportion of the volume, and particle-size data are based on the <2-mm fraction), argillic criteria are met in younger soils in gravelly than in nongravelly materials. Late Holocene (<2.5 ka) soils have 5YR Bt horizons with 12% clay in gravelley materials (Fig. 8.15), but a Bt seems to require a middle Holocene age in nongravelly materials, and it is 5–7.5YR with 9% clay. A hue of 2.5YR, the reddest in the area, is reached in the late Pleistocene, given by them as 10–250 ka. By late middle Pleistocene (250 ka) the clay content has reached 34%.

The Las Cruces area is the area in which the carbonate accumulation stages were developed (Appendix 1, Fig. A1.5). As with the argillic properties, gravel concentration determines the morphology of the early stages of accumulation, but the two pathways converge to a similar morphology by stage III. Higher stages are reached in less time in the gravelly materials, because with gravel taking up so much volume,

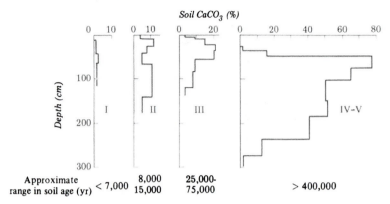

Figure 8.25 Carbonate distribution with depth for nongravelly soils of different ages, Las Cruces, New Mexico. Roman numerals refer to carbonate morphology stages (Appendix 1). *(Data from Gile and Grossman, 1979, and Gile and others, 1981.)*

there is less matrix space available for the carbonate to occupy, and therefore for morphology to be expressed in. Hence, with a constant influx of eolian carbonate, the relatively low volume of the matrix of gravelly soils accumulates carbonate faster than does the matrix of soils without gravel. As the carbonate stages increase, so does the amount of $CaCO_3$ in the soil (Fig. 8.25), and at the highest stages the amount of $CaCO_3$ is close to that of limestone.

In chronosequence studies and in studies using soils to estimate ages of Quaternary deposits, we would like to know the rate and processes associated with carbonate accumulation. The rate at which carbonate builds up in soil depends on the process of formation as well as the rate of leaching in the soil. If the necessary Ca^{2+} comes from mineral weathering, buildup will be controlled by the rate of weathering of the calcium-bearing minerals. If $CaCO_3$ was originally present in the parent material, the rate of buildup in the soil is a function of the rate at which it can be dissolved and translocated by leaching waters in the soil profile. If, however, the $CaCO_3$ is of atmospheric origin, buildup is a function of the Ca^{2+} in the rainfall, the carbonate content of dust, and the amount of rainfall to both move Ca^{2+} into the soil and dissolve carbonate in dust, and to translocate it to some depth where precipitation occurs.

Provided the $CaCO_3$ was inherited and can be shown to have been distributed uniformly in the soil parent material, one can roughly calculate how long it might take to redistribute the $CaCO_3$ by solution, translocation, and reprecipitation. Arkley (1963) has presented a method for these calculations, using estimations on the volume of water passing various lev-

els in the soil, as determined from soil available-water-holding capacity and climatic data, and data on $CaCO_3$ solubility. One such calculation is shown in Fig. 8.26 and, although it and other ages he calculated are fairly rough and only a minimum, the calculated ages are reasonably consistent with the geological evidence on age.

Another major discovery of Gile and others (1981) was that much of the carbonate in the soil is of atmospheric origin. Many workers in the southwestern

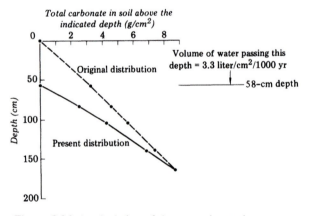

Figure 8.26 A calculation of the approximate time necessary to redistribute carbonate in a soil. *(Data from Arkley, 1963.)* Carbonate is assumed to have been uniformly distributed in the soil, subsequently leached from the uppermost 58 cm, and precipitated in the 58–165 cm interval. Arkley calculates that about 3.3 liters water/cm²/ky moves by the 58–cm-depth level in the soil and uses a soil carbonate solubility of 0.1 g/liter for the prevailing conditions at that depth.

United States agree with this origin. Airborne sediment traps placed in the Las Cruces area collected from 9.3 to 26.3×10^{-4} g/cm²/year. Silt commonly exceeded clay (the latter made up 20–40% of the sample), organic carbon is 3–7%, and $CaCO_3$ is 1–3%. The Ca^{2+} in the precipitation is considered to contribute much more to the pedogenic carbonate than does the carbonate in the dry dust, perhaps by a factor of 2 to 3. An estimate is that Ca^{2+} from all sources is sufficient to form about 0.2 g pedogenic carbonate/cm²/ky, assuming all the available Ca^{2+} enters the soil and is deposited as pedogenic carbonate.

McFadden and Tinsley (1985), Mayer and others (1988), and McFadden and others (1991) have suggested an improvement on the Arkley model that takes into account such things as $CaCO_3$ solubility as a function of the partial pressure of CO_2, solution ionic strength, and the rate of atmospheric Ca^{2+} influx into the soil. The method was tested by trying to match the calculated carbonate profile with that for two Holocene soils, and the match was fairly good (Fig. 10.24).

Another way to rank areas is by the total carbonate in the soil for soils of various ages. This is the preferred method because gravel content is taken into account. Amounts are calculated using the following equation (Machette, 1985):

$$C_s = (C_3 P_3 - C_1 P_1)d$$

C_s = weight of secondary carbonate in a soil horizon per unit area (g/cm²)
C_3 = present total $CaCO_3$ content (g $CaCO_3$/100 g oven-dry soil)
C_1 = initial (primary) $CaCO_3$ content
P_3 = present oven-dry bulk density (g/cm³)
P_1 = initial oven-dry bulk density
d = thickness (cm) of the sampled interval

Here C_1 and P_1 have to be estimated either from Cu horizons at depth or from Cu horizons of younger soils in the area. The C_s values for each sampled interval can be summed to give C_S, the total secondary carbonate for the soil. This is expressed as grams of $CaCO_3$ per square centimeter of surface area for the soil.

One way to estimate ages of calcareous soils in the latter example is to divide the present-day influx rate into the C_S value. If this is done for soils in the Las Cruces area (data given above), to estimate age in thousands of years, divide the C_S value of the soil in question by 0.2.

It is difficult to estimate ages of soils by the above method because the system is very complex. Some complications are (1) part of the water runs off and therefore does not translocate $CaCO_3$ in the soil; (2) some of the eolian $CaCO_3$ may be blown away before the next rainfall moves it into the soil; and (3) the Ca^{2+} that goes to making additional $CaCO_3$ may come from other sources, such as calcium-bearing salts (e.g., gypsum) and the Ca^{2+} on the exchangeable sites of the colloidal fraction.

Machette (1985) added to the carbonate accumulation work of Gile and others (1981) in several ways. First, he added two more stages (V, VI) to represent development beyond stage IV (Appendix 1, Table A1.5). Second, in a regional study he demonstrated that the time for attainment of a particular carbonate stage varies between areas; low rates are found in either mountains or basins where atmospheric dust influx is low and/or the dust has a low carbonate content, and high rates under the opposite conditions (Fig. 8.27). For example, New Mexico sites with rapid development of higher stages are closer to carbonate dust sources than are the Colorado sites where attainment of higher stages takes longer. That is, stage equates to age.

Third, by examination of stable soils of known age (parent materials dated by associated volcanic ash), Machette could estimate the rate of mass accumulation of carbonate by dividing C_S by the age of the deposit. He found a positive correlation between the accumulation rate and the rapidity with which successive carbonate morphological stages are attained (Fig. 8.27). This method could be preferred over one that determines accumulation rate by dust traps placed in the landscape (Gile and others, 1981; Reheis and Kihl, 1995), because the former is a measure of what actually gets into the soil in the long term. In contrast, the dust trap measures the total possible, and probably does not take into account that dust moved from the site before a rainfall event can move it into the soil, or carbonate that dissolved and was removed from the soil during times of higher than average rainfall.

Finally, Machette (1985) diagrammed long-term carbonate accumulation rates, taking into account climate change (Fig. 8.28). Two main fields are depicted in the diagram, moisture limited and influx limited. In moisture-limited areas there is a high influx of Ca^{2+} from all sources relative to the rainfall. Potentially, therefore, not all the available Ca^{2+} is carried into the soil. In contrast, in influx-limited areas there is a relatively low supply of available Ca^{2+} rel-

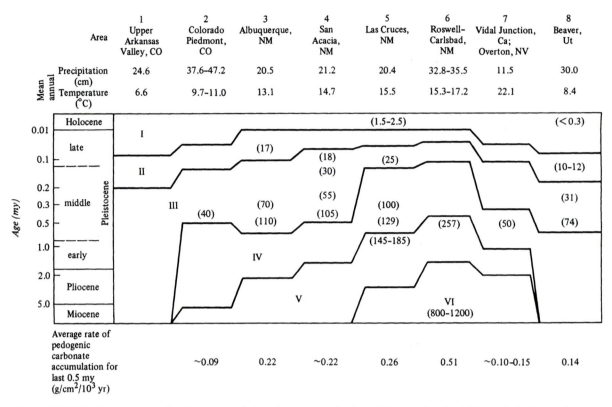

Figure 8.27 Maximum stages of carbonate morphology in surface soils formed in gravelly alluvial deposits in the southwestern United States. Numbers in parentheses are the average grams of $CaCO_3/cm^2$/soil column in soils of the indicated ages. *(Redrawn and modified from Machette, 1985, Table 2, with additional data from Machette, 1982, and McFadden, 1982.)*

ative to the rainfall. Potentially, therefore, all of the Ca^{2+} is carried into the soil, but some could be leached out. In both these extreme cases, the eventual amount of Ca^{2+} residing in the soil is less than that delivered to the surface. Along the boundary between the two fields, all of the available Ca^{2+} is moved into the soil and resides there as pedogenic carbonate. Finally, as climate fluctuates, so does the rate of accumulation of carbonate. One should be aware of these pathways when evaluating dust-trap data, as an efficient trap should retain everything delivered to it.

There are some well-established carbonate influx rates for other parts of the western Cordillera. Using the soil-estimated-age method of calculation, Harden and others (1985) obtained a range of rates from 0.14 to 0.26 g/cm²/ky near the La Sal Mountains, Utah, and Scott and others (1983) estimated a rate of 0.5 g/cm²/ky

for the eastern Lake Bonneville area. Data from across parts of California and Nevada provide a regional dust-trap data set, and allowed Reheis and Kihl (1995) to calculate a range of potential carbonate influx rates of 0.07–0.39 g/cm²/ky. Their study also shows the importance of alluvial fan surfaces as dust sources, particularly because these are a far more extensive part of the landscape than are the more localized playas. They also found that playa-lake sources provide dust richer in salts than do alluvial-fan sources. In a follow-up study relating the atmospheric dust influx to the actual soil properties, Reheis and others (1995) obtained average rates of <0.01–1.2 g/cm²/ky. In northern Mexico the average rate is 0.09 g/cm²/ky (Slate and others, 1991).

As an example of dust influx on both silt plus clay and carbonate distribution in soils, we can compare

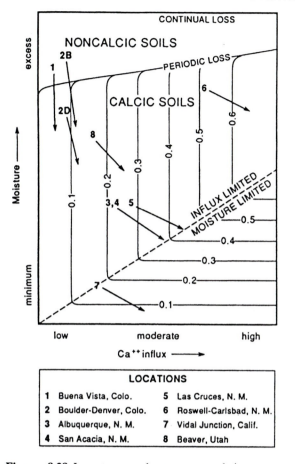

LOCATIONS

1 Buena Vista, Colo.	5 Las Cruces, N. M.
2 Boulder-Denver, Colo.	6 Roswell-Carlsbad, N. M.
3 Albuquerque, N. M.	7 Vidal Junction, Calif.
4 San Acacia, N. M.	8 Beaver, Utah

Figure 8.28 Long-term carbonate accumulation rates, shown by vertical and horizontal lines (in grams of $CaCO_3/cm^2/ky$) under varying conditions of moisture and Ca^{2+} influx. The localities here as the same as in Fig. 8.27. Each arrow depicts conditions and rates during two climate modes: the point of the arrow is for interpluvial (interglacial) episodes, and the end of the arrow for pluvial (glacial) episodes. Influx-limited and moisture-limited parts of the diagram are defined in the text. *(Redrawn from Machette, 1985, Fig. 5, from the Geological Society of America Special Paper 203, p. 1–21, © 1985, with permission from the Geological Society of America, Boulder, Colorado.)*

soils on Sierra Nevada tills (area of low dust influx, and previously discussed) with those on lake shoreline deposits in the western Great Basin. Recall that in 20 ky, the soils on tills have no accumulation bulge of silt or clay, and clay reaches 14% maximum in a Bt horizon in 140 ka in one place (Fig. 8.15). Chad-

wick and Davis (1990) sampled a downwind transect of soils formed on pluvial-lake beach deposits of close to the same ages in the Lake Lahontan Basin, east of the tills, where dry lake beds are a ready source of eolian material. The upwind site has <5% clay in a 12-ka soil, and this increases to close to 30% downwind of the eolian source (silt follows clay and is more abundant). In contrast, the upwind 130-ka soil has close to 30% clay and the downwind soil close to 50%. Hence, the downwind 12-ka clay distribution is similar to the upwind 130-ka clay distribution, confusing the use of soils in correlating deposits in the region. In contrast, the carbonate profiles for both age soils are similar, being 5% or less at upwind sites, and 15–20% at downwind sites.

Areas of exposed playa sediments and paleoclimate explain the above trends. The 12-ka soil has received influx during the Holocene. However, the clay content of the 130-ka soil is only twice that of the younger soil, even though it is 10 times older. It is thought that the time of introduction of fines was restricted mainly to the following two interpluvial periods, the Holocene and O-isotope stage 5, as the playa source did not exist when pluvial lakes flooded the basins. Thus the clay content of the older soils is only twice that of the younger soil. The reason the two different-age soils have similar carbonate profiles is that the carbonate in the older soil was probably leached from the profiles during the subsequent pluvial climates (O-isotope stages 4–2), a time of greater soil moisture, so a Holocene carbonate profile was overprinted on an older clay profile.

Soils formed on alluvial deposits derived from rhyolitic ash-flow tuffs at the Nevada Test Site (MAP 15; MAT 15) accumulate secondary silica in addition to carbonate (Taylor, 1986; Harden and others, 1991; Reheis and others, 1995). Pedogenic opal-silica stages are defined, modified after the carbonate morphology stages (Harden and others, 1991b; Appendix 1, Table A1.8). Atmospheric dust seems to account for the accumulation of clay and carbonate, whereas weathering seems to have released sufficient silica to form the pedogenic opal silica. Soils of about 10 ka lack a distinct Bt horizon but have a 10YR Bw horizon, stage I–II carbonate morphology and stage I silica morphology. The youngest deposit with a Bt horizon is 30–47 ka. By 270–430 ka, the Bt is 7.5YR, a K horizon is present, the carbonate and silica stages are III–IV and III, respectively, and various properties attain higher values at higher elevation. The long-term rate of carbonate accumulation is 0.1–0.3 g/cm²/ky, a moderate rate compared to those elsewhere in the southwestern United States (Fig. 8.27), and that of pedogenic silica is 0.08 g/cm²/ky.

It is difficult to separate some pedogenic from parent material properties when desert soils form from carbonate parent materials. Such is the case for soils formed from alluvial fans consisting mainly of limestone and dolomite clasts near the Nevada Test Site at Kyle Canyon (MAP 21–23, MAT 13–18; Reheis and others, 1992). Standard techniques for analysis could not be used because of the high carbonate content of the parent materials; hence, percentage estimates were made by point counting the constituent of interest. The progression of profiles with time is Av/7.5YR Btk/Bk (stage II carbonate, stage I silica) in 15 or 75 ka; Av/7.5YR Bt/Kmq (stage IV carbonate, stage III–IV silica) in 130 ka; and Av/Km (stage IV carbonate, stage I silica) in >730 ka.

Thin-section study indicates some of the processes going on in the Kyle Canyon soils. First, foreign grains (eolian) increase in content toward the surface. Second, despite the highly calcareous nature of the system, which should flocculate clay particles, clay coats on grains, indicative of clay translocation, are present in Bt and Btk horizons. Third, carbonate precipitation has displaced grain fragments. Fourth, parts of grains have been replaced by carbonate. There is both thin-section and field evidence for the solution of carbonate grains and clasts, which adds to the carbonate introduced by dust influx.

Chemical data at Kyle Canyon mainly reinforce the idea of dust influx in the upper part of the soil. Major oxides match the contents in dust more than they do those in the parent material. Also, Fe_d increases in the surface horizons, and there are few trends with time, suggesting Fe_d is introduced with dust and little is added by weathering.

Soil development at Kyle canyon is relatively rapid. PDI data show that development is more rapid than at the nearby Nevada Test Site (see above), no doubt helped by the additional source of parent-material carbonate at Kyle Canyon (Harden and others, 1991a). Pedogenic carbonate and noncarbonate accumulation rates depend on which date one assumes for the younger soil: 1.3 and 0.6 g/cm²/ky, respectively, if the younger date is used, or 0.3 and 0.1 if the older date is used. The lower rates are more in line with the rates determined for the next-older soil, and for the regional rates.

There are two studies in the western Cordillera using dated basalts as a substrate for soil development on desert eolian fines. In the Cima volcanic field of SE California (MAP 12–25, MAT 16–18) soils are formed from loess and are grouped into four phases (McFadden and others, 1986). Before burial by loess, however, a rubble zone forms on top of the flows. Phase 1 soils form in <140 ka, have a A/Bk (stage I) profile, and some have a thin Btk horizon (Fig. 8.29). Most horizons are 10YR but Btks are 7.5YR, and maximum clay is 12%. The weighted mean PDI value is 0.23 or less. As the loess is deposited a thin layer of stones remains at the surface as desert pavement (McFadden and others, 1987). Phase 2 soils are on flows between 140 and 700 ka, and consist of the phase 1 soil–loess couplet over a thick 7.5YR Btk/Bk (stage II; rarely K with stage III) profile with a maximum clay content of 43% and maximum weighted mean PDI of 0.42. Phase 3 soils, on flows of 700 ka to 1 Ma, also can occur beneath the phase 2 soil–loess couplet. The characteristic morphology is a carbonate-engulfed Bt (hue no redder than 7.5YR)/Bk or K (stage III). Abundant clay (maximum of 40%) and clay films in the Bt horizon indicate strong development prior to engulfment by carbonate. Maximum weighted mean PDI is 0.71. Phase 4 soils are on Pliocene flows and are highly eroded, consisting mainly of fragments of stage IV–V K horizons. Clay mineral and Fe oxide analyses indicate greater chemical alteration of the phase 2 soils relative to the phase 1 soils. This is a good example of welded soils with pedogenic carbonate being the material that welds one soil to the other.

The existence of four phases of soil development is taken to mean that loess deposition has been episodic, and separated by long intervals of soil formation (McFadden and others, 1986). For example, the fact that phase 1 occurs on flows as different in age as 16 and 140 ka indicates that the loess and soil post-date the 16-ka flow. The main source of loess would be upwind playas and fans. The phase 2 soils mark a long interval of soil formation, with higher rainfall at times. By phase 3 time, the flows and soils are beginning to show signs of erosion, some of which could be generated by soils with less infiltration capacity and therefore greater runoff. By phase 4, erosion is dominant over either loess deposition or soil formation.

Another soil chronosequence involving eolian deposits/basalt (age range 140 ka to 1.1 Ma) occurs in northern Mexico (MAP 10, MAT 21; Slate and others, 1991). These deposits have more sand and less silt than the Cima soils. The soils differ markedly from those in the Cima area in that buried soils are not present, reddening and clay films were not observed, and Bt horizons are recognized on the basis of subsurface relative clay accumulation. Because both clay and carbonate contents have significant increases with time, it appears that deposition has been slow and continuous, with pedogenesis going on in a dry environment that perhaps was no wetter than present. The soils, therefore, are cumulic.

Chronosequences Involving Both Surface and Buried Calcic Soils

Two chronosequences involving both surface and buried soils illustrate how this record can be pieced

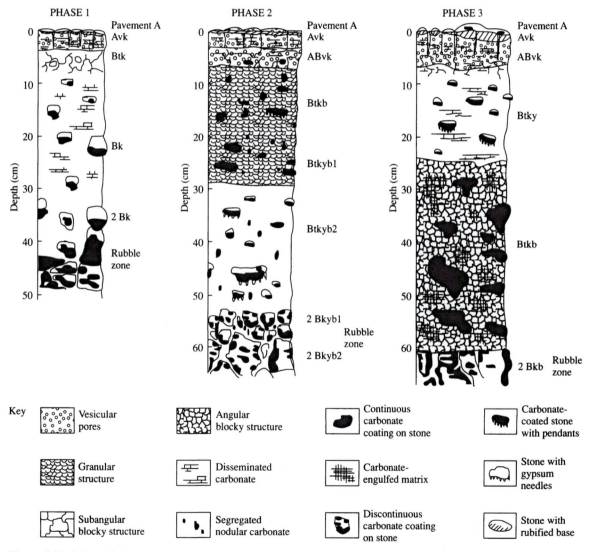

Figure 8.29 Schematic diagrams showing typical horizons and morphologies of three phases of soil development for desert loess overlying dated basalt flows, Cima volcanic field, California. *(Redrawn from McFadden and others, 1986, Fig. 2, Influence of Quaternary climatic changes on processes of soil development in desert loess deposits of the Cima volcanic field, California. Catena 13, 361–389, © 1986, with kind permission from Elsevier Science, The Netherlands.)*

together. Good dating control is essential, such as that done for archeology sites, as the interval over which each soil was at the Earth's surface is needed. Furthermore, one has to be sure diagenesis during burial has not altered the soils to any great extent; because these soils are in a desert with minimal soil moisture, this does not appear to be a problem.

A fairly well-dated chronosequence consisting of both surface and buried soils exists at the Lubbock Lake Archaeological Site, Texas (MAP 47; MAT 15) (Holliday, 1985a,b, 1988). The youngest soil formed at the surface in the past 100 years and, where best expressed, has an A/Bwk profile with stage I carbonate morphology. The intermediate-age soil formed over 450 years, where it is at the surface, and 200 years where buried; the best expressed profile is A/Bt/Bk with stage I carbonate morphology. In contrast, the oldest soil had 4500 years to form where not buried and 3500 years to form where buried, and the best expressed profile is A/Bt/Bk with a maximum carbonate morphology of stage II. Parent-material hue is 10YR, and the oldest soils have 7.5YR to 5YR hues. An adequate amount of dust and moisture combined to produce the exceptionally rapid development of the Bt horizon, and a rate of carbonate accumulation that could be as much as 10 times faster than the rapid rates for soil development in areas of eastern New Mexico (Fig. 8.27).

The other surface and buried soil chronosequence is at Black Mesa, Arizona (MAP 30, MAT 9; Karlstrom, 1988). Horizons that meet the criteria of Soil Taxonomy form rapidly: cambic in 0.2–0.5 ky and argillic in 1.0–1.5 ky. Clay accumulation index mean values increase progressively from 20 in 0.5 ky, to 131 in 1.3 ky, 518 in 3.7 ky, >976 in 15 ky, and 2049 in 28 ky. All clay minerals seem to be inherited with little pedogenic alteration. Carbonate stages progress as follows: stage I in 1.5 ky and stages II and III in 15 ky.

The rates of formation at Black Mesa are rapid when compared to other areas. For example, the clay accumulation index values are much higher than they are in Pennsylvania (Levine and Ciolkosz, 1983), and the attainment of stage III carbonate morphology is one of the most rapid for the western United States (compare with Fig. 8.27).

Chronosequences Involving Gypsic Soils: Israel, Western United States, and Peru

Rapid soil development takes place in the extremely arid environment of southern Israel (Dan and others,

1982), where much of the pedogenic salt, silt, and clay are considered to be of atmospheric origin (Gerson and others, 1985; Amit and Gerson, 1986; Gerson and Amit, 1987). An example is soils on a flight of river terraces that range from recent to 14 ka (MAP 6; MAT 23) (Amit and Gerson, 1986; Gerson and Amit, 1987). Maximum profile thickness, 40–50 cm, is achieved in about 1 ka. In the oldest soil, coarse-clay content approaches 10% and fine clay 15%, fine silt accumulation is a maximum of 25% near the surface, and gypsum is about 8% (Fig. 8.30). Although dating is poor, Bt horizons are present in late Holocene soils. While the above accumulate, stones within the soil profile shatter due to salt weathering.

For older soils in the region with estimated ages of several hundred ky, maximum silt content is 43% near the surface and 24% at depth in the B horizon, and the gypsum maximum of 43% produces a Bym (petrogypsic) horizon (Dan and others, 1982; Gerson and Amit, 1987). Bt horizons are nongravelly in spite of the presence of both an overlying well-developed gravel pavement and underlying gravelly By and Bz horizons. In places such Bt horizons can form in 50 ka. The formation of the desert pavement with the above variations of gravel with depth has been addressed by McFadden and others (1987); the original surface clasts remain at the surface while atmospheric silt and clay continue to infiltrate to shallow depths. Thus the surface clasts rise and the gravel-free B horizon is created.

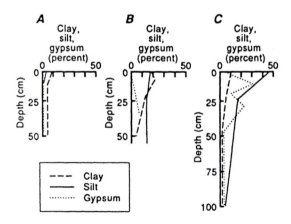

Figure 8.30 Data on clay, silt, and gypsum contents in soils of various ages in southern Israel. Ages are (A) late Holocene, (B) 14 ka, and (C) Middle Pleistocene. *(Data from Dan and others, 1982; Amit and Gerson, 1986.)*

The rates of accumulation of silt, clay, and salts in these dry deserts can change with time for various reasons (Gerson and Amit, 1987; Reheis and others, 1995). Some are extrinsic and include factors in the source area such as wind direction and degree of protective vegetation cover. Others are site specific and include the efficiency of the surface to trap dust, with high efficiencies associated with vegetation and/or a rough gravelly surface. Most stratigraphic studies (e.g., Wells and others, 1987) involving alluvial-fan deposits have shown that there is a progression with time from an initial rough surface marked by bar and swale topography strewn with many gravel sizes to one that is smooth with a tightly packed desert pavement of nearly uniform small sizes. This change can happen in about 10 ky. In this case the trapping capability of the surface diminishes with time. Yet other site factors are the formation of a surface crust and the plugging of near-surface voids with silt and clay; both of these diminish the capacity of the soil to vertically transmit water and thereby carry solids and salts to depth in soils. If runoff increases because of the latter, the landform, with its soils, could undergo erosion. Wells and others (1987) have shown clearly how closely linked both geomorphic and pedogenic processes are in the desert.

A chronosequence of alluvial-fan deposits in north-central Wyoming (MAP 7; MAT 7) suggests rates of clay and gypsum accumulation somewhat similar to those of southern Israel (Reheis, 1987b). Marked clay accumulation (27%) and some gypsum (0.2%) are noted in 5-ka soils, and gypsum exceeds 50% in some horizons of soils estimated to be 410 ka. The accumulation of gypsum is about 9 g gypsum/cm^2/100 ky for 250-cm-thick soils. Accumulation of atmospheric gypsum in dust traps placed in the field is sufficient to account for this rate of pedogenic gypsum. Reheis warned, however, that not all trapped gypsum necessarily gets into the soils, as some may be carried off by wind to other sites before a rainfall event moves it into the soil.

The latter study brought attention to several processes in gypsic soils that are important to the predicted progressive pedological variations in chronosequences. A common observation is that even though dusts of both CaCO$_3$ and gypsum are available to accumulate at depth, and are together in many soils, in extremely arid environments gypsum accumulation dominates. For example, the gypsic soils of Israel, mentioned above, are in an area of abundant limestone outcrops. One explanation for this preference of gypsum in the soils is that although both may be deposited at the surface of the soil, the solubility differences are such that primarily gypsum is dissolved and moved into the soil; although some of the CaCO$_3$ also does, most of the remaining CaCO$_3$ could be removed by wind erosion. Also shown to be important is the common ion effect, with the higher Ca^{2+} concentrations associated with the higher gypsum contents markedly inhibiting CaCO$_3$ solubility. Still another factor could be that with low precipitation, vegetation is nonexistant, so soil CO$_2$ is low (Brook and others, 1983), perhaps even near that of the atmosphere (McFadden and others, 1991); if sulfate is present, this could favor precipitation of gypsum over that of carbonate. Gypsum buildup could also inhibit infiltration of clays of atmospheric origin by increasing Ca^{2+} concentrations to levels at which clay particles are flocculated and their downward movement is greatly diminished. Such flocculated clays at the surface could be removed from the site by winds. Finally, although present climate and climate change are important in explaining the position of gypsic horizon(s) in soils, gypsum, being hydroscopic, can, if in high enough concentrations, absorb some increases in effective soil moisture related to a climate change toward more moist conditions. The result could be that the By horizon, the one used to estimate depth of wetting and climate, is not shifted to greater depths.

The coastal desert of Peru is so dry (MAP 1, MAT 18) that one would expect soils similar to those of driest Israel, but this is not the case, according to the chronosequence study by Noller (1993). Soils in parts of the northern coast are characterized by deep (>1 m) translocation of atmospheric silt and clay, and slight salt accumulation at depth (Fig. 8.31A). In contrast, some southern coastal areas have shallow gypsic soil chronosequences similar to those of southern Israel. Because of the contrasts in morphology, the northern soils have higher PDI values by about an order of magnitude (Fig. 8.31B).

The above variation in soils along one of the driest coastal deserts is attributed to the periodic El Nino event, which brings high rainfalls to the northern but not to the southern coast (Noller, 1993). The rains occur every 3–7 years and last less than 6 months; the amounts range from 4 to 40 cm, and can go as high as 300 cm. When the rains occur they translocate airborne fines deep in the soil, while leaching salts from the soil. Because the rains do not hit the

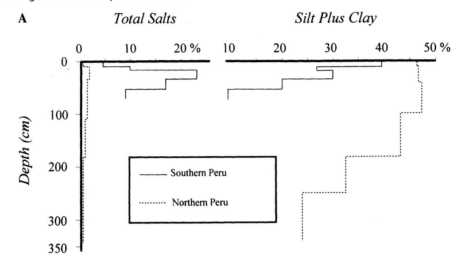

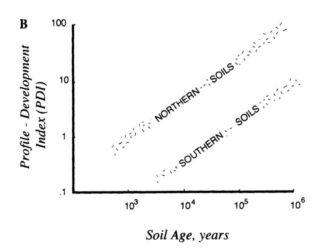

Figure 8.31 Data on (A) total salts and silt + clay, and (B) PDI with age for soils characteristic of northern and southern Peru. *[Data from Noller, 1993; (B) modified from his Fig. 3.8, and (A) courtesy of him.]*

southern coast, the soils there form the usual hyperarid profile.

A very involved diagram was developed to explain the movement of clay, silt, carbonate, gypsum, halite, and soda niter (order of increasing solubility) into the soils along the coast (Fig. 8.32). The diagram is mod-

eled after that of Machette (1985; see Fig. 8.28). For example, southern coastal soils (SC1) are in the accumulation area for all salts, but as rainfall increases (to SC2) all salts except soda niter accumulate. Silt and clay move into the soils to shallow depths, but there is far too little rainfall to move all of the poten-

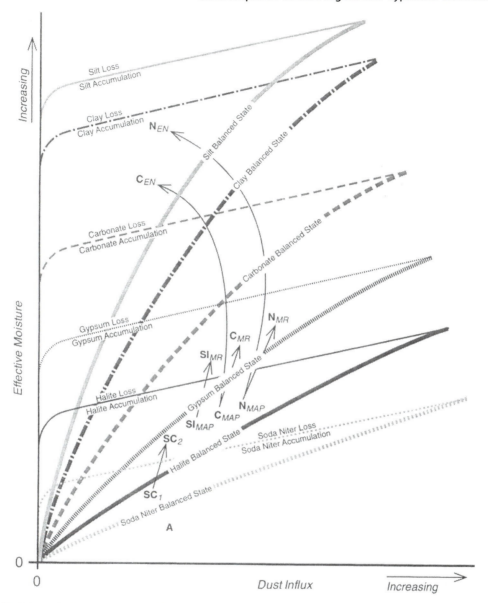

Figure 8.32 Schematic diagram showing effective moisture, dust influx, and fields of accumulation and loss of various soil components in the Peruvian and Atacama (A) deserts. Locations are SC, southern coastal; SI, southern inland; C, central coastal; and N, northern coastal. For SC, 1 is for drier conditions, 2 for wetter. For the other areas, MAP is the mean annual precipitation, MR is for a mud rain, and EN is for El Nino events. This diagram is patterned after Fig. 8.28, but includes more salt and solid phases and climatic events. *(Modified from Noller, 1993, Fig. 6.1 and courtesy of him.)*

tial fines into the soils. In contrast, southern inland (SI-MAP), central (C-MAP), and northern (N-MAP) soils normally are in the halite accumulation part of the diagram, meaning they collect halite as well as all materials higher in the diagram (gypsum, carbonate, etc.). During mud rain events (SI-MR, C-MR, N-MR), these areas move up to the gypsum-accumulation part of the diagram, meaning that halite is preferentially dissolved. Finally, during El Nino events the central and northern soils move into the area where all salts can be dissolved and leached from the soil, but clay and silt accumulate (C-EN, N-EN).

Summary of Desert Soil Development with Time

The work of Machette (1985), McFadden and others (1986, 1987), Gerson and Amit (1987), Wells and others (1987), Chadwick and Davis (1990), Harden and others (1991a,b), Reheis and others (1995), and Reheis and Kihl (1995) can be combined to suggest the following long-term soil-geomorphic evolution of gravelly fluvial deposits in arid lands. Shortly after a floodplain is abandoned, bar-and-swale topography is present, and such a rough surface is an ideal dust trap. In time, weathering reduces the size of surface rocks, colluviation and eolian influx erode the bars and bury the swales, and the bar-and-swale topography is replaced by a smooth surface with desert pavement. The desert pavement does not form by deflation, as claimed in many textbooks, but forms when atmospheric silt and clay infiltrate beneath the surface stones to form a gravel-free Bt horizon, and at the same time lift the stones so that the latter always remain at the surface. The smooth surface is a less efficient dust trap than the rough one, so that with time a greater proportion of the atmospheric solids may not become translocated to depth but can bypass the site. However, if cushiony mosses are present, the surface can remain an effective dust trap (Danin and Ganor, 1991). The formation of a surface crust, primarily in the Av horizon, and the plugging of voids with fines diminishes infiltration and the capacity of the soil to transmit solids vertically in suspension and salts in solution; high concentrations of Ca^{2+} could flocculate clay particles and limit the depth to which they are carried.

That dust is a major source for pedogenic silt, clay, and carbonate, and more soluble salts is shown by dust-trap data and chemical and mineralogical data. In time, a Bw horizon is transformed to a Bt horizon as the underlying carbonate-accumulation horizon (or that containing more soluble salts) attains increasingly higher stages of morphological development. Commonly, the older soils, those that have gone through at least one wet–dry climatic cycle, display signs of chemical weathering and superimposed properties of both moisture regimes. Alteration is such that parent material clay minerals can alter within 10 ky, and, under the correct conditions, palygorskite and sepiolite form.

Dust influx seems to be episodic, with Holocene rates much higher than rates averaged over longer time spans. The presence of playa sources for dust is one explanation, but substantial dust is also derived from alluvial surfaces. Pre-Holocene soils could have experienced similar short episodes of rapid influx, but the evidence for this is partly masked by subsequent pedogenesis. Another factor in dust availability is the variation of vegetation cover with climatic change, and this would tend to enhance high influx during the dry climates. The very dust that led to rapid soil formation eventually brought about soil destruction. One result of increased fines in the soil is a decreasing soil-infiltration capacity, which results in relatively greater runoff, the development of a drainage network on the abandoned terrace, and eventual evolution of the landform to the ballena stage and loss of the soil (Fig. 8.10).

■ Leaching of Parent-Material Carbonate

Parent materials in some humid regions contain original $CaCO_3$ and, because of the prevailing climate, this material is removed slowly in solution by the percolating waters. Some figures on the rate of $CaCO_3$ depletion from soils are as follows: removal from the top 5 cm of the soil in less than 50 years at Glacier Bay, Alaska (Crocker and Major, 1955), and from the top 2 m in 1 ky in sand dunes along Lake Michigan (Olson, 1958). Stratigraphic studies in the midcontinent have tended to use the depth of $CaCO_3$ leaching as one criterion for the approximate age of the parent material. Thorp (1968) points out, for example, that O-isotope-stage 2 drifts in northern Illinois are leached to 0.5–1.5 m depth, whereas 140-ka drifts in the southern part of the state are leached to a 2.5–3.5 m depth. This distinction can be incorporated into the soil description. For example, Follmer and others (1979, Appendix 3) have suggested subdivisions of the C horizon if formed from calcareous materials, such as till. In addition to carbonate leaching, the criteria include alteration of mineralogy (including clays) and staining (oxidation). Their C4 horizon probably corresponds to the Cu horizon used here. Thus, for any region with a uniform parent material, there should be a depth of $CaCO_3$ leaching corresponding to the age of the parent material. Many workers, however, warn that this criterion for age should be used with caution because many variables can alter the rate of $CaCO_3$ removal. Furthermore, even if environmental conditions remain constant for long periods of time, the rate of leaching would decrease with depth in the soil because less water is available for leaching at greater depths, and the water there could be saturated with respect to $CaCO_3$. Hence, one would not expect a linear relationship between depth of leaching and duration of soil formation.

■ Summaries on the Time Required to Reach Various Levels of Pedological Development

We can summarize some of the above data to show the time needed to reach a particular level of soil development or attainment of a particular amount of some soil material. The data are meant to give both an approximate figure, as well as show how one might go about making such comparisons. Comparisons are fraught with problems, however, as not all workers address the problem of inherited vs pedogenic properties, nor do they always present the data on the same basis so that comparisons can be made easily. Carbonate properties have been covered above and will not be repeated.

A-Horizon Organic Matter and Associated Properties

Organic matter probably reaches a steady state more rapidly than any other property of the soil. Schlesinger (1990) reviewed data from a wide variety of environments and concluded that organic matter increases relatively rapidly for the first 3 ky of soil development; after that the rate is less and may continue for several ky. His review gives the range in long-term accumulation from 0.2 g C/m^2/year in a polar desert to 12 g C/m^2/year in temperate and boreal forests. Eventually a steady state has to be achieved in which gains are balanced by loses.

The organic matter steady state is reached when the curve of organic matter vs time is flat. The time required to reach the steady state varies with environment. Not too many curves are available, but Harden and others (1992) present several from a variety of environments (Fig. 8.33). Buildup in the Spodosol is the most rapid, and still is increasing at 1 ky; Schlesinger (1990) presents a curve for New Zealand sand dune

soils that also is rapid, reaching ca. 20 g C in 10 ky, and the curve has not flattened. I believe Figure 8.33 is representative, and most soils reach the steady state between <5 and 20 ky.

The dynamic nature of the organic-matter system and the rapidity at which steady state is reached suggest that a new steady state could be achieved quite rapidly if conditions change. Hence, it is one soil property that is amenable to the functional use of factors because it can be quantified, and polygenesis is not a great problem. The A horizon properties of most soils are probably in a steady state with prevailing conditions and therefore are of limited use in stratigraphic studies except for very young deposits or surfaces.

Several other soil properties develop as rapidly as the trends of the organic-matter constituents of the A

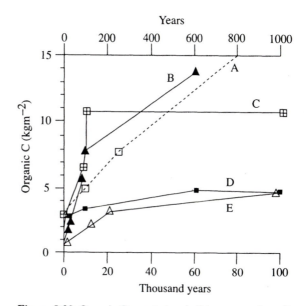

Figure 8.33 Organic C versus time buildup curves for soils in various environments. Each symbol is for one or more soil profiles of a given age, and all points of a chronosequence are connected by a straight line. Localities are (A) Alaskan Spodosols at Glacier Bay, with age in years (top of diagram); one point is off the scale (27 kg/m^2 at 1000 years) (all other ages are in thousands of years shown at the bottom of the diagram); (B) Michigan Spodosols; (C) Baffin Island, NWT, Mollisols; (D) Baffin Island Inceptisols; and (E) Pennsylvania Alfisols. *(Redrawn from Harden and others, 1992, Fig. 1, Dynamics of soil carbon during deglaciation of the Laurentide Ice Sheet. Science 258, 1921–1924, © 1992, with permission from the American Association for the Advancement of Science.)*

horizon, and they probably respond to these trends. pH commonly becomes more acid quickly and could reach a steady state before or at about the same time as does the A-horizon organic matter content (Fig. 8.34). Carbonates, if present, will influence the pH, unless they are leached as rapidly as the organic constituents build up. The rate of leaching will depend on the local climate. The eventual pH at the steady state will depend on some combination of climatic, vegetational, and parent-material influences.

Comparative Data on Clay and Iron Buildup in Soils of Different Climates

From some of the above data we can compare the mass of clay and iron buildup in soils in various environments. However, for much of the data used here, the parent-material amounts have not been subtracted from the totals in the soils so that pedogenic amounts cannot be reported. This distinction is important because pedogenic amounts should plot at the origin for 0 years. In contrast, if parent-material amounts are not subtracted from the data, it is difficult to know the shape of many curves in the first 10–20 ky of soil formation. Data for the clay-accumulation index could also be used here but they are not plentiful.

The bulk of the clay-accumulation data seems to fall into two artificial groups (Fig. 8.35). High amounts (group A) are characteristic of such diverse areas as the coast and Great Valley of California, Cowlitz River area, Washington, and Barbados. In contrast, low amounts (group B) are characteristic of the Sierra Nevada, the arid and semiarid soils of McFadden's (1982) southern California transect, Rock Creek, Montana, and the Rocky Mountains. The coastal plain soils of the eastern United States and the

xeric soils of McFadden's (1982) transect lie between the two groups, and the rate of increase in the coastal plains soils is rapid enough to intersect group A at about 1 Ma. Some desert soils with high influx of fines fall within group A, such as the Kyle Canyon soils (Reheis and others, 1992).

One application of the above is that the clay content in Bt horizons can be used to help estimate the ages of some Quaternary soils. For example, Boardman (1985) notes that the soils in the midcontinent of the United States that presumably formed during pre-Holocene interglacials cannot be adequately explained by assuming a Holocene-length interglacial of 10 ky. He obtains an unrealistically high rate of clay accumulation if the early Sangamon (O-isotope stage 5) soil of Guccione (1982) is assigned a formation duration of 10 ky, and gets a more reasonable rate if the duration is 100 ky. He suggests that some of the interglacial soils formed either during a longer interglacial or over several glacial–interglacial cycles.

The Fe_d-accumulation data also fall into two artificial groups (Fig. 8.36). Greater accumulation (group A) is present in Ventura, California, the northern California coast, Cowlitz River area, and the xeric soils of McFadden's (1988) transect. Lower accumulations (group B) are present in the Great Valley of California, the semiarid and arid soils of McFadden's (1988) transect, and the Rocky Mountains (Swanson, 1984, 1985). The soils of the coastal plain of the eastern United States have intermediate values at about 100–200 ka, and trend into group A at about 1 Ma. In contrast, the New Zealand data (note that it is for Fe_o, not Fe_d) fall into group B for the younger soils, but then the curve reverses as Fe is lost from the system.

Factors involved in explaining the above trends in clay and Fe_d are the parent-material amount, atmos-

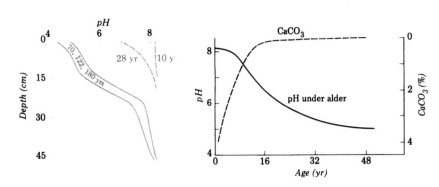

Figure 8.34 Variation in pH and carbonate content (top 5.1 cm) with time, Glacier Bay, Alaska. *(Redrawn from Crocker and Major, 1955, Figs. 4 and 11.)*

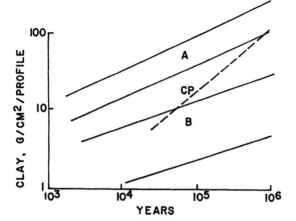

Figure 8.35 Clay accumulation with time in soils in various environments. Group A includes San Clemente Island, California (Muhs, 1982), Ventura, California (Harden and others, 1986), San Joaquin Valley (Harden, 1987), Cowlitz River area, Washington (Dethier, 1987), Missouri (Boardman, 1985), and Barbados (Muhs and others, 1987). Group B includes the Rocky Mountains (Swanson, 1984; Colman and Pierce, 1986), Rock Creek area, Montana (Reheis, 1987a), the arid and semiarid areas of McFadden's (1982) southern California transect, and the Sierra Nevada (Berry, 1994). The intermediate plots include the eastern United States coastal plain (CP) (Markewich and others, 1987), and the xeric soils (not shown) of McFadden's (1982) transect. Only data for Ventura, Cowlitz River area, Rocky Mountain data of Colman and Pierce (1986), Sierra Nevada, and that of McFadden (1982) have been corrected for parent material amount of clay. *(Reprinted from Birkeland, 1990, Fig. 6 Soil-geomorphic research—a selective review. Geomorphology 3, 207–224, © 1990, with kind permission from Elsevier Science, The Netherlands.)*

glosses over the individual accumulation rates and any fluctuations driven either by climatic changes or by episodic pulses of atmospheric dust.

Comparison of PDI Values with Time in Different Climates

PDI values also can be used to rank the various areas. Values increase in most areas with time. Intuitively, one would think that plots of PDI versus age would be different in different climatic regimes. However, in one study that included climates that range from arid to humid, most of the data cluster broadly together (Harden and Taylor, 1983). The clustering was best when only the four best field properties in each area were used (Appendix 2, Fig. A2.6). This has important implications when field soil data are used to pro-

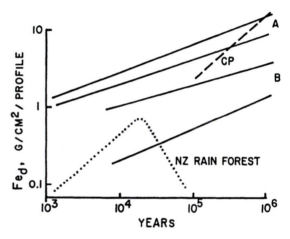

Figure 8.36 Fe_d accumulation with time in soils in various environments. See caption for Fig. 8.35 for references. Group A includes Ventura, California, Cowlitz River area, Washington, and xeric soils of McFadden's (1988) southern California transect. CP is the coastal plain soils. Group B includes the San Joaquin Valley, the arid and semiarid soils of McFadden's (1988) transect, the Rocky Mountains (Swanson, 1984, 1985), and Barbados. New Zealand data are for the west coast of the South Island and are for Fe_o (Campbell, 1975); Fe_o probably makes up a major proportion of the pedogenic iron in these soils. Only the data of McFadden (1988) have been corrected for parent material values. Data for Slovenia (not shown), corrected for parent material, plot just left of the CP line (Vidic, 1994, Fig. 5.10). *(Reprinted from Birkeland, 1990, Fig. 7 Soil-geomorphologic research—a selective review. Geomorphology 3, 207–224, © 1990, with kind permission from Elsevier Science, The Netherlands.)*

pheric influx, and weathering of various primary minerals. The relative importance of these factors varies with the area, and is not always addressed by workers, probably because in places pedogenesis mutes the signal. Three additional comments can be made about these data. First, in general the soil chronosequences with high clay accumulation also have high Fe_d accumulation; this supports the findings of McFadden (1988; see Fig. 8.23), only here the data are for a wider range of environments. Second, one would intuitively expect that soils in wetter environments (eastern United States, New Zealand) should have much higher amounts of Fe_d compared to the relatively dry soils of the western United States, but this does not appear to be the case. Third, grouping into envelop curves vide broad age estimates for deposits. Vidic and Lobnik (1997) added two more data sets to the that of Harden and Taylor (1983). The chronosequence Vidic studied, near Ljubljana, Slovenia (MAP 140–170, MAT 10), on river-terrace deposits made up of limestone and dolomite clasts plots higher than many other areas, many of which are in drier environments (Fig. 8.37). Interestingly, if one were to make an envelope curve of the data, data for the two wettest areas would form the upper and lower limits of the envelope. Furthermore, although these data on aridic carbonate-accumulating soils are not included in Fig. 8.37, data for the aridic soils of McFadden's (1988) transect, the

Cima soils (McFadden and others, 1986), the Kyle Canyon soils (Reheis and others, 1992), and the Black Mesa soils (Karlstrom (1988) all would plot in the upper half of the plotted data set.

Summary of Bt (Argillic) Horizon Development

The formation of a Bt horizon that meets argillic criteria is time dependent because eolian influx, weathering, clay formation, and translocation all proceed at different rates in different climates. Here I will take the stratigraphic approach and compare clay-content

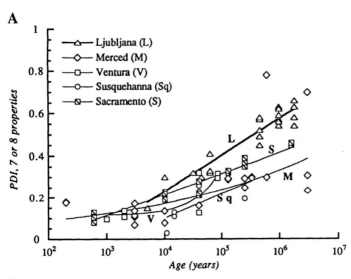

Figure 8.37 Plots of PDI versus age for soils in various environments. (A) is for all properties measured (seven or eight), and (B) is for the four best properties. Data for Merced (San Joaquin Valley) and Ventura, California, and Susquehanna, Pennsylvania, are from Harden and Taylor (1983), that for Sacramento, California from Busacca (1987), and Ljublana, Slovenia from Vidic and Lobnik (1997). *(Reprinted from Vidic and Lobnik, 1997, Fig. 5 Rates of soil development of the chronosequence in the Ljubljana Basin, Slovenia. Geoderma 76, 35–64, © 1997, with kind permission from Elsevier Science, The Netherlands.)*

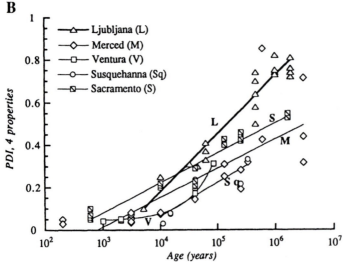

profiles of deposits of different age from different regions.

A plot of clay content vs depth usually shows an increase in the amount and thickness of B-horizon clay with time. Data on soils from a variety of regions are given in Fig. 8.15, many of which have been discussed above; these soils are thought to be representative of the rate of clay buildup for the respective region and for the indicated parent material, and more similar diagrams are included throughout the text. The plots emphasize that the rate of clay buildup varies remarkably from region to region. This is important in soil-stratigraphic studies because clay content is one of the basic properties used in the correlation of unconsolidated deposits.

There are several reasons for the regional differences in rates of Bt-horizon development. Both central and northern California coasts have sufficient moisture for Bt horizon formation, but the minimum time needed to form such horizons is not known. High Na^+ contents, even if seasonal, associated with coastal environments (Chapter 5) could help accelerate the translocation process. Dust from mainland California helps speed up Bt development on the offshore San Clemente Island of southern California (Muhs, 1982, 1983). Relatively low soil moisture, combined with sandy parent materials and a relatively low dust influx, could limit the rate of Bt development in the San Joaquin Valley and on the flanks of the Sierra Nevada and the Rocky Mountains. Of interest for stratigraphic studies is the lack of a Bt horizon in soils 10–20 ka in these localities, as well as in soils of similar age in Pennsylvania, and the gross similarity between clay accumulation plots for the Sierra Nevada and Rocky Mountains for soils about 140 ka. The Bt horizon in a 10-ka soil in the Rocky Mountains above treeline is a bit of a surprise as the soils are frozen half of the year, but they probably formed primarily from mechanically infiltrated atmospheric dust, subsequently bioturbated. In contrast to the surrounding mountain flanks, soils in parts of the Great Basin and the Las Cruces area display rapid Bt-horizon development, even in Holocene soils. Part of the reason for such rapid development in these aridic soils is that the translocated clays are eolian dust derived from abundant sources and concentrated in a relatively thin zone. And, if the parent materials are gravelly, Bts form even faster. If carbonate content is sufficiently high, however, Bt-horizon formation is inhibited and may require all of Holocene time to form (Gile, 1975). In

Southern California, McFadden (1982, 1988) reports Bt horizons in early Holocene soils in both semiarid and arid climates. Perhaps dust influx is less there than in New Mexico. Maximum time for the formation of a Bt horizon is 40 ka for the San Joaquin Valley and Pennsylvania and about 140 ka for the mountain-front tills of the Sierra Nevada and Rocky Mountains. Thin Bt horizons, in places, have formed sooner in the Rocky Mountain mountain-front tills, such as in the Wind River Mountains (Hall and Shroba, 1993). In Pennsylvania, Bilzi and Ciolkosz (1977) cite other work to suggest that occasionally Bt horizons are present in soils formed from 12-ka alluvium.

The impression was given above that Bt horizons take a long time to form in the eastern United States, but this is not necessarily so. In West Virginia (MAP 112) a Bt is recognized in artificially built mound material, constructed 2.1 ka (Cremeens, 1995). Soils in northern Georgia (MAP 152, MAT 13) take about 5 ky before a Bt horizon is recognizable (Leigh, 1996).

Finally, Bt horizons form rapidly in some midcontinent deposits and environments. Soils formed on the late Wisconsin loess (O-isotope stage 2) have such well-differentiated clay profiles (Fig. 7.14) that Bt horizons are expected to form within the Holocene (Ruhe, 1969, 1983). Perhaps this rate results from some combination of high-clay-content parent materials and a leaching soil-moisture regime. They also form rapidly in stream deposits in Ohio. In contrast, Bt horizons require at least Holocene time to form in sanddune deposits (Miles and Franzmeier, 1981).

A few studies have reported Bt-horizon formation within the latter part of the Holocene. Clay translocation is recognized in a soil formed in loess spoil in 0.1 ka in Iowa, but the amount falls short of argillic criteria (Hallberg and others, 1978). One of the more rapidly formed Bt horizons formed in 0.2–0.4 ka near Lubbock, Texas (Holliday, 1985b); recall, however, that this is an area in which mud rains are a contributing factor (Chapter 5). In the Nevada desert, a Bt horizon has formed in a soil that might be less than 2 ka, and this is attributed to both dust and the dispersive effects of Na^+ (Peterson, 1980). The world record, however, could be 41 years under pine trees in a field controlled experiment in California (MAP 68, MAT 14; Graham and Wood, 1991).

In contrast to the above, Bt horizons are not common in exceptionally dry and cold environments. The review above showed that in the cold desert of Antarctica, Bt horizons are not present in soils several million years old. In the hyperarid desert of southern Israel, however, they are present in Holocene soils, and the difference in development pathway is airborne dust vertically translocated by the little rainfall that exists in Israel.

The development of Bt horizons with time is somewhat different in very sandy materials, such as dune sand. Discrete clay bands are the first indication of translocated clays, and these increase in number, lateral continuity, and thickness to eventually merge into a Bt horizon, surely within 10 ky.

A cursory look at many data sets suggests that the maximum amount of Bt clay in the oldest soils of chronosequences lies in the 60–80% range. This is reached in about 2 Ma in the Colorado Piedmont

(Birkeland and others, 1996), but takes five times as long in the more humid eastern United States.

Summary of Development of Red Color

A common observation in many areas is that older soils are redder than younger soils (Ruhe, 1965, 1968). This is readily shown by the color of the B or Cox horizon, and by the dry rubification; data in Table 8.5 can be used as an illustration. The offshore California colors show little progression with time, probably because the soils are clay rich, the older ones are Vertisols and thus mix, and in the dry climate, chemical weathering might be minimal. In the wetter environment of the central California coast (not plotted),

Table 8.5 Color of the B Horizon, or Cox Horizon if No B Horizon is Present, for Soils at Various Localities, and Rubification Values (in Parentheses)

Location[a]	Parent Material	0	10³	10⁴	10⁵	10⁶	Moist (m) or Dry (d)
				Approximate Age (years)[b]			
1. California offshore island	Eolian over marine sediment	—			10YR 3/3	10YR 3/3	m
2. San Joaquin Valley, California	River sediment	10YR 7/2	10YR 5/4 (20)	10YR 5/4 10YR 6/6 (20) (40)	5YR 4/4 2.5YR 3/6 (40) (70)	10R 3/6 (80)	d
3. Sierra Nevada, California	Till	5Y 7/2		10YR 5/4 (40)	7.5YR 5/4 5YR 4/4 (50) (60)		d
	Till[c]	5Y 7/1		10YR 5/3 (40)			d
4. Lower Colorado River, California	River sediment	10YR 7/2		7.5YR 6/6 5 YR 5/8 (50) (80)	5 YR 5/6 2.5YR 5/6 (60) (70)		d
5. Las Cruces, New Mexico	River sediment	—	5YR 3/4	5YR 5/4 5YR 5/5	2.5YR 4/6		d
6. Rocky Mountains, Colorado	Till	2.5Y 7/2		7.5YR 6/3 (30)	7.5YR 5/4 7.5YR 6/6 (40) (60)		d
	Till[c]	7.5Y 6/1	10YR 4/4 (60)	7.5YR 4/5 (80)			
7. Southern Alps, New Zealand	Till	5Y 6/1	10YR 6/4 (50)	7.5YR 4/8 (100)			d
8. Colorado Piedmont	River sediment	10YR 3/2		7.5YR 4/6 (50)	5YR 5/5 (50)	10R 4/6 (80)	m,d

[a]Location reference: (1) Muhs (1982); (2) Harden (1987); (3) Burke and Birkeland (1979); Birkeland and others (1980); and unpublished data of author, R.M. Burke, and J.C. Yount; (4) McFadden (1982); (5) Gile and others (1981); (6) Shroba (1977); Birkeland and others (1987); (7) Birkeland (1984); (8) Machette and others (1976a,b).

[b]Scale is arithmetic between tick marks.

[c]Tills in alpine environment.

2.5YR hue occurs in 100-ka soils (written communication, J.W. Harden and others, 1996), and the older soils do not have redder hues. For many of the other areas, data are shown for the parent-material colors ($t = 0$ years), so one can better appreciate both the color change with time and the value of obtaining a color as close to the parent material as possible. Redness seems to increase more rapidly in the San Joaquin Valley relative to the Sierra Nevada, and redness develops in the Rocky Mountains at a rate that is close to that for the Sierra Nevada. The Colorado Piedmont has a rate that could be slightly higher than that for the San Joaquin Valley, and the 10R hue for the oldest soils in both areas is about as red as soils get. It might take three to five times longer to reach the 10R hue in the wetter soils of the eastern United States (Fig. 8.18). Some of the highest rates are those in arid areas, such as Las Cruces and the Lower Colorado River. Also of interest is the relatively rapid increase in redness of alpine soils, especially those in New Zealand and the Rocky Mountains. The rubification values more or less parallel the redness rates, but for any detailed study, all colors should be taken in the same manner by one observer. Although data are not given in the table, the rate of increase in redness is very slow in cold, dry arctic conditions of Baffin Island, and even slower in the cold desert of Antarctic.

One problem in linking color development with age of parent material is the role of paleoclimate in producing redness. As shown above, red soils form fairly rapidly in areas characterized by high temperatures. Hence, if soil redness increases with age of deposit, one or two factors may be responsible. One is time, and the other is warmer past climates. For example, many people link climate change with soil properties in desert areas (examples in Bull, 1991); a common horizon is a Btk, with Holocene carbonate overprinting the deeper leaching of the late Pleistocene (see Chapter 10). It is difficult to separate the effects of these two factors (Ruhe, 1965). Red Bt horizons in England, termed paleoargillic, are ascribed to formation under former warm climates over a relatively long interval of time, and are key stratigraphic markers (Rose and others, 1985). Paleoclimatic interpretation of soils, therefore, should rest only partly on color, and should include information gleaned from other soil properties, as well as stratigraphic and topographic setting.

Red Btk horizons may suggest climate change. In young soils that probably have not undergone a cli-

mate change, I have observed that the red color is restricted to the Bt horizon, and that the latter overlies a Bk horizon with little rubification. Soils with a Bt/Btk morphology that are old enough to have formed over several climatic cycles can have red Btk horizons. This could mean that the red color is imparted on the soils mainly during times of greater leaching, and that little additional red color is produced when the carbonate is overprinted on the red Bt.

Summary of Clay Mineralogy

The clay mineralogy of many soils may change with time, and, if it does, this may confound efforts to use clay minerals as paleoclimatic indicators. Two cases will be explored. In one, the initial parent material contains little inherited clay, and clay formation from alteration of nonclay minerals predominates. In the other, the parent material contains clay minerals of diverse types, and many of the clay minerals that form in the soil do so by alteration of preexisting clay minerals. Whether a clay mineral will alter is a function of its stability in the soil solution, as shown by stability diagrams (Figs. 4.7 and 4.8).

Clay minerals that form from the weathering of nonclay minerals should be stable in that environment, or else they would not have formed. If, however, the properties of a soil change with time so that the internal environment of the soil changes, the originally formed clays may alter. For example, the buildup of clay-sized particles in a soil may change the leaching environment enough that (1) existing clay minerals are unstable or (2) clay minerals that form in the future differ from those that formed in the past, yet both were in equilibrium with the water chemistry at the time of their respective formation. For example, chronosequences involving old red, clay-enriched soils commonly display mottling with time, suggesting that drainage becomes impeded. This change from well drained to poorly drained should be reflected in the clay mineralogy. Another example of a pure time control on clay-mineral alteration would be the weathering of the nonclay minerals that was so intense that certain minerals that delivered weathering products and ions to the solution were depleted, and their depletion so altered the water chemistry that the previously formed clays then became unstable. This may happen in areas of intense weathering, such as the tropics, areas of podzolization, or stable landscapes in which weathering dates back to the early Quaternary and Tertiary. For example, in the study summarized in Fig. 8.18, the interpretation is that two thresholds were crossed in time, each marked by changes in mineral products, and each related to depletion of various minerals. In contrast, some areas, such as the much drier Sierra Nevada–Great Basin transect, show little age control on soil clay mineralogy (Birkeland, 1969; Birkeland and Janda, 1971; Fig. 10.16).

In places where a variety of clay minerals are present in the original parent material, changes in mineral species may take place for any mineral not stable in the environment, as shown by the stability diagrams. Jackson and others (1948) ranked the clay minerals by relative resistance to weathering and assigned a number to each mineral characteristic of a weathering stage (Fig. 8.38). Three of the more stable minerals are not given in Fig. 8.38; they are allophane (stage 11), hematite (stage 12), and anatase (stage 13). For most of the minerals, a decrease in iron, magnesium, and silicon is associated with an increase in mineral stability. The arrows in the figure indicate some possible paths for alteration from one mineral to another (for others see Fig. 4.9). In soils with a combination of original clay minerals in the parent material, any alteration due to weathering will be regulated by the water–mineral equilibria. The extent of the alteration, or in other words the weathering stage reached, will be a function of time, because the alterations involve slow reactions and water–mineral equilibria; that is, alteration takes place until an equilibrium mineral assemblage is formed, after which time further alteration probably does not occur.

The midcontinent is a good region in which to study the effect of time on clay mineral alteration. Tills and loesses of several ages are present in many environments, and parent materials contain a variety of clay minerals derived from bedrock units over which the glaciers advanced. Clay-mineral transformations are judged from the vertical sequence of clay minerals in a soil and by comparison of soils of different ages. Clay minerals formed in soils from tills in Illinois demonstrate the changes possible with greater age (Fig. 8.39). The soil on 40+ ka till shows only an alteration of chlorite to vermiculite–chlorite in the B and leached Cox horizons. Some soils on 20-ka till display similar alteration (Frye and others, 1960, 1969). This similarity between clay-mineral alteration in soils formed from two tills of these ages does not always hold true. In some places, illite depletion is recognized in soils on the older till, but not in soils on the younger till (Willman and others, 1966); in addition, some of the soils in the older tills with original montmorillonite show an increase in montmorillonite in the soil. In contrast to the soils on the 40+ ka tills, soils formed from 140-ka till have more altered clays, and the alteration goes to greater depth (Fig. 8.39). Chlorite is altered through a vermiculite–chlorite stage to material termed "heterogeneous-swelling material," which probably is either montmorillonite or a mixed-layer mineral. Illite alters to illite–montmorillonite and finally to montmorillonite. Older soils seem to be altered to about the same stage. Soil formed from 20-ka loess in Illinois shows slight clay-mineral alteration (Frye and others, 1968). The major alteration is that of montmorillonite to the heterogeneous-swelling material. Other changes are slight depletion of illite, modification of chlorite, and possibly formation of vermiculite and kaolinite.

In Illinois, therefore, the stable clay mineral seems to be stage 9 montmorillonite, but of the heterogeneous-swelling material variety. This material can form in the time span represented by the weathering of 20-ky-old deposits. More intense alteration to great depths, however, requires weathering durations on the order of several tens to 100 ky. Forsyth (1965) used

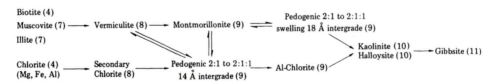

Figure 8.38 Various clay-mineral reactions in soil. Parenthetical numbers are weathering stability numbers for the clay minerals; mineral stability increases with number. (*Redrawn from Jackson, 1964, in F.E. Bear, ed. Chemistry of the soil, p. 124. by © 1964 by Litton Educational Publishing, Inc., Reprinted by permission of Van Nostrand Reinhold Company.*)

Figure 8.39 Clay-mineral variation with depth in soils of different ages formed from till, Illinois. Early Wisconsin till could be O-isotope stage 4, and Illinoian till stage 6. The expandable vermiculite in the original publication subsequently has been shown to be heterogeneous-swelling material (Willman and others, 1966); here it is shown as montmorillonite in the Sangamon soil. Note the difference in vertical scale for the two profiles. *(Redrawn from Frye and others, 1960, Fig. 3.)*

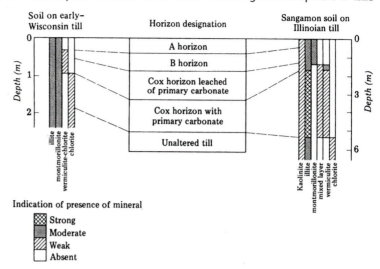

similar relationships to help estimate the age of a buried soil in Ohio. A different clay-mineral pattern is seen in the accretion-gley deposits (Chapter 9).

Guccione (1985) calculated profile mass losses and gains of various clay minerals in a till chronosequence in north central Missouri (MAP 97, MAT 12). The approximate age of soil formation of the chronosequence is 400 ky for the oldest soil, whereas the intermediate-age and young soils formed in early and late parts, respectively, of the interval 130–20 ka. She was also able to suggest pathways of alteration in dominantly oxidized and dominantly reduced profiles. The oxidized profiles (the two in the 130- to 20-ka interval) display gains in vermiculite (31–71 g/cm^2/profile) > smectite (18–31g) > kaolinite (5–9 g), and loss of illite (58–70 g). In contrast, in the dominantly reduced profiles (the older of the soils in the 130- to 20-ka interval and the oldest soil) the pathway was still greater losses of illite (65+ to 79 g), greater gains in smectite (26–108+ g), no change in kaolinite, and a gain followed by a loss in vermiculite. This study quantifies the gains of vermiculite, smectite, and kaolinite at the expense of illite in oxidized profiles and, as is usual for soils, the respective gains and losses increase toward the surface.

The extensive data set of Reheis and others (1995) dispels the notion that clay alteration is minimal in the deserts. The regional dust is fairly homogeneous and dominated by smectite and mica. This can be taken as the initial clay mineralogy from which transformations take place. Once on the ground and/or infiltrated

into the soil, alteration takes place, and is time dependent. Generally, the Av horizons of Holocene soils are similar, and the main alteration seems to take place at depth. Pleistocene soils generally have a different mineralogy in all horizons, and because this can be seen in some late Pleistocene soils, transformations can be rapid. The trend with time is for greater smectite and mixed-layer clay, perhaps at the expense of mica in places, as well as increases in palygorskite and some sepiolite in carbonate-accumulation horizons.

The fibrous clay minerals show some time dependency for formation. In carbonate soils, montmorillonite transforms to palygorskite in about 120 ky, and it takes more than 300 ky for palygorskite to dominate (Bachman and Machette, 1977). Sepiolite then forms by conversion from palygorskite.

Even in aridic soils the rapidity of the clay-mineral transformation depends on bioclimate, just as do many chemical trends. Greater leaching with warmer temperatures should increase the rate of change, and this is seen in those hot deserts with sufficient moisture. At the other extreme are the cold deserts. Bockheim (1980b) reviews data for the Arctic and Antarctic and concludes that little or no change takes place in the Antarctic, even though some chronosequences extend beyond 1 Ma. The most common alteration is a broadening of the 10 Å mica peak (Bockheim, 1997). In contrast, in the Arctic, the alteration goes to vermiculite, and if leaching is sufficient, to kaolinite. Isherwood (1975) recorded the almost complete illite ⇒ vermiculite transformation on Baffin Island in

100–200 ky. Evans and Cameron (1979) report slight changes in clay mineralogy on Baffin Island in soils that vary in age from 8 to 115 ka. Illite seems to transform to vermiculite at depth and to Al-vermiculite toward the surface, and the alteration is most extensive in the older soil.

Holocene soils in the high mountains display clay-mineral changes with time, and the degree of alteration and direction depends on the climate. In the alpine tundra of the Colorado Rocky Mountains, the main alteration in 10 ka is a slight alteration from mica and illite–montmorillonite to mixed layer (10–18 Å) clays (Shroba and Birkeland, 1983). In a still wetter and leaching alpine environment in New Zealand, the transformation is (Table 8.1):

$$\text{mica} + \text{chlorite} \Rightarrow \text{interlayered}$$
$$\text{mica-14Å} + \text{mica–smectite}$$
$$(3 \text{ ka}) \qquad (4 \text{ ka})$$

A recent study in the Southern Alps shows how complicated the clay-mineral transformations can be (R. Harrison and others, 1990). Two chronosequences have oldest soils of 20+ ka. The drier soils (MAP 83) alter to vermiculite, dioctahedral chlorite, interstratified minerals, and gibbsite. The wetter soils (MAP 145) appear to show greater alteration to some of the these products, but include smectite and vermiculite–chlorite, and greater amounts of gibbsite. Important factors besides time in explaining these trends are past changes in bioclimate.

Other New Zealand clay-mineral studies at lower altitude support rapid clay-mineral change in areas of high rainfall. One chronosequence (MAP 495, MAT 11) that goes back to 22 ka detected the transformation mica ⇒ vermiculite in 0.25 ka, with chlorite and montmorillonite increasing in soils of 1–12 ka (Mokma and others, 1973).

Clay mineral formation in the tropical areas should be rapid. In the Pacific Ocean chronosequences mentioned above, even the youngest soils have boehmite and gibbsite. If the soils of Maré come from Australian dust, the inherited composition probably would have been illite, kaolinite, and chlorite (McTainsh, 1989). Without younger surfaces, we cannot determine how long it takes to make these changes, but it is less than 200 ka. If fact, if the hypothesis presented earlier is correct, that of extreme weathering accompanied slow dust deposition, the alteration may have been nearly instantaneous.

Atlantic Ocean carbonate island soils may not have had as rapid a clay-mineral conversion as above (Muhs, in preparation). The dust is mainly from the Sahara and the conversions are as follows:

Barbados (drier):
(MAP 110, MAT 26)
$$\text{illite} \Rightarrow \text{kaolinite–smectite} + \text{smectite} \Rightarrow$$
$$(<125 \text{ ka})$$
$$\text{kaolinite–smectite} \Rightarrow \text{pure kaolinite}$$
$$(460 \text{ ka}) \qquad (700 \text{ ka})$$

Jamaica (wetter):
(MAP 114–324, MAT 26)
$$\text{illite} \Rightarrow \text{kaolinite} \Rightarrow \text{boehmite} + \text{kaolinite}$$
$$(<125 \text{ ka}) \qquad (<860 \text{ ka})$$

This sequence matches that of others in that boehmite forms in the wetter area.

Clay minerals that form from volcanic glass are neoformed and in places the formation is quite rapid. Much work along these lines has been done on the North Island of New Zealand (Lowe, 1986; Lowe and Percival, 1993). Tephras of many ages are present, and they weather in a udic moisture regime. Clay content increases with duration of weathering, being <5% at 3 ka, 5–10% at 3–10 ka, 15–30% at 10–50 ka, and >70% at >50 ka. The most abundant secondary minerals are allophane and halloysite. Allophane, under the correct chemical conditions, can form in 3 ka and is stable enough to persist for several 100 ky. Although halloysite can form in less than 0.1 ka, the conversion of allophane to halloysite is more typical of tephras that weathered 10–15 ka. In contrast, Wada (1989) reports allophane formation in 1 ka, allophane to halloysite in 6 ka, and allophane to gibbsite in 10 ka.

In summary, it seems that the following generalizations are valid. Clay minerals that form by weathering of primary grains in soils from parent materials low in clay content probably form mineral assemblages stable in that environment, and therefore variation in the minerals with age may not be found, unless either major depletion of minerals with contrasting chemical composition occurs, or the internal soil environment is drastically altered (intrinsic threshold). Soils formed from parent materials with a high content of inherited clay minerals, including mica, are another matter. Some clays in the assemblage may be unstable and gradually change to more stable forms. The change will progress from the surface downward, and complete change to a stable assemblage may take considerable time, and much of this is dependent on the amount and chemistry of the percolating soil waters. In these cases, the presence of particular clay-mineral assemblages with depth may aid in estimating either the age of a surface soil or the occurrence of a period of weathering and/or soil formation during the accumulation of a thick section of sediments.

■ Relationship between Diagnostic Horizons, Soil Orders, and Time

The examples presented above help put limits on the time needed to form the diagnostic horizons (Fig. 8.40A). It is difficult to generalize as factors such as parent marterial and climate will alter the rate at which horizons attain the properties necessary to become diagnostic. A horizons form relatively rapidly, and those that qualify as diagnostic epipedons probably require less than 5 ka to form. Cambic horizons form rapidly and by 10 ka should have good expression. Calcic (Bk in Fig. 8.40A) horizons might form as rapidly as the cambic horizons, and many environments would have them by 100 ka. Argillic (Bt) horizons probably take between 10 and 100 ka, with maximum development between 100 ka and 1 Ma. Kandic horizons would take longer than argillic horizons to form in a particular environment, because a longer duration of leaching is required to meet the key properties. Petrocalcic horizons of the K horizon variety take 100+ ka to form. Oxic horizons take the longest to form from the average parent material, and they are put close to 1 Ma.

There is a fairly good correlation between the soil orders and the age of the underlying deposits or landscapes in some regions (Fig. 8.40B). A striking example in the United States is the widespread occurrence of Ultisols in the midcontinent and in the east (Fig. 2.8). These soils have formed on deposits and landscapes of >100 ka, and some are several Ma (Markewich and Pavich, 1991; Howard and others, 1993). The Oxisols of the world also seem to have required long periods of time to form, perhaps 1 Ma or more (Tardy and Roquin, 1992). Although the climate in some regions, such as Australia, was different during the time of Oxisol formation, these soils are so highly weathered that maximum formation times of

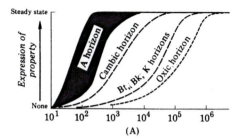

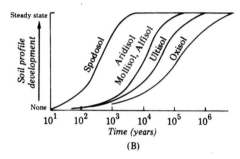

Figure 8.40 Schematic diagrams showing the variations in time to attain (A) the steady-state or maximum expression for various soil properties that would be associated with various soil horizons (e.g., organic C in the A horizons, depletion of all but the most resistant oxides in the Oxic horizon) and (B) the steady-state soil profile development for various soil orders.

several Ma do not seem unreasonable (Division of Soils, CSIRO, 1983).

In contrast to soil orders that seem to require a long time to form, some can form in a short time. Entisols are so little developed that material would be classified in that order shortly after exposure to soil-forming processes; perhaps a century is all that is required. Others, such as Histosols and Vertisols, require so little pedogenesis that formation under appropriate conditions could be very rapid, surely between 1 and 5 ky. Inceptisols could form in that time, or slightly longer. Podzolization occurs very rapidly, and McKeague and others (1978) believe that soils that qualify as Spodosols probably require several ky to form. Andisols could form in the same time. Both Mollisols and Aridisols would require variable times to form because they include both cambic and argillic horizons; those with cambic horizons probably could form in 5 ky and those with argillic horizons about 10 ky. Alfisols, because an argillic horizon is required, probably also require about 10 ky in wet climates and perhaps the same time in dust-influenced dry climates. The time to reach maximum development is more difficult to plot, and those given are broad estimates.

From this it can be seen that soil maps at the order level (Fig. 2.8) can be used to suggest broad groupings of deposits or landscapes on the basis of age.

■ The Steady-State Condition and Time Necessary to Attain It

In the two previous editions of this book time was spent discussing the attainment of the steady-state condition in soils. Curves for the buildup of most properties initially are fairly steep, but after some time the curves flatten, indicating less or no visible change thereafter with time. There can be several reasons for the flattening of the curve. For the A horizon, the gain of organic matter at the surface and its conversion to humus is balanced by loss due to decomposition. For Bt, Bk, and K horizons, I doubt if growth of a particular constituent is balanced by its loss. Rather, the lack of further change with time (steady state) could be due to the balance between slow removel of surficial material by erosion and the slow extention of the soil profile at depth into unweathered material (Nikiforoff, 1949). Or, perhaps, regressive processes (next section) are at work to limit further soil development. For many Ultisols and Oxisols, the steady state might be reached only when most of the weatherable minerals have been depleted by weathering, thus rendering further change unlikely.

Because of the attainment of the morphological steady state, earlier soil-stratigraphic studies concluded that it was difficult to differentiate pre-100-ka soils from 100-ka soils on most field criteria (Richmond, 1962; Morrison, 1964). Hence, the best hope of differenting pre-100-ka deposits using soils was if they were buried and the sequence was complete. We now know that a soil can reach a morphological steady state while particular properties continue to change (Tonkin and Basher, 1990), so we should be sure to state the basis on which steady-state condition is defined.

The above statements on steady state are intended only as broad guidelines. In detail, for specific areas, they may not always be acceptable. Rodbell (1990), for example, shows many curves for various properties of New Zealand soils flattening in 10–20 ka. In any similar study one has to be certain there are suf-

ficient constraining data points for the older part of the curves, so that a steady state can be argued. In the latter example, it would be good to have more data on older stable soils, but these are not abundant in that area. Many plots of soil data demonstrate that some curves have not flattened off even after long durations of pedogenesis (1–10 My) (Bockheim, 1980a; Markewich and others, 1987; Howard and others, 1993). Furthermore, plots of the PDI for a variety of environments have not flattened at 1 Ma or more (Fig. 8.37). At any rate, the flattening of the curves means that for the time span during which the curve is relatively flat, correlation of deposits on soil data will be less precise. Finally, some aspect of regressive pedogenesis could be the reason some soils do not evolve beyond a particular state.

■ Regressive Pedogenesis

This chapter has concentrated on examples of progressive pedogenesis. As defined by Johnson and Watson-Stegner (1987), this is the progressive buildup of various profile forms and constituents with time. They contrast it with regressive pedogenesis, defined as processes that result in soils that no longer change or evolve in the same direction with time.

Regressive pedogenesis can follow progressive pedogenesis, with the result that either the soil morphology no longer changes or properties and constituents no longer increase with time (steady state), or the soil reverses its morphological evolution or pathway, or loses properties or constituents. Torrent and Nettleton (1978) called some of these examples negative feedback, the return to a previous stage in the soil development sequence. One way to do this is to mix soils, causing them to lose some or all contrast between horizons; these processes are reviewed by Johnson and Watson-Stegner (1990) and are given the general term, pedoturbation. They give many subdivisions of pedoturbation. Regressive soils could be part of a chronosequence, as long as it is recognized that the direction of pedogenesis has changed.

There are several ways in which progressive pedogenesis can be halted or reversed, giving regressive pedogenesis.

1. With time the landform erodes, eventually to the ballena form, and as it does surface erosion progressively removes the upper part of the profile. Depending upon the opposing forces of pedogenesis and erosion, the profile could remain the same or it could lose constituents (e.g., maximum clay in upper Bt) and become less developed.

Fire can accelerate surface runoff and rapid soil loss. Kutiel and others (1995) list three reasons for this: (1) destruction of the protective vegetation cover, (2) creation of a less permeable surface crust, and (3) formation of a hydrophobic layer (DeBano, 1980; Wells, 1987) that drastically reduces water infiltration, although fire is not always necessary for its formation (Doerr and others, 1996). Certain soil features may help in detecting if fire is the culprit (Ulery and Graham, 1993; Ulery and others, 1996).

2. Wind erosion could be the process that causes a soil to switch from a progressive to a regressive mode. Climate change with a change to less vegetation cover and increasing wind speed could strip a surface of its soil as long as it did not have sufficient gravel to form an armor and stop the process. Such a process has been called on in New Zealand to explain some of the loess–soil relations (Bruce, 1973). Frost action could help with the above process. One way in which this effects soils is illustrated in another example from New Zealand (Soons, 1968, Soons and Greenland, 1970). Needle ice can form readily in fine-grained material, such as loess. This can begin to reduce the vegetation cover, and when the ice melts on a daily basis, leave a rather puffy, exposed surface. High velocity winds can then remove the loess and soil and expose the underlying material, such as gravelly scree or till.

Burrowing animals can bring fine-grained materials to the surface and, if left exposed, be removed by erosion. Burns (1979) studied gopher activity in the alpine region of the Colorado Front Range. Surface mounds are built in the summer, whereas in the winter the gophers fill tunnels in the snow with soil, which, on melting, form "eskers" (casts) on the surface. These unvegetated materials are prime locales for erosional removal by wind and water. If removal exceeds the influx, the soils thin in time.

3. Exceptionally cold climate with accompanying cryoturbation could disrupt the morphology, homogenize the profile, or erode the soil. A dramatic example of this is periglacial gelifluction features impressed on an older, highly rubified Bt horizon formed under a temperate climate near London (Whiteman and Kemp, 1990).

4. The development of a soil itself (e.g., increasing clay content or formation of an iron pan) could so reduce the movement of soil water vertically that runoff increases and the soil and landform erode. This link between soil development and erosion was clearly indicated by Wells and others (1987) for an area in the Mojave Desert.

5. Accumulation of constituents in the soil can change the directions of pedogenesis in several ways. As Alfisols form, the increasing degree of development is shown by increased clay content and increased soil anisotropy; the smectite content of the clay can also increase. In time, and in the right climate, the soil can have so much clay that it becomes progressively more vertic, eventually forming a Vertisol (Muhs, 1984). The result is a loss of soil anisotropy, and this is presented by Muhs as an example of a pedogenic threshold. Total clay content could continue to increase with time, either through weathering or from an atmospheric source, as the cracked surface could trap most incoming dust.

Carbonate accumulation can have several effects. One is that as a Km horizon builds upward it can engulf all or part of the overlying Bt horizon (Fig. 8.41). Another effect of carbonate accumulation is that as carbonate content increases, the rate of infiltration decreases. In one study, as carbonate content increased to >90%, infiltration decreased from 15 to 0.13 cm/hour, a decrease of two orders of magnitude (Gile and Grossman, 1979, p. 220). Changes such as this could increase soil moisture in the overlying horizons to the point that surface runoff increases and the horizons overlying the Km are removed.

6. Another potential process for arresting progressive soil development is the uprooting of trees, usually by wind (Schaetzl and others, 1989; Johnson and Watson-Stegner, 1990; Schaetzl, 1990; Schaetzl and Follmer, 1990). This creates a pit and mound microtopography and the soil horizons can be quite disturbed. Horizon sequences can be reversed, some horizons are missing, and lateral continuity is disrupted (Johnson and Watson-Stegner, 1990, Figs. 7 and 9). Subsequent pedogenesis works to reorganize these materials with respect to the present-day surface, and it might be some time (several ky) before the original profile form is reconstituted. The return to the former morphology takes less time than the time to first reach that morphology as many constituents already are in place; they just have to be rearranged into proper order again. Another complication is that the pit and

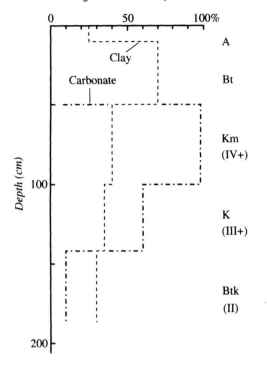

Figure 8.41 Engulfment of a Bt horizon by pedogenic carbonate, as shown by the clay bulge (analysis on carbonate-free portion) extending down into the Km horizon. The soil has formed from gravelly Rocky Flats alluvium, south of Boulder, Colorado (described in Machette and others, 1976a,b), and the alluvium could be 2 Ma (Birkeland and others, 1996).

mound microtopography could set up micro-environments so that opposing soil processes (e.g., oxidation vs reduction) occur side by side over short distances.

7. Ferrolysis (Chapter 5) can also reverse the direction of pedogenesis. In a New Zealand example, above, pedogenic Fe buildup (Bw or Bs) is followed by its removal (Fig. 8.36), leading to formation of a prominent E horizon overlying a thin Fe-pan (Bsm). In time, the progressively lower pH solubilized the Fe, some of which went to form the Fe-pan and, combined with reducing conditions on top of the pan, much of the Fe in the upper part of the profile was depleted. In another example, a soil can progress as follows: Inceptisol ⇒ Udalf ⇒ Glossic Aqualf; as internal processes reverse, an E horizon engulfs the Bt horizon, and gleying increases (Fig. 8.42).

8. Formation of pans can lead to their demise (Torrent and Nettleton, 1978). In the case of a Km hori-

zon (petrocalcic), the laterally flowing waters can seek out avenues for downward movement, such as cracks, dissolve the carbonate, and form a window in the Km horizon. In time a steady-state number of windows per unit area could have formed. The overall result is a decrease in the grams of carbonate per unit area.

It is important to keep these, and other, processes in mind when trends in chronosequences that should be predictable are not found in the field. An example of this is the absence of many trends with soil age near Laramie, Wyoming, in spite of some deposits being over 500 ka (Reider and others, 1974). The reasons for this are not clear, but strong erosive winds and paleopermafrost conditions (Mears, 1981) might be contributing factors.

■ Estimating Ages of Deposits

Trends such as many shown in this chapter demonstrate that soils can be used to estimate the ages of the associated deposits. There are not many data on the error limits (±) of such studies, but one in New Zealand put it at ~50% (Knuepfer, 1988). In a statistical study of PDI values, Harden (1987) put the error limits at ~70%. For particular parts of the time scale, recalling the logarithmic nature of many curves, such

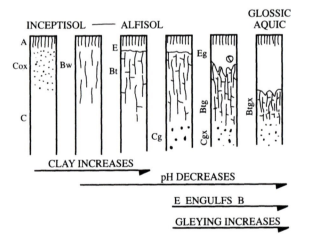

Figure 8.42 Soil profile changes with time illustrating progressive pedogenesis followed by regressive pedogenesis in a calcareous loamy parent material. *(Redrawn from Pedro and others, 1978, Fig. 2. Two routes in genesis of strongly differentiated acid soils under humid, cool-temperate conditions, Geoderma 20, p173–189, ©1978, with kind permission from Elsevier Science, The Netherlands.)*

estimates are still quite valuable, and perhaps no worse than age estimates by some other techniques (i.e., stream incision rate, rock varnish cation ratios, ^{14}C dating of rock varnish, cosmogenic nuclide dating of surface boulders, weathering-rind data, etc.). As an example of this, one can cite the weathering and soil study of Burke and Birkeland (1979) in trying to estimate ages of moraines in the eastern Sierra Nevada. We ended up lumping moraines together that had pre-

viously been subdivided, and in places different classification on age. This created a roar, which promoted much high-quality work many fronts. Commonly, the later work usual changed that of Burke and Birkeland (1979). However, more recent research using cosmogenic dating of boulders showed that the age classification of Burke and Birkeland compares favorably with the more modern techniques (Chapter 11).

Chapter 9

Topography–Soil Relations with Time in Different Climatic Settings

Topography, or local relief, controls much of the distribution of soils in the landscape, to such an extent that soils of markedly contrasting morphologies and properties can merge laterally with one another and yet be in equilibrium under existing local conditions (Fig. 9.1). Many of the variations in soils with topography are due to some combination of microclimate, pedogenesis, and geological surficial processes, and sorting out the effects of each on soil distribution is difficult. The fields of pedology and geomorphology probably overlap here more than with any other pedologic factor, as discussed by Gerrard (1992) and Hall (1983).

At a much grander scale, one can compare soil landscapes of various continents, incorporating climate change and plate tectonics. This was done by Patton and others (1995), in comparing Australia and New Zealand. Australia has low relief, exceptionally old landscapes, both a function of being located far from converging plate margins. The old landscapes have survived the erosional effects of climate change as the extent of glaciation has been minimal. Hence, old deep, red soils are common (Division of Soils, CSIRO, 1983). Contrast that with New Zealand, where uplift along a converging plate margin produces high mountains subject to large-scale glaciation and rapid erosion (e.g., Tonkin and Basher, 1990; Bull, 1991). Late Quaternary moraines, river and marine terraces are plentiful, and loess provides yet another parent material once it is laid down across the landscape. Hence, New Zealand soils are young and there are many opportunities for chronosequences and toposequences in environments in which the soil-forming factors are rather well known.

Soil properties vary with topographic setting. One reason for this is the orientation of the hillslopes on which soils form; this affects the microclimate (i.e., N- vs S-facing) and, hence, the soil. Another is the shape of the slope and how it influences the redistribution of moisture along the slope; this affects soil properties because the rates of surface-water runoff influence erosion and of soil-moisture variation with slope position. In areas of rolling terrain, the simple landscape depicted in Fig. 9.2 (a catena, defined later) helps explain the soil-moisture redistribution in an area of uniform MAP. Higher parts of the landscape could experience greater evaporation, and the soils would have lower soil-moisture contents and wet to shallow depths. In cold areas, snow could blow off these sites, leaving them relatively dry. The redistribution of water downslope, both at the surface and subsurface (throughflow), gives soil properties downslope indicative of moister conditions, and moves solutes laterally. If snow redistribution occurs, these could be the sites of greater snow accumulation. Also, low areas can be influenced by a high water table, which could have a considerable effect on the soil (mottling and gleying). If sediment is eroded and deposited, it

Figure 9.1 Variation in soils with topographic position in Indiana. Poorly drained, dark-colored Aquolls occur in the lowlands, and somewhat poorly drained Aqualfs occur in the higher parts of the landscape. These latter soils are light colored here because erosion has exposed the light-colored B horizon. *(Reprinted from Marbut Memorial Slide Collection, prepared and published by the Soil Science Society of America, Madison, Wisconsin, in 1968.)*

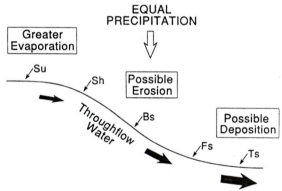

Figure 9.2 Schematic diagram of some geomorphic and pedologic processes related to position on a slope. Sites commonly studied are the summit (Su), the convex part or shoulder (Sh), the straight section or backslope (Bs), the concave part or footslope (Fs), and the relatively flat part beyond the Fs, the toeslope (Ts). The slope at the Ts site is still to the right and the only source of surficial or throughflow water, as well as sediment, is from the landscape upslope (to the left). Although precipitation is equal at all sites, evaporation is greatest at the Su. In more humid environments, water is redistributed downslope as surface runoff and throughflow water, so that the lower parts of the slope receive more water (indicated by width of arrow for throughflow water). Soil properties should correlate with erosion–deposition patterns and relative amounts of water. *(Reprinted from Birkeland and others, 1991, Fig. 5.1.)*

will show up in soil and sediment patterns along the slope.

Young (1976) depicts the multitude of possible horizons and horizon positions downslope (Fig. 9.3a–i). The horizons may remain the same (a) or thicken (b) or thin (c) downslope. Some get deeper (d) or shallower (e) downslope or terminate downslope, or a new one commences (f). Finally, a horizon might change laterally to another downslope and the contact have different geometries (g, h, i). All of these relationships have different soil-geomorphic interpretations. Understanding these soil–slope relations is critical for the correct identification of buried soils (Valentine and Dalrymple, 1975).

Although there are many successes with catena studies, Gerrard (1990) shows many of the complicating factors and concludes that "these examples cast doubt on the applicability of the catena concept to slopes in humid temperate environments" (p. 241). Tighter factorial control on catena studies, such as many of the examples used here, is necessary to understanding pedogenic processes operating on slopes over time. Perhaps part of the problem leading up to Gerrard's statement is that some of the slopes he was discussing were bedrock overlain with many different surficial materials. In fact, using the soil-catena ap-

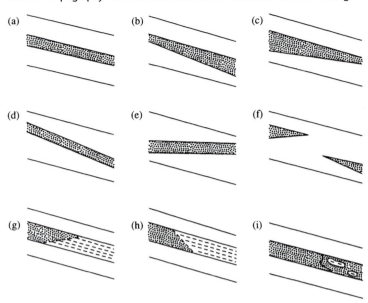

Figure 9.3 Possible variations in soil horizon thickness and type (different patterns) with slope position in catenas. See the text for explanation of (a)–(b). *(Redrawn from Young, 1976. Tropical soils and soil survey. Cambridge Univ. Press, Fig. 20.)*

proach can lead to some surprising results, such as that of Frolking (1989), who indicates that slopes that range from 2–55% have such similar soil profiles that surface erosion has been minimal in the last 12 ka.

Geomorphologists, soil scientists, and archeologists all study soil erosion. A book edited by Bell and Boardman (1992) shows how soil properties, coupled with agricultural practices and climatic events, help explain present-day erosion. Examination of older soils, surface and buried, is essential in understanding ancient erosion episodes, whether influenced by humans or not. Soil–slope relations offer the best hope of understanding these.

■ Influence of Slope Orientation on Pedogenesis

Slope orientation results in microclimatic and vegetation differences, and thus in soil differences. Jenny (1980) argues that topography is the primary factor in explaining soil variation in these field situations. Here we are concerned with topographic relief of meters, or tens of meters, and therefore regional climate can be considered a constant. Hunkler and Schaetzl (1997) argue that the influence of aspect on microclimate and therefore soils should be greatest between 40 and 60° latitude, and less important in both equatorial and polar latitudes.

An instructive example of the effect of slope orientation on soil properties is the study of Finney and others (1962) in southeastern Ohio (Fig. 9.4). Several NW–SE-trending valleys were studied. The parent material is mostly colluvium derived from sandstone. The microclimate varies quite markedly with orientation, with the SW-facing slopes displaying higher temperatures on the leaf litter, as well as greater annual fluctuations. Temperatures beneath the leaf litter show the same trends with orientation, but the differences between maximum and minimum values are less. Soil-moisture values follow the temperature differences, with soils on NE-facing slopes generally being more moist than those on SW-facing slopes. Vegetation correlates with the moisture–temperature trends. A mixed-oak association is dominant on the SW-facing slopes, whereas a mixed mesophytic plant association is dominant on the NE-facing slopes. The microclimatic–vegetation differences produce soil suborder differences. The NE-facing slopes are characterized by Ochrepts with thick A horizons on slopes with gradients over 40% and Udalfs with relatively thick A horizons on the rest of the slopes. However, Udalfs with thin A horizons predominate on the SW-facing slopes, and Ochrepts with thin A horizons are present locally (H.R. Finney, written communication, 1972).

Slope orientation greatly affects soil organic carbon distribution with depth, the presence or absence of an E horizon, pH, and percentage exchangeable bases (Fig. 9.4). Organic matter differences probably result from greater moisture and vegetation cover on the NE-facing slopes, combined with greater organic matter decomposition rates on the SW-facing slopes, along with some loss due to surface runoff. E horizons are more common on SW-facing slopes, and the reason for this is not clear. If anything, they should be more common on the moister, NE-facing slopes. Base saturation and pH trends are somewhat parallel with depth, and these can be explained in terms of vegetation differences and fire. The vegetation assemblage on the NE-facing slopes has a higher base content, and bases are returned to the soil surface as litterfall. However, the mixed-oak association has a lower base content and probably a higher incidence of fire. Fire not only burns off some of the soil organic matter, but also brings about the loss of bases released from the organic matter by either runoff or deep leaching.

Figure 9.4 Topographic, vegetative, soil, and microclimatic data for slopes of different orientation, southeastern Ohio. Temperatures are maximum and minimum monthly averages on leaf litter, 1956. *(Redrawn from Finney and others, 1962, Figs. 1, 3, 5, 6, and 7.)*

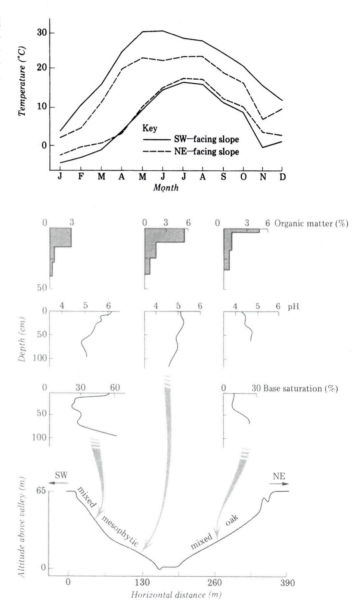

A somewhat similar relationship of soils to slope orientation is seen in the rolling hills of eastern Washington (Lotspeich and Smith, 1953). Parent material is 20-ka loess deposited in NW–SE-trending ridges that have about 65 m of relief (Fig. 9.5). Although the regional MAP is 53, most of which occurs in the winter, distinct microclimates are produced by the orientation of the hillslopes. It is estimated that the S-facing slopes receive close to the average annual precipitation. However, because about one-fifth of the precipitation is snow, the moisture can be easily redistributed in the landscape. It is estimated that the hilltops might receive only 25 cm of precipitation, with losses attributed to the removal of snow by wind as well as high evaporation rates on the exposed ridges. In contrast, the N-facing slopes have more effective moisture than do the S-facing slopes due to a combination of accumulated drifting snow and lower evaporation rates. The virgin vegetation follows these trends, with the more mesic shrubs, grasses, and forbs on the N-facing slopes and the more xeric forbs and grasses on the S-facing slopes.

Soils are closely related to the microclimate (Fig. 9.5). The S-facing slopes are characterized by non-calcareous Mollisols that are thought to be normal for the region. In contrast, the N-facing slopes, largely because of the additional moisture, have more strongly developed Mollisol profiles and greater amounts of organic matter in the A horizon. The soils on the ridgetops are thin Mollisols, characterized by weak B-horizon development and carbonate accumulation at depth, both properties that are clearly the result of a more arid microenvironment. Thus, it is seen that fairly large differences in soil profiles and properties can be closely associated in space because of microclimatic differences associated with slope orientation.

A soil-geomorphic study in a small watershed in northern California (MAP 200) also demonstrates the relationship of soil development to aspect (Marron and Popenoe, 1986). Five soil-topography map units were identified, and the soils of the map units show a progression in development from A/Bw Inceptisols with 10-7.5 YR hues and up to 30% clay, to A/Bt Ultisols with 5-2.5YR hues and up to 57% clay. Weathering

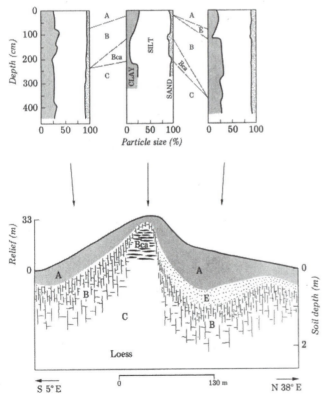

Figure 9.5 Variation in soil properties with slope orientation in loess hills, eastern Washington. *(Redrawn from Lotspeich and Smith, 1953, Figs. 2 and 3, © 1953, The Williams & Wilkins Co., Baltimore.)*

of the schist bedrock in an area of high rainfall probably accounts for the high clay content of the soils. The N-facing soils have a higher proportion of Ultisols than do the S-facing slopes, where Inceptisols dominate. Various geomorphic processes help account for these soil-aspect differences. Creep and slow, deep block sliding are thought to occur on N-facing slopes, and soils are slowly translated downslope as they form. In contrast, on S-facing slopes fluvial processes and shallow landsliding are rapid and episodic, remove soil, and expose bedrock or saprolite. With less residence time to form soils, Inceptisols are more common. Here, geomorphic processes, dictated by microclimate, help create the N–S soil contrast.

Graham and others (1990) and Graham and Buol (1990) show in detail how mass movements interrupt soil-process pathways at different stages on slopes in the mountains of North Carolina. At stable, upslope sites soils form in place as various minerals weather in succession to clays, and there is a time succession of Bw $\Rightarrow$ Bt horizon with clay films. At any time mass movements can remove part or all of the above soil, and form a colluvial deposit downslope composed of the homogenized transported soil materials. Once the slope is again stable, the upslope soil restarts its developmental sequence on a truncated profile, and a soil begins to form from the colluvial material, which also goes through a Bw $\Rightarrow$ Bt succession if residence time is sufficiently long. Presumably, a study such as this could be combined with a study such as that of Marron and Popenoe (1986) for greater detail of aspect and soil–mass movement interactions.

Spodosol formation in forested northern Michigan is influenced by aspect, as demonstrated in the statistical study of Holocene sandy soils by Hunkler and Schaetzl (1997). N–NE-facing slopes are generally cooler and have greater amounts of infiltrating water relative to S–SW-facing slopes. The result is that soils on the N–NE-facing slopes show field (POD index, eluvial–illuvial color contrast; Chapter 1) and laboratory (extracts of organic matter, Fe and Al) properties more indicative of podzolization than do the S–SW facing soils. Of the soils studied, the Spodosol:Entisol ratio is 9:1 for the N–NE-facing slopes, but only 3:7 for the opposite slopes.

One of the more dramatic influences of aspect on soils occurs in those parts of Alaska in which ice-rich permafrost occurs at shallow depths on N-facing slopes, but is absent on S-facing slopes. The soils of interest are forested and have formed in loess/schist (Krause and others, 1959). Soils on the S-facing slopes are well drained, relatively deep, and have an O/A/Bw/R morphology, whereas those on N-facing slopes are poorly drained, shallow, with an O/A/Bg morphology. The latter soils are more acid and leached of bases relative to the S-facing soils. We can predict that this contrast would be less pronounced farther north under continuous permafrost, as well as farther south with no permafrost.

■ Soil Catenas

Numerous studies have shown that many soil properties are related to the gradient of the slope as well as to the particular position of the soil on a slope. Milne (1935a,b) proposed the term catena to describe this lateral variability on a hillslope, and emphasized that each soil along a slope bears a distinct relationship to the soils above and below it, for a variety of geomorphological, pedological, and hydrological reasons. The word comes from the catenary curve, that depicted by a rope when held by two people. A journal devoted to these topics bears this name. These also are called toposequences. Yaalon (1975) reviewed many of the topofunctions derived from catena studies in various areas in the world.

Several models have been proposed to describe how landscapes are related to soil catenas. Conacher and Dalrymple (1977) discuss soil–slope relations in considerable detail and propose a nine-unit landsurface model defined on process and response. For many studies, however, it is sufficient to use a simple five-unit model based on slope form (Ruhe and Walker, 1968). Going from the top to the base of a slope (Fig. 9.2), these units are the flat summit (Su), the convex portion or shoulder (Sh), the more or less uniform slope of the backslope (Bs), and the concave portion at the base, which is divided into an upper footslope (Fs) position and a lower toeslope (Ts) position. Soil catenas can be further divided into open and closed systems (Ruhe and Walker, 1968; Walker and Ruhe, 1968). In the former, drainage is such that sediment can leave the area; the latter is characterized by a closed depression and all sediment is trapped in the topographic low.

Although most soil catenas, including many of those in this chapter, are depicted as a two-dimensional cross section, Huggett (1975) suggests that a better approach would be to consider a three-dimensional model, such as a small watershed. The watershed's boundary is the divide with the adjacent watershed. He then makes the point that properties of soil catenas in such a setting are partly explained in terms of lateral flow of soil waters, called through-flow water, from the higher parts of the watershed to the lower parts (Whipkey and Kirby, 1978). Referring to the three-dimensional soil chemistry that might result in a humid environment within such a watershed, three zones can be identified: (1) in the upper reaches, an eluvial zone in which soils show a net loss; (2) along the lower parts of the watershed, an illuvial zone in which the same constituents accumulate; and (3) separating these, a transluvial zone. The degree to which the soil properties match the zones will be mainly a function of the climatic regime as well as the solubilities of various soil constituents in the particular environment. Clay particles also can move laterally by throughflow waters (Huggett, 1976). As soil profiles develop through time, the increased profile anisotropy downslope results in increased anisotropy of vertical permeability, which leads to lateral flow (throughflow) of soil waters (Zaslavsky and Rogowski, 1969). The latter workers use this relationship to talk of diminished pedogenesis at convex positions, and increased pedogenesis at concave positions, with the result that B horizons thin upslope.

It also is important to recognize the results of both divergent and convergent soil water flowlines as these influence local soil moisture (Hall, 1983). In simple terms, soil water moves at right angles to the landform contours, and downslope soils can receive additional water as dictated by the flowlines. If the contours are straight, the flowlines are parallel. If, however, the soils are on a topographic nose (contours are convex downslope), the flow is divergent off the nose and soils downslope on the nose do not receive much surface or subsurface moisture from positions higher on the nose. In marked contrast, in a topographic hollow (contours are concave downslope) soils waters converge in the lower parts of the landscape, adding considerably more water than that of the mean annual rainfall.

Divergent and convergent flow can occur on various parts of the catena, and influence depths of various soil properties. In southern Saskatchewan, Canada, for example, Pennock and de Jong (1990) recognize regions of aridic, ustic, and udic Borolls. A-horizon thickness and depth to $CaCO_3$ should be related to local soil-moisture conditions, with both

thickness and depth greater with greater soil moisture. Comparing divergent–convergent pairs for Sh, Bs, and Fs positions, in most places the soil at the convergent site has a thicker A horizon and a greater depth to $CaCO_3$ when compared to the soil at the divergent site. The largest differences between the two sites take place at the Fs position, no doubt because the Fs collects soil water from a larger part of the catchment. Hence, in describing soil-site characteristics, it is important to describe the local landscape configuration (Appendix 1, Fig. A1.1).

Huggett (1975) has produced some useful diagrams to illustrate soil properties in the three-dimensional model (Fig. 9.6). The watershed is depicted as bounded by a headwaters and two noses, with a hollow down the middle. At $t = 0$, a particular constituent has a uniform concentration across the watershed. As time goes on ($t = 2$, $t = 4$, etc.), the constituent is lost in the upper reaches of the watershed and accumulates in the hollow to produce a three-dimensional concentration diagram that is almost the opposite of the topographic diagram. Profile weighted mean percentage clay in a humid climate is an example of this. In another case, that of excessive leaching, and depending on the solubility of the constituent of interest, the entire watershed could be in the eluvial zone, and the diagram could mimic the topographic diagram, but with a concentration less than that of the parent material; some of these concentration diagrams could have less relief than that of the topographic diagram. Profile weighted mean percentage Ca is an example of this. If on the other hand, the moisture regime is aridic, leaching and throughflow water movement are minimal and the particular constituent is introduced from outside the watershed (i.e., airborne carbonate), the watershed could be entirely in the illuvial zone with the concentration greater than parent material values, and the resulting diagram could be relatively flat across the area. Add to this the localized divergent and convergent flowlines and one can appreciate the complexities of the soil-geomorphic system. Not only soil properties, but saprolite thicknesses can follow these topographically controlled soil-moisture contents, as Graham and others (1990) report thicker saprolite in the lower slope positions. This diagram of saprolite thickness would be the reverse of the topographic diagram.

It is common to find markedly different soils in juxtaposition in rolling topography in which rounded hills rise above very gentle slopes or closed depressions

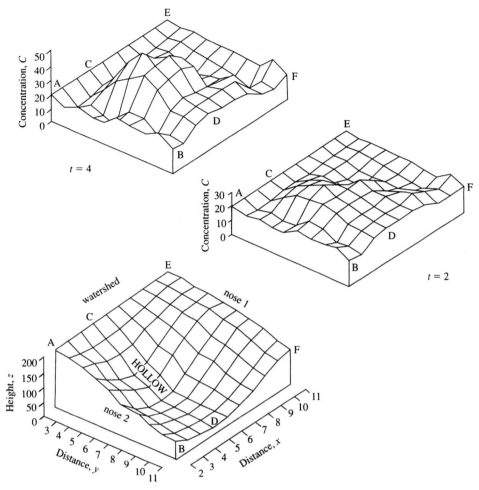

Figure 9.6 Schematic diagram of concentrations of material (e.g., percentage clay or some element such as Ca^{2+}) in soils at two times ($t = 2$, $t = 4$) in an idealized basin (topography in bottom diagram). Here the concentration maps are a reversal of the topographic map. Each environment and each pedogenic constituent probably will have a different topographic versus concentration relationship. *(Redrawn from Huggett, 1975, Fig. 1.5, Geodema 13, 1–22 © 1975, with kind permission of Elsevier Science, The Netherlands.)*

(e.g., Ollier, 1973). Soils on the uplands commonly are well drained, whereas those in the depressions are poorly drained and rich in clay and organic matter, with signs of various degrees of gleying. In Arctic climates, well-drained A/Bw profiles can grade downslope to cryoturbated poorly drained soils. Tropical landscapes can have red kaolinitic soils upslope and dark smectitic soils downslope. In dry climates, saline and alkaline soils occupy the depressions, better-leached soils the slopes, and the less-leached soils the summits. Different soil taxonomic units can be found at different positions. In aridic areas, one could find, for example, Orthids on the summits, Argids on the backslopes, and Natrargids in the footslopes and toeslopes. In more humid environments, the summit Argiustoll can give way (in the downslope direction) to Haplustoll (shoulder), Argiustoll (backslope), and finally Aquoll (toeslope). The differences in soil properties with position could be due to pedogenesis in place, resulting from differences in moisture, leaching, and vegetation over the rolling landscape. In this case, the various parts of the landscape are assumed to be approximately the same age, and soil differences are attributed solely to the topographic factor. Recent work, however, has cast some doubt on this simple model, because it fails to take into account the fact that some material in the depressions could be derived from erosion of the landscape that slopes into the depression. In this latter case, it is unlikely that the parent material of the soils formed on the slopes is the same as that in the depressions (i.e., Dahms, 1994). The soil differences, however, have their primary origin in topographic position.

In addition to the strictly pedologic redistribution of materials on a slope, one also must consider the geomorphic processes operative on slopes that cause erosion in some places and deposition in others (Gerrard, 1992). Important processes include rainsplash, overland flow of water, and mass movements such as creep.

The K-cycle model of Butler (1959) attempts to place slope events in a time framework. Start with a simple slope. For a variety of reasons, the steeper, upper part of the slope might be unstable and erode fairly rapidly; material of this event can be deposited in the footslope–toeslope positions. Stability follows and is accompanied by soil formation along the entire slope. One K-cycle encompasses the erosional–depositional interval, as well as the soils that subsequently form over the entire slope. A subsequent period of instability could result in a second K-cycle. The position and extent of both the erosion and deposition zones can change with time. The end result could be that erosion keeps soils relatively poorly developed in the upper parts of the slopes, and episodic deposition produces a sequence of buried soils in the lower parts of the slopes. Some parts of the slope can be beyond the areas of major erosion and deposition. If, instead of the periodicity of K-cycles, erosion and deposition are more gradual through time, the cumulative and noncumulative profiles depicted in Fig. 7.10 could be produced.

One could categorize catenas as relatively stable and unstable (Fig. 9.7). The unstable ones have a K-cycle history, eroded parts have weakly developed soils, stable parts better developed soils, and buried soils are present in the lower slopes. The transition from one type of soil to the other could be abrupt. In contrast, stable catenas have not been subject to appreciable erosion, even though creep is an ongoing process, soil properties along the slope are mainly explained in terms of pedological process governed by

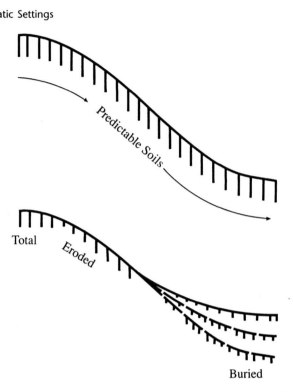

Figure 9.7 Schematic diagrams of soil–slope relations. If the slope is stable (upper diagram), one might find a predictable sequence of soils from Su to Ts sites, and the properties would be dictated by the climate. In contrast, if the slope episodically erodes (lower diagram), the soil formed over the total available time would be present at the Su, younger and thinner soils would be present at the eroded Sh and Bs sites, and a sequence of buried soils would be present at the Fs and Ts sites. The number of surface and buried soils at the latter sites would correspond to the number of episodes of erosion.

topographic position, and so are predictable. On an essentially stable slope, A-horizon characteristics are essential in evaluating the recent erosion history of the slope. Thinner than average A horizons suggest areas of erosion, and thicker than average A horizons denote areas of deposition, and the changes laterally can be gradational. Of course, one first has to decide if the A-horizon properties instead reflect microenvironment, that is greater soil moisture downslope. In contrast, B-horizon properties could be used to suggest longer-term overall slope stability, because, relative to A-horizon properties, many of these properties take longer to form.

Tonkin (1994) and Tonkin and Basher (1990) have used much of the above to explain in considerable detail the soil–landform relations in the hills and mountains of New Zealand. They argue that New Zealand is an excellent laboratory for such studies as geomorphic activity is great, soils form at a relatively fast rate, and human interference has been minimal. Morphological development is used as a proxy for the time span of stability between erosion–depositional events. However, to best estimate the time factor, it is necessary to know the soil property variations along catenas of various ages in different climatic settings. These soil patterns can be used to identify patterns of hillslope erosion and deposition, with subsequent delivery to the fluvial system, and the total can be put into a time frame. No other methodology could do this in the detail presented.

Geomorphologists classify slopes according to the thickness of the unconsolidated material on the slopes (Summerfield, 1991). Weathering-limited slopes are those in which erosion is limited by the rate at which material is made available by weathering; the result is thin soils/bedrock. In contast, transport-limited slopes are those in which there has been a longer residence time on the slopes to form colluvial deposits and soils; erosion here is controlled by the capacity of the transport process. A simple way in which to recognize such slopes is by study of soils. In the example of Tonkin and Basher (1990) in New Zealand, the former slopes have A/R profiles and the latter have A/Bw/R or C, with the soil thickening downslope and perhaps containing gley features.

Several kinds of catenas are recognized (Fig. 9.8). One is formed on bedrock, but these are often unpredictable because there is no certainty that either parent material or time is constant (equivalent) at all sites. For example, there could be pockets of old soils at different places along the slope. We see this in the granitic slopes around Boulder, Colorado, where soils with red, strongly argillic horizons depict portions of slopes that must have been stable for several 100s of ky.

A second kind of catena is formed in surficial materials that when deposited created a sloping landform; in other words the landform was created relatively quickly and all sites are close to the same age. Examples are moraines or sand dunes. This type of catena has an advantage over the bedrock catenas in that parent material and time are nearly constant at all sites. Furthermore, by studying moraines of different ages

we can track catena development over various time spans.

A third catena is formed in surficial deposits related to fluvial scarps within a flight of river terraces. The terrace flat and the terrace scarp (riser) immediately above it are of approximately the same age, and the scarp evolves by erosion in the upper part and deposition in the lower part (Pierce and Colman, 1986). If the fluvial materials of the terrace and the scarp colluvium are of similar texture and lithology, then par-

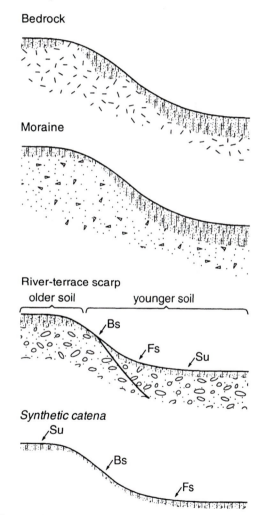

Figure 9.8 Soil development on several kinds of catenas. Depth and degree of soil development are shown by line and stipple patterns. The synthetic catena is derived from the river-terrace scarp catena. *(Reprinted from Birkeland and others, 1991, Fig. 5.2.)*

ent material can be considered to be nearly constant. The next higher terrace, at the top of the scarp, is older and can have older soils related to it. Depending on the properties and depth of the latter soil, the scarp colluvium can have varying degrees of inherited soil properties, but this can be corrected for in the scheme presented by McCalpin and Berry (1996). Because in assessing catena processes, we start at the summit and progress downslope, I sometimes rearrange the catena positions in river-terrace scarps to reflect this arrangement, thus producing a synthethic catena (Fig. 9.8). In this way, all soils are approximately the same age.

The selection of sites for catena studies requires great care. Because water-flow pattern is a major process connecting the properties of one catena site to the next, the soils on the slope should be aligned so that both surface and throughflow waters at an upslope site have the potential for passing through all downslope sites. In other words, the soil pits are aligned at right angles to the contours. Only with this placement strategy are the soils linked in both a physical and geochemical sense, and there is some predictability in soil properties from one site to the next.

Catena properties have been shown to vary with climate (Ollier, 1973; Tonkin, 1994), so the following examples of soil catenas are organized according to major climatic regimes. The order of the time chapter will be followed: first we deal with cold-climate soils, then look at noncalcic soils in areas of progressively greater leaching, and finally those soils of progressively lesser leaching with accumulations of carbonate and more soluble salts. Within each climatic regime my field strategy has been to find catenas of different ages; these will be presented where data are available.

Soil geochemistry will be included where data are available. Geochemistry varies with slope position if there is sufficient rainfall to bring about significant redistribution of soil water along the slope. Elements released by weathering may or may not be redistributed along the slope as a function of their mobility in the constant or changing geochemical environment along the slope (Table 9.1). For example, under slightly acid-oxidizing conditions, Fe and Al remain close to the point of release, whereas the more mobile cations (Ca, Mg, Mn, K, and Na) can move to the lower parts of the slope. McDaniel and others (1992) suggest using the dithionite-extractable Mn/Fe ratio as a pedochemical tracer for water movement in some of these environments. In their field area of North Carolina (MAP 120, thermic) with Kanhapludults, Fe_d is strongly correlated with percentage clay, suggesting in situ formation of both or comigration. Mn, being more mobile than Fe, can behave independently of both clay and Fe. Low Mn_d/Fe_d ratios indicate areas of sufficient water movement in areas of divergent flow to deplete the soil in Mn, whereas high ratios indicate accumulation of Mn in areas of convergent flow. In contrast, under acid-chelating conditions, both Fe and Al are solubilized and can move to lower parts of the slope along with the more mobile cations. If conditions in the lower part of the latter catena are reducing with an increase in pH, it is possible that Fe remains mobile and leaves the system, but Al accumulates. Finally, under aridic conditions with associated high pH, all constituents, including the most soluble, accumulate where introduced into the soil with little redistribution downslope. Catena studies, therefore, can answer questions such as (1) at what MAPs does reduction become an important down-

Table 9.1 Relative Mobility of Various Elements with Different Soil Conditions

Relative mobility	Mobility phase	Soil Conditions		
		Slightly-Acid Oxidizing	Acid Chelating	Reducing
↑ Increasing	I	Cl^-, SO_4^{2-}	Cl^-, SO_4^{2-}	Cl^-, SO_4^{2-}
	II	Ca^{2+}, Na^+, CO_3^{2-} Mg^{2+}, K^+	Ca^{2+}, Mg^{2+}, Na^+, CO_3^{2-} K^+, Fe^{3+}, Al^{3+}	Ca^{2+}, Mg^{2+}, Na^+, CO_3^{2-} Fe^{2+}, Mn^{2+}
	III	SiO_2, P	SiO_2, P	SiO_2, P
	IV	Fe_2O_3, Al_2O_3		Al_2O_3

Modified from Young et al. (1977).

slope process, or (2) what MAPs provide the geochemical setting for specific geochemical separation on a slope (Fe, Ca, etc.)?

Tardy and others (1973) have shown clearly how catena studies explain one product of slope geochemistry along a climatic transect from hot, aridic to hot, humid. Clay mineralogy was used, and it reflects the overall soil geochemical environment in the long term. As a generality, upslope clay-mineral suites will be more depleted in Si relative to downslope clay-mineral suites, which will be characterized by Si accumulation (Fig. 9.9). Soluble salts follow a pattern of the more soluble ones accumulating in the downslope positions, just as they do vertically in a soil profile. In the driest catenas, smectite, gypsum, and halite might be present at all positions. At slightly greater MAP, however, smectite is still the clay mineral present, carbonate is present along the slope, but gypsum (and maybe halite) is present only in the lower slope positions. At higher MAP, leaching is sufficient at the upslope sites to produce kaolinite and pedogenic Fe, whereas the translocated Si and Ca accumulate downs-

lope to give rise to smectite and carbonate. At still higher MAP, gibbsite forms upslope and the downslope transfer of Si produces kaolinite. Finally, at sufficiently high MAP and old soils, leaching is such that gibbsite is present at all sites. Thus, clay mineralogy gives great insight into geochemical processes in catenas along a climatic transect.

■ Catenas in Polar, Arctic, and Alpine Areas

One would suspect that catenas would develop slowly in landscapes near the poles and at high altitides. Soils are frozen all year or at least for parts of the year, thus inhibiting weathering. Limited soil moisture would limit the downslope movement of both solid and dissolved materials. Cold, dry climate inhibits vegetation growth and therefore A-horizon development. Finally cryoturbation could mix up the soil so much that trends could be difficult to decipher.

Catenas involving saline soils have been reported in the extremely cold, dry environment of Antarctica (Campbell and Claridge, 1987, p. 140). Undrained depressions are present and some are damp or saturated with water. The reason for this lies in the high salt content, some being so high that the freezing point is lowered to the point that the soils do not freeze. The soils in these depressions can be sticky and clay rich, with some clay contents approaching 30%. In contrast, soils on the surrounding slopes are the usual dry, sandy soils.

In one Antarctic catena, color, clay content and mineralogy, and position of the salt maximum vary with catena position in a till landscape (Campbell and Claridge, 1987, p. 191). The soil at the dry site has 10YR oxidation to 23 cm and a vertical decrease in clay from 6 to 1%; this overlies a 2.5Y horizon with 1% clay. In contrast, the hollow soil is 2.5Y, with a clay decrease with depth of 11–5%. The contrast in hue is thought to be due to reducing conditions in the hollows. Salts are deposited at depth in the dry soil, but at the surface in the hollow soil. At another place (Campbell and Claridge, 1987, Table 6.4) workers report that in water extracts for the top 10 cm, Na^+ is the most abundant cation, and it is 1.5 times greater in the hollow soil. NO_3^- and SO_4^{2-} are the main anions in the dry soil, whereas NO_3^- and Cl^-, at slightly higher concentrations, dominate at the wetter site. Thus some of the more soluble ions are accumulating

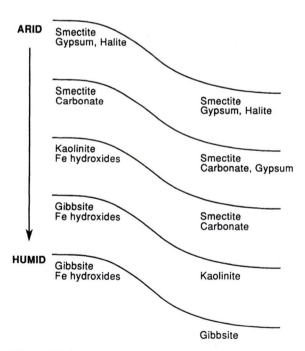

Figure 9.9 Schematic diagram showing the variation of clay minerals and other pedogenic products with position in catenas along a climatic gradient. *(Redrawn and modified from Tardy and others, 1973, Fig. 1.)*

in these soils. Finally, the original mica is 80–90% at the dry site and has altered to vermiculite, Al-chlorite, and smectite, whereas the hollow soil has 70–85% mica with a wider variety of alteration products: vermiculite, Al-chlorite, chlorite, and mica–chlorite. In other sites in Antarctic hollows, mica alteration can be extreme enough to produce 60% smectite. Relative to the alteration at dry sites (Chapter 8), alteration at these locally moist sites is accelerated.

In one study on bedrock and colluvium in central Alaska (MAP 42; MAT −6), Swanson (1996a) demonstrates the influence of the various aspects of the topographic factor on soil formation. Relatively warm, dry mineral soils are associated with coarse-textured materials, convex slopes, steep slopes, and S-facing aspect. In contrast, cold, wet organic soils occur with fine-textured materials, concave slopes, gentle slopes, and N-facing aspect. Depth to permafrost is deeper under convex terrain (hills) than under concave terrain (valleys), and soils containing relatively shallow permafrost are cryoturbated.

Moraines in the discontinous permafrost zone of northern Alaska (MAP 42, MAT −6) display catena trends with respect to slope position (Swanson, 1995, 1996a,b, personal communication, 1997). For the catena on 10- to 30-ka moraines, loess is absent, and the relatively dry Su soils are Typic Cryochrepts (A/10YR Bw/2.5Y C). Only two Fs + Ts soils have been described, and one is a Cryorthod, probably the result of ponded water, and the other is a Cryaquept. Permafrost is >1.5 m depth or absent on the upper slopes, and about 0.5 m depth on the lower slopes. Burning increases the proportion of the upper slope that is underlain by deeper permafrost, and this is reflected in the soil properties. The next older set of moraines (Ambler) might be as old as 140 ka, and is characterized by a cap of loess. The transition zone at the Bs site is narrow and there are few soil data on it. Upslope soils are Cryochrepts (O/A/Bw in loess/2C in till). Colors are similar to those in the younger moraines and loess at the Su is <0.5 m. Downslope soils are entirely in loess (>0.5 m thick) and are Pergelic Cryaquepts (O/A/2.5Y Bg/Bgf). O horizons thicken on average downslope, from 3 to 21 cm. Still older moraines (Kobuk) have lost most of their morainal topography and have soils similar to the latter ones.

Unusual catenas are present in the Alaskan Arctic Coastal Plain (Walker and others, 1996). There, pingos of two morphological types are present, and these

rise as small, dry hills with dry tundra vegetation above the saturated Pergelic Cryaquepts of the flat coastal plain. Pingos are ice-cored mounds that form when thaw-lakes drain; the saturated sediments then begin to freeze, and as water is concentrated and under pressure, the overlying material is mounded up, forming a pingo. In this case, the catenas form from a landscape produced by forces pushing upward from beneath; for contrast, many catenas are original depositional features (e.g., moraines), or formed by erosional processes (e.g., river-terrace scarps). Two pingo types are recognized, younger (>5 ka) steep-sided ones with slopes >10 degrees, and older (>14–22 ka) broad-based ones with slopes <5 degrees. Although the purpose of the study was to use soils to estimate ages of the pingos, we can use their properties as an example of soil-catena development in an extreme environment (MAP 20 cm, MAT −13). Parent material is complex, and consists of a thin layer of lake sediment over river gravelly alluvium.

The common soil profile form is A/Bk, and there are a few trends with slope position and age. The Bk is oxidized and identified mainly on carbonate coats on the bottoms of clasts (stages I and II, subdivided according to Forman and Miller, 1984). Of the laboratory data, clay and organic carbon contents show significant downslope increases in both catenas (Fig. 9.10). Mean profile clay content is <10%, and the greatest profile increase downslope is about three times; the contrast downslope is better for the younger catena than for the older one. The older catena has higher clay content. Mean profile organic carbon content stays under about 3%; it also increases downslope in most transects, but the more consistent trends are in the younger catena.

Several sets of laboratory and field data indicate pingo catena trends with time (Walker and others, 1996). IPA, which measures profile anisotropy (Chapter 1), was calculated for all laboratory data and values for clay and Fe_d contents, and pH showed significantly higher values for the older catena. PDI data also shows higher values with age. They modified the property indices thusly: the total texture index included eolian additions of silt; the carbonate index reflected subdivisions of stages I and II for gravelly materials as suggested by Forman and Miller (1984); and an index for silt-cap morphology was added. Eight of nine weighted mean property indices for the older catena have greater values relative to those for the younger catena, and five of the differences are significant. Four measures of PDI have higher values for the older catenas. The maximum weighted mean PDI values are between about 0.15 and 0.24; of interest for age-estimate studies, these values are similar to those for weighted mean catena PDI values for 20-ka moraines under a much warmer climate in the western United States (Fig. 9.16).

Alpine soil catenas can show marked contrasts in profile horizonation with slope position. Those in extremely cold and windswept environment (MAP 102, MAT −4) above treeline in the Colorado Rocky Mountains display morphological variation with position on rolling bedrock terrain (Burns and Tonkin, 1982). Topography controls soil moisture by limiting much of the snowfall to the lee sides of hills; these same sites also trap windblown silt and clay (mixed loess). Soil distribution follows trends in snow-cover duration and in thickness of mixed loess (Fig. 9.11). Summit and shoulder sites have little snow cover, and the resulting profiles have thin A horizons over thin Bw horizons. There is no loess at the most windblown sites, and patches of mixed loess are as much as 8 cm thick at protected sites. Soils lower on the slopes show

a relation to duration of snow cover. In late winter, snow covers much of the area, but with spring and summer melting the soils become exposed; those higher on the slopes first and those lower on the slopes last. Below the shoulder are the minimal snow-cover sites, which are characterized by the best-developed soils: thick organic-matter-rich A horizons formed from mixed loess over thick Bw horizons with the best color development in the catena. Next downslope are soils in which snowbanks melt early. Mixed loess is not as thick as in the minimal snow-cover soils, but the A horizons are the thickest of all (probably due to pocket-gopher activity) and overlie a Bw horizon. In areas in which the snowbanks melt late, soils have poorly developed A/Bwj profiles with no loess. Finally, at the base of the slope are soils with poorly drained, mottled, and gleyed O/Bg profiles, which are characterized by high concentrations of silt and clay that have been transported during snowmelt from upslope positions.

Soil laboratory data on an alpine slope catena near the latter site provide insight into the processes operating along a catena over about 10–15 ka (Litaor,

Figure 9.10 Mean clay and organic carbon percentages for two pingo types in relation to slope position (Sh, Bs, Fs) and aspect (N-facing slopes on left, S-facing on right, and leeward slopes in center). *(Redrawn from Walker and others, 1996, Fig. 1.)*

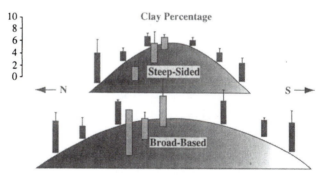

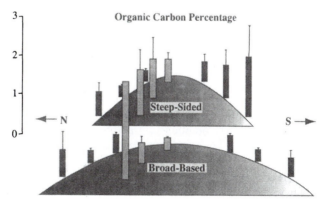

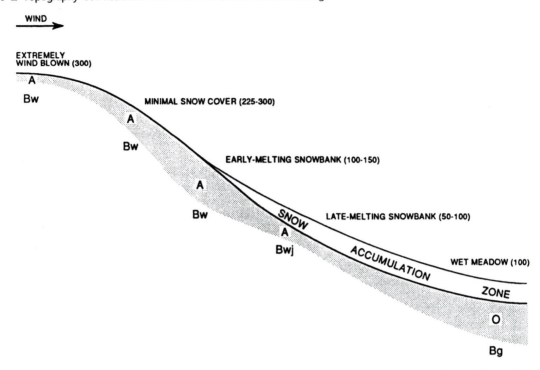

Figure 9.11 Schematic diagram of soil catena relations in an alpine tundra environment, Colorado Front Range. Depths of A and (or) O horizons are schematic, and numbers in parentheses are the estimated numbers of snow-free days per year. *(Data from Burns and Tonkin, 1982; Figure reprinted from Birkeland and others, 1991, Fig. 5.4.)*

1992) (Fig. 9.12). The collection of soil solutions at various depths helps elucidate the processes operating. Morphologies along the catena are generally O/A/Bw/BC, in thin mixed-loess A horizon/till parent material; a buried A horizon occurs at shallow depth at the toeslope. Vertical clay translocation is insignificant, whereas vertical silt translocation seems to have taken place at the summit position, as shown by both the increase in silt downprofile and the abundance of silt caps . However, both size fractions increase in content in the surface horizons downslope. Most of this is attributed to enhanced lateral flow of water within the surface horizons during the snowmelt season (May and June) while the deeper parts of the soil are still frozen. The dense alpine tundra vegetation is thought to inhibit downslope movement of fines by surface runoff. The relatively dry summer conditions are not conducive to either vertical or lateral movement of fines except during high precipitation events.

Marked geochemical trends occur along the catena. Organic carbon content doubles downslope (Fig.

9.12). This is attributed to the above-mentioned enhanced lateral flow during snowmelt, and this is supported by the sixfold increase in dissolved organic carbon in solutions collected from the surface soil horizons when the summit and backslope are compared to the toeslope. Greater vegetation cover due to greater moisture at the toeslope also could be a factor. The same lateral flow could explain the increase in exchangeable Al downslope and that, along with the lowering of pH by about 1 unit (5 to 4) downslope, could partly explain the decrease in base saturation downslope. Analyses of soil solutions indicated that vertical and lateral transport accompanying extreme summer precipitation events (recurrence interval >100 years), when the soils are not frozen, also could play a role in explaining some of the trends.

Fe extracts of the above soils do not reflect great mobility (Fig. 9.12, M.I. Litaor, written communication, 1996). In the mixed-loess O and A horizons, Fe_d and Fe_o increase about 30% downslope, and Fe_o/Fe_d is about 0.3, expected for the alpine tundra environment. Relative to the latter, in the Bw horizons, Fe_d

increases less downslope, and Fe_o stays relatively constant.

Mahaney and Sanmugadas (1983, 1989) incorporated time into their studies of alpine catenas on moraines of the Wind River Mountains, Wyoming. The area probably has a climate close to that of the Colorado Front Range. The main trends in the 10-ka catena are as follows: (1) the greatest organic carbon content is upslope, and because the lower slopes do not have cumulic profiles, there does not seem to be much downslope redistribution of surficial materials; (2) pH values are slightly lower downslope; and (3) clay and Fe_d contents and depth to most intensive oxidation (10YR 4/4) are greatest downslope. Perhaps greater soil moisture, as indicated by soil mottling, help explain both (2) and (3), but (1) seems odd.

Two alpine grassland catena transects on moraines of about 12 ka in the Peruvian Andes demonstrate that greater mobility of Fe takes place at >100 MAP. The relatively dry catena has MAP <100 cm and the relatively wet area, located just west of the Amazon rainforest, has an MAP of about 300; MAT is 6–12 (Miller and Birkeland, 1992). Mixed-loess A/till B horizons are the characteristic profile form. PDI data indicate greater profile development downslope in both areas, and the wetter area has a weighted mean catena value 60% greater than that in the drier area (0.16 vs 0.10). In the wetter area profile weighted mean Fe_d and Fe_o

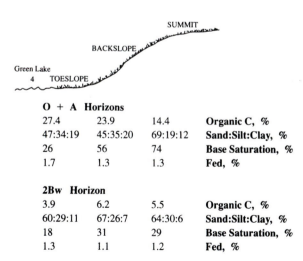

O + A Horizons

27.4	23.9	14.4	**Organic C, %**
47:34:19	45:35:20	69:19:12	**Sand:Silt:Clay, %**
26	56	74	**Base Saturation, %**
1.7	1.3	1.3	**Fed, %**

2Bw Horizon

3.9	6.2	5.5	**Organic C, %**
60:29:11	67:26:7	64:30:6	**Sand:Silt:Clay, %**
18	31	29	**Base Saturation, %**
1.3	1.1	1.2	**Fed, %**

Figure 9.12 The distribution of various soil properties along an alpine catena, Colorado Front Range. *(Redrawn from Litaor, 1992, Figs. 2 and 3, and written communication, 1996.)*

Table 9.2 Properties of Soils in a Catena in Northern Norway[a]

Soil Property	Landscape Position		
	Summit/Shoulder	Backslope	Footslope
A horizon thickness (cm)	15	25	30
Bh horizon thickness (cm)	10	30	55
Humus in Bh (%)	2.1	2.2	4.4
Fulvic acid:humic acid in Bh	1.8	—	9.3
Fe_2O_3 in Bh (%)	6.9	6.9	7.8
Al_2O_3 in Bh (%)	18.4	22.2	33.4

Taken from Glazovskaya (1968, Fig. 2).

[a]The author is not clear on the extraction procedure used for Fe and Al.

are 0.5 and 0.4% upslope and increase to about three times these values downslope. In contrast, the drier catena upslope values are 30–60% less than those of the wetter site, as are those in the downslope soils, being about 50% less. Greater influence of throughflow waters is thought to be responsible for the much greater Fe mobility at the wetter site.

Marked geochemical trends are apparent in humid subarctic areas of northern Norway (Glazovskaya, 1968). Horizon thicknesses increase downslope, as do humus and Al_2O_3 contents in the Bh horizon; in contrast, Fe_2O_3 content changes little with slope position (Table 9.2). The chemistry of these soils seems to follow the direction of movement and the amount of throughflow water, with possible organic matter complexes moving Al in preference to Fe.

■ Catenas for Noncalcic Soils in a Temperate ⇒ Humid Transect

Western United States, Israel, and New Zealand

Muhs (1982) described a catena formed on andesite bedrock on an offshore California island (MAP 17, MAT 16), and reviewed others formed in mediterranean climates. The soils are thin (<0.5 m) but thicken downslope. The Su soil is an Alfisol with a large clay increase from the A to the 7.5YR Bt horizon, whereas the Ts soil is a uniformly dark (10YR 3/1), high-clay Vertisol with cracks and slickensides.

The clays seem to be from a combination of bedrock weathering, eolian additions from the Mojave Desert in mainland California, and downslope transport. Illuvial processes upslope, due to vertical and through-flow waters, give way to vertical mixing (pedoturbation) downslope. Muhs suggests that thresholds also might be involved in the soils along the catena. Early in the history of the catena, Alfisols may have been the dominant soils. In time as the toeslope soil thickened and became more clay rich, an intrinsic threshold was crossed, pedoturbation became the dominant process, and a Vertisol formed.

Nettleton and others (1968) described three soils formed in a bedrock catena from tonalite in southern California (Fig. 9.13). They assumed that the three soils formed on slopes of about the same geomorphic age, and therefore the differences in the soils could be attributed to topographic position. Because of the warm climate (MAP 38, MAT 13) and precipitation mainly in the winter and early spring, the soils have A horizons with low organic-matter content. However, the properties of the B horizon vary markedly downslope. The Vista soil has only a Bw horizon, whereas the Fallbrook has a Bt horizon, and the Bonsall has an Na-enriched Bt horizon. Soil-moisture measurements were taken at various times of the year, and from these data and the 1965–1966 precipitation records, the moisture in the soils was estimated on a monthly basis (Fig. 9.13A). These data show that the downslope soils (Fallbrook and Bonsall) receive more soil moisture and retain their moisture longer than the upslope soil (Vista). A marked feature of these soils is that the clay content increases downslope (Fig. 9.13C). Weathering of the primary minerals is greater downslope; consequently, most of the clay present can be attributed to weathering of the underlying rock at that site. The differences in weathering and clay-formation downslope are most likely due to soil moisture, which is controlled by slope position. Soils in lower slope positions can receive more moisture than those in upslope positions because of lateral movement, either at the surface or within the soil. Furthermore, clay formation enhances a soil's water-holding capacity, which in turn further accelerates clay formation.

Sand dunes along the Mediterranean coast in Israel (MAP 52; MAT 20) also demonstrate soil variations with landscape position (Dan and others, 1968). The soils on the tops and sides of the sand dunes have well-developed Bt horizons and color hues of 2.5YR, overlying a C horizon of low clay content (Fig. 9.14). In the depressions, however, the soil at 3-m depth has nearly 60% clay and meets the color criteria for a gleyed horizon. Other notable differences below the surface from ridgetop to depression are a decrease in content of pedogenic Fe, a change in clay mineralogy from predominantly kaolinite to montmorillonite, and an increase in pH.

Variation in soil development on the Israel sand dunes is explained by eolian influx combined with slope processes and pedogenesis. Much of the clay in these soils is derived from eolian influx, some of which, after reaching the surface, is translocated downward in the soil. During pedogenesis, the clay minerals are thought to alter from montmorillonite to

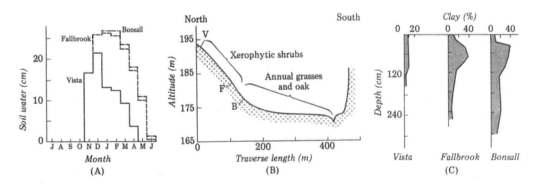

Figure 9.13 Data for a toposequence of soils formed from tonalite, southern California. (A) Estimated soil water for November 1965 through June 1966. (B) Topographic and vegetative relations at the sampled sites; slopes are 8% (Vista), 12% (Fallbrook), and 8% (Bonsall). (C) Distribution of clay with depth at the three sites. *(Redrawn from Nettleton and others, 1968, Figs. 1 and 2, and data from Table 3.)*

Figure 9.14 Generalized distribution of clay with depth in soils developed from sand-dune deposits. Topographic relief is 5 m. Munsell colors are for Bt horizons of soils located above the depression, and for soil materials at a comparable depth within the depression. *(Redrawn and modified from Dan and others, 1968, Fig. 9.)*

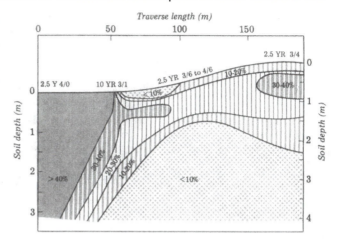

kaolinite under favorable chemical and leaching conditions. The soil in the depressions is not considered to have formed in situ, mainly because there are few mineral grains in the dune sand that can alter to clay. It is also suggested that some clay transfer could take place within the soil by laterally moving soil water at the top of the strongly developed Bt horizon (about 20 cm deep). Transfer of dune sediment from the slopes to the depression is not considered important because so little sand is found in the depression. Leaching conditions in the depressions are slight and thereby favor retention of montmorillonite as the main clay mineral. Perhaps redoximorphic conditions in the depression explain the low pedogenic Fe.

Catenas formed on moraines at the foot of mountains located across the western United States illustrate development under relatively dry conditions. A forested one in Idaho represents average conditions (Berry, 1987; MAP 90, MAT 2). As was mentioned in Chapter 8, Bw or minimal Bt horizons are typical of 20-ka moraines and Bt horizons of 140-ka moraines; in these catenas B horizons thicken downslope, as does the A horizon of the older catena (Fig. 9.15). Index values for various properties follow suit; rubification, clay-film development, and PDI values are about 1.5–2 times greater in most of the older catena sites, and the lowest values for each catena are either at the Sh or Bs positions. Clay profiles are typical for the region for the ages, and comparison is best made using weighed mean percentage values. These values for clay are rather constant downslope in the younger catena, but increase 1.5 and more downslope in the older catena. Comparing the Su and Fs sites for

younger vs older catena, the increases are two to three and three to four times greater, respectively. In contrast, Fe_d almost doubles downslope in the younger catena and almost triples downslope in the older one; the Su and Fs increases with age are two and three times, respectively. In short, the Su ⇒ Fs increase in both data sets increases with age.

Somewhat similar catena relations exist in California and Wyoming. Like the Idaho example, the parent materials are gravelly sands of granitic composition. For the eastern Sierra Nevada, California, catenas in sagebrush (MAP 40–45, MAT 7–8; Birkeland and Burke, 1988), a sandy colluvial wedge typically is present at the footslope sites, A/Bw/Cox profiles have formed at most sites on the 20-ka moraines, and there is little change in clay and Fe_d downslope. In contrast, soils on 140-ka moraines may not be any better developed at the Su sites than the 20-ka summit soils, but some soils downslope have Bt horizons with the highest amounts of both clay and Fe_d for the catena. Catenas can be compared by calculating weighted-mean catena values in this manner (Birkeland and others, 1991): multiply the values at each site by the fraction of the catena estimated to be represented by each soil in the catena (e.g., 0.33), and sum all such values for the catena. In the location with the greatest catena-property differences with age, the catena weighted-mean values for the younger vs the older moraine are PDI (0.14 vs 0.37), clay (4.7 vs 10.1), and Fe_d (0.45 vs 0.69). Thus, values for the older catena are about 1.5–3 times greater than those for the younger catena. Because the older deposits are about seven

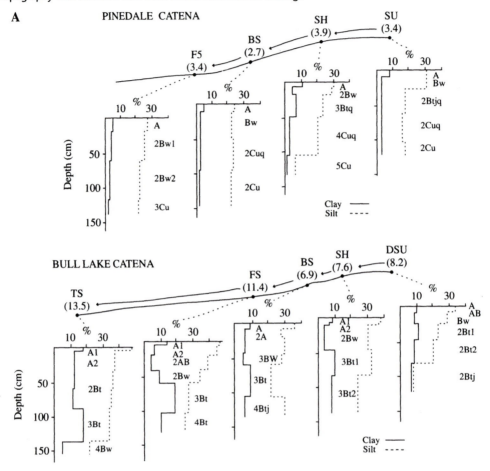

Figure 9.15 Distributions of (A) silt and clay and (B) Fe$_d$ with depth for soils of moraines of different age (Pinedale = 20 ka; Bull Lake = 140 ka), Salmon River Mountains, Idaho. Numbers in parentheses at the top of each profile are the weighted mean value for each profile. *(Redrawn from Berry, 1987, Figs. 5 and 8.)*

times older than the younger deposits, all plots vs age will have the characteristic logarithmic shape.

In another California moraine catena study (Pine Creek: MAP 20–28, MAT 9–11) Berry (1994) calculated the mass of pedogenic clay in catena soils. Here the Su vs Fs masses in g/cm^2/profile are compared. Mass amounts systematically increase downslope and with age: (1) for 20-ka soils, 1.2 vs 1.6; (2) for 140-ka soils, 1.4 vs 3.8; and (3) for >140-ka soils, 3.2 vs 6.0. In this data set, the amount of clay in the downslope soil of one catena closely matches that in the upslope soil of the next older catena.

In the Wind River Range, Wyoming (MAP 40; MAT −1), Swanson (1984, 1985) found a rather sim-

ilar relation under sagebrush. For 20-ka moraines, the main morphology for all soils is A/Bw, both horizons thicken downslope, and (in the same direction) pH increases slightly, as does Fe$_d$. In contrast, the soil catena on the 140-ka moraine has a Bt horizon at most sites and increases in thickness, percentage clay, and Fe$_d$ downslope. Catena soil data for clay and Fe$_d$ were compared on a mass basis for a 130-cm-thick soil. The highest values are at the downslope sites, and values for the older catena (22.5 g clay, 1.1 g Fe$_d$) are two to four times greater than those for the younger catena.

Swanson (1985) regressed several soil properties against slope curvature, specifically the second derivative of the slope profile (d^2y/d^2x) at the sample site.

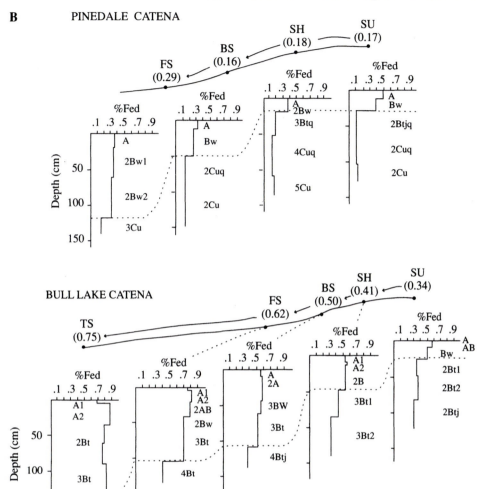

Figure 9.15 (continued)

The regressions confirmed that soils on concave slopes are better developed than those on convex slopes. It was also shown that the slope and intercept of the 140-ka-soil regression line is greater than those for the 20-ka-soil line, meaning that the difference between soils on both concave and convex slopes increases with increasing age (the older ones are better developed). Swanson uses these relations to suggest that soil creep is a major geomorphic agent on these slopes.

There are some trends in weathering and clay mineralogy with slope position in the above moraine studies. Soils at the drier site in Wyoming display little

signs of either differential weathering or differences in clay mineral type and abundance with slope position. In contrast, soils at the wetter site (Idaho; Berry, 1987) have a greater degree of plagioclase weathering downslope, but only in the older catena. As regards clay mineralogy, mica decreases downslope in the younger catena, whereas in the older catena both mica and vermiculite decrease downslope while a mixed-layer clay mineral (10–17Å) increases.

From the above examples, it can be seen that soil data casted in a variety of ways from the above three localities in the western United States give quite sim-

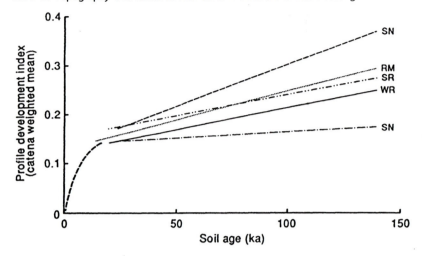

Figure 9.16 Variation with time in the catena weighted mean PDI for soils on moraines for several localities in the western United States. Localities are SN, Sierra Nevada, California; RM, Ruby Mountains, Nevada; SR, Salmon River Mountains, Idaho; and WR, Wind River Mountains, Wyoming. Dashed line estimates the trend for all areas from 0–20 ka. *(Reprinted from Birkeland and others, 1991.)*

ilar results. We are not sure about the reasons for the trends, but erosion can result in thinner than expected profiles, and both deposition and greater moisture downslope result in thicker profiles with higher values in various soil properties.

The above catena studies were conducted to determine (1) trends in catenas with time and (2) the usefulness of the soil catenas as a field tool for differentiating moraines. For example, Burke and Birkeland (1979) clearly illustrated the problems in using summit soils alone for recognizing moraines of different ages. In some valleys, summit soils on both 20- and 140-ka moraines have nearly identical A/Bw profiles, so other relative-age methods (e.g., rock-weathering features) must be used to differentiate the moraines in the field. If similar weakly developed soils are present at summit sites of moraines of different ages (owing to erosion), the next soil-sampling strategy is to either study multiple soils along the crests, or study soils downslope in a catena. The above three studies

demonstrate that the second strategy could be the best one, because strongly developed Bt horizons commonly are present at some or most downslope sites (Fs or Ts) in the older catenas. This is interesting from both paleoclimatic and erosional points of view, because temperature reduction during the 20-ka glaciation in Wyoming was at least 10°C (Mears, 1981). Despite these extreme conditions, soils were not entirely stripped from the slopes of the 140-ka moraines during the 20-ka glaciation.

PDI data for soil catenas can be used for making broad correlations of moraine sequences. Here data are compared for catena weighted-mean PDI values for four locations in the western United States, including the three discussed above (Fig. 9.16). There is a rather tight grouping of values for the 20-ka moraines, and a spread of data points for the 140-ka moraines. The tight grouping at 20 ka is expected as so many of those soils are similar. The spread at 140 ka is less easily explained, but could be due as much to the degree to

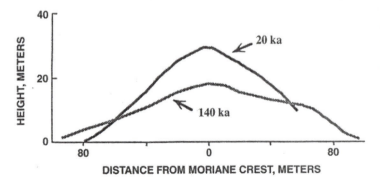

Figure 9.17 Typical cross-moraine profiles for 20-ka and 140-ka lateral moraines in the upper Colorado River basin. Each profile is averaged from 10 profiles surveyed in the field. *(Modified from Meierding, 1984, Fig. 4.)*

which some slopes have been eroded as to pedogenic reasons.

One major geomorphic problem with the PDI values (individual and catena-weighted mean) described above is that, in places, the values are high even though the moraine might have undergone long-term erosion. Meierding (1984), for example, has constructed cross sections of 20- and 140-ka moraines (Fig. 9.17) that suggest considerable erosional lowering of the older moraines. The soil data described above fit with this erosional story if either (1) erosional adjustment (that is, moraine flattening) takes place relatively soon after moraine deposition and soil formation encompasses a subsequent longer time interval; (2) slower erosion upslope keeps the soils there poorly developed and the combination of continual deposition and soil formation downslope results in overthickened cumulic profiles; or (3) erosion and the resulting poorly developed soils upslope grade downslope to relatively stable, well-developed soils, and the materials eroded from upslope bypass the downslope sites.

Although we attributed the above Su vs downslope soil differences to erosion, the latter was never quantified. One moraine-catena study in northern Colorado was able to do this. McMillan (1990) studied soils across a 20-ka moraine that had been affected by a recent forest fire. The Su A/Cox soil (weighted mean PDI = 0.04) grades downslope to much better developed soils with Bw and Btj horizons (PDI increases progressively downslope to reach a maximum of 0.20); all soil properties display the marked contrast between the Su and catena slope soils. Because a fire 10 years before the study had burned much of the forest vegetation, postfire erosion could be evaluated. It was concluded that fire-induced erosion of the nearly flat Su site could help explain the striking contrast between weakly developed soils at the Su site relative to better developed soils on adjacent slopes.

Catenas in a relatively dry area in New Zealand (MAP 57, MAT 9) duplicate rather well the above and show how total soil chemistry can aid in such studies (Birkeland, 1994). The soils are A/Bw/Cox profiles formed in loess/till. Not all soils thicken downslope, and in places the Ts soil is thinner than the Fs one. Fe_d increases downslope, but ratios of mobile-to-immobile elements are essentially the same from site to site down catena, as well as from one catena to the next. These data suggest that (1) the slopes are quite stable and (2) the moraines are close in age. I estimate

the moraine ages as either O-isotope stage 2 or stages 2 and 4, counter to estimates by others that put the older moraine as stage 8, based on other criteria.

With an increase in the amount of water leaching through the system, aided perhaps in places by organic matter complexes, quite marked pedologic changes take place over short distances. McKeague (1965) presents data on a catena involving Spodosols in eastern Canada and includes information on both the depth to water table and the redox conditions. Cryorthods are present in the summit, shoulder, and backslope positions. At these latter slope positions, the top of the water table is always below about 1 m upslope, but is at the surface for several weeks every spring downslope; measured redox conditions are more oxidizing upslope than downslope, where gleyed soil properties are present. In contrast, the footslope soils have the water table within 30 cm of the surface, and redox conditions vary from high to low Eh in the shallow zone of a fluctuating water table. The resulting soils are Cryaquents. Extractable Fe in the B horizon doubles from the summit (1.6%) to the footslope (3.2%). Clay minerals above the C horizon are smectite and chloritized vermiculite in the upslope soils and vermiculite, smectite, and mixed-layer clays in the footslope soils.

New Zealand workers have obtained soil catena data somewhat similar to that in the relatively humid area above, and they have compared them to catenas in somewhat drier conditions (Tonkin and others, 1977; Young and others, 1977). Under an MAP of about 100 cm the sequence is Inceptisols at and above the backslope position and Aquepts at the footslope position (Fig. 9.18). In contrast, under about 200 cm of precipitation, the system is more leached, with Spodosols at the summit, Aquods at the backslope, and Histosols at the footslope. As expected, the more humid catena displays a greater elemental loss relative to the drier catena. Clay-mineral alteration is more advanced for the more humid catena, and for both catenas, alteration is more advanced for the more leached upslope positions than for the less leached downslope positions. The geochemistry of the waters at the downslope sites, specifically the relatively high accumulation of bases and Al, seems to inhibit attaining higher clay-mineral alteration stages of Jackson (Fig. 8.38).

Pedogenic Al and Fe can be used to demonstrate geochemical conditions along a catena somewhat similar to the wetter one above in Canada, formed

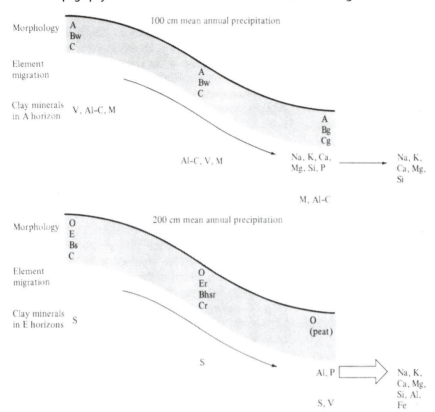

Figure 9.18 Schematic diagrams showing soil morphology, element migration, and surface clay mineralogy for two catenas in New Zealand, in different precipitation regimes. Key to clay minerals: V, vermiculite; Al–C, Al-chlorite; M, mica; S, smectite; "r" with horizon symbols indicates reduced horizon. For element migration, the width of the arrow is proportional to the amount of element moving through that portion of the catena in throughflow waters. *(Redrawn and modified from Tonkin et al., 1977, Fig. 1, and Young et al., 1977, Fig. 1 and Table 2.)*

on a 20-ka moraine (Young, 1988; MAP 300). Al_o is the Al extract with the highest concentration; it accumulates at all catena sites, commonly reaching 2–3%. Fe_d is the Fe extract with the highest concentration, it accumulates to a high value of 1–2% at backslope and upper footslope sites, but is depleted at the lower footslope and toeslope sites. Reducing conditions can explain this geochemical separation of Fe and Al in the lower sites; Fe is solubilized but Al is not.

The above catena development in a wet environment can be tracked through time on moraines on the west coast of the South Island, New Zealand (Birkeland, 1994; MAP 300, MAT 11). The 15-ka catena is characterized by an A/Bw/Cox form, with increasing thickness of the Bw horizon downslope (Fig. 9.19). An E horizon is present between the A and Bw horizons in the 65-ka catena, and the sub-A horizons thicken downslope. In contrast, the >100-ka catena has an A/E/B/Bsm/Cox profile, with the dominant E horizon thickening downslope. PDI for the 15-ka

catena increases downslope, but the older ones do not, probably because of the increasing dominance of the light-colored E horizon with time. We can also compare the catenas on the weighted mean catena PDI. Here, with increasing age the values are 0.26, 0.17, and 0.21. The lack of a progressive trend with time is attributed to the dominance of the E horizon over the Bw horizon with time; this produces a reversal in key properties with time (rubification and structure), and increases in other key PDI properties with time, such as clay buildup and associated properties, are minimal.

Soil chemistry follows the above morphological trends (Fig. 9.20). Fe_d increases downslope in the younger catena, but displays massive depletion downslope in the older catenas. Total chemical ratio data (element of interest vs more insoluble element) show depletion trends downslope, and in general relative depletion follows relative age. One of the most depleted horizons (A horizon of oldest Su soil) has 93% SiO_2, 4% Al_2O_3, and 0.5% Fe_2O_3, in other words it is close to a quartz sandstone in composition (initial composition was greywacke). These chemical and morphological trends are explained by the high leaching rates in an acid regime in which reduction (both downslope and above the Bsm horizon) becomes a progressively more important process with time.

Midcontinent United States and the Gumbotil Problem

Many catenas have been investigated in the midcontinent, in both loess and till. Loess of uniform particle size of about 20 ka is the parent material for a soil catena in southeastern Nebraska (Al-Janabi and Drew,

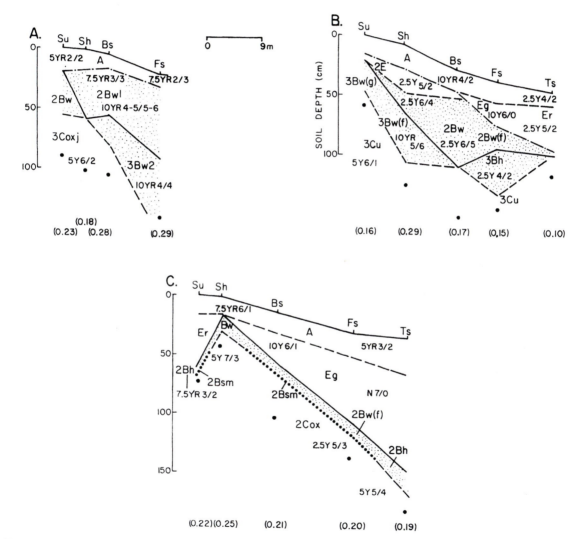

Figure 9.19 Generalized soil profiles and colors for catenas of different ages, Hokitika area, New Zealand. Estimated O-isotope stages are (A) stage 2, (B) stage 4, and (C) stage 10. Scale in the upper right is for both horizontal and vertical dimensions of the landscape. Dot–dash line is the base of loess, solid line is the top of the till, and if both lines are present, material between them is mixed loess. Dashed line depicts horizon boundary within one parent material. Row of dots depicts a thin (1–2 cm) Bsm horizon. Stipple pattern denotes a B horizon. Solid circle at the bottom of each profile is the depth of hand-dug hole, and the number in parenthesis at the base of the diagrams is the weighted mean PDI for that soil. *(From Birkeland, 1994, Fig. 4.)*

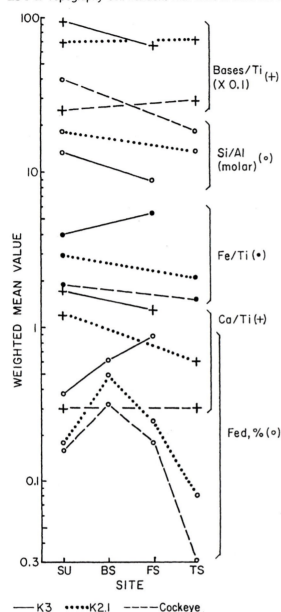

1967). Soils formed in the downslope positions are more strongly developed than are soils in upslope positions and on interfluves (Fig. 9.21). Resistant to nonresistant mineral ratios also vary markedly with slope position, and this suggests that much of the clay content variation is due to mineral weathering within each profile. Soil-moisture values in the C horizon for the summer of 1966 averaged 21% (by volume) at site A compared with 25% at site B. Although this slight difference in moisture might explain the differences in weathering and clay formation, it is possible that past conditions of greater moisture in the downslope soils, suggested by gleyed features, may have accelerated the trends. The quartz to feldspars ratio with depth can be compared with those of Ruhe (Fig. 8.5). The ratio at site A is typical for 20-ka deposits, but that at site B compares with those obtained from soils formed on much older Yarmouth–Sangamon surfaces. Thus, in this case, there seems to be a severalfold acceleration of weathering and clay formation with slope position.

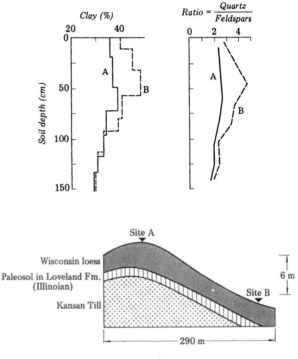

Figure 9.20 Weighted mean profile chemical data for catena soils of different ages in the Hokitika area, New Zealand. Estimated O-isotope ages are K3, stage 2; K2.1, stage 4; and Cockeye, stage 10. All ratio data (e.g., Fe/Ti) are for element oxides (e.g., Fe_2O_3/TiO_2), and bases = $CaO + MgO + Na_2O + K_2O$. *(Reprinted from Birkeland, 1994, Fig. 6.).*

Figure 9.21 Relations between topography, stratigraphy, and soils, southeastern Nebraska. Approximate O-isotope stages are Wisconsin, stage 2; Illinoian, stage 6; Kansan, >stage 6. *(Redrawn and modified from Al-Janabi and Drew, 1967, Figs. 3 and 4, Table 2.)*

Ruhe and Walker (1968) show that many soil properties correlate well with the slope steepness for rolling loess topography in Iowa (Fig. 9.22). In particular, with increasing steepness the soils are thinner, and there is less organic matter in the A horizon; also the depths to a pH of 6, to base saturation of 80%, and to a greater than 1 percent carbonate all decrease. Thus, the soils are shallower and less well developed on steeper slopes.

These soil trends with slope gradient may result, in part, from surface erosion, because, as Ruhe and Walker (1968) and Walker and Ruhe (1968) point out, the slopes are not always the same age. The summit surfaces, for example, are more than 14 ka, whereas the slopes adjacent to the summit areas have undergone more recent erosion and are less than 7 ka. This is not an isolated case as Dahms (1994) documents mid-Holocene erosion of moraine slopes in the mountains of Wyoming. Age of surface, therefore, a factor not always appreciated in soil–slope studies, could explain many of the soil differences in a comparison of summit vs slope soils.

Slopes are difficult to date. However, alluvium derived from slope erosion would be deposited at the foot of slopes. Therefore, careful radiocarbon dating of the alluvium produces dates on the times of stability and erosion of the slopes from which the alluvium came.

Soil development can be tracked over both time and topographic position in an undulating landscape in central Louisiana (MAP >100, MAT 18; Schumacher

and others, 1987). Here moderately well-drained Udalfs on broad slightly convex landscapes and poorly-drained Aqualfs of the nearly flat lows are compared. Both sequences are formed on loess, with the younger catena being about 10 ka, and the older about 75 ka.

There are differences between the catenas, some attributed to catena position, some to time, and some to both. A + E horizon thicknesses about double from the better to the poorly drained sites, but the thicknesses for both age groups are the same. Bt horizon thicknesses are about 1 m greater in the older catena, and in both catenas the thickness at the poorly drained site is about 50–65% that of the better drained site. Both maximum Bt clay content and weighted mean solum clay content are the same at both drainage sites, and the older catena contents are about 1.5 times greater. The clay-mineral suite of the better drained younger soil is smectite = mica > kaolinite = vermiculite and this changes to smectite > mica = kaolinite = vermiculite at the poorly drained site. In contrast, the older catena has kaolinite > mica = smectite > vermiculite at the better drained site and smectite = vermiculite > kaolinite > mica at the poorly drained site. In short, kaolinite increases with time at the better drained site at the expense of both mica and smectite, whereas age trends at the poorly drained site are an increase in vermiculite and kaolinite at the expence of mica, and perhaps smectite. These soil relations with drainage perhaps would be better expressed in terrain with higher relief and steeper slopes.

The till landscapes of the midcontinent are good examples of the interplay of pedologic, sedimentologic, and erosional processes in areas characterized by closed depressions. We will first consider processes operative on the present landscape and then discuss the problem of gumbotil.

Walker (1966) has made a detailed study of closed depressions and associated hillslopes in Iowa. The parent material is till of about 13 ka. The soils vary from Udolls in well-drained positions to Aquolls in poorly drained, closed depressions. Major differences in the soils are an increase in clay and organic matter toward the center of the depression, along with a decrease in the gravel content (Fig. 9.23).

The stratigraphy and radiocarbon dating of several depressions indicate that the sedimentation rate there and the erosion rate of the surrounding slopes have varied in the past. The uppermost 60 cm of material

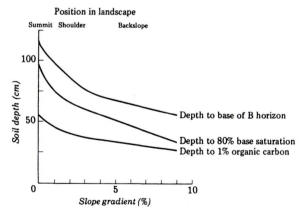

Figure 9.22 Relation between slope gradient and several soil properties, Iowa. (*Data from Ruhe and Walker, 1968, Table 1.*)

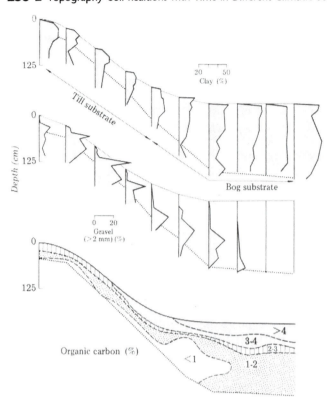

Figure 9.23 Relation of particle size and organic carbon with topographic position in a closed depression and on surrounding hillslopes. The dotted line separates hillslope sediments from till on the hillslopes, and younger from older deposits in the low-lying area. *(Redrawn from Walker, 1966, Figs. 19, 30, and 31.)*

in the depression shown in Fig. 9.23 is rich in organic matter and was deposited during a period of relatively slow slope erosion dating from about 3 ka to the present. Material below 60 cm, however, dates back to 8 ka; it is relatively low in organic matter and was deposited during an interval of relatively rapid slope erosion. Older deposits, indicating additional periods of slope erosion and stability, are found deeper in the depression.

The depression, therefore, contains mostly postglacial sediment, and properties of the latter directly influence the soil properties. These sediments were derived from the adjacent slopes, and the operative processes brought about the lateral separation of size fractions, so that gravel sizes were not moved into the centers of the depressions, whereas clay sizes were. Thus, the major lateral differences in soil-particle sizes are sedimentologic, not pedologic, in origin. Add to this primary textural control the variation in soil moisture due to variation in internal drainage, and most of the lateral variation in soils is adequately explained. Finally, the stratigraphic relationships indicate that

most of the soils in the area formed over the past 3 ka under grassland vegetation. Hillslope soils that formed over a previous period under postglacial forest vegetation were essentially removed by erosion before 3 ka.

Vreeken (1984) introduced a diagram, called a chronogram, that depicts well the interplay of deposition, erosion, and soil formation to form the complex soil–till landscapes of the midcontinent.

The arguments regarding the formation of the well-known gumbotils of the midcontinental Pleistocene record involve aspects of both the topographic factor in an undulating landscape and the time factor. Kay (1916) defined gumbotil as "a gray to dark-colored, thoroughly leached, non-laminated, deoxidized clay, very sticky and breaking with a starchlike fracture when wet, very hard and tenacious when dry, and ... chiefly, the result of weathering of drift." Kay (1916, 1931) and Kay and Pearce (1920) considered gumbotil to be primarily the result of prolonged chemical weathering of pre-O-isotope-stage-2 tills on flat, poorly drained plains; it is a gleyed soil. That chemical weathering had occurred was shown by chemical analyses of the gumbotil, compared with the unweathered till, and by the concentration in the gumbotil of siliceous pebbles that are resistant to weathering. A lively debate on what was included and what was excluded in the original definition of gumbotil developed. These arguments are reviewed by many workers (Frye and others, 1960a,b; Trowbridge, 1961; Leighton and MacClintock, 1962; Ruhe, 1965) and the details will not be repeated here. Basically, it was agreed that there are two contrasting origins for gumbotil. One origin is weathering in place under conditions of poor internal drainage; the other is that they are deposits laid down in depressions on the original till surface. Although Kay considered chemical weathering to be the main factor in gumbotil origin, he did agree that other processes, such as slope wash, were operative. Ruhe (1956), in his comparison of gumbotils located on swells and in swales in slightly undulating topography (Fig. 9.24), also recognized both processes of gumbotil formation. He thought that either weathering and sedimentation could go on contemporaneously in the swales, or that the two processes could be separated in time, and demonstrated from mineral ratios that material in the swales is more weathered than is material on the swells.

Frye and others (1960a) called attention to these two different origins of gumbotil and suggested that,

if the term is kept, it be restricted to profiles that have weathered in place from till. They suggest that gleyed material that accumulates in depressions be called accretion gley. It is important to recognize both of these origins and probably to use separate terms for both because the profiles formed in place attain their main characteristics by pedogenesis in poorly drained areas, whereas accretion gleys attain their main characteristics by sedimentary processes, as well as pedogenesis, in poorly drained areas. Some criteria found to be helpful in differentiating gumbotil from accretion gley are listed in Table 9.3.

The relationship of the gumbotils to the adjacent accretion gleys is analogous to that of deposits and soils with topography described by Walker (1966) mentioned above. This relationship has been pointed out by Ruhe (1965) and the lateral variation in gumbotils and accretion gleys appears to be due to the topographic factor. Time is also involved here, however, because gumbotils are found only on prelast-glaciation surfaces. Thus, given enough time, the materials and soils described by Walker might progress to the gumbotil and accretion gley stage of development (compare Figs. 9.23 and 9.24 and Table 9.3).

Follmer (1982) incorporates many of the above concepts in a detailed discussion of the Sangamon soil (began to form in O-isotope stage 5) in southern Illinois. The soil varies with topographic position on the Illinoian (0-isotope stage 6) till plain, and in places its recognition as a separate entity is obscured by the overlying younger deposits. Clues are given on how to correctly interpret these soils and stratigraphic relations.

The above discussion is over a rather old argument, and the terms could be modernized. However, the approaches used are important in understanding how to distinguish between in situ pedogenic processes and slope sediments, perhaps masked by pedogenesis. These problems are widespread and not only a midcontinent problem.

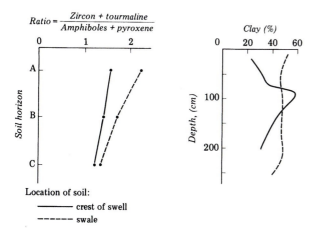

Figure 9.24 Comparison of pre-O-isotope stage 4 soils formed on different parts of an undulating landscape, Iowa. Parts of both profiles would be classed as gumbotil. *(Redrawn and modified from Ruhe, 1956, Figs. 4 and 5, Table 1, © 1956, The Williams & Wilkins Co., Baltimore.)*

Warm, Humid Environments

Catenas in which all soil-forming factors are reasonably well known may be more difficult to locate in strongly developed soils of warm, humid climates because the landscapes are so old and their histories more difficult to decipher. Such complex landscapes are common to parts of Australia (Division of Soils,

Table 9.3 Some Criteria for Differentiating Gumbotil from Accretion Gley

Criterion	Gumbotil (Formed in Place by Weathering)	Accretion Gley (Deposited in Lows in the Landscape)
Stratification	Massive, not stratified, leached of carbonates	Stratification can be seen as humus-rich layers toward the top of the deposit and as layers of contrasting particle size
Contact with fresh till	Gradational contact with fresh till includes a leached and oxidized zone over an unleached and oxidized zone	Contact with underlying fresh till can be sharp; configuration of the contact might suggest that the deposit occupies a depression on an old paleolandscape
Distribution of resistant pebbles	Pebbles of resistant lithologies show an orderly increase in relative abundance upward	Some deposits may have few resistant pebbles, relative to the till, due to sorting on adjacent hillslopes
Silicate mineral weathering	Silicate minerals are decomposed in order of weatherability; decomposition shows an orderly increase toward surface	Weatherable silicate minerals may be depleted; depletion toward the surface need not be systematic
Clay-mineral alteration	Clay minerals show an orderly arrangement of weathering stage with position in the profile	Clay minerals are a mixture of minerals of different stages of weathering; mechanical mixing is suggested

Taken from Frye et al. (1960a,b) and Trowbridge (1961).

CSIRO, 1983), where landscapes and soils date back to the Tertiary. Tardy and Roquin (1992, Fig. 16.4), for example, illustrate five interpretations for laterites that crop out at different elevations, only one of which might qualify as a toposequence. Ollier (1973) and Gerrard (1992) discuss some of these catenas. In contrast, many of the catenas previously mentioned have formed on landscapes formed of young glacial or loess materials, and, therefore, their histories are easier to interpret.

In spite of the problems, there are trends common to many soil catenas in warm humid climates (Nye, 1954; Radwanski and Ollier, 1959; Mohr and others, 1972; Hussain and Swindale, 1974; Kantor and Schwertmann, 1974; Young, 1976; Macedo and Bryant, 1987). Internal soil drainage is a major factor responsible for many of the trends. Soils in the higher positions of the landscape are well drained and oxidized red. Progressing to the lower parts of the landscape, the soils are mottled if under the influence of a fluctuating water table, and those at the base of the slope are usually gleyed. Iron and manganese concentrations and concretions can be present in the more poorly drained soils. Clay mineralogy contrasts commonly match the soil drainage contrasts, with minerals indicative of greater leaching associated with the upper slopes positions.

As an example, soils of catenas in Sri Lanka (MAP 147; Panabokke, 1959) and Rhodesia (MAP 84, MAT 18; Watson, 1964a,b) have hues of 10R and 2.5-5YR with high chromas at well-drained upslope sites, whereas colors commonly are 10YR 6-7/1-2 at poorly drained sites, where soils can also be gleyed and mottled. Kaolinite is the common clay mineral in the well-drained soils and smectite in the poorly drained ones. Both minerals form as a function of the geochemical environment of their respective soils, which is related to the leaching and groundwater conditions. If kaolinite has been eroded and transported to the topographic lows, however, there has to be some mechanism for its conversion to smectite. In one place, however, kaolinite is common and smectite not present in the gleyed soils (Watson, 1964b) so either such a kaolinite $\Rightarrow$ smectite alteration does not take place, the geochemical environment inhibits the alteration, or it is time dependent and more time is needed for the alteration to occur.

An Oxisol catena in Brazil (MAP 130, MAT 24) shows predictable trends for some of the more weathered landscapes (Curi and Franzmeier, 1984). Comparisons here are for the Bo horizon at 1-m depth. Clay content ranges from 58 to 71% with no slope-position relationship. When the fine-clay fractions are compared, these are the downslope trends: (1) soils become progressively less red: 2.5YR ⇒ 5YR ⇒ 7.5YR ⇒ 10YR, (2) Fe_d shows a slight decrease from 14 to 11%, (3) hematite is present only in the upper part of the slope where the hues are 2.5 and 5YR, and decreases in content downslope from 8% to 2%, (4) goethite ranges from 5 to 17%, with little relationship between content and slope position, but is the only crystalline Fe mineral present in the downslope sites where the hues are 7.5 to 10YR, (5) the kaolinite/gibbsite percentage ratio changes progressively from 22:22 to 42:6, and (6) magnetic susceptibility, used as a measure mainly of the presence of magnetite, decreases markedly. In this highly leached landscape, the wetter, lower slopes experience greater weathering (loss of magnetite) and loss of Fe in soil waters, but what pedogenic Fe is there forms goethite, and Si is redistributed downslope to form kaolinite. This gibbsite ⇒ kaolinite downslope distribution is repeated in the Ivory Coast (Tardy and Roquin, 1992, Fig. 16.9).

Tardy and Roquin (1992, Fig. 16.6) show several ways in which tropical toposequences form. One has a ferricrete on the slope that undergoes solution, called chemical dismantling, and the Fe is precipitated downslope, much as it is in the last example. A second one involves the work of termites. Termite mounds are an important feature of ferricrete plateaus, and they can be as much as 4 m high (Fairbridge, 1968, p. 778). Termites bring materials to the surface, commonly from the mottled weathered zone beneath the ferricrete. Distance of vertical transport can be as much as 13 m. The mound materials, dominantly silt and clay, are then subjected to lateral movement by surface erosion processes. The toposequence result is ferricrete upslope grading downslope to a profile in which a soft silt–clay horizon, the reworked mottled zone, overlies ferricrete. This surface horizon can then retain soil waters, which can bring about the chemical dismantling of the underlying ferricrete.

The association of red soils in the highs and grey soils in the lows is reversed in a special situation in an old landscape in Queensland, Australia (semiarid tropical, MAP 59 ; Coventry, 1982; Coventry and others, 1983). Depth to bedrock is shallow (0.4–2.5 m) in the upslope area, and increases downslope (3–20+ m). With the high water table perched on bedrock, the upslope soils are reduced and either goethite forms or Fe^{2+} is removed with the laterally flowing soil waters. In contrast, the well-drained deep and permeable soils of the lower slopes have soil waters that are continually resupplied with O_2 so that the O_2 level is not

lowered to the point of producing reducing conditions. The result is an oxidized profile and hematite formation in the lower positions of the landscape. A key point here is whether the O_2 in the soil water is depleted with time to produce gley features, or is continually resupplied (see also Evans and Franzmeier, 1986).

Daniels and co-workers (Daniels and Gamble, 1967; Daniels and others, 1967, 1968, 1971) recognize some complications in soil–slope relations in the Ultisols of the flat Coastal Plain of the eastern United States. These are old soils, with some beginning development during the early Pleistocene; in others, initial development was still earlier. Water table has an important influence on the soil distribution. Where divides are wide (about 3 km), the water table is high and within 50 cm of the surface about one-half of the time. Subtle depressions (relief ~2 m) within these exceptionally flat areas have the usual gleyed soils in contrast to the yellowish brown soils of the slightly better drained surrounding terrain. In the better-drained soils at the upper edge of the depressions there is a marked increase in texture between the A and Bt horizon (approximately 20–40% clay). Soils in the topographic lows have about 20% clay, again with an abrupt increase at the A/B boundary. Between these two locations, the soils have much less clay and are loamy sands or sandy loams. The indication is that the clays have been eluviated from these sandy soils because the water table fluctuates there more than at the other two sites.

Soils adjacent to the major terrace escarpments in the coastal plain also display the influence of the water table, because the water table is deeper at the edge of the escarpment than it is a short distance inland from the scarp. Where the water table is relatively high, B-horizon colors are 10YR 5/6. Toward the escarpment, the water table is deep enough not to influence the soils much, and the colors can change laterally to 7.5YR 5/6 and to 5YR 6/5 within several meters. Greater amounts of pedogenic Fe accompany the increase in redness. As one approaches the edge the E horizon thickens and tongues into the Bt horizon, the E-horizon clay content decreases, and the B-horizon clay content increases. In the areas of fluctuating water table, small patches of clay eluviation occur within the Bt horizons. Thus, a major change in these old highly weathered soils comes only when scarps retreat and the drainage of the soils at the edge improves, thus paving the way for a marked morphological change through eluvial–illuvial processes if a sufficiently long time has elapsed. In addition, soils in the middle of the divide will be influenced by a high water table until the divide becomes less than about 0.6 km wide. Finally, with greater geomorphic age and widespread landscape dissection, the water table lowers, and this will have a marked effect on the morphological trends over the region. Prior to dissection, the soils, except for those at the scarp edges, underwent little change with time.

In places in the tropics, Spodosols are part of catenas, forming at the lower parts of the slopes to produce what some call areas of white sand. In Brazil (MAP 240, MAT 26; Bravard and Righi, 1989, 1990), for example, Oxisols form at Su, Sh, and upper Bs sites, Ultisols at the Bs, and Spodosols (white sands) at the Fs (Fig. 9.25). Clay content shows a progressive decrease downslope, from 90 to 1%. Total Fe_2O_3 and Al_2O_3 are 4 and 35%, respectively, at the Su and decrease downslope to <0.3% in the Spodosol; over the same distance total SiO_2 increases from 57 to 99%, forming a quartz sand. Element-oxide mobility ranking upslope is Si > Al > Fe = Ti as quartz is dissolved there, whereas downslope Si reverses position as quartz accumulates giving a ranking of Al > Fe = Ti > Si. Kaolinite and gibbsite are the main clay minerals, and their contents at the Su (84 and 2%, respectively) change to 54 and 24 at the Fs. Fe_o and Al_o are <1% at all sites except the Spodosol, where they each make up less than 3% of the Bh horizon. One in-

terpretation is that during Oxisol formation quartz is resistant enough to weathering that it persists as other minerals weather away and convert to the resistatant oxides of Fe, Al, and Ti. Furthermore, in the mid-slope position, progressive clay illuviation increases the A/B horizon contrast and that, perhaps combined with downslope movement and accumulation of quartz sand into a thicker layer in the lower slope position, brings about a change in the dominant pedogenic process from vertical clay movement to podzolization. In this case, podzolization does not produce the sands; instead it is a dominant process in the lower slopes because of the presence of nonreactive quartz sand. Once podzolization takes over, it can lead to destruction of clays as well as the migration of organometal complexes.

Finally, a still wetter environment is a swamp on the coastal plain of northern Florida (Coultas and others, 1979). Sandy Aquents are present on the better-drained higher parts of the area. On the more poorly drained slopes are Aquods with Bh horizons; those on the lower parts of the slopes can be submerged for parts of the year. Finally, humus-rich Aquepts are present in the lowest parts of the topography. Trends from the topographic high to low position are clay content increasing from <10 to >20%; organic carbon in the surface increasing from 0.5 to 37%, and staying generally above 10% in the upper 81 cm of the Aquept; pH decreasing from near 5 to 3–4; and extractable cations showing a general increase in the surface horizons, from 0.3 to 9.7

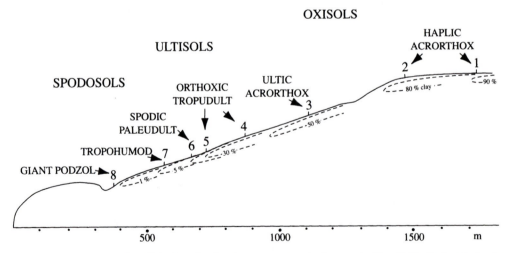

Figure 9.25 Schematic diagram of an Oxisol ⇒ Spodosol toposequence in Amazonia, Brazil. Vertical scale is about 50 m. *(Redrawn from Bravard and Righi, 1990, Fig. 1, Podzols in Amazonia (Brazil). Catena 17, 461–475, © 1990, with kind permission from Elsevier Science, The Netherlands.)*

meq/100 g. Since the environment is sufficiently leaching, most horizons have no bases.

■ Catenas Involving Calcic and Gypsic Soils

Soil catenas in progressively more arid environments are characterized by lateral translocation of only the more soluble compounds, the salts. For cold deserts, Glazovskaya (1968) noted the presence of the more soluble salts in soils at progressively lower parts of the landscape. Thus, $CaCO_3$ can exist in all the catena soils, but in the downslope direction, the more soluble salts appear in order of increasing solubility. For example, if Midslope soils accumulate gypsum and Na_2SO_4, chlorides of Ca, Mg, and Na could be present in the lowest parts of the landscape. Salts can move laterally if sufficient throughflow water is present. Iron released by weathering, which can be rapid in saline environments, would stay at the site of release and, along with aluminum, probably not be translocated laterally. The most likely clay mineral would be montmorillonite, but conditions appropriate for formation of fibrous clays such as palygorskite and sepolite could also exist.

At the wetter end of an aridic transect, one might find carbonate leached from the upslope soils but accumulated in the downslope soils. In drier climates, however, carbonate should accumulate at all sites.

A soil catena formed from late Pleistocene to early Holocene loess in eastern Colorado (MAP 40; MAT 11) illustrates the conditions under which carbonate accumulates in all catena soils (Honeycutt and others, 1990a,b). The vertical distribution of organic carbon is similar at all sites (Fig. 9.26), suggesting little recent downslope movement of surface material. The clay distribution shows a slightly greater clay bulge at the footslope site, due either to greater clay production there or to the slow accumulation of clay at the footslope derived from upslope sites. The carbonate distribution curves are similar at all sites, but the carbonate bulge is at greater depth at the footslope site. Perhaps the microclimate at the footslope is slightly wetter and carbonate is leached to greater depths.

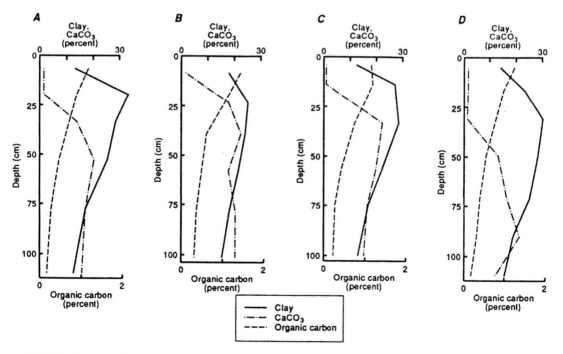

Figure 9.26 Distribution with depth of organic carbon, clay, and carbonate in a catena formed from loess in eastern Colorado. Catena positions are: (A) Su, (B) Sh, (C) upper Bs, and (D) Fs. *(Redrawn and modified from Honeycutt and others, 1990a, Figs. 2 and 5, and Honeycutt and others, 1990b, Fig. 1.)*

Birkeland (unpublished data, 1990) studied a series of river-terrace scarp soils arranged into synthetic catenas near Las Cruces, New Mexico (MAP 20; MAT 17), where Gile and others (1981) conducted their classic soil studies (Fig. 9.27). Catenas of three ages were studied: <10 ka, ca. 20 ka, and ca. 75 ka (J.W. Hawley, personal communication, 1990). The Bs soil of the intermediate-age catena has bedrock at shallow depth, so data for it are not comparable with that at the other sites. Catena soil profiles change with time from A/Bt or Btk/Bk with a maximum carbonate stage of II, to A/Btk or Bk/K with a maximum carbonate stage of IV. Soil hues are all 7.5YR and values and chromas are within the ranges of 5–8 and 2–4, respectively, with little trend with time except the K horizons are 8/2.

There are trends within and between catenas. Limited particle-size data suggest that the parent materials have <10% silt and <7% clay. The younger catena has a silt:clay percentage ratio of 10:9 in the Su upper Bk horizon, whereas the Fs Btk horizons have ratio ranges of 21–30:13–19. The oldest catena has a ratio of 25:13 in the Su Bk horizon vs ratio ranges of 21–43:15–19 in the Fs Btk horizons. From this limited data set the percentage fines increase downslope, but there is little increase with time. Gravel percentage decreases toward the surface in the Su soils, and downslope in all catenas. Organic-carbon accumulation index increases downslope in the younger two catenas (6.8 ⇒ 18.2 in the <10-ka catena, and 7.3 ⇒ 20.7 in the 20-ka catena; data gravel corrected) but remains constant in the older one (15.0 ⇒ 15.2), suggesting some contemporary downslope movement in the younger two catenas. The graphic carbonate profiles have a great deal of within-catena similarity, suggesting similar accumulations (Fig. 9.27; recall these are for the <2-mm fraction). However, the estimated mass of carbonate accumulation indicates a two- to fourfold increase downslope. The gravel-corrected values (g/cm^2/profile) for Su ⇒ Bs ⇒ Fs are

<10 ka: 3.4 ⇒ 6.4 ⇒ 12.8
20 ka: 20.5 ⇒ 21.8 ⇒ 29.4
75 ka: 17.3 ⇒ 32.1 ⇒ 43.1

PDI values reflect the above trends, and are as follows for Su ⇒ Ba ⇒ FA:

<10 ka: 0.07 ⇒ 0.19 ⇒ 0.22
20 ka: 0.22 ⇒ 0.25 ⇒ 0.36
75 ka: 0.24 ⇒ 0.31 ⇒ 0.37

There is a consistent increase downslope in all catenas, and there is a consistent increase with age when each catena position is compared. When these data are plotted by position vs age and compared with data for moraines in the western United States (Fig. 9.16), the Las Cruces catenas plot above the moraine average, and the highest values (Fs positions) are near the highest values for the moraines.

Geomorphic and pedologic events combine to produce the catena relations at Las Cruces. Once the original scarp forms, it erodes at the top, deposits colluvium at the base, and assumes a more or less stable form in about 10 ky. Most workers feel that the Holocene was a time of relatively rapid dust influx (e.g., Machette, 1985), so this accounts for the relatively high values of fines, carbonate, and PDI. Pedogenesis was sufficient to form profiles as geomorphic

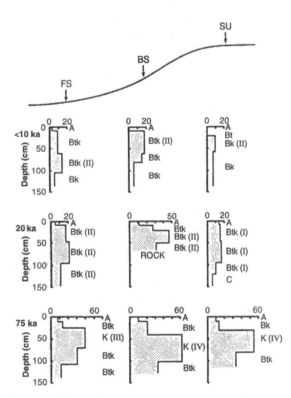

Figure 9.27 Morphology and carbonate content (percentage on horizontal axes) versus age for synthetic catenas on river–terrace scarps, southern New Mexico. Maximum stage of carbonate morphology is given in parentheses. *(Reprinted from Birkeland and others, 1991, Fig. 5.9.)*

activity took place, and downslope transport of both solid and dissolved materials seems to have occurred. Pedogenesis has dominated over geomorphic activity in the two older catenas, and because at least the older one formed for a long time under relatively low long-term dust influx (e.g., Machette, 1985), the amounts of carbonate are not proportional to the age differences, but are less. For the youngest and oldest catenas, the sevenfold difference in age is somewhat matched by soil properties at the Su and Bs sites, but falls way short of this at the Fs site, for some unknown reason. Finally, the influence of dust is seen in several ways: (1) it introduces fines and carbonate to the system, (2) the fines dilute the gravel content of the soils in place (Su) as well as dilute the gravel in the downslope colluvium, and (3) the carbonate is dissolved and moved both vertically in the profiles and downslope parallel to the landscape surface.

Soil catenas formed on river-terrace scarps have been described in the hyperarid climate (MAP 3; MAT 25) of southern Israel (Birkeland and Gerson, 1991). The youngest catena is 5 ka (SII, Fig. 9.28) and the soils are characterized by weak Bt and By horizons. The next older catena (SI) is 10 ka, and soils have a thin Bt/By/Byzmt profile at most sites, in strong contrast with the youngest catena. The progressively older catenas (AII, about 20 ka; AI, about 100 ka) have thicker Bt horizons, and salts have accumulated in some Bt horizons. The weighted mean PDIs for the catenas display a good trend between the two youngest catenas, but values for the three oldest catenas are

quite similar. The best trends are for horizon PDI data summed to the base of the clay accumulation horizon. Depth plots of electrical conductivity, a surrogate for total salt content (most likely gypsum and halite), also indicate similar values for the three older catenas, and much less for the youngest catena (Fig. 9.29). If the gravel-corrected weighted mean values are compared, the SI and AII catenas show an increase in salt content downslope. Another trend with slope position is that the Bt thickens from the backslope to the footslope position in the oldest catena, suggesting slight erosion and deposition, and the base of the Bt at the footslope is cemented by salts. In general, in these hyperarid catenas, only the most mobile components (salts) have been moved downslope.

Paleoclimate could have played a role in the development of the above soils in Israel. Salt content, for example, does not display a linear increase with time, as the values are quite similar beyond 20 ka. This could be explained (1) if influx of airborne dust was much less in the interval 20–100 ka, (2) if a wetter climate >20 ka leached all salts from all the soils and accumulation began anew near 20 ka, or (3) if surface crusts developed in time that reduced the ability of the soils to transmit water (and included solubilized salt) vertically. Under (2), one could envisage crossing the line between gypsum accumulation and depletion as shown in Fig. 8.32).

A N-facing catena changing from bedrock upslope ⇒ colluvium downslope in southern Israel (MAP 9) shows one aspect of the influence of shallow bedrock

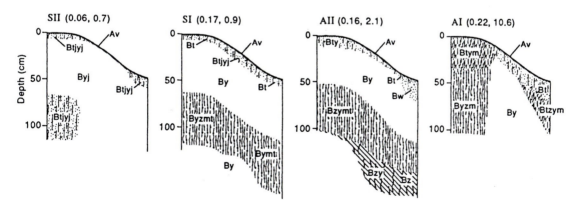

Figure 9.28 Soil catenas developed from river-terrace scarps in southern Israel. There are two numbers in the parentheses above each soil; the first is the catena weighted mean PDI, and the second is the nonnormalized PDI summed to the base of the clay-accumulation horizon. *(Reprinted from Birkeland and others, 1991, Fig. 5.10.)*

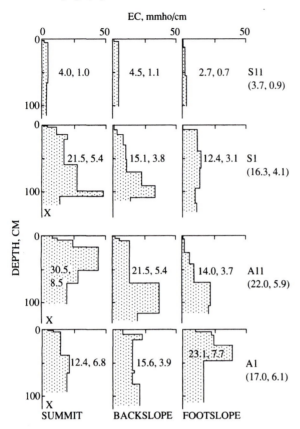

Figure 9.29 Plots of electrical conductivity (EC) with depth for soils of the four catenas in Fig. 9.28, southern Israel. Ages are younger from top (SII) to bottom (AI). "X" indicates EC value for present-day river bed sediment. Two numbers are adjacent to each profile plot; the first is the profile weighted mean for the <2-mm fraction, and the second is the same but corrected for gravel content. To the right of each catena are two numbers in parentheses; the first is the catena weighted mean for the <2-mm fraction, and the second is the same but corrected for gravel content. *(Redrawn from Birkeland and Gerson, 1991, Fig. 4 Soil-catena development with time in a hot desert, southern Israel—field data and salt distribution. J. Arid Environments 21, 267–281 © 1991, by permission of Academic Press Ltd., London.)*

in a catena–soil relations (Weider and others, 1985). The upper slope is either bedrock or shallow A/R or A/C profiles, whereas the lower slope has soils with subsurface Bk and Bky horizons in colluvium >1 m thick. Leaching by soil waters is indicated by the vertical salt distribution. Just downslope of the bedrock–colluvium contact the salts are low in con-

tent with no pronounced increase with depth. Farther downslope in the colluvium, however, salt content is not only greater but there is a dramatic increase with depth, and the soil farthest downslope has the greatest salt content.

The above relations have to do mainly with bedrock-induced runoff. Because the upper slope underlain by bedrock encourages runoff and little infiltration, water runs off downslope and infiltrates where it first meets the colluvium, thus producing a leached soil profile. Farther downslope, however, soil moisture content is lower, being mainly from atmospheric sources and not from runoff, and pedogenic salts accumulate in a manner predictable in a hyperarid climate. Throughflow water movement might account for some of the abrupt increases with depth.

Many soil properties in other environments might be similarly influenced by shallow depth to bedrock. Hence, to keep most factors constant in a catena, it might be best to avoid ones involving bedrock. This was the rationale in selecting the catenas that have been discussed in this chapter.

The Boulder, Colorado area (MAP 47; MAT 11) is characterized by Ustolls, but leaching is such that in the lower parts of some poorly drained rolling landscapes are soils with gypsic horizons. A toposequence on colluvium/shale studied by M.A. Siders (1991 unpublished report) is rather unique in that the total geochemistry of the soil solutions was examined, and computer programs used to suggest the pedogenic salt minerals that are expected, given the ionic concentrations. The soils at the site are calcareous and are Torriorthents (A/C) at the Su and Bs sites, and Halaquepts (A/Bky/Bkyg) in the poorly drained Fs site (Moreland and Moreland, 1975). Salt concentrations (exclusive of carbonate), best shown as total dissolved salt, increase down all profiles to a maximum, and then decrease; the ranking of the depth to the maxima is Fs > Su > Bs. Almost all ions are in greater concentration downslope. This is best shown by the profile averaged data, which show that the Su and Bs soils are rather similar, but that the Fs values are one to two orders of magnitude greater (Fig. 9.30). Na > Ca > Mg are the dominant cations, and SO_4 is the dominant anion. Using the WATEQB program to identify equilibrium concentrations and species by calculation of the saturation index, only gypsum was shown to be in equilibrium in the Fs soil. It also can be readily seen with a hand lens. Both halite (NaCl) and thenardite (Na_2SO4) are

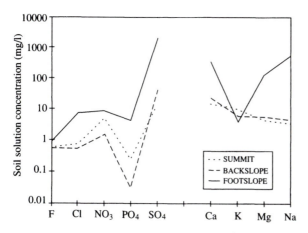

Figure 9.30 Data for soluble salts in a catena on shale near Boulder, Colorado. *(Data from M.A. Siders, 1991, unpublished class report; diagram courtesy of John Drexler.)*

undersaturated at all sites, but less so at the Fs site, suggesting they are not present.

■ Concluding Remarks about Catenas

The catena relations described above and others reviewed by Yaalon (1975) allow soil ages for discrete portions of slopes to be estimated. For example, a strongly developed soil with a red Bt horizon on a colluvial mountain slope in the western United States suggests geomorphic stability surely for 100 ky. Menges (1990) presented similar soil arguments for the duration of stability of tectonic mountain-front bedrock facets in New Mexico. With these data, one can also map drainage basins, and assign general ages to various parts of the basin. Well-developed soils would indicate relatively stable sites, whereas poorly developed soils would indicate sites of active erosion or deposition. Although creep undoubtedly goes on at the stable sites, pedogenesis proceeds at a rate much faster than erosion. An example of this is the work of Tonkin and Basher (1990) in New Zealand. This is an application of the K-cycle model of Butler (1959), where soil morphology at different slope positions was placed in a time framework. With this the areal position and extent of both the erosion and deposition zones with time can be estimated. One end result could be a mosaic of soils in which erosion keeps soils relatively poorly developed in the upper parts of the

slopes and episodic deposition yields a sequence of buried soils in the lower parts of the slopes. If, instead of the periodicity of K cycles, erosion and deposition are more gradual through time, the soil distribution should reflect this. In this case, instead of buried soils, one would see overthickened, cumulic profiles at the downslope sites.

Several generalized models can be depicted to demonstrate how interconnected catenas are with climate and time.

The first model is one in which the soil or specific soil properties that are seen downslope in one climate or at one time are present upslope in a wetter climate or in an older catena. The model for this, a climate model, is the clay mineralogy vs slope position one of Tardy and others (1973) (Fig. 9.9) in which higher-silica-content clay minerals are found downslope, and these same clay minerals occur in the upslope soils of the catenas in the next drier environment. Once it gets so dry that carbonate and more soluble salts are involved, the amounts progress with slope position in concert with their relative solubilities.

A generalized trend similar to the above has been shown for other soil properties in catenas of different ages. For example, in many environments PDI and clay, Fe, and carbonate amounts all are greater downslope, and in those places where the ages between the catenas are fortuitously about right, the amounts of pedological constituents (e.g., Fe_d in Fig. 9.15B, or carbonate in Fig. 9.27) in the relatively better developed downslope soils of a particular-age landform are somewhat similar to the summit soils of the next older landform. Continuing, the same constituent increases downslope from the Su soil to the Fs soil of the latter catena, and the content there is similar to that in the Su soil of the yet older catena, and so forth. This is akin to a pedological forecast with time.

The second model depicts soil catenas along a precipitation gradient in which all soils become more acid, and downslope soils become first gleyed and then reduced at progressively higher precipitations; this model is an extension of that of Tonkin and others (1977) (Fig. 9.31). At low precipitation, similar gypsic soils are found at all slope positions, such as those discussed in southern Israel. At slightly higher precipitation, gypsum does not accumulate but carbonate does, and although the soils can be morphologically similar at many slope positions (e.g., the New Mexico area; Fig. 9.27), there is some movement of carbonate downslope, shown by the mass accumulation

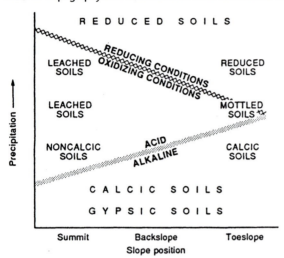

Figure 9.31 Schematic diagram depicting type of soil versus catena position along a precipitation gradient in which pH becomes progressively lower and reducing conditions become more extensive (laterally) at higher precipitations. *(Modified from Tonkin and others, 1977; Reprinted from Birkeland and others, 1991, Fig. 5.11.)*

data. At still greater precipitation, calcic soils may form only at the downslope sites. The next step is to those higher precipitation levels at which soils are leached, no carbonate accumulates, and surface and throughflow accumulation of moisture in toeslope sites results in mottling. With even greater moisture and low enough pH, leached and oxidized upslope soils may grade downslope to reduced soils. Finally reduced soils might form across the whole landscape. In the New Zealand example (Figs. 9.19 and 9.20), this happens at about 3 m MAP in soils >100 ka.

The third model depicts the pedological contrast down a catena with precipitation or with time. The horizontal axis of the plot is precipitation, and the vertical axis is the degree of contrast in the catena (Fig. 9.32). Degree of contrast could be recorded as morphological contrasts (e.g., Bt horizon downslope but not upslope, or PDI increases drastically downslope), geochemical contrasts (e.g., carbonate or Fe contents double downslope), or clay mineralogy contrasts.(e.g., kaolinite upslope and smectite downslope). The greater the change downslope, the greater the degree of contrast within the catena. For the soils discussed

in this chapter, the gypsic soils are so similar within catenas that the degree of contrast is slight. Although the morphology of the New Mexico calcic soils is similar within each catena, carbonate has been redistributed downslope as shown by the mass data, and because carbonate is less soluble than gypsum, the contrast for the New Mexico soils is deemed greater than for the Israel soils. Catenas formed on moraines in the western United States have a still higher degree of contrast because of the Bw $\Rightarrow$ Bt contrast downslope. The marked increase in clay downslope in the midcontinent, as well as the difference in thickness of clay accumulation, places those soils at a still higher degree of contrast. Soils in the hot tropics have the highest degree of contrast as the red, leached, kaolinitic soils of the uplands are markedly different from the dark, lesser leached, smectitic soils of the lower slopes.

The wet area of New Zealand (Figs. 9.19 and 9.20) illustrates a situation in which time might be the factor on the horizontal axis of Fig. 9.32. The catena contrast first increases with age of catena, but in the oldest catena, where reduction processes dominate, the result is a dramatic lessening of the within-catena contrast, hence the bell shape of the New Zealand curve for contrast vs time. Another time factor would be the

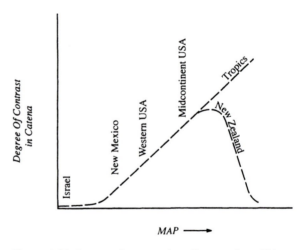

Figure 9.32 Degree of contrast in soil properties within a catena versus MAP (or for some properties in some environments, greater time).

moraine soils of the western United States: the 20-ka soils have low contrast downslope, whereas the 140-ka soils have much greater contrast (Fig. 9.15).

An excellent review of catenas by Sommer and Schlichting (1997) has recently appeared. They classify catenas into three principal types: (1) transformation catenas, those in which transformation processes operate but there are no gains or loses of elements or soil components; (2) leaching catenas, with loses of elements and no gains; and (3) accumulation catenas, which show gains in elements or components. They also have a translocation catena, characterized by upslope depletion of some elements or components, and downslope accumulation of the latter. I suggest that the translocation catena be included as a major type, for a total of four. Examples of the correlation of some catenas discussed in this chapter with the four Sommer–Schlichting types are as follows:

1. Transformation catenas: southern California catena in which most of the clay weathers from bedrock (Fig. 9.13); also the dry area in New Zealand (Tekapo) where pedogenesis products form but they do not migrate downslope (Birkeland, 1994).

2. Leaching catenas: the wet catenas of New Zealand characterized by widespread loss of elements (Fig. 9.20).

3. Accumulation catenas: the aridic catenas that accumulate $CaCO_3$ and more soluble salts (Figs. 9.27 and 9.29).

4. Translocation catenas: some catenas on moraines in the western United States (Fig. 9.15).

Sommer and Schlichting also point out that catenas can shift from one type to another over time. For example, the New Zealand example in Fig. 9.32 (also Fig. 9.20) starts out as a transformation and translocation catena, and in time becomes a leaching catena.

Summary of Climate–Soil Relations, and Paleoclimatic Interpretations

Climate is considered by many to be the most important factor in determining the properties of many soils [e.g., Hunt and Sokoloff (1950), but this particular case was challenged in Chapter 8]. This climate–soil relationship can be seen by comparing the world soil map (Fig. 2.8) with maps of soil moisture and temperature (examples for Pennsylvania in Waltman and others (1997); most soil orders and suborders are restricted to certain climatic regions. In Chapters 8 and 9 the format presents soil relations along a global climatic gradient, so many of the details are in those chapters. Here we will look into larger-scale relationships, mention specific soil properties that are related to climate, and discuss those properties that might be most useful in reconstructing past climates. Because vegetation varies with climate, these might best be termed bioclimate–soil relations.

The intent here is not to belittle the significant impact of vegetation on pedogenesis (reviewed in Birkeland, 1984, Chapter 10, plus other soil textbooks). Most of the trends with vegetation have to do with organic matter, base saturation, and pH. If climate is kept constant, these trends include the following: (1) organic matter remains at higher levels with depth under grassland than under forest, and base saturation is higher under grassland; (2) all three properties vary with tree species; and (3) all three properties, as well as others such as clay and Fe content, vary with distance from a tree, due to the influence of rainwater coming through the tree canopy, and the amount and composition of stemflow water. Finally, any soil classification scheme reflects the combination of climate and vegetation, so soil morphologies of surface and buried soils that are not expected under present-day conditions could indicate a former, and different, vegetation perhaps under the same general climate or in combination with a different climate.

Moisture and temperature are the two aspects of climate most important in controlling soil properties. Moisture is important because water is involved in most of the physical, chemical, and biochemical processes that go on in a soil, and the amount of moisture delivered to the soil surface influences the weathering and leaching conditions with depth in the soil. Temperature influences the rate of chemical and biochemical processes. Jenny (1980) demonstrates how to derive separate functions for precipitation and for temperature in some areas, provided other aspects of the climatic factor and the other soil-forming factors can be considered constant.

A point discussed in parts of this chapter is that in comparing some soil properties in quite different climates, pedologic differences that might be predicted from the climatic data are not always apparent. This causes some problems in interpretation, but it is good for stratigraphic studies in which soils are used in correlation.

■ Climatic Parameters

A numerical value for the climate has to be used to demonstrate, quantitatively, the functional relationships between climate and the various soil properties. In places, the mean annual values of either precipitation or temperature can be used as an approximation of the climate. The use of mean annual values, however, fails to take into account the monthly distribution of precipitation and temperature (Fig. 10.1). It is important to know, for example, whether precipitation is seasonal, and if it is, whether the precipitation maximum coincides with the annual temperature maximum or minimum, because these climatic variables strongly influence soil leaching and soil-water chemistry, both of which are important in determining key soil properties. For example, some areas have strong seasonal contrasts with relatively wet winters and dry summers. Under such a fluctuating moisture regime, it could be that kaolinite is the clay mineral stable with

the chemistry of the winter soil waters, whereas smectite is stable with the summer soil waters. If, in studying clay-mineral stability, we collect soil waters that are representative only of the mean annual conditions, we might err in our interpretation. Few such detailed analyses have been made, but the point is that many of the details of the soil–climate connection still have to be worked out.

Water balance can be used to describe some of the climatic characteristics of a region that are important to soil formation (Arkley, 1963, 1967). Geomorphologists have spent considerable effort in trying to estimate the balance (an excellent treatment is Dunne and Leopold, 1978). This balance represents the gains and losses of soil moisture over a certain interval of time (week, month, year). Gains are from precipitation (P); these data come from climatological stations. Losses, however, are from evapotranspiration; values for potential evapotranspiration (ET_p) can be calculated by the method of Thornthwaite and Mather (1957). If

Figure 10.1 Monthly variations in precipitation and temperature for various climatic stations around the world. Precipitation value on each graph is MAP in millimeters. Climate and altitude for each station as follows: (A) tundra, 7 m; (B) continental, 285 m; (C) mediterranean, 47 m; (D) steppe, 1613 m; (E) tropical savanna, 16 m; (F) marine, 43 m; (G) humid subtropical, 33 m; (H) desert, 111 m; (I) tropical rain forest, 5 m. Upper curve, temperature; histogram, precipitation. *[Reproduced by kind permission of John Bartholomew & Son Ltd. from the Times atlas of the World (The Times of London, 1967, XXVIII–XXIX).]*

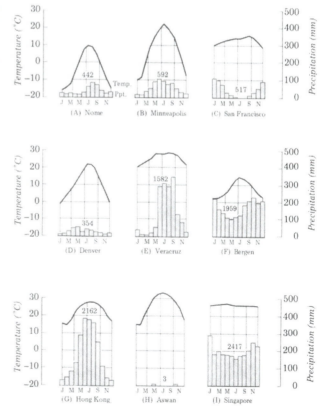

Manhattan, Kansas is used as an example (Table 10.1), the water balance shows the approximate monthly variation in soil moisture. For those months with positive values for $(P - ET_p)$, water is stored in the soil at moisture values above the permanent wilting point; this water is available for plant growth and weathering reactions; and if the water is moving, it can translocate dissolved or solid materials within the soil. In contrast, months with negative values of $(P - ET_p)$ are marked by the removal of water from the soil. Water removal can go on until the soil reaches water contents approaching the permanent wilting point. If the water-holding capacity of the soil is known, the amount of water percolating to depths within or beneath the B horizon can be calculated.

Plots of water balances for several different climatic regions indicate significant soil-moisture differences of pedologic importance (Fig. 10.2). For example, both the California coast (Half Moon Bay) and the California Great Valley (Sacramento) are characterized by soil-moisture buildup for about the same months, but a greater amount of water leaches through

the soil at the coastal than at the inland site. Fallon, Nevada, located in the rain shadow of the Sierra Nevada, is characterized by a short period of soil-moisture storage and slight leaching; thus, soil profiles there are shallow and slightly leached and contain Bk horizons. However, Manhattan, Kansas is in an area of summer precipitation and thus is characterized by fairly high soil-moisture contents throughout much of the year, including some of the warm summer months. These data point out the basic differences between the climate of the midcontinent and that of California and the Basin and Range Province; that is, in the midcontinent, the soils are moist during much of the warm season, whereas the western areas are winter wet and summer dry. From this it can be predicted that the rates of biological and chemical processes will vary between these regions. Compared with western sites, sites in the midcontinent should produce more aboveground organic matter annually, and they should have higher rates of organic matter decomposition, mineral weathering, and clay-mineral transformation.

Table 10.1 Water Balance at Manhattan, Kansas, for a Soil with 10.2-cm Available Water-Holding Capacity

	Month												Annual (cm)
---	J	F	M	A	M	J	J	A	S	O	N	D	
Potential evapotranspiration (ET_p)	0.0	0.0	1.9	5.1	9.3	13.9	16.4	15.0	10.1	5.2	1.4	0.0	78.3
Precipitation (P)	2.0	3.0	3.8	7.1	11.1	11.7	11.5	9.5	8.6	5.8	3.8	2.2	80.1
Soil-moisture gains, $P - ET_p = (+)$	2.0	3.0	1.9	2.0	1.8					0.6	2.4	2.2	15.9[a]
Soil-moisture losses, $P - ET_p = (-)$						2.2	4.9	5.5	1.5				14.1
Soil-moisture storage at end of month (A and B horizons)	7.2	10.2	10.2	10.2	10.2	8.0	3.1	0.0	0.0	0.6	3.0	5.2	
Possible deep percolation below B horizon[b]				1.9	2.0	1.8							5.7

[a]This annual value is the leaching index of Arkley (1963).

[b]In this case it is assumed that the water-holding capacity of the A and B horizons is 10.2 cm. Therefore any surplus of water over 10.2 cm can move to greater depths in the soil.

Figure 10.2 Monthly values of precipitation (*P*), potential evapotranspiration (*ET*p), soil-moisture storage in the A and B horizons (10.2-cm available water-holding capacity), and soil-moisture percolation below the B horizon for various climatic stations.

Station	Mean Temperature (°C)	
	January	July
(A) Manhattan, Kansas	−1.6	26.6
(B) Fallon, Nevada	−1.2	22.9
(C) Sacramento, California	7.5	23.2
(D) Half Moon Bay, California	9.9	17.0

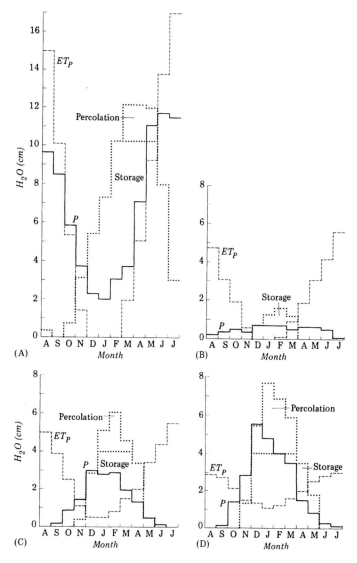

Water-balance data were used to define the soil-moisture regimes for Soil Taxonomy. The first published edition of the latter (Soil Survey Staff, 1975) has diagrams similar to Fig. 10.2 with all the described soils. In addition to data as in Fig. 10.2, they identify the months of recharge, surplus, utilization, and deficit. From a pedological point of view, we would want to calculate the amount of available water for each specific soil, rather that use some seemingly arbitrary value, such as the 10.2 cm in Fig. 10.2. Sometimes what seems to be an arbitrary value from a pedological standpoint, however, is

critical for crop production and the timing of irrigation.

The data on water balance and water-holding capacity of the soil can be combined to approximate water movement with depth in soils (Arkley, 1963). Knowing the frequency of wettings per year, and assuming that downward movement of water takes place only when field capacity for that part of the soil has been reached and that moisture can be removed from the soil by evapotranspiration down to the permanent wilting point, one can construct curves depicting water movement (Fig. 10.3).

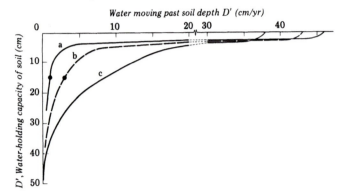

Figure 10.3 Amount of water moving past various depths of the soil, calculated from water-balance data. Depth is in water-holding capacities, but these can be converted to soil depths. Filled circles, depth of the top of the Bk horizon. Key for curves: a, Clovis, NM; b, Boise City, OK; c, Santa Monica, CA. *(Some data from Arkley, 1963, Fig. 4, © 1963, The Williams & Wilkins Co., Baltimore.)*

These curves then can be compared with soil data, such as clay-mineral variation with depth, or the top of the Bk horizon, to determine which soil features can be attributed to present-day water movement and which may have formed under some past water-movement regime. For example, one such curve superim-posed on clay and carbonate data shows a distinct clay bulge over a distinct carbonate bulge (Fig. 10.4A). The transition from one to the other is about 5 cm thick, and takes place where the water movement values are very low. The reason for the sharp demarcation be-tween the two bulges probably is because not much

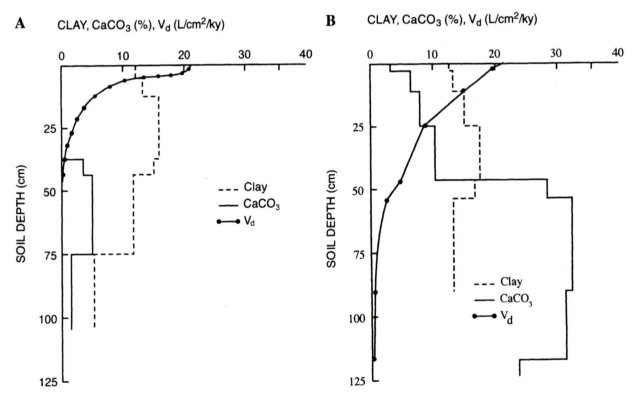

Figure 10.4 Vertical distribution of clay and carbonate compared with the volume of water passing each depth increment (V_d, in liters/cm^2/ky) for two soils near Las Cruces, NM. (A) Onite sandy loam; (B) Dona Ana sandy loam. *(Unpublished data of R.J. Arkley, written communication, 1997.)*

water moves deeper (on average), and any clay moving deeper probably would be flocculated. In contrast, another soil in the same area shows an overlap of both clay and carbonate, and the carbonate accumulation zone extends to the surface (Fig. 10.4B). This diagram is difficult to interpret, but one interpretation could be that the superimposed Bt and Bk horizons indicate climate change toward a drier climate as the Bk horizon engulfs Bt, followed by erosion of the carbonate-free upper part of the soil that should overlie the zone of carbonate accumulation.

Although a single number cannot be derived from the water-balance data to characterize a particular climate adequately, water balances do provide data from which one can rank soils from various regions by the amount of water leaching through the soil. Arkley (1963), for example, derived a number called the leaching index, which is the mean seasonal excess of $(P - ET_p)$ for those months of the year in which $P > ET_p$ (Table 10.1). Soils located in areas characterized by a high leaching index commonly have properties associated with large amounts of water percolating through the soil, such as 1:1 clay minerals, whereas soils in regions with a low leaching index have properties associated with slight leaching, such as 2:1 clay minerals or a Bk horizon close to the surface. Mc-Fadden (1982), McFadden and Tinsley (1985), and Reheis (1987a,b) use the leaching index as the climatic parameter in chronosequence studies in several climatic regimes in dealing both with present climate–soil relations and paleoclimate–soil relations.

Because most data in the literature on climate–soil relations use mean annual values of precipitation and temperature, these will be used here. Water-balance data might be more meaningful, however. Mather (1964) produced a useful data set of water balances for stations around the world.

■ Regional Soil Trends Related to Climate: Pedogenic Gradients or Climosequences

Many soil properties show distinct trends with regional climate in going from the equator to poles, as is shown by the often-reproduced diagram of Strakhov (1967) (Fig. 10.5). Such variations in the soils originate in such processes as organic matter influx and decomposition, presence or absence of chelating agents, soil-water chemistry, and the amount and rate of leach-

ing of water through the soil. These processes, in turn, are controlled by the climate, but we have shown that sometimes parent material can alter the particular pathway taken. The tropical forest regions are characterized by intense, deep weathering, with iron and aluminum oxyhydroxides predominant close to the surface. With depth, clay minerals of the 1:1 and finally 2:1 varieties are found. Organic matter in these soils is relatively low because even though the amount of organic matter annually added to the soil is high, the decomposition rate of organic matter also is high. The above trends diminish to the north in the savanna region. The deserts are characterized by low organic matter input relative to the rate of decomposition, and low organic matter content in the soil results. Slight leaching produces 2:1 clays, pedogenic carbonate at depth, and gypsum at depth under extreme aridity. Increased precipitation and decreased evapotranspiration characterize the steppes compared to conditions in the deserts; the result is a fairly thick cover of vegetation and thick A horizons rich in organic matter, with moderately leached subsurface horizons. We find $CaCO_3$ in soils in the more arid part of the steppes. North of the steppes is the forested taiga, a region of fairly high soil leaching. The low temperatures of the taiga result in fairly low rates of organic matter decomposition and prominent O and/or A horizons form. Where conditions favor iron and aluminum movement, an E horizon forms. The tundra lies north of the taiga, and the combination of fairly low precipitation, low temperature, and permafrost close to the surface in places results in a moist soil with relatively slow rates of organic matter influx and decomposition. These conditions produce a soil in which an A horizon with a fairly high organic content commonly overlies a gleyed horizon. Continuing northward, the tundra gives way to the polar desert, a region of low precipitation and temperature (see Tedrow, 1977). Because vascular plants are nearly absent in the drier parts of this desert, the organic matter content in the soil is fairly low and is provided mainly by lichens, algae, and diatoms. Although these polar desert soils generally are permeable, the absence of appreciable moisture and water movement in the soil causes them to be saline and alkaline, with very little weathering. In many respects, these soils are similar to those of some hot deserts, as discussed in Chapter 8.

The above variation in soil properties and processes forms what Tedrow (1977) calls a pedogenic gradient. Tedrow's examples come from the northern latitudes,

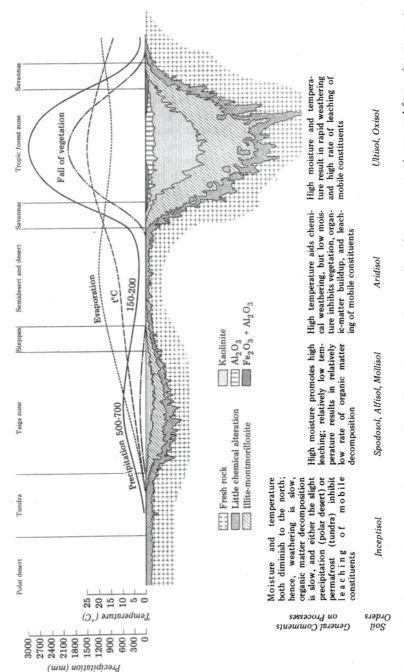

Figure 10.5 Diagram of relative depth of weathering and weathering products as they relate to some environmental factors in a transect from the equator to the north polar region. One could add Gelisol to the predicted soil orders at the polar side of the diagram. (*Taken mostly from Strakov, 1967, Fig. 2.*)

and Fig. 8.12 depicts a pedogenic gradient for the southern latitudes. The term is appropriate for any pedologic variation associated with climatic gradients. For example, the work of McFadden and Tinsley (1985) and McFadden (1988) is an example of a pedogenic gradient along a climatic gradient from aridic to xeric (Fig. 8.22), that of Birkeland and others (1989) is one that includes the high mountains where cold climate (cryic and frigid) is an important factor (Fig. 8.14), and that of Reheis (1987a,b) is an example of one from aridic to ustic-udic.

Schaetzl and Isard (1996) present an example of a pedogenic gradient involving Spodosols over about 400 km in the Great Lakes area (Fig. 10.6). The strength of development of Spodosols increases from south to north, as shown by the Soil Taxonomic classification to the subgroup level, and by the increase in the podzolization (POD) index of Schaetzl and Mokma (1988; in Chapter 1). They modeled the soil

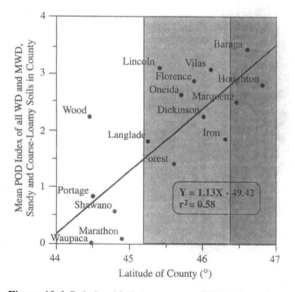

Figure 10.6 Relationship between mean POD index values and latitude of the center of the county, northern Wisconsin and Northwest Michigan. Names are those of the counties. Only moderately drained (or drier) soils are included. Shading (white, light gray, dark gray) corresponds with areas of weak, moderate, and strong podzolization, respectively. (*Redrawn from Schaetzl and Isard, 1996, Fig. 3, Regional-scale relationships between climate and strength of podzolization in the Great Lakes region, North America. Catena 28, 47–69, © 1996, with kind permission from Elsevier Science, The Netherlands.*)

climate and found that areas of strong podzolization are characterized by soil temperatures rarely >16–17°C, the thickest snowpack, and a steady and seasonally large influx of cold snowmelt waters. The main contrast with the area of weak podzolization is the latter area has only one-third the snowmelt flux, and there is a small infiltration peak in the autumn.

Another pedogenic gradient has been studied in the New Zealand Southern Alps, over a MAP gradient from 64 to 200 cm (Webb and others, 1986). Soils are <20 ka and formed in loess. General trends with increasing rainfall are (1) change in profile form from A/Bw to A/E/Bs, (2) subsoil clay content about doubles, (3) percentage base saturation of >>50 to <<50 that accompanies increasing acidity, (4) total P at least halves, (5) Fe_d increases over two times, and (6) Fe_o increases about four times, but there is no trend in Fe_o/Fe_d ratio, as all profiles have ratios > 0.6.

R.V. Ruhe was one of the foremost soil geomorphologists in the United States, and his last work was on pedogenic gradients in the central part of the United States, transects of about 1500 km north–south and east–west. Prior to this work, Jenny and Leonard (1934) and others had collected soils in the region along climatic gradients and obtained good data on climate–soil relations. Ruhe shows that the system is more complex that the early workers thought, mainly because of more recent work on Quaternary stratigraphy and paleoclimate.

The north–south transect extends from Minnesota to Mississippi, and all soils formed over about 14 ka from Peoria Loess, the regional younger loess along the Mississippi River system (Ruhe, 1984a,b) (Fig. 10.7A). The transect includes 15° of latitude; climatic parameters are MAP of 74 and MAT of 7 in the north, and 140 and 18, respectively, in the south. Generalized trends going southward (wetter) are (1) solum thickness increases from about 150 cm to >200 cm; (2) although all the soils have Bt horizons, the weighted mean percentage clay decreases from about 25–35 to about 20–25, nearly parallel to trends in weighted mean clay-free percentage fine silt in the Bt and weighted mean percentage clay in the C horizons; (3) illite and expandable 2:1 clay minerals dominate, illite decreases with depth, whereas the expandables (mostly montmorillonite) increase, and the contents of surface illite increase from about 40 to 50–60% at the same time as the subsurface expandables decrease from 80 to 60%; (4) Bt-horizon hue reddens from 10YR to 5YR, yet pedogenic Fe remains near constant at a weighted mean percentage near 0.8; and (5) Bt-horizon total bases halve, base saturation nearly halves, the mean depth of the mean base saturation almost doubles, and acidity increases.

Ruhe argues that many of the above properties and trends are more related to inherited sediment properties than to pedogenic trends along a climatic gradient. In particular, some trends are opposite to that expected for the pedogenic gradient: the clay percentage should increase to the south, as should the weathering of illite. In contrast, the trends in color and bases follow the climatic gradient, and overprint the parent material control.

The east–west transect goes from Iowa to the Colorado border, and the climate gets drier in that direction from an MAP of 86 to one of 46; MAT is constant (Ruhe, 1984c). The parent materials are loess again, but the ages differ, with soils having formed over 14 ky east of the Missouri River and over 9 ky west of the river. Argiudolls to the east give way to Haplustolls to the west. Generalized trends going westward (drier) are (Fig. 10.7B) (1) solum thickness

decreases from about 160 to 40 cm; (2) Bt horizons thin and the weighted mean percentage clay abruptly increases near the river, and decreases farther west; (3) the abrupt change is not duplicated in the C horizon weighted mean percentage clay, but the latter also decreases west of the river; (4) abrupt changes in silt fractions are not seen; (5) base saturation increases in both C and B horizons, from <80% to the east to >90% to the west, with Ca the dominant cation, particularly to the west; (6) pH increases over one unit (from 6 to 7–8); and (7) clay mineralogy is relatively constant and trends are similar to the north–south transect. Again Ruhe argues for a parent material influence, points out a paradox as the soils with the greater clay contents not only are the younger soils, but are also the lesser weathered soils as shown by relatively high base saturation, high exchangeable Ca, and a lack of illite weathering. He concludes that atmospheric dust has added clay to the western soils. In contrast, solum thickness, base saturation, exchangeable Ca, and pH form the pedogenic gradient. Climate change also plays a role, as the eastern soils, which partly be-

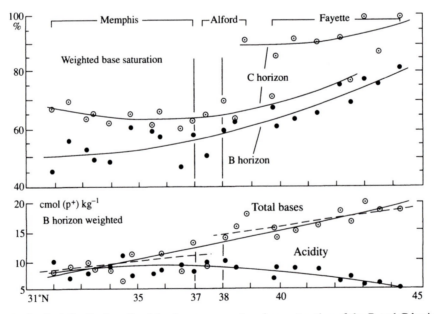

Figure 10.7 (A) Latitudinal distribution of weighted mean percentage base saturation of the B and C horizons (upper diagram) and weighted mean total bases and extractable acidity in B horizons (lower diagram) of Hapludalfs from Minnesota (right) to Mississippi (left). *(Redrawn from Ruhe, 1984b, Fig. 1.)*

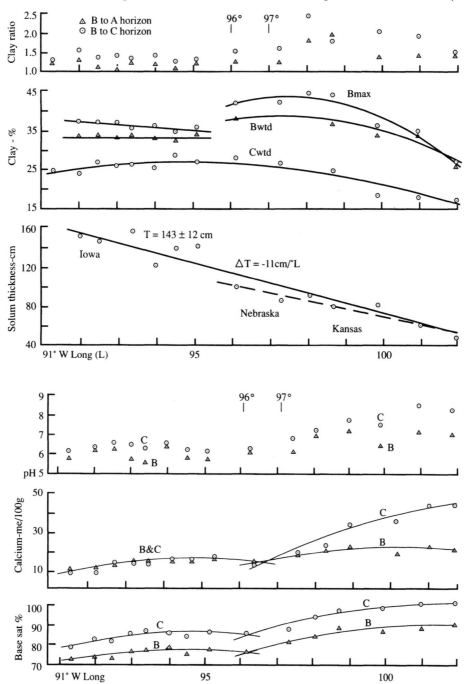

Figure 10.7 (continued) (B) Longitudinal distributions of various soil properties: clay ratios; percentage clay (B_{max} is the maximum clay in B horizon; B_{wtd} is the weighted mean clay in the B horizon; C_{wtd} is the weighted mean clay in the C horizon for a thickness equivalent to that of the associated B horizon); solum thickness (T); and weighted mean values for B and C (equivalent thickness to associated B horizon) horizons in pH, exchangeable Ca, and base saturation. *(Redrawn from Ruhe, 1984c, Figs. 4 and 6, Soil-climate system across the prairies in midwestern U.S.A. Geoderma 34, 201–219, © 1984, with kind permission from Elsevier Science, The Netherlands.)*

cause they are older, have formed under more and different bioclimates than have the western soils.

The above examples show the sampling strategies and potential for pedogenic gradient studies. One important lesson is that if working in a rather small area, one can consider all the important factors and perhaps keep them more or less constant. Broadening the scale of the research introduces problems, as the system may not be known so well. In his study, however, Ruhe relied on the extensive knowledge of soil scientists in each of the states included for site selection, he was very knowledgeable of the properties inherited from the regional variation in loess properties, and there has been much work done on paleobioclimate in the region. There are not many large areas in which work with this detail could be done. In these transects, it is likely that with increasing time for soil formation, more soil properties should follow climatic parameters as they overprint the inherited properties.

Little has been done on the relation between climatic parameters and large-scale soil-classification units, probably because the relationship is an extremely complex one. Arkley (1967) attempted such a correlation for the western United States, in which soils shown on a regional soil map were related to water-balance climatic parameters obtained from climatic stations located within the mapped soil units (Fig. 10.8). The parameters chosen were calculated actual evapotranspiration, based on a soil-moisture storage capacity of 15.2 cm, leaching index, and MAT. These parameters were considered to relate best to the major processes responsible for the various soil-classification units. The plot of data indicates that although there is considerable overlap of some soil orders and although some orders range widely in values for the climatic parameters, most soil orders seem to fall within well-defined values for the climatic parameters. In some places, the overlap might occur because the climatic parameters chosen were not those most highly correlated with the particular orders; in other places, the overlap might occur because both orders are stable in that climatic regime, and one soil order may grade into another with time. Or some other factor explains the overlap. Recall how parent material differences can determine the genetic pathway of a soil sequence. These plots also bring up a major problem in working with soil morphology for paleoclimatic reconstruction: each order is represented by a large variation in climatic parameters, and thus soil classi-

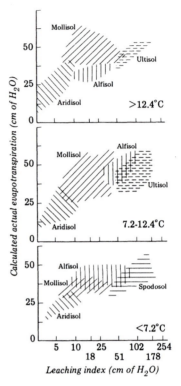

Figure 10.8 Relationship among soil orders, leaching index, calculated actual evapotranspiration (soil-moisture-storage capacity 15.2 cm), and MAT for the western United States. Horizontal axis scale is linear between recorded values. *(Data from Arkley, 1967, Fig. 5, © 1967, The Williams & Wilkins Co., Baltimore.)*

fication units do not give very precise data on paleoclimate, except in those transitional regions that separate soils of markedly different morphologies. One apparent success story is that of Karlstrom (1991), because the modern soils contrasted so much with the soils of interest (discussed later). For more precise information it might be best to construct figures for soil suborders rather than orders.

■ Variation in Specific Soil Properties with Climate

The main soil morphological and mineralogical properties that correlate with climate are organic matter content, clay content, kind of clay and iron minerals, color, various chemical extracts, the presence or absence of $CaCO_3$ and more soluble salts, and depth to the top of salt-bearing horizons. Many of these trends were included in the discussion of soil chronosequences (Chapter 8) and soil toposequences (Chapter 9). As an example, one could go through Chapter 8 and obtain data on soils of a particular age, say 100-ky soils, and relate the data with some parameter of climate in an *X–Y* plot (pedogenic clay content versus MAP). Ideally, one would want to keep parent material constant, or if parent material varies it potentially would not influence the properties being studied. One cannot stick with the same vegetation, however, as it changes with climate; so one really is comparing soils along a bioclimatic gradient. Only the highlights will be summarized here.

Organic Matter Content

There are several forms of soil organic matter distribution with depth that are associated with climate and the accompanying vegetation. Common to many soils is the distribution that is low (should approach zero eventually) at depth and content increases toward the surface. The maximum value/profile is at the surface, and this value varies with soil order: common values taken from tables in Soil Taxonomy are <0.5% in Aridisols (some of these soils have a subtle maxima at shallow depth), 1% in Alfisols, and 2–5% in Mollisols, Oxisols, and Ultisols. Grassland soils have higher contents to greater depths relative to forested soils, in concert with root density with depth. Spodosols have a much different depth distribution; a surface O horizon overlies an E horizon of relatively low organic-matter content, which, in turn, overlies a B horizon of high content due to the concentration of translocated organic matter.

The long-term accumulation of soil organic carbon varies with bioclimate. Schlesinger (1990) gives data from which we get this general ranking (maximum values for each group in g C/m^2/year): Boreal forest = temperate forest (10–12) > tundra sedge moss (6) > polar desert = temperate grassland (2) > temperate desert (<1). At the steady state, loses balance gains, and the amount in the soil remains constant thereafter.

The amount of organic carbon in soils at the steady state varies with bioclimate and soil order. Post and others (1985) have contoured the mass of carbon versus the world life zones and get these generalized values: grasslands have about 10 kg/m^2 of soil organic

carbon, most forests and tundra > 10 kg/m^2, and deserts < 10 kg/m^2 (Fig. 10.9A). The global ranking to 1 m depth by soil order is as follows (kg/m^2): Histosols (206) >> Andisols (31) > Inceptisols (16) = Spodosols (15) = Mollisols (13) > Oxisols (10) = Entisols (10) = Ultisols (9) > Alfisols (7) = Vertisols (6) > Aridisols (4) (data from Eswaran and others, 1993). Similar data for the contiguous United States result in a rather similar ranking, but Inceptisols and Entisols have a lower ranking and Vertisols a higher ranking (Kern, 1994); mass of C for the orders is comparable. The data set of Kern (1994) also lists mass of C by suborder. Intuitively, one would expect the ranking within an order to follow climatic parameters, such that aquic = frigid or cryic (bor-) = udic > ustic = xeric. With some reversals, this general ranking is seen for Alfisols, Mollisols, and Oxisols. Eswaran and others (1995) have a map depicting the global organic carbon, which one can compare with a global soil classification map (Fig. 2.8).

Jenny (1941a, 1961, 1980) studied many climatic transects to determine the trends in organic matter constituents in the soil (organic carbon and total nitrogen). In general, he finds that soil nitrogen increases logarithmically with increasing moisture and decreases exponentially with rising temperature. These relationships hold for such diverse climatic regions as the Great Plains (Jenny, 1941a, Fig. 92), India, and California (Fig. 10.10).

The above organic matter–climate trends are related to yearly gains and losses of organic matter. Jenny (1950, 1961) compared the dynamic nature of several tropical and temperate ecosystems to find some basic differences (Table 10.2). The tropical soils he studied contain more organic carbon and total nitrogen than the temperate soils, but the latter have a higher proportion of organic constituents in the forest floor. This is the case in spite of data indicating that litterfall is much greater in tropical forests. The annual rates of decomposition of the litterfall and forest floor material explain this apparent discrepancy, for the rates are much higher in the tropical region. In California, the rates of decomposition also have been shown to decrease with elevation and thus with climate (Table 10.2).

To access the influence of climate on organic soil properties, workers commonly undertake elevation transects up a particular mountain or range. One such transect in northern California varies from 50 MAP

A

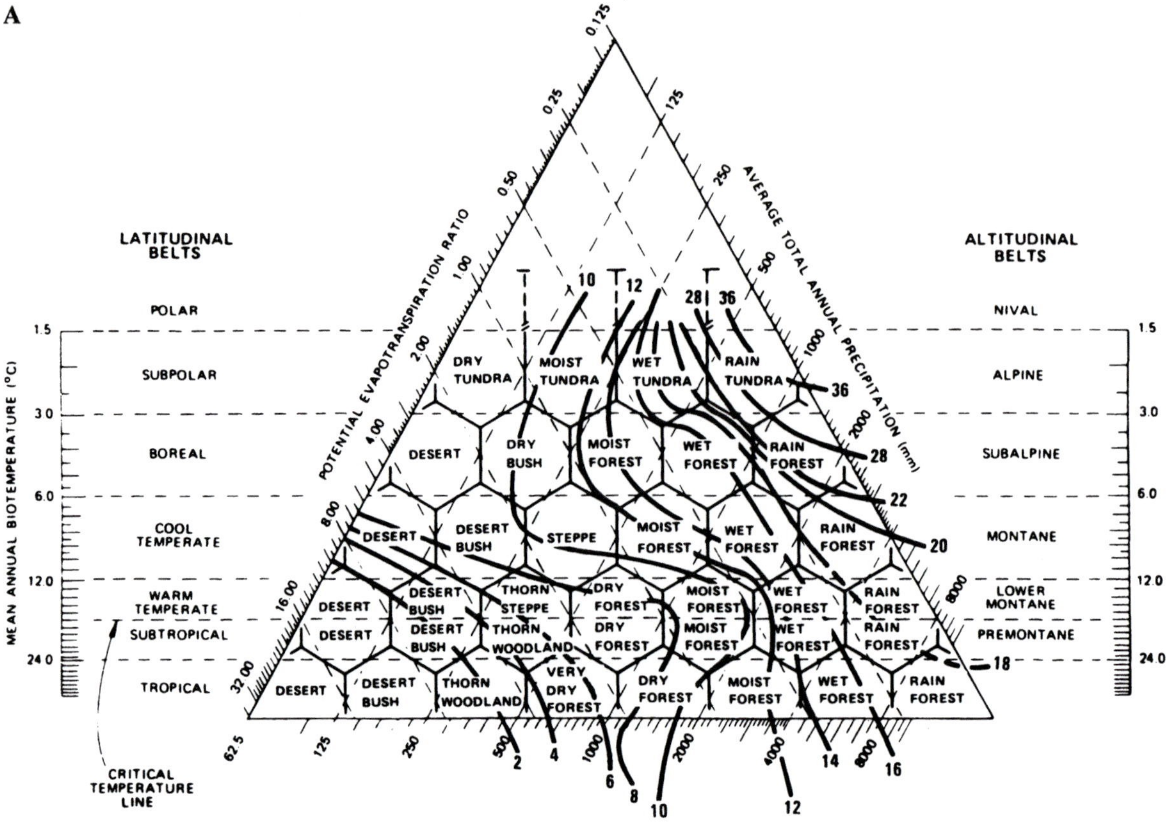

LATITUDINAL BELTS

POLAR
SUBPOLAR
BOREAL
COOL TEMPERATE
WARM TEMPERATE
SUBTROPICAL
TROPICAL

MEAN ANNUAL BIOTEMPERATURE (°C)

POTENTIAL EVAPOTRANSPIRATION RATIO

AVERAGE TOTAL ANNUAL PRECIPITATION (mm)

ALTITUDINAL BELTS

NIVAL
ALPINE
SUBALPINE
MONTANE
LOWER MONTANE
PREMONTANE

CRITICAL TEMPERATURE LINE

DRY TUNDRA MOIST TUNDRA WET TUNDRA RAIN TUNDRA
DESERT DRY BUSH MOIST FOREST WET FOREST RAIN FOREST
DESERT DESERT BUSH STEPPE MOIST FOREST WET FOREST RAIN FOREST
DESERT DESERT BUSH THORN STEPPE DRY FOREST MOIST FOREST WET FOREST RAIN FOREST
DESERT DESERT BUSH THORN WOODLAND DRY FOREST MOIST FOREST WET FOREST RAIN FOREST
DESERT DESERT BUSH THORN WOODLAND VERY DRY FOREST DRY FOREST MOIST FOREST WET FOREST RAIN FOREST

B

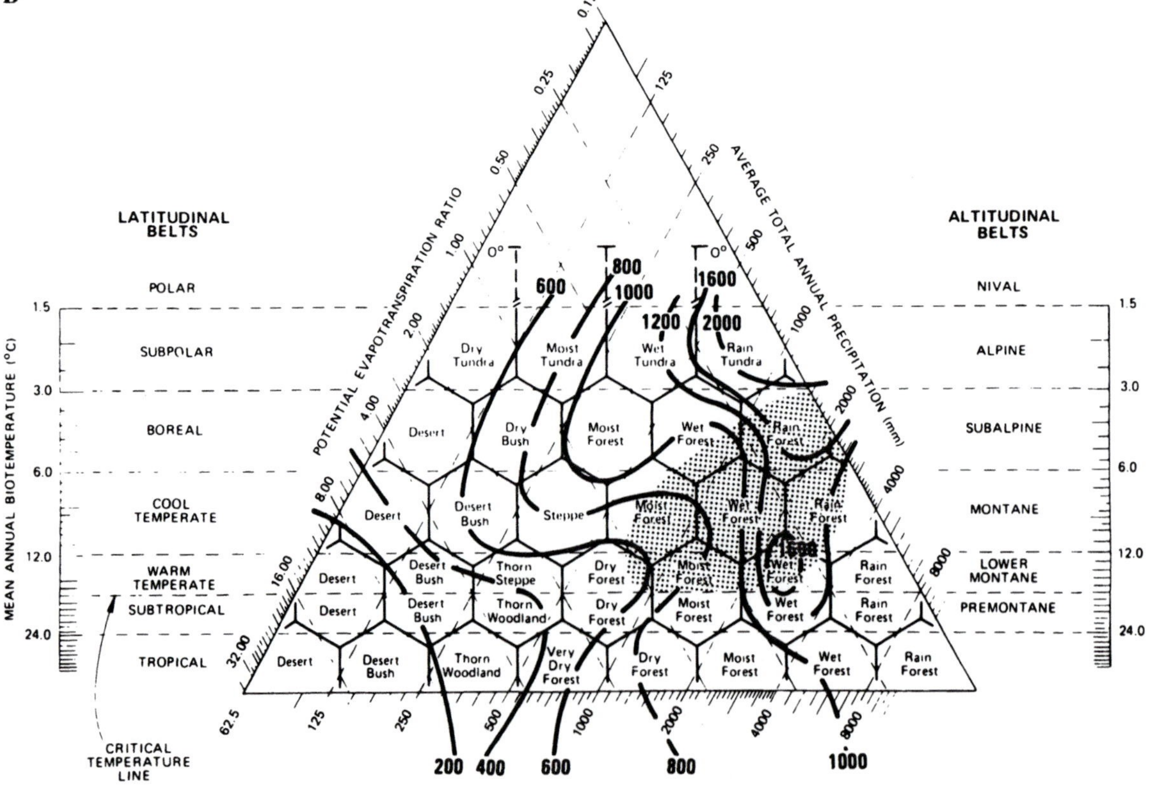

LATITUDINAL BELTS

POLAR
SUBPOLAR
BOREAL
COOL TEMPERATE
WARM TEMPERATE
SUBTROPICAL
TROPICAL

MEAN ANNUAL BIOTEMPERATURE (°C)

POTENTIAL EVAPOTRANSPIRATION RATIO

AVERAGE TOTAL ANNUAL PRECIPITATION (mm)

ALTITUDINAL BELTS

NIVAL
ALPINE
SUBALPINE
MONTANE
LOWER MONTANE
PREMONTANE

CRITICAL TEMPERATURE LINE

Dry Tundra Moist Tundra Wet Tundra Rain Tundra
Desert Dry Bush Moist Forest Wet Forest Rain Forest
Desert Desert Bush Steppe Moist Forest Wet Forest Rain Forest
Desert Desert Bush Thorn Steppe Dry Forest Moist Forest Wet Forest Rain Forest
Desert Desert Bush Thorn Woodland Dry Forest Moist Forest Wet Forest Rain Forest
Desert Desert Bush Thorn Woodland Very Dry Forest Dry Forest Moist Forest Wet Forest Rain Forest

C

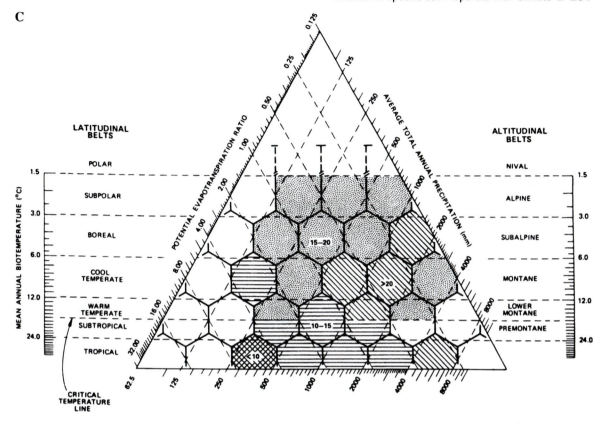

Figure 10.9 (A and B on facing page) Isopleths of amounts of carbon (A, kg/m³), nitrogen (B, g/m³), and C/N ratio (C, patterns delineate similar ranges in ratio), for various Holdridge world life zones. *(Reprinted from Post and others, 1985, Fig. 1, with permission from Nature 317, 613–616, © 1985, Macmillan Magazines Ltd.)*

and 17 MAT at the base (200 m elevation) to 150 and 5, respectively, at the top (2000 m) (Alexander and others, 1993). In going up the range, organic carbon in the top cubic meter increases from about 10 to 25 kg, as the range in C/N ratio of the surface layer increases from 11–14 to 25–35.

Global patterns of soil nitrogen and C/N ratio also follow the life zones (Post and others, 1985). Nitrogen values are about 800 g/m³ in grasslands, and this value is greater in the forests and less in the deserts (Fig. 10.9B). C/N ratio partly reflects the amounts of N sequestered in both the microbial biomass and what are termed recalcitrant materials in the soil (Fig. 10.9C). Grassland C/N values are in the range of 10–15 whereas deserts, where low precipitation limits primary productivity and leaching, have similar or lower values. In contrast, many forests and tundra are

characterized by ratios of 15–20 and >20 because of the composition both of the litterfall and the lower decomposition rates in the cooler temperatures. Finally in some wet tropical forests, in spite of high litterfall rates, decomposition rates are high, resulting in C/N ratios that can be 15–20 with some <10.

These studies can serve as models for further study of ecosystems in other climates. Thus, to understand the relationship between climate and the organic matter constituents in a soil to the extent that Jenny has, we must know the gains and losses involved, the decomposition constants, and the total weight of the organic matter constituents, as well as the climatic parameters. Compared to other soil properties, those related to the organic constituents are most amenable to functional analysis as most of the other factors can be kept constant. This is because most soils reach an

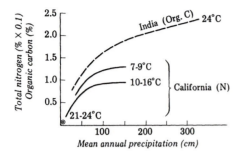

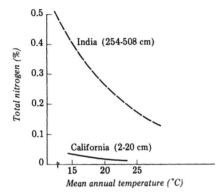

Figure 10.10 Idealized trends of organic carbon and nitrogen with MAP and MAT, India and California. Temperatures in upper graph are MATs, and numbers in lower graph in parentheses are MAPs. *(Redrawn from Jenny, 1961, Missouri Agri. Exp. Sta. Res. Bull. 765, and Harradine and Jenny, 1958, Figs. 5 and 6, © 1958, The Williams & Wilkins Co., Baltimore.)*

organic property steady state in about 5 ky, so soils in any climatic setting can be included as long as they are about this old. Soils of this age should be relatively easy to recognize in most settings, as they have the maximum content and proper depth distribution.

Clay Content

It is commonly reported that clay content in soils varies with climate, but precise data on this relationship are not often available because, in many places, factors other than climate have influenced clay production, and their influence is hard to quantify.

Two of the early quantitative relationships between soil clay content and climate are the moisture and temperature functions of Jenny (1935), in which he found a linear relation with moisture and an exponential relation with temperature (Fig. 10.11). Sampling fol-

lowed a classic Jenny format of sampling along both temperature and precipitation gradients. Thus, if all other factors are constant, one would expect low rates of clay production in cold-dry, cold-wet, and hot-dry environments, increasing rates with increasing precipitation, and highest rates in hot-humid environments. These data were obtained before we knew much about Quaternary stratigraphy, the relationships between soil morphology and age, and the influence of atmospheric dust on soil properties, so, although Jenny's generalized trends may still hold for some areas, the absolute amounts of clay for any climate might not coincide too precisely with the data of Fig. 10.11.

To determine the precise relationship between clay production and climate, it would be best to have data on total clay present per unit area of the ground surface (g/cm²/soil column). Percentage values, although based on the <2 mm fraction, do not reflect the true amounts of clay in the profile because variations in gravel content and bulk density have a marked effect on absolute clay content (see Fig. 1.7). An alternative would be to calculate the clay accumulation index of Levine and Ciolkosz (1983; see Fig. 1.8). If one could, one should differentiate clays of different origin; for example, some clay might be formed by weathering of primary grains, and other clay might be part of the atmospheric influx into the soil. Both processes produce pedogenic clays, yet the rates of clay accumulation by each process could be quite different.

The relations between climate and pedogenic clay are complex, as described in Chapter 8, and shown in the selected soils of Fig. 8.35. In general, the paradigm of increasing clay with increasing precipitation (as in the New Zealand example of pedogenic gradient, above, and in Fig. 8.22) seems to hold. But examples were given in which the amount of clay in desert soils due to dust influx matches or exceeds that produced by weathering. The result is that some desert soils plot high in Fig. 8.35, quite the opposite of what Jenny's initial attempt showed (Fig. 10.11). Another example of where the clay–rainfall relation does not work is in comparing the oldest soils of the eastern United States (Figs. 8.17 and 8.18) with the oldest one in the Colorado Piedmont (Fig. 8.41), where about 60% clay is reached in about one-fifth the time in the drier environment of Colorado.

Clay production can be qualitatively related to climate by comparing stratigraphically dated deposits and their respective soils in many different environments (specifics in Chapter 8). One useful age datum

Table 10.2 Annual Gains, Losses, and Decomposition Constants for Several Forest Ecosystems

	Tropical Forests of Columbia		California		
			Temperate Forest (1640 m)		Cold Forest (3280 m)
	Sealevel	1540 m	Oak	Pine	Pine
Litterfall (g/m²)					
Weight[a]	730	935	149	305	101
Total nitrogen	10.4	15.7	1.27	1.54	—
Organic carbon	391	510	74.6	164	—
Forest floor[b] (g/m²)					
Weight[a]	432	1,455	2,517	12,635	11,081
Total nitrogen	8.8	35.4	31.1	117	—
Organic carbon	225	795	1,224	6,463	—
Decomposition constant of litterfall and forest floor (annual percent loss)					
Weight[a]	62.8	39.1	5.6	2.4	0.90
Total nitrogen	54.2	30.7	3.9	1.3	—
Organic carbon	63.5	39.1	5.7	2.5	—
Total weights in soil profile[c] (g/m²)					
Total nitrogen	2,502	3,521	633	650	—
Organic carbon	36,681	45,196	11,606	17,317	—

Taken from Jenny (1950, Table 1), © 1950, The Williams & Wilkins Co., Baltimore, and Jenny (1961, Table 2), Missouri Agri. Exp. Sta. Res. Bull. 765.)

[a]Volatile weight only, because the forest floor was contaminated with sand grains.

[b]Forest floor is the fresh and partially decomposed plant debris overlying the mineral soil.

[c]Includes forest floor and amounts in mineral soil.

is soils formed on O-isotope stage 2 tills. Such tills in many parts of the Cordilleran Region of the western United States, in areas characterized by igneous and metamorphic rocks of granitic composition, commonly have only A/Bw or A/Cox soil profiles; Bt horizons usually are absent. This relation holds for many soils, in spite of the wide range in MAP values (>120 in the Cascade Range of Washington, <40 in the Sierra Nevada) and the presence of volcanic clasts in the high-precipitation areas of the Puget Sound Lowland, Washington and on Mt. Rainier, Washington.

Soils with Bt horizons occur on O-isotope stage 6 tills at some of the localities mentioned in the above summary. It is difficult to generalize, but it is not unusual for soils of this age in tills in the western United States to double their clay content from the Cu to the Bt horizon (approaching 15% absolute), and for the Bt to be about 50 cm thick. Soils on tills of O-isotope stage 4 are not commonly recognized, but one in Idaho has a percentage clay close to that of the soil about twice as old, but clay mass is less (Colman and Pierce, 1986). Similarly, the youngest soil with a Bt horizon in western Washington is estimated to be about the same age as the latter soil (Dethier, 1988).

Soils related to O-isotope stage 6 deposits probably formed during drastic regional climatic changes; for example, they were forming when the O-isotope stage 2 glaciers advanced to positions quite near the older soils. What was the influence of that climate on the soils, and was the influence the same everywhere? Recall that the intermontane basins of Wyoming were about 10–13°C colder then (Mears, 1981), and amino-

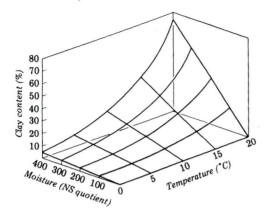

Figure 10.11 Idealized relationship between percentage clay in the top approximate 1 m of soil derived from granites and gneisses and mean annual moisture (see Jenny, 1941a, p. 109, for definition of NS quotient) and MAT. *(Redrawn from Jenny, 1935, Fig. 8, © 1935, The Williams & Wilkins Co., Baltimore.)*

acid analysis suggests that parts of the northeastern Great Basin could have been equally as cold (McCoy, 1981). Richmond (1965) estimated that these soils are preserved on 60–80% of the stage 6 moraines of the southern Rocky Mountains, but only on 10–20% of the same age moraines of the northern Rocky Mountains, and ascribed the difference to glacial-age (O-isotope stage 2) erosion. The main point is that soils from quite different areas formed under different present and paleoenvironments across the Cordilleran Region are grossly similar, where not eroded. Cremaschi (1987) contrasts northern Italy with the rest of Europe in this regard. Much of Europe underwent severe periglacial erosion during the glacials, removing previously formed soils, whereas these conditions did not exist in northern Italy where soils are better preserved.

Comparison of the above western United States soils with Bt horizons formed in O-isotope stage 6 tills in the humid climate of Pennsylvania (Fig. 8.15) shows that the latter have over twice as much clay and clay accumulation to greater depths. This results in a clay accumulation index of 1299–1387. One important factor in explaining these differences is parent material, because tills in the western states are formed from crystalline rocks, whereas those in Pennsylvania are formed from sedimentary rocks.

Still older soils are hard to locate and difficult to date. However, one can compare the Bt horizons of about 0.5-my-old or older soils in the Rocky Mountains (Shroba, 1977) and the Sierra Nevada (Birkeland and others 1980). Such soils in the former area have a 92-cm-thick Bt, a maximum clay content of 28%, and an absolute increase relative to the Cu of at least 18%. In contrast, those soils in the latter area have a 145-cm-thick Bt, a maximum of 40% clay, and probably at least a 20% absolute increase in clay. It is tempting to ascribe the differences to the warmer climate in California, but too many unknown factors could be responsible, not the least of which is erosion.

The midcontinent is an area of seemingly rapid clay production. Soils formed on 20-ka and older tills have more strongly developed Bt horizons than their Cordilleran counterparts (e.g., Forsyth, 1965). Although part of this variation could result from parent materials with relatively high clay content in the midcontinent (Ruhe, 1984a,b,c), at least some of the greater clay production of the midcontinent could result from the climate, which is characterized by significant rainfall during the warm summers. Soils suitable for a comparison of soil development on parent materials of somewhat similar textures include those on outwash deposits of the eastern Sierra Nevada (Fig. 10.12) and of Ohio (Fig. 8.15).

A comparison of midcontinent soil formation with that in the western United States can be made with sandy parent materials (dune sand) in the former area (Miles and Franzmeier, 1981). Climate is udic and probably mesic. Soils formed from the low-clay (<5%) sands progress through clay + Fe-enriched lamellae to a Bt stage at 20 ka. The latter have 10–20% clay, 0.7–1.4% Fe_d, and 7.5YR hue; the clays are considered to have formed from feldspar weathering. In this comparison, the midcontinent shows up as a region of rapid clay production, relative to that in the low-dust-influx areas of the western United States.

So far we have compared semiarid soils with little dust influence to more humid soils. If the diagram of Pye (Fig. 7.12) is used, dust influence on the wetter soils should be minimal, as the dust presumably is rained out of the atmosphere before reaching the wetter sites. No doubt some atmospheric fines get into the humid-climate soils, however, as shown by the soils on carbonate islands in the tropics (Chapter 8).

Various climatic transects illustrate the influence of dust on pedogenic clay in drier environments. A pedogenic gradient in Bt-horizon properties is seen in the transect from wetter and colder to drier and warmer in the western Great Basin (Fig. 10.12; see also Chad-

Figure 10.12 Clay distribution in soils formed from stream terrace deposits and a lacustrine sand (younger soil at Mustang), Truckee River, California and Nevada. Truckee is at the eastern foot of the Sierra Nevada with MAP 80, MAT 6, and a leaching index of 65.8 cm. Comparable climatic data for Reno, Nevada (between Verdi and Mustang) are 18, 9.5, and 7.4, respectively. Summers are dry. *(Redrawn from Birkeland, 1968, Fig. 3.)*

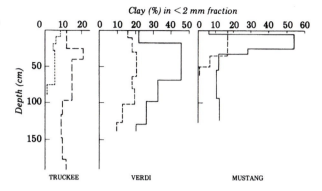

wick and Davis, 1990). Compared to the wetter soils, the more aridic Bt horizons are thinner and have a greater buildup of clay relative to that of the presumed Cu horizon. Whereas this relation used to be perplexing, almost all workers now would attribute the clay increase in the drier area to a notable eolian component (Chapter 8). Other examples of this are the Las Cruces area (Fig. 8.15) and southern Israel (Fig. 8.30). The mass data of McFadden's transect (Fig. 8.22) did not show this, but the clay versus pedogenic Fe data (Fig. 8.23) provided a clue on recognizing the dust component of pedogenic clay in deserts (i.e., the combination of high clay with low pedogenic Fe).

Finally, arctic and polar regions are characterized by virtually no clay accumulation. Many of the soils are formed in sandy tills made up of igneous and metamorphic rocks of granitic composition, so the parent materials are somewhat similar to those of the Cordilleran Region. The lack of Bt horizons, no matter what the age (some soils in Antarctica are several My), could be ascribed to low values for MAP and MAT, which keep weathering and pedogenic processes at a minimum. These environments also have low atmospheric dust. Hence, I suspect the lack of pedogenic clay, at least in that part of the arctic with summer rainfall, can partly be the result of low levels of atmospheric dust. The Antarctic would never produce a Bt horizon via dust, as there is no rainfall for translocation.

In summary, we are obtaining better data on variation of clay content with climate, particularly when soil studies are combined with Quaternary strati-

graphic studies so that age can be kept constant. Bulk densities should be taken so that the rates can be based on weight of clay formed or accumulated per unit surface area for the entire soil. Inherited clay should be subtracted from the total to produce pedogenic clay. If of interest, the pedogenic clay can subdivided into that fraction derived from weathering and that from atmospheric sources. Once this is done, we can attempt a newer version of Fig. 10.11; my suspicion is that it would look much different than the original, for at least we know now that some desert regions are areas with high amounts of pedogenic clay.

Clay Mineralogy

Clay minerals formed in the soil vary with the water chemistry and the rate of leaching, and thus with the climate. Many transects relating clay mineralogy to climate have been described, but only a few will be mentioned here. In any climatic transect, however, the clay mineralogy seems best correlated with precipitation or leaching index, and, in going toward areas of greater precipitation, the clay mineralogy trend usually is toward species containing progressively less silica. The reason for this trend is that soil leaching increases with increasing precipitation, and more silica is lost from the soil during the course of weathering (Fig. 10.13). Data are given here for well-drained soils so that the mineralogy reflects regional climate, not topographic influence (Fig. 9.9). The discussion here will focus mostly on soil clay minerals formed from crystalline-rock parent materials, or materials derived

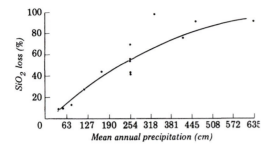

Figure 10.13 Trend of silica loss with MAP in a 10- to 17-ky volcanic ash, Hawaii. *(Redrawn from Hays and Jones, 1972, Fig. 2, published by the Geological Society of America.)*

from them (e.g., till and loess). Clay minerals formed from sedimentary rocks or deposits derived largely from sedimentary rocks will not be discussed, because some of their clay minerals are inherited. Finally, clay-mineral associations can be quite complex, with several clay minerals and clay–mineral interlayering, so this discussion will be simplified and include the common minerals and no interlayered ones.

Clay-mineral associations with climate first will be made on a broad scale. Allen and Hajek (1989) do this by tabulating the mineralogy families in soil orders (Table 10.3). A mineralogy family has over one-half of the indicated mineral present; mixed means a mixture of many minerals. Because orders commonly are

associated with different climates, some broad trends are apparent. The best relationship is kaolinite with Oxisols and Ultisols. Smectites are dominant in Vertisols, but the latter also characterize several climatic regimes. Although smectites are common in the Aridisols, they are equally as common in many other orders, especially the Mollisols. Mixed mineralogies are found in many orders and appear to be dominant in the Alfisols and Inceptisols. Spodosols represent a rather unique pedogenic environment, and the clay mineralogy corresponds to this, as reviewed by Ross (1980). The clay mineralogy of the C horizon, for the soils he reviews, consists mainly of chlorite and mica. Vermiculite is common in the B horizon, in addition to the above two minerals. The A or E horizon is characterized by an absence of chlorite and abundant Al-rich 2:1 clay minerals, specifically, beidellite and dioctahedral vermiculite. One problem with broad relationships at this level is that many orders include such a wide variation in the soil-forming factors that a good correlation is not possible. Correlation could be improved, perhaps, by isolating some of the factors.

Even though tropical regions typically are used to characterize the end products of intense weathering (aluminum and iron compounds), there is a definite mineralogical trend with climate. Commonly, montmorillonite forms at low amounts of precipitation, kaolinite at higher amounts, and oxides and hydrox-

Table 10.3 Clay-Mineralogy Families for Soil Orders in the United States, Puerto Rico, and the Virgin Islands

Order	Clay-Mineralogy Families (%)					
	Kaolinitic[a]	Illite	Smectitic	Vermiculitic	Mixed	n[b]
Alfisols	5	4	37	2	52	351
Aridisols		2	70		28	141
Entisols	1	4	62		33	121
Inceptisols	5	10	26		59	125
Mollisols	2	1	62	1	34	523
Oxisols	83				17	7
Spodosols						
Ultisols	25	4	2	1	68	102
Vertisols	2		86		12	90

Data are from Allen and Hajek (1989, Table 5.4), and are for clayey particle-size class; see table for other classes.

[a]Kaolinitic family includes halloysitic family.

[b]n is the number of families represented in each order.

ides of iron and aluminum at still higher amounts. The relations found in an early study in Hawaii still generally hold. Sherman (1952) reports that montmorillonite predominates below about 100-cm MAP, kaolinite between about 100 and 200, and the iron and aluminum compounds above 200 (Fig. 10.14). Other areas with other combinations of the other factors could have a different set of MAPs for the various clay minerals.

Barshad (1966) reported on the clay mineralogy of several hundred surface-soil samples (0 to 15 cm depth) in California and found a close correspondence of clay mineral with precipitation (Fig. 10.15). Generally, montmorillonite is found only at less than about 100 cm MAP, and gibbsite at greater than about 100. Kaolinite and/or halloysite are present over a wide range of precipitations and are predominant above about 50 cm MAP. Illite and vermiculite are also present, the former only in felsic igneous rocks and the latter in both felsic and mafic igneous rocks. The abundance of the various minerals varies somewhat for the same amount of precipitation between soils formed from felsic igneous rocks and those formed from mafic igneous rocks. The main reason for this shift in clay-mineral abundance with parent material is the cation content of the parent material. Thus, for example, montmorillonite can form under higher precipitation from mafic igneous rocks than from felsic igneous rocks because the former have a higher cation content, and high cation content in the soil solution favors formation of montmorillonite.

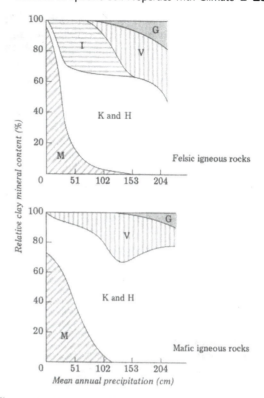

Figure 10.15 Relative clay-mineral content as a function of precipitation and major rock type. MAT ranges from 10 to 15.6. Key to symbols: M, montmorillonite; K, kaolinite; H, halloysite; I, illite; V, vermiculite; G, gibbsite. *(Redrawn from Barshad, 1966, Figs. 1 and 2.)*

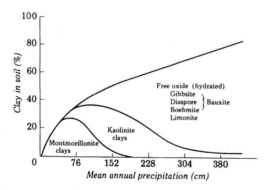

Figure 10.14 Clay mineralogy as a function of MAP in a continuously wet climate, Hawaii. Sherman (1952) presents data for an alternating wet and dry climate with a somewhat similar relation of MAP and soil-clay minerals. *(Redrawn from Sherman, 1952, Fig. 3 in Problems of Clay and Laterite Genesis, published by AIME.)*

The fibrous clay minerals, palygorskite and sepiolite, are associated with aridic soils, alkaline pH, and relatively high concentrations of both Mg and Si (Singer, 1989). Although they were originally thought to be inherited clay minerals, recent work leaves little doubt as to their pedogenic origin. Conditions favorable for carbonate accumulation also favor these minerals. In one study he cited, palygorskite was reported to be stable at less than 30 cm MAP. As mentioned earlier (Chapter 8), sepiolite can form from palygorskite, and the transformation is time dependent.

Clay-mineral studies of soils from a large number of till and outwash deposits of many ages along the east side of the Sierra Nevada in California and Nevada substantiate the conclusions of Barshad (Birkeland, 1969; Birkeland and Janda, 1971). In general, in the northern part of the range in deposits with a mixed lithology of granitic, andesitic, and basaltic rocks, the change from high halloysite content to high montmorillonite content takes place at the pine-sagebrush vegetation boundary at about 40-cm MAP (Fig. 10.16). However, near the southern end of the study area, soils formed from granitic till presently located below the pine–sagebrush boundary, in what is probably a drier climate, do not contain montmorillonite. Again, this variation with parent material can be predicted from the data of Barshad, although the exact climatic values at the transition from one predominant clay mineral to another may vary with the precise environmental conditions, including any paleoclimatic influences.

Clay mineral assymblages will vary with local climatic conditions along various transects. For example, those along the aridic ⇒ semiaridic ⇒ xeric transect of McFadden (1988) are (in approximate order of abundance) illite–smectite–palygorskite–kaolinite, vermiculite–illite–kaolinite–smectite, and kaolinite–vermiculite–illite–smectite, respectively. To simplify, high palygorskite and smectite are at the dry end, and high kaolinite at the moist end.

Folkoff and Meentemeyer (1987) examined the relation between A-horizon clay mineralogy and climate for a still larger area: 99 soils considered to be representative of the contiguous United States. Several regional dominant clays are kaolinite in southeastern soils, vermiculite in northeastern soils, montmoril-

lonite or montmorillonite + mica in the Great Plains, and mica or montmorillonite + mica in the southwestern soils. They found that several water-balance variables, calculated for nearby climatic stations, correlated well with clay mineralogy, especially soil-moisture surplus (leaching), water deficit (concentration of ions in solution during the dry season), and seasonality. Their model was able to predict the dominant clay mineral correctly for 80% of the sites.

To sum up, clay minerals have a fairly good correlation with climate. Correlation is often best and most specific at either end of the climatic spectrum, that is, 1:1 clay minerals in warm, humid environments; biedellite in cool, humid conditions characterized by podzolization processes; and montmorillonite, palygorskite, and sepiolite in the more aridic environments. Clay-mineral associations in other environments may be more difficult to predict. In time, inherited clay minerals (e.g., from dust in deserts) should convert to forms more representative of and stable in the environment.

Soil Redness and Pedogenic Fe Minerals

The color of a soil, in particular its degree of redness (see Blodgett and others, 1993), is generally related to climate. One has to be cautious in this regard because at least part of the redness of old soils is a function of

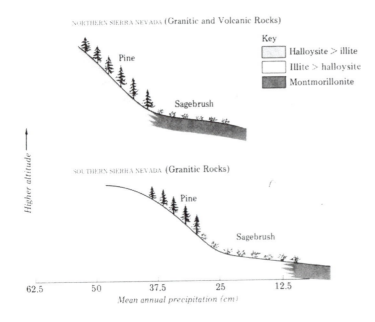

Figure 10.16 Relationship of clay mineralogy and climate with different parent materials along the eastern Sierra Nevada. (*Data from Birkeland and Janda, 1971, Table 3.*)

time of soil formation (Table 8.5). Part of the differences in soil color with climatic region, however, is a function of climate. Temperature probably is more important than precipitation in producing redness, because red soils are common in the humid southeastern United States, the humid tropics, and hot deserts of the southwestern United States and Baja California, Mexico; red soils are not common in areas of low temperatures and varying amounts of precipitation (e.g., along the northern border of the United States and in polar regions).

The degree of redness of soils has been shown to be a function of the species of the pedogenic Fe-bearing minerals (Schwertmann, 1993). The association with color is as follows: hematite with hues of 5R–5YR, goethite with 7.5YR–2.5Y, and ferrihydrite with 5YR–7.5YR and values <6 (Table 4.3). Presumeably, the Fe extracts that are the surrogates for the pedogenic Fe minerals also correlate with the color. Schwertmann and co-workers have done much of the work. One was an east–west transect in soils formed in glaciofluvial gravels in the foreland north of the European Alps (Schwertmann and others, 1982). The data suggest that the more hematitic soils are redder and that these occur in the warmer, drier climatic regimes (Table 10.4). The C horizons are 10YR hue and contain goethite. They also recognized a parent-material effect, in that silty till of the same age as the glaciofluvial gravels has no hematite and has 10YR hue (Table 10.4). With these data and those

from soils in Spain, they obtained the following relationship between soil redness (Y), calculated from the Munsell notation, and hematite content (X):

$$Y = 0.81 + 8.4X - 0.75X^2 \qquad (r = 0.97; \ n = 21)$$

In another transect involving Oxisols and Ultisols in southern Brazil, hematite is favored over goethite at mean annual temperatures ∼17°C, mean annual precipitation minus mean annual evapotranspiration values ca. 100 cm, and a low content of organic matter (<3%) (Kämpf and Schwertmann, 1983; Fig. 10.17).

Fragipan Occurrence

The presence of fragipan in the United States indicates a climatic control. The book edited by Smeck and Ciolkosz (1989) reviews their occurrence by region, and here are some generalities related to climate. First, however, they occur in surficial materials that are silty or loamy (loess, till, colluvium, alluvium, lacustrine deposits, and some bedrock units), but loess seems to be preferred, and clay content is usually <35%. Sometimes the occurrence of fragipan seems controlled by a contact of contrasting materials, such as underlying bedrock or a buried soil. Most are found in either udic or aquic moisture regimes, and mesic or thermic temperature regimes. Some workers see an association with a summer soil-moisture deficit. Most are under forest vegetation.

Table 10.4 Variation in Soil Color and Iron Extracts and Minerals for Bt Horizons of Alfisols Formed from Late-Wisconsin Glaciofluvial Gravels, European Alps

MAT[a] (°C)	MAP[b] (cm)	Color		Clay Fraction (%)[c]			Ratio of Hm:Hm + Gt
		Soil	Clay	Fe$_o$	Fe$_d$	Fe$_o$/Fe$_d$	
10.3	57.8	5YR 3/4	5YR 4/6	0.80	4.94	0.16	0.29
8.0	75.0	5YR 4/8	5YR 3/6	0.38	3.83	0.10	0.09
6.5	137.9	7.5YR 4/4	5YR 4/4	0.40	4.64	0.09	0.08
7.5	106.7	7.5YR 4/4	7.5YR 4/6	0.81	5.27	0.15	0.05
7.6[e]	118.7	2.5Y 4/4	10YR 4/4	0.40	3.43	0.12	0.0

Data from Schwertmann and others (1982, Table 1).

[a]Mean annual temperature.

[b]Mean annual precipitation.

[c]Fe$_o$ is the oxalate extract, Fe$_d$ the dithionite extract.

[d]Hm is hematite, Gt goethite.

[e]Parent material is silty till.

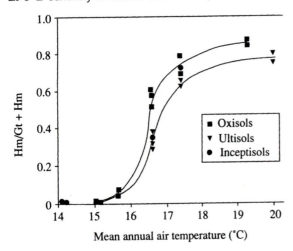

Figure 10.17 Relationship of hematite/hematite + goethite ratio and MAT, southern Brazil. *(Redrawn from Kämpf and Schwertmann, 1983, Fig. 3, Goethite and hematite in a climosequence in southern Brazil and their application in classification of kaolinite soils. Geoderma 29, 27–39, © 1983, with kind permission from Elsevier Science, The Netherlands.)*

A climosequence in the southern part of the South Island, New Zealand, shows a relation of fragipan with MAP (Malloy, 1988, Fig. 13.2). Parent material is loess of O-isotope stage 2. Although fragipan occurs at MAP 55, the soil is sufficiently dry that there is little gleying in the overlying B horizon. Soils in the MAP range of 70–90 have fragipan and enough water movement to form a perched water table above the fragipan. Mottling is strong, and gleying is moderate both in the overlying horizon and along gamma-shaped veins that penetrate deeply into the fragipan. At >90 MAP fragipan is absent, the soils are well drained, and there is little evidence of gleying.

Calcium Carbonate, and More Soluble Salts

Several features of $CaCO_3$ accumulation in soils are related to the climate, more specifically to that fraction of the precipitation that leaches downward in the soil (Fig. 10.4). Much of this was covered in Chapter 8. Here the amount of precipitation at which $CaCO_3$ begins to appear in soils and the relationship between the amount of precipitation and the depth to the top of the $CaCO_3$-bearing horizon are discussed.

In any climatic transect running into a sufficiently dry region, $CaCO_3$ usually appears in the more arid soils. The climatic value at which it first appears is obviously a function of the distribution of moisture and temperature throughout the year. In the midcontinent, for example, an area of significant summer precipitation, the boundary between the calcic and noncalcic soils (Fig. 10.18) is at about 50 MAP and 5 to 6 MAT in northern Minnesota and at about 60 MAP and 22 MAT in southern Texas. However, in going from the northern Sierra Nevada to the Basin and Range Province, $CaCO_3$ first appears in the soils near

Figure 10.18 Approximate position of calcic–noncalcic boundary for the midcontinent. *(Redrafted from Jenny, 1941a, Fig. 99.)*

Reno, Nevada with an MAP of 18 and an MAT of 9.5. The calcic–noncalcic soil boundary occurs at a lower precipitation at Reno relative to the midcontinent partly because soil leaching per unit of rainfall is more effective in a winter-wet, summer-dry climate such as Reno's. Another factor that might explain the moister climate at the midcontinent calcic–noncalcic soil boundary is texture, since midcontinent parent materials commonly are finer textured than are the parent materials near Reno. Still another factor to be considered in explaining this boundary is the calcium content of the parent materials, because Jenny (1941b) has shown that $CaCO_3$ horizons can persist to higher values of precipitation in soils with calcium-rich parent materials relative to parent materials low in calcium.

Although Fig. 10.18 accurately depicts the calcic–noncalcic soil boundary for the midcontinent, it is not accurate for the western United States because the mountains and intermontane basins have markedly different climates and soils. Nearly every mountain–basin transect is characterized by a relatively moist-cool climate in the mountains and a relatively dry-warm climate in the basins. Under appropriate conditions, therefore, soils on the lower slopes of the mountains and in the basins will contain $CaCO_3$. These relationships are repeated over and over again in the western United States. Richmond (1962) made a particularly detailed study of these relations in the La Sal Mountains of Utah, and finds that the cal-

cic–noncalcic soil boundary is related not only to present altitude and climate (Fig. 10.19), but also to past conditions of pedogenesis. He also found that with lower altitude the Bk horizons of all soils of a particular age get thicker, their pH shows a slight increase, and mixing of B and Bk horizons is more common.

Depth to the top of the Bk or K horizon is closely related to the amount of precipitation. This is qualitatively shown by the La Sal Mountains data (Fig. 10.19). Depth relationships have been quantitatively assessed in two regions of the United States. Jenny (1941a) worked on a transect from the relatively dry climate of the Colorado Great Plains eastward to the more humid regions of Missouri and found the following relationship:

$$D = 86.74 - 0.2835P$$

where D is the depth (centimeters) to the top of the Bk or K horizon, and P is the MAP (millimeters; metric conversions from Retallack, 1994). Arkley (1963) analyzed the same relations for a number of California and Nevada soils and got a different relationship:

$$D = -2.734 - 0.1508P$$
(metric conversion as above)

The difference between the two relationships is ascribed to climate. In California and Nevada, with a

Figure 10.19 Relationship of soil order and morphology with altitude, La Sal Mountains, Utah. The Spring Draw soil is close to 100 ky old. The numbers on the right side of the profiles are boundary depths in centimeters. (Redrawn from Richmond, 1962, Fig. 18.)

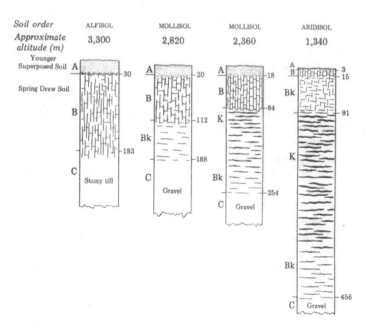

predominant winter rainfall, a unit increment of rainfall is much more effective in leaching $CaCO_3$ to a particular depth than it is in the midcontinent, where significant rainfall comes during the season of high evapotranspiration. Retallack (1994) expanded the data set drastically, included 317 soils from around the world, and in many different climatic settings, to get this relationship:

$$P = 139.6 - 6.388D - 0.01303D^2$$

The above equations predict the top of the calcic horizon at ca. 1 m under an MAP ~75, whereas the relations derived from southern Israel would suggest an MAP of half this (Fig. 8.24). All of the data sets have a wide scatter of data points, no doubt due to local factors. For example, subsequent accumulation of surficial material, or locally more coarse material might result in the top of the carbonate horizon being deeper than predicted, or erosion of surficial material may result in the top being on the shallow side.

One climate-related influence on the rate of carbonate buildup with region (Fig. 8.28) is the chemical composition of the precipitation, for it has been shown to vary with region. Many of the calcic soils in the western United States lie in areas of relatively high concentration of Ca^{2+} in the rainwaters, and this is important to regional rates of carbonate buildup. Recall that over one-half of the Ca^{2+} involved in the carbonate buildup at Las Cruces is ascribed to that dissolved in rainwater (Chapter 5).

In still drier climates, gypsum and more soluble salts are present in well-drained soils lying above the influence of groundwater (Eswaran and Gong, 1991). If pedogenic carbonate also is present, the top of it will be shallower than the top of the By horizon. The origin of the salts is mainly atmospheric, as solid wind-blown particles or dissolved in the rainwater. The salts occur at the depth to which soil waters penetrate during the wetter years (Nettleton and others, 1982), and because of their high solubility, are very sensitive to climatic changes that might alter this depth of penetration. The climates in which gypsum accumulate vary. Page (1972) studied gypsum soils in Tunisia and found them to be more common in areas with less than about 17.5 MAP and a leaching index of <12 mm. In the northeastern Bighorn Basin, Wyoming, gypsum is accumulating in a climate with about 16.5 MAP and a leaching index of 26 mm (Reheis, 1987b). The salt–climate relations in southern Israel have gypsum appearing at <30 MAP (Fig. 8.24). As usual these trends are general and affected by local factors, an important one being soil texture.

Of interest is the lack of pedogenic carbonate in areas of extreme aridity where pedogenic gypsum is forming. Apparently the lack of much vegetation results in such low soil CO_2 values (Brook and others, 1983) that the common Ca salt is the sulfate rather than the carbonate.

Trends in Iron, Aluminum, and Phosphorus

Extracts of Fe, Al, and P display rather predictable trends with climate. Much of this was discussed and referenced in Chapter 8 (see Fig. 8.14). Soils in relatively humid environments have an accumulation of Fe_o and Al_o and a depletion of acid-extractable P that seem to correlate fairly well with precipitation. Arid environments, cold or hot, either do not show trends, or rates of accumulation and depletion are so slow that only the very old soils display trends.

■ Soil-Elevation Transects

Many workers have used changes in soil properties with elevation to depict and combine the above influences of climate and vegetation on soils. In the Basin and Range Province of the western United States the basins are dry and the adjacent mountains are moist. Amundson and others (1989a) found these changes in soils in Nevada formed from alluvium consisting of limestone and dolomite clasts, from lower (MAP 16, MAT 18, leaching index = 2 cm/year) to higher elevation (MAP 55, MAT 9, leaching index 24 cm/year): organic carbon increases (0.8 kg in the top m^3, to 9.4 kg); B-horizon development increases as does Fe_d; pedogenic carbonate content decreases; and palygorskite gives way to smectite. A transect in northern California in soils formed from basaltic and andesitic materials gives these trends from lower (MAP 50, MAT 17) to higher elevations (MAP 150, MAT 5) (Alexander and others, 1993): organic carbon increases (9 kg in the top m^3 to >25 kg), as does the C/N ratio; pH decreases about 1 unit as percentage base saturation decreases; argillic horizons are more distinct at the lowest elevation, increase in clay content and thickness with elevation, but then have lesser clay and become thinner at still higher elevation with the highest soils lacking a Bt; redness decreases; and the trend in predominant clay mineral is kaolinite $\Rightarrow$ halloysite $\Rightarrow$ allophane. All of these trends are fairly predictable, but before all trends can be attributed to climate and vegetation, any major influence of the other factors has to be discounted.

■ PDI as One Example of Poor Correlation between Climate and Soil

It should not be assumed that all soil properties will show definite and predictable associations with present-day climate, for some do not. The examples used in the last sections were selected to demonstrate trends. However, trends with climate are not always obtained.

Harden and Taylor (1983) calculated the PDI of chronosequences in four climatic regions and concluded that, depending on how the index is calculated, the change in the index with time can be independent of climate (Appendix 2, Fig. A2.6). The soil-moisture regimes and soils compared are aridic in the Las Cruces, New Mexico area, xeric inland and xeric coastal in California, and udic in Pennsylvania. In one analysis, the profile index was calculated using the four properties that best correlate with age for each chronosequence; these were not always the same property for each chronosequence. Although there is scatter in the plots, the index data from the four soil-moisture regimes when plotted against time are all very similar. When eight soil properties are used to calculate the index, the index data for the aridic soil chronosequence plotted at lower values relative to the rest, but those for the other three climatic regions again overlapped.

The general lack of separation of index data for such contrasting climates is surprising and difficult to explain from a climatic point of view. Perhaps a different manipulation of the data would better reflect the climatic differences. However, the fact that all four areas did provide such similar plots against time is important for workers who use the index to assign broad ages to soils in contrasting climatic regions.

■ Reconstruction of Past Climates from Pedologic Data

Morphology and various other properties of soils can be used to infer past climates. This is important to Quaternary research because in many places the soils represent hiatuses in the depositional record and may be the only record left of certain time intervals. Before paleoclimatic interpretations can be made, however, it is necessary to be certain that the observed feature was truly imparted on the soil by a past climate and is not related to other factors of soil formation. It has commonly been recognized, for example, that the effect of a longer interval of soil formation can sometimes give the same pedologic result as a climatic change. Moreover, although many workers have used soil features to infer past climates, such studies must be backed up with quantitative soil data. It may also be difficult to separate the effect of greater precipitation from that of lower temperature, as both increase the amount of soil moisture and leaching, and many soil properties reflect the latter. Pawluk (1978) correctly urges caution in these interpretations and points out that there may be several genetic pathways for the formation of soil features and that a different former climate may be only one such pathway.

Soil properties vary in their usefulness as tools for paleoclimatic interpretation. If the soil has remained at the surface since the climatic change, the property or properties used to decipher changes in climate must have been resistant enough to persist in the soil, and not be altered entirely during subsequent pedogenesis. Obviously, those soil properties that alter readily with changing environmental conditions, such as pH, cannot be used as indicators of past conditions. The same is true for buried soils. Here, however, the properties imparted to the soil during pedogenesis must be resistant to subsequent diagenetic alteration to be useful in paleoclimatic reconstruction. Morphological features can be ranked by their persistence with changing environmental conditions or with burial (Table 1.5). For example, those horizons or features ranked as relatively persistent or persistent are most liable to carry a legacy of former conditions.

The concept of pedological thresholds (Chapter 8) has to be addressed in any paleoclimatic reconstruction using soils. Some of the same soil trends that are used to explain a threshold can be used to suggest climatic change. In some cases the threshold causes a change in pedogenic rate, just as do some climatic change scenarios. In the example discussed by Mc-Fadden and Weldon (1987), the accumulation of fines preceded the increase in accumulation of pedogenic Fe. The threshold requires about 8 ky to occur, and the transition that accompanies it (recognized by a change from Bw ⇒ Bt horizon) takes about 3 ky. The strategy for recognizing the threshold lies in soils (and associated deposits) close in age, good dating, and independent information on climatic change in the region. In this case, if climatic change was the driving force, an increase in weathering would take place, and this probably would show as fines and Fe increasing together, rather than the former leading the latter.

The same arguments as above could help sort out threshold versus climate change in the threshold examples of Muhs (1984) and the positive feedback examples of Torrent and Nettleton (1979). In these examples, progressive changes in the soil chemistry, and accumulation contents of clay, carbonate, and Fe dictate the future pathway of soil development. As thresholds are crossed, Bt horizons can become mottled, Bs horizons convert to Bhs horizons, stage III carbonate morphology convert to stage IV morphology, and Alfisols convert to Vertisols. Some of these same changes could take place under changing climate, and the challenge to the researcher is to devise a strategy to determine which is controlling the pathway.

Success in deciphering paleoclimate from soils evidence will depend partly on the field area chosen. The most promising areas are those of rather sharp transition from one climatic region to another. As an illustration, if working in the center of a large region of Mollisols that are characterized by a relatively wide range in climate, a past change in climate may have left the area of interest still within an area of Mollisol formation (Fig. 10.8). It is possible that no diagnostic soil feature would record this climatic change. In contrast, if working at the margins of the Mollisol area, where they grade into either the Aridisols or the Alfisols, and if a past shift in climate were accompanied by a shift in the geographical position in the soil orders, then marked changes in the soil properties would have accompanied the change in climate, and one could hope to read these properties as due to a past different climate in either surface or buried soils (see Busacca, 1989).

An example of the above is the work of Janda and Croft (1967) on soil formation and Quaternary climates in the northeastern San Joaquin Valley of Cal-ifornia. Soils on terraces of all ages are Xeralfs, and the field evidence suggests that they all have formed under a climate conducive to Xeralf formation. Janda and Croft then looked at the west coast distribution of Xeralfs and noted that they presently lie in areas characterized by 16- to 84-cm MAP, falling mainly in the winter, and 11.3 to 18.3°C MAT. They concluded that this wide range in climatic parameters probably would result in Xeralf formation, and thus, rather large changes in the climate could take place without those changes being registered in soil morphological features. About all they could conclude was that the Quaternary climates in that area probably had not deviated beyond the above precipitation and temperature values.

McFadden (1988) came to a similar conclusion in the xeric ⇒ semiarid ⇒ arid transect he studied. In the xeric climate, there is little unambiguous evidence for climatic change in the noncalcic soils. In contrast, evidence for climatic change is fairly clear in the calcic soils formed in the semiarid and arid climates. The reason the latter soils show the evidence is that they contain pedogenic material (carbonate) that is readily recognized and that readily shifts position vertically with changing soil-moisture conditions.

Overall Soil Morphology

Some soils in a region, either at the surface or buried, might have a soil morphology quite different from those currently being formed in the region. If they have, one might be able to reconstruct the past climate by comparing the present-day climate and morphology of soils in the region with the climate and morphology of the soil or soils in question. As mentioned earlier, one might use data such as that in Fig. 10.8 to determine how much of a change is suggested by the differences in soil morphologies, but for this approach to be successful, the differences have to be great, and success could be best along the boundary between two contrasting soil orders.

Thick loess sections in central Europe contain buried soils with contrasting morphologies that are interpreted as due to climatic change (Kukla, 1970, 1977). Soils formed during interglacials are mainly leached of carbonate and strongly weathered, with Bt horizons; these formed under forest vegetation. In contrast, soils formed during the glacials under a cold continental climate and steppe vegetation are less leached and weakly developed.

Another soil-loess sequence in the western United States illustrates the use of soil morphology in climate reconstruction. The Columbia Plateau of Washington, Idaho, and Oregon has a loess cover that in eastern Washington is as much as 75 m thick, and the oldest loesses could be 1–2 Ma (Busacca, 1989). Multiple buried soils are the key to unraveling the stratigraphic succession. Key soil horizons seem to be restricted to particular soil-moisture regimes, along a transect in which the vegetation changes from sagebrush ⇒ grasses ⇒ forest. Cambic, calcic, petrocalcic, and duripan horizons occur at <45 MAP, albic, cambic, and argillic horizons between 45 and 70 MAP, and albic, cambic, argillic, and fragipan horizons in a relatively narrow zone (20 km wide) at >70 MAP. In this case, 45 and 70 cm MAP are areas of pedogenic tension for they separate soils of markedly contrasting morphologies. One hypothesis suggests that fragipans form at the steppe–forest transition. Properties of mul-

tiple buried soils can be used to suggest limits on past climate change. Many of the buried soils in a section at ca. 30 MAP lack well-developed argillic horizons, leading to the conclusion that MAP probably was never twice that of the present. At the wetter end of the transect, one buried soil with fragipan is 20 km west of the average, suggesting a shift in the steppe–forest boundary. Finally, soils in the area of both calcic and argillic horizons have Btk horizons. All of these examples suggest lateral shifts in key climatic parameters with time. Many of the soils near the base of the loess have properties indicating greater soil moisture then.

Some areas have experienced exceptional climatic change, as shown by the morphologies of different-aged soils. In southwestern Australia, thick Oxisols lie on old dissected plateau remnants (Fig. 10.20), whereas younger soils in the river valleys are much less leached, and some contain $CaCO_3$ and are alka-

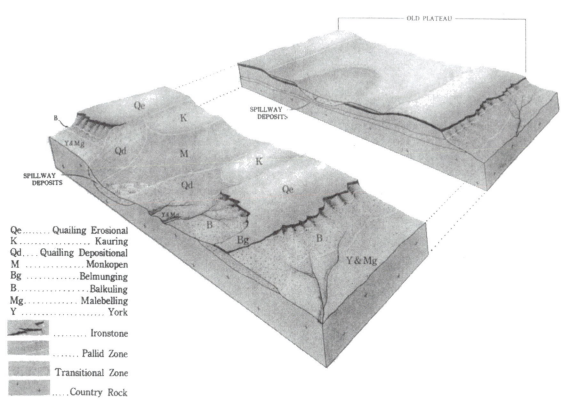

Qe........ Quailing Erosional
K.................. Kauring
Qd.... Quailing Depositional
M .... Monkopen
Bg ....Belmunging
B.................Balkuling
Mg............ Malebelling
Y .... York

.... Ironstone
.... Pallid Zone
Transitional Zone
.....Country Rock

Figure 10.20 Diagram of a lateritic plateau in Western Australia. Patterns refer to the vertical profile, as defined. *(Reprinted from Mulcahy and Hingston, 1961, Plate 1.)*

line (Mulcahy, 1967; see also Division of Soils, CSIRO, 1983). The area in question is characterized by <50 MAP. Clearly, the Oxisols could not form in the present climate because it is doubtful if much water wets as deeply as their thickness. A much wetter former climate, perhaps extending back into the Tertiary, is postulated to explain their presence.

At a larger scale, that involving plate tectonics and the time associated with plate movement, some of these deep and intensely weathered soils may have obtained some of their properties when Australia was far south of its present position (Division of Soils, CSIRO, 1983). Northern Australia was in a latitude close to present-day Tasmania at 65 Ma, so southern Australia was much farther south where present climate is cold (Veevers, 1984, Figs. 20–26 and 38). For the formation of some Oxisols, sufficiently warm and wet conditions seem to have extended far south of present-day Tasmania.

Other areas have old soils that formed under much different climates. An interpretation similar to that for Australian Oxisols has been made in California, where the only Oxisol in the conterminous United States seems to have formed between 45 and 25 Ma (Singer and Nkedi-Kizza, 1980). Present climate in the area is 51 MAP and 16 MAT, and the interpretation is that a wet, hot tropical climate (see Eswaran and Tavernier, 1980) was responsible for the thick, highly weathered soil. Other examples of buried soils with morphologies indicative of much wetter conditions are a highly weathered red soil formed from basalt during the Cretaceous in the Negev Desert (Singer, 1975), and some with fragipan in Oligocene terrestrial deposits in the

Western Desert of Egypt (Bown and others, 1982) (Fig. 10.21). In the last example tree vegetation is suggested, in contrast to an interpretation by a previous worker that the area at the time was semiarid and almost treeless.

One study used soils to show that a marked climatic change had not occurred. Pollen data have been used to indicate widespread pine and spruce boreal forest across the Southern High Plains in the late Pleistocene. Forests should have imparted diagnostic properties on surface and buried soils alike, but instead the soils are all analogous to those expected in grasslands (Holliday, 1987). This is a powerful argument against the forest hypothesis.

Organic Matter Distribution and Content

Soil properties associated with the organic matter content in the soil as well as distribution with depth are not too useful in reconstructing paleoclimate. Because organic matter properties of soil reach a steady state quite rapidly (Fig. 8.33), with a change in environmental conditions, the organic matter properties quickly adjust to values in equilibrium with the new conditions. This is especially true for surface soils, and if climatic change is to be deciphered from these properties, the change would have to have taken place in less time than the time necessary to reach the steady state for that environment. An A horizon in a buried soil might be useful, but the organic matter is usually so depleted that the A horizon is seldom recognizable on either color or content of organic matter constituents, particularly in soils buried for longer than

Figure 10.21 Fragipans (f) within Oligocene sediments in the western desert of Egypt. *(Photograph courtesy of M.J. Kraus.)*

about 10 ky. A paleoclimatic reconstruction might be possible if some of the organic matter properties could be shown to have been inherited from a former vegetation cover (Birkeland, 1984, Chapter 10) and if the change in vegetation was due to a change in climate.

Bt-Horizon Properties

The position of the Bt horizon, its color, and the amount of clay all might be used to interpret past climates, but unambiguous interpretations are hard to make.

The position of the Bt horizon should relate to the climate under which the soil formed. It is commonly reported that aridland soils are shallow and that soils in progessively wetter climates are progressively thicker; the Bt-horizon position would parallel these trends. One might be able to relate the position of the Bt horizon to the water-holding capacity of the soil and to water movement within the soil (Fig. 10.4A).

With past climate change to a wetter climate, the soil-moisture curve is altered so that water penetrates to greater depths, and clay translocation should follow. Curves such as those in Fig. 10.4 would be shifted to the right with a climate change favoring both greater soil moisture and greater leaching index, and to the left with a change toward decreasing soil moisture and leaching. Curve A (Fig. 10.22) depicts water movement for a particular time and climate and profile form could be A/Bt/Bk. With a shift to a drier climate (curve B), most parts and properties of the former profile are preserved, but the top of the Bk horizon rises, giving a profile form of A/Bt/Btk/Bk. With a shift to a wetter climate (curve C), however, the entire profile is in a leaching regime and the carbonate of the previous profile (Bk horizon) removed from the soil; thus, only the more resistant parts and properties of the previous profile are preserved. As an example, if kaolinite formed in the soil (curve A) and the climate turned aridic (curve B), the kaolinite would be preserved. In contrast, if montmorillonite formed in the soil (curve A) and the climate became more humid (curve C), the montmorillonite could convert to kaolinite and the soil lose the clay-mineral record of the former drier time. In summary, a present drier climate preserves soil features from a former wetter climate, but a present wetter climate may not preserve soil features of a former drier climate.

In a transect along a river in Montana, Reheis (1987a) calculated these changes in leaching index for

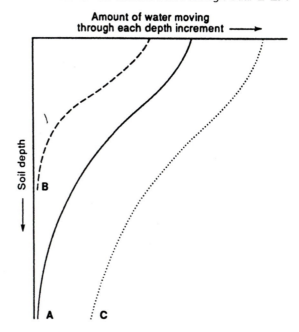

Figure 10.22 Schematic diagram showing the movement of water with depth, according to the model of Arkley (1963). Curves as follows: (A) is any one time and the top of the Bk horizon is close to where the curve becomes asymptotic with the vertical axis; (B) denotes a change to a drier climate, the top of the Bk horizon is much closer to the surface, and most persistent properties formed during the time of curve A are preserved; (C) change to a wetter climate, which would result in removal of the more soluble soil constituents, greater depth of clay formation and translocation, and eventual formation of a noncalcic soil.

an assumed glacial-climate change of 10°C lowering of MAT, with MAP remaining the same: at the wetter end of the transect (noncalcic soils), the index increased from 26 to 44 cm, whereas at the drier end (calcic soils) it increased from 8 to 24 cm. This change resulted in clay accumulation to greater depth for those soils older that about 20 ka (pre-last glacial), in comparison to soils less than that age, at most localities (Fig. 10.23). The fact that many of the clay-accumulation depths for the older soils are somewhat similar is taken to mean that the soil-leaching regimes during the older glacials were somewhat similar.

A surface soil that is redder than younger surface soils in the area might be used as an indication of a former warmer climate. This can be done, however, only with equivalent parent materials and duration of soil formation.

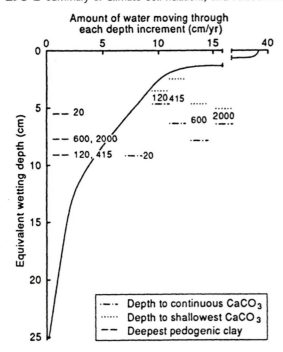

Figure 10.23 Relation between water-movement curve and depth of pedogenic clay and carbonate for soils in a transitional area between the mountains and basin, southern Montana. Equivalent wetting depth is the depth of soil expressed in centimeters of available-water-holding capacity (see Arkley, 1963). Numbers adjacent to depth symbols are age estimates in ka. *(Redrawn from Reheis, 1987a.)*

The amount of clay present in a soil can be used to indicate past climatic change if the duration of soil formation is known. Here, if the clay content of an old buried soil is greater than that in a soil formed under the present climate, and if both soils formed over a similar time span, it can be concluded that the older soil formed in an environment conducive to a more rapid production of clay. To increase the rate of clay production, one might postulate an increase in precipitation, an increase in temperature, or both, but a greater rate of eolian dust influx also has to be considered.

The above approaches to paleoclimate reconstruction can be demonstrated with several field examples. It took some time for red clay-enriched deposits in England to be recognized as soils. Some of these formed during interglacial times on relative old landscapes, and the B horizons have been termed paleoargillic (Avery, 1985). Kemp (1985) and Kemp and others (1993) review many detailed studies of strati-graphically important deep, red, clay-enriched profiles in eastern England, called the Valley Farm Soil. This soil was a key stratigraphic marker, as it was initially thought to have formed during one interglacial time, with the accompanying temperate climate. Further field and laboratory work, of which thin-section analysis was an important component, shows that some of these soils may have formed during at least two temperate interglacial climates and one intervening periglacial climate.

Although the Valley Farm Soil preserves many properties of temperate climate pedogenesis, in places it is crosscut by periglacial features reminiscent of Arctic soils, collectively called the Barham Soil (Rose and others, 1985; Kemp and others, 1993). Common features are involutions, ice-wedge casts, frost cracks, sand wedges, as well as disrupted clay films in the host Valley Farm Soil. Some of these features suggest MATs as cold as $-12°C$, which, when compared to the climates prevailing during the formation of the Valley Farm Soil, provide pedological evidence for one of the greater drops in temperature.

Another example of a paleoclimatic interpretation of old red soils is along the United States–Canada border, on the east side of the Rocky Mountains (Karlstrom, 1991). The modern soils are A/Bw or Bt, and some have an underlying Bk horizon. At least five, and perhaps 10 surface and buried soils are present, and they date from the late Tertiary to the late Quaternary. The soils have what are termed paleoargillic horizons—thick (0.3–5 m), reddish (at least 5YR, and 1.5–3 times increases in Fe_d), with 1.5–5 times the clay content of the C horizons, and the calcic soils have stage I–III+ carbonate morphology and are 0.3–15 m thick. Carbonate leaching varies from 2.6 to 10 m. Included clasts are more weathered than they are in the modern soils. Duration of soil formation is estimated at 100 ky to 1 My. Comparison is made with the properties of soils in other climatic regimes to suggest that the main soil properties were mainly formed under MATs 6–8°C higher than present, and 40 cm greater MAP.

A final example illustrates the difficulty of separating the climate and time factors, and comes from the moderately developed Churchill Geosol in the Lake Lahontan Basin, which Morrison and Frye (1965) initially suggested formed in a soil-forming interval of not much more than 5 ky under a climate wetter and warmer than the present, during the mid-Wisconsinan. This soil formed from a sand (see soil formed on O-isotope stage 6 deposit at Mustang, Fig. 10.12). Because in reconnaissance studies in many areas, we had not seen a 5-ky soil anywhere as well developed as that one, given a similar parent material, it was suggested that more time was required to form the soil (Birkeland and Shroba, 1974). Later data bear this out, for the revised Quaternary history of Lake Lahonton suggests that as much as or more than 10 times the 5 ky suggested above were available for formation of the soil (reviewed by Morrison, 1991). One final aspect of this soil was brought up by Chadwick and Davis (1990), and that is that the soil-forming interval was characterized by high atmospheric dust influx, high enough to form the soil in a relatively short time.

If dust-infiltrated properties are used as the climate-change proxies, one should know if most available dust gets into the soils. Gerson and Amit (1987) show that this is not necessarily so in southern Israel. As infiltration proceeds, both the capabilities of the gravelly surface to trap dust and the subsurface horizons to transmit dust decrease. The result is a decrease in dust influx into the soil unrelated to climate change. Furthermore, this decrease in infiltration with time can lead to greater runoff (e.g., Wells and others, 1987), which could start the erosional destruction of the landform that harbors the soil needed for study.

Therefore, in any desert soil study one should factor in the potential episodic nature of atmospheric clay (and silt) influx, and their distribution with depth. These can be due to factors such as regional climate change or the progressive changes in internal drainage of the soils.

Clay Mineralogy

Clay-mineral formation and transformation in the soil are slow processes in some environments, and therefore clay mineralogy may be a useful tool in assessing past climatic influences on the soil. Here I will simplify the situation and mention only the interpretations that can be made for kaolinite, montmorillonite, and palygorskite.

The stability of a clay mineral in a changing soil-leaching environment is a function of the clay mineral originally formed and the direction of the climatic change (Pedro and others, 1969; Singer, 1979/1980). As mentioned earlier, kaolinite and halloysite formation is favored by a high leaching environment, and

montmorillonite by a relatively low leaching environment, in well-drained parent materials. A change in soil-leaching environment triggered by a change in climate may bring about a change in the clay mineralogy, but the climatic change usually has to be in one direction to be read. For example, many studies have shown kaolinite to be a stable mineral, one that will persist for long intervals of time in a neutral to alkaline environment. Thus, if kaolinite is formed in a soil under a fairly high leaching and relatively acid environment, and the climate becomes arid, kaolinite should persist in the soil and serve as an indicator of a former wetter climate. Montmorillonite, on the other hand, is less stable than kaolinite and will change to the latter if greater leaching conditions are obtained. In summary, therefore, a climatic change from humid to arid could be read in the clay mineralogy, but one from arid to humid most likely could not. One final point should be made and that is that clay-mineral transformation is a slow process, and therefore the climatic change would have had to persist long enough to allow time for slow reactions to take place. A climatic change of short duration may not be registered in the clay mineralogy.

A transect from the relatively humid east side of the Sierra Nevada into the dry western Great Basin can serve as an example of the use of clay mineralogy to help define Quaternary climatic change (Birkeland, 1969; climatic data are given in the caption for Fig. 10.12). The clay mineralogy of soils developed from tills, outwash deposits, and lacustrine deposits is closely correlated with the present climate. Halloysite is the dominant clay mineral in the humid environment, and montmorillonite is dominant in the arid environment (Fig. 10.16). The change from one mineral assemblage to the other takes place at the present position of the pine–sagebrush vegetation boundary, at about 40 cm MAP. The interesting thing about this study is that the clay mineralogy of soils formed on river-terrace deposits of all ages changes at about the same location under similar present-day environmental conditions. This is taken to mean that long-term past climatic change toward a wetter soil-moisture regime in the western Great Basin has not taken place, for if it had, it probably would be seen as dominant kaolinite or halloysite in the older soils far east of their easternmost dominant occurrence in the younger soils. The above conclusions seem to be valid for a large number of sites along the eastern Sierra Nevada (Birkeland and Janda, 1971).

In a more detailed and site specific study, Chadwick and others (1995) examined soil clay minerals in a sequence of soils in eastern Nevada. Glacial–interglacial climates were estimated, and climatically sensitive soil properties were related to specific climatic regimes along a basin-to-mountain elevation transect (increasing effective moisture with elevation). Smectite is the main pedogenic clay mineral at <16 cm/year effective moisture, and kaolinite at >22 cm/year. Because the low-elevation soils have both clay minerals, the indication is that these soils had the requisite increase in effective moisture to produce kaolinite during glacial climates (times of greater leaching).

For clay mineral–climate change interpretations, one has to consider both clay-mineral stabilities in changing environments as well as the time needed to convert original minerals to new species. For example, if the major portion of clay formation takes place in a relatively warm and dry interglacial interval, and the subsequent glacial climate is wetter and cooler, the cooler temperatures may inhibit alteration to a new clay-mineral species if the rate of alteration can be shown to be strongly temperature dependent. If this is the case, the clay mineralogy of old soils will carry more of the record of the interglacials than of the glacials. Data presented in Chapter 8, however, and that of the above example from eastern Nevada, suggest that alteration can be fairly rapid in relatively cold environments as long as sufficient moisture (MAP > 50?) is present.

The presence of palygorskite and sepiolite in soils may be an excellent indicator of long-term aridity, as shown by their common occurrence with carbonate horizons (Bachman and Machette, 1977; McFadden, 1988), and the calculated stability field for palygorskite with respect to the soil-water chemistry (Singer, 1989). Calcareous soils in the Texas High Plains contain both of these minerals as a parent-material component, but under the present pedogenic regime, both minerals are being degraded (Bigham and others, 1980).

Rather than rely solely on specific clay-mineral species for paleoclimatic interpretation, one can use the depth of pedogenic clay-mineral alteration. The depth of such alteration should be closely related to the depth of soil-moisture penetration, provided sufficient time was available for alteration to be registered in the clay minerals. This approach has been used in the central Yukon Territory, Canada, to determine paleoclimates (Foscolas and others, 1977). Soils have formed from glacial and glaciofluvial deposits of several ages. Under the present pedogenic environment, which is subarctic and semiarid, it is calculated that water penetrations are about 20 cm depth. The soil on the O-isotope stage 2 deposits probably formed under that climate, and the pedogenic alteration of clay minerals in glaciofluvial deposits extends to about that depth. For the progressively older soils, the depth of pedogenic clay-mineral alteration is 93 and 190+ cm, respectively; a more humid climate in the past seems to be required for alteration to extend to those depths.

Mineral Etching

Mineral etching has been used to infer past climate in only one study (Locke, 1979). The etching of hornblende grains has been shown to vary with depth and with age of parent material on Baffin Island (Fig. 8.8). It was noted, however, that the rate of etching changed markedly, with some time intervals characterized by low rates and others by high rates. It is suggested that low rates of etching are related to limited soil moisture and high rates to excessive soil moisture, and these are related to climate. A cold, dry climate is inferred for the times of limited soil moisture, and a relatively mild, moist climate for the times of excessive soil moisture.

Position of Horizon of CaCO₃ Accumulation and Morphology of the K Horizon

Because the presence or absence of $CaCO_3$ in a soil is related to soil-water movement, and hence to climate, the position and morphological features of $CaCO_3$ accumulations may provide insight into past climates. The position of the top of the zone of $CaCO_3$ accumulation is related to the regional climate, as long as the water-holding capacity of the soil is taken into account. In those places where the tops of Bk or K horizons plot close to the sharp break in the calculated soil-water-movement curves that represents a rapid decrease in water movement with depth (Fig. 10.4A), climatic change does not have to be used to explain the position of the tops of such horizons. However, before the lack of such correspondence between climate and the top of the Bk or K horizon can be considered paleoclimatically significant, it is necessary to investigate the possibility of surface erosion or deposition as reasons for the top of such a horizon being either closer to or farther from the surface, respectively, than predicted.

The position of the Bk and/or K horizons is related to that of the Bt horizon; in most places, they lie directly beneath the Bt. If, however, they lie some distance below the base of the Bt, and the position and origin of the carbonate can be shown not to be due to groundwater, there is the possibility of a shift to a wetter climate, driving the $CaCO_3$ to greater depth. I have seen this in one locality in Wyoming, where the parent material is a well-drained gravel of a high river terrace. About 20 cm of unweathered Cu material separates the base of the Bt horizon from the top of the Bk horizon. One interpretation is that the top of the Bk was 20 cm higher in the past and that weathering of iron-bearing minerals was inhibited in the carbonate-accumulating environment. A subsequent climatic change resulted in deeper water penetration to force the top of the Bk horizon to greater depths, but the change has been recent enough so that the silicate grains in the former Bk horizon have not altered to form either a Cox or a Bw horizon.

Btk horizons can provide information on climate change toward aridity as long as no other reasons for the upward movement of the top of the Bk can be demonstrated. There are two reasons for nonclimatic upward movement of the top of the Bk. One is that as clay accumulates in the Bt horizon, the water-holding capacity of the horizon increases. The result of this is that more water is held in the Bt horizon than previously was the case. Hence, the water carrying Ca^{2+} and HCO_3^- through the Bt does not move so deeply, and these ions precipitate to form a Btk horizon. Chadwick and others (1995), for example, show how atmospheric dust influx in a 150-ka soil increased the soil-water-holding-capacity by 25%, which, in turn, decreased water penetration by some 65 cm. Such a change in any aridic area would move the depth of carbonate accumulation toward the surface, where it would overlap with the Bt horizon. The other, nonclimatic reason for upward movement of the top of the $CaCO_3$ accumulation layer involves the K horizon. As the K horizon forms, the pores of the soil are progressively plugged so that deep water penetration is inhibited. The downward-moving waters, therefore, are held at the top of the K, and eventually ionic concentrations in the soil solution build up to the point where $CaCO_3$ precipitates out. In this way, the top of the K moves toward the surface, engulfing the base of

the B horizon in the process (Gile and others, 1966, 1981; see Fig. 8.41).

Several examples of Btk horizons resulting from climatic change in the aridic parts of the western United States can be given. Changes in soil-water movement with depth, such as those shown in Fig. 10.22, would accompany the changes. In Southern California soils <10 ka have a shallow accumulation of carbonate, whereas those in older deposits are three to four times as deep (McFadden, 1988). In fact, late Pleistocene soils have two carbonate accumulation maxima with depth—a shallow one related to the present Holocene climate and a deeper one related to the greater effective soil moisture of the late Pleistocene. A similar relation of overlapping pedogenic clay and carbonate has been reported for southern Montana (Fig. 10.23) (Reheis, 1987a). In thin section, overlapping films of clay and carbonate are present, and the number of overlapping pairs of films is greater on older deposits (Reheis, 1987c). Here, glacials were times of deeper water movement and accompanying clay translocation, and interglacials were times of shallower movement and deposition of carbonate films on previously deposited clay films.

One other aspect of climatic change with respect to carbonate soils is possible variation in accumulation rate with climatic change. For southern New Mexico, Machette (1985) has evidence that the rate of carbonate accumulation during interpluvials could be twice that of the pluvials (these and others are depicted in Fig. 8.28). Reasons for this are changes in vegetation type and percentage cover, and exposure of dry lake beds. During pluvials, for instance, the source areas may be more restricted relative to those of the interpluvials. In his regional comparison, only one locality (Vidal Junction) experienced a lower carbonate influx rate during the interpluvials; the explanation is that interpluvial soil moisture was insufficient to move all of the relatively abundant carbonate into the soil.

McFadden and Tinsley (1985) and Mayer and others (1988) modeled the important parameters of calcic soils (soil CO_2, dust flux, water movement, etc.) to determine carbonate profiles in aridic climates that have undergone climate change. The approximate pluvial-to-interpluvial (late Pleistocene to Holocene) change in climatic parameters is a decrease in leaching index from about 2.3 cm to 1 cm, with an accompanying rise in MAT of >3. The model data mimic the double bulge in pedogenic carbonate common in the Pleistocene soils in the area of interest (Fig.

10.24), with deep accumulation attributed to the pluvial climate, and shallow accumulation to the Holocene interpluvial climate. In one case, however (Silver Lake), the modeled curve predicts much less carbonate than is present. Perhaps the model underestimated either (or both) the past carbonate-influx rates and the amount of effective moisture.

Lateral and altitudinal displacements of the calcic–noncalcic soil boundary along climatic transects in soils of different age can also be used as indicators of climatic change. Richmond (1962, 1965) reports such displacements in many parts of the Rocky Mountain region (Fig. 10.25), and they are also seen in deposits along the eastern side of the Sierra Nevada near Reno and in the Colorado Piedmont near Boulder. In all these transects, the calcic–noncalcic soil boundary lies at a higher altitude, and therefore in a wetter present climate, on the older deposits. For the Rocky Mountain example, it is believed that the older soils formed during an interval of time that was warmer than the present, or possibly that the combination of precipitation and temperature at that time resulted in $CaCO_3$ accumulation in soils at a higher altitude than would occur under the present climate. These explanations are plausible, if the position of the calcic–noncalcic soil boundary is not related to the different water-holding capacities of the different-aged soils. For example, the pre-Illinoian (pre-O-isotope stage 6) soils commonly have Bt horizons of high water-holding capacity, whereas the soils formed from Wisconsin (O-isotope stage 2) deposits often lack Bt horizons, and thus, their water-holding capacities are less. The alternative explanation is that under certain climates, it might be possible for Ca^{2+} and HCO_3^- to be leached from the soils with low water-holding capacity, yet be retained to form a Bk horizon in an adjacent older soil with high water-holding capacity.

Morphological features of K horizons may provide important data on solution and reprecipitation of $CaCO_3$ that perhaps owe their origin to climatic change (Bryan and Albritton, 1943; Bretz and Horberg, 1949). Some K horizons, for example, have solution pits on their upper surface. If it can be shown that the solution pits are not due to deeper water penetration accompanying ground-surface lowering by erosion, this evidence suggests a change toward a wetter climate. Other soils show evidence of formation of laminated K horizons, brecciation of these horizons, and subsequent cementation of the breccia (carbonate morphology stages V and VI). Although the processes responsible for this history are not clearly understood, they could be due to climatic change; perhaps brecciation occurs during intervals of drier climate and cementation of the fragments during intervals of wetter climate.

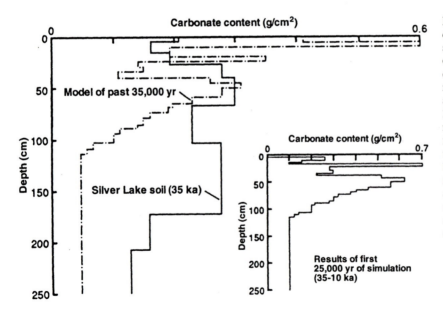

Figure 10.24 Plot of carbonate distribution for latest Pleistocene soil from Silver Lake, California, compared to modeled carbonate distribution. The model incorporates two distinct climates: a relatively wet one for the period 35–10 ka (see inset diagram and note that surface is carbonate free), and a relatively dry one for the past 10 ka, during which carbonate accumulated at the surface. *(Redrawn from Mayer and others, 1988, Fig. 4, © 1988, with permission of the Geological Society of America, Boulder, Colorado.)*

Figure 10.25 Altitudinal relationship of calcic–noncalcic boundary with age of parent material, Rocky Mountains. Approximate soil ages: short-dash line, 20 ka; long-dash line, 140 ky; solid line, >140 ka. *(Redrawn from Richmond, 1972, Fig. 2.)*

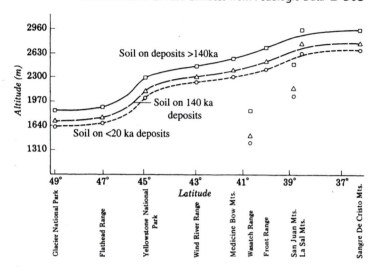

Position of Gypsic and Salic Horizons, and Associated Weathering Features

The position of soluble salt accumulations with depth could be one of the better indications of climatic change in arid regions. The reason for this is that because of their high solubility, a change in soil-moisture regime could be reflected in a rapid change in the depth of salt accumulation. If the change is toward a drier soil-moisture regime, it should show as an accumulation of salts high in the profile where it would be disjunct with respect to the rest of the profile; an example is salts in the Bt horizons of some soils in Arizona (Nettleton and others, 1975). In this case, the climatic change was toward diminishing soil moisture with time. In other places, there might be several positions of salt accumulation maxima within a profile, as exemplified by a soil in the central Sinai (Dan and others, 1982). If localized permeability control on distribution can be discounted, the interpretation could be one of stepwise increasing aridity with time.

Because of the high solubility of these salts, there are some problems in using them in paleoclimatic reconstruction. One is that if the younger climate is the more humid one, salts could be quickly leached from the profile and the record of previous aridity eradicated. Another is the effect of wetter-than-normal years that occur every century or even less frequently. These latter might result in complex salt-accumulation patterns that should not be interpreted as due to climatic change. Finally, surface erosion and deposition could influence the depth of salt accumulation, and may result in complex accumulation patterns.

There is one problem with paleoclimate reconstruction using gypsum. Reheis (1987b) cautions workers to consider the hygroscopic qualities of gypsum, since it can absorb and adsorb larger quantities of water than can silicate mineral grains. She suggests, therefore, that horizons of high gypsum content may persist under a climatic change toward greater soil moisture as long as the excess soil moisture is taken up by the gypsum. Perhaps a soil-moisture threshold has to be exceeded for the level of gypsum accumulation to move to greater depth.

In spite of the possible ephemeral nature of salt-accumulation horizons with fluctuating climates, associated weathering features may be used to infer former positions of salt accumulation. In the northern Sinai Desert and in the northeastern Bighorn Basin, Wyoming, some horizons of salt accumulation in gravelly parent materials are marked by mechanically cracked clasts (Gerson, 1981; Gerson and Amit, 1987; Birkeland and Gerson, 1991; Amit and others, 1993) (Fig. 10.26). The cracking is due to the salt weathering (Chapter 3) and presumably depends both on salt content and time. Levels of cracked stones in soils have the potential for marking former horizons of salt accumulation, which indicate former soil-moisture regimes and therefore former climates; the cracked stones would persist as easily recognized features even if a younger, more humid climate leaches the salts from the horizon or soil.

A

Bw

Bk

By

⊢━┤
cm

Figure 10.26 Salt-weathering-shattered clasts in the By horizon of an Aridisol, Sinai Desert, Egypt.

Isotopic Evidence in Soils for Climate Change

Stable isotopes of carbon and oxygen in soils can be used to suggest some combination of vegetation and climate change. These isotopes (^{13}C, ^{18}O) are present in the pedogenic carbonate, and their ratios to the more common ^{12}C and ^{16}O are used in isotopic studies (Cerling, 1984; Cerling and Quade, 1992, 1993; Ronald Amundson, unpublished manuscript, 1997). $\delta^{18}O$ in the carbonate is determined by several climate variables, most importantly the $\delta^{18}O$ of rainfall and the MAT for continental stations with <100 cm MAP. There is close to a 1:1 relationship between the $\delta^{18}O$ of the pedogenic carbonate of Holocene soils and that estimated for the associated meteoric waters, with the more negative sites corresponding to colder climates (e.g., higher altitude and latitude). The relationship is weaker for stations dominated by monsoonal climates. In contrast, $\delta^{13}C$ of the soil carbonate is best related to the isotopic composition of the biomass, and a strong correlation with the photosynthetic pathway used by the associated plants. There are three photosynthetic pathway plant groups: C3 plants, C4 plants, and CAM plants. Of importance to these studies, C3 plants include trees, most shrubs and herbs, and cool season grasses. In contrast, C4 plants are adapted to high light and water-stressed conditions; prairie grasses and many desert plants of the United States, and savanna grasses of the tropics are common examples. $\delta^{13}C$ values vary with the photosynthetic plant group, averaging $-27\%_{oo}$ for C3 plants and $-13\%_{oo}$ for C4 plants. In general, however, trees, shrubs, and some grasses contribute to the more negative values, whereas those in the less negative group are the product mainly of grasses. Cerling (1984) plots $\delta^{13}C$ of Holocene soil carbonate versus the proportion of C4 plants in the local flora, and gets a general trend from near $0\%_{oo}$ (or slightly positive) for 100% C4 plants in the local flora to near $-10\%_{oo}$ or lower for 0% C4 plants.

Isotopic data for pedogenic carbonate in a continental climate follow a distinct trend, whereas those for coastal, monsoonal, and periglacial regions fall outside the trend (Fig. 10.27). The data for continental climate trend come from Holocene soils to avoid problems with climate change, and the more negative values on either axis correlate with colder climates. Several elevation transects also show the trend of more negative values in both $\delta^{13}C$ and $\delta^{18}O$ at progressively higher elevations (Quade and others, 1989a). Obvi-

ously a mixing of isotopic signals would be obtained if a calcic soil remained at the surface through several climatic cycles with associated shifts in C3 and C4 plants. Buried soils isolated from carbonate additions after burial, and lacking evidence for diagenesis, should be best for obtaining the clearest climatic record. Amundson and others (1989b) address some of these issues, as well as the problem of inherited bedrock carbonate.

Several climate-change applications have been made using isotopic signals from soil carbonates. In the Negev Desert of Israel, buried soils containing calcic nodules have been dated at about 13, 28, and 37 ka (Goodfriend and Magaritz, 1988). The soils are considered to have formed during relatively wetter times. $\delta^{13}C$ data are used to intrepret precipitation trends with time. All soil data reflect a N-to-S gradient of lesser precipitation to the south that parallels the present gradient. There may have been a steeper precipitation gradient during the time the older soil formed, and more negative values suggest that the time of formation of the 28-ka soil was wetter than the time of formation of the 13-ka soil. Furthermore, the precipitation source seems to have been to the north, with little southern monsoonal influence from the Indian Ocean.

Numerous buried soils are present in sediment of the last 2.2 my at the archeologically significant Olduvai Gorge in East Africa (MAP 57; MAT 22; Cerling and Hay, 1986). Several assumptions were used in the analysis of isotopic data, including a direct relation between the proportion of C4 grasses in the flora and MAT. Some of the older paleosols lack pedogenic carbonate, suggesting that MAP exceeded 75–85 cm, and some calculations put MAT as low as 12–15°C. Between 1.6 and 1.7 Ma MAT could have risen as high as 22–25°C, followed by a return to the former, cooler conditions. Much of the time between about 1.3 and 0.6 Ma seems to have had MATs of 15–18°C. Another short (ca. 50 ky long) aridic period with MAT >20°C took place, followed by a cooling, and finally the warming to the present dry conditions.

The Siwalik Group sediments of northern Pakistan contain many calcic floodplain paleosols formed over the last 18 Ma (Quade and others, 1989b). Prior to about 7 Ma, $\delta^{13}C$ data plot between -13 and $-9\%_{oo}$, the $\delta^{18}O$ data average -10 to $-9\%_{oo}$, and these are taken to indicate a dominance of C3 plants, probably closed canopy forests (Quade and others, 1995). In the interval 7–5 Ma, both $\delta^{13}C$ and $\delta^{18}O$ shift toward more

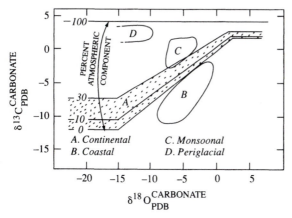

Figure 10.27 Relation between $\delta^{13}C$ and $\delta^{18}O$ in pedogenic carbonate in a continental environment, assuming 0–30% admixture of atmospheric CO_2 with soil CO_2. Fields for carbonates in monsoonal, coastal, and periglacial areas also shown. PDB refers to the PeeDee Belemnite, used for reference in the stable isotope analysis. *(Redrawn from Cerling, 1984, Fig. 5, The stable isotope composition of modern soil carbonate and its relationship to climate. Earth Planetary Science Letters 71, 229–240, © 1984, with kind permission from Elsevier Science, The Netherlands.)*

positive values, and from 5 Ma to the present $\delta^{13}C$ ranges between -2 and $+2\%_{oo}$, while the $\delta^{18}O$ ranges between -7 and $-5\%_{oo}$. The interpretation is that C4 grasses dominated the flora after ~7 Ma. The climatic change at 7 Ma coincides with the oldest documented C4 grasses, and could have been the result of an intensification of the regional monsoon, or a global decrease in atmospheric CO_2.

In a study of relict Quaternary calcic soils on the east side of the Wind River Mountains, Wyoming, the isotopic data are complex (Amundson and others, 1996). The Holocene soils seem to have formed under the present continental climate, with a westerly flow from the Pacific Ocean. However, all pre-Holocene soils have a significantly more positive $\delta^{18}O$ value, contrary to other isotopic evidence for pedogenic carbonate formation during cooler-than-present temperatures. This discrepancy is used to suggest a different source for paleosoil waters, one from the Gulf of Mexico via northward monsoonal flow.

Application of Soils to Geomorphological, Sedimentological, and Environmental Studies

There are many applications of soils to geological studies. One common one is to help decipher or put limits on possible changes in climate and/or vegetation during the Quaternary and older periods of geological time, topics covered in Chapter 10. Another is to aid in the subdivision and correlation of unconsolidated sediments, as these are of great interest to geologists, geographers, and archeologists. Soils are useful in neotectonic studies, for they can be displaced by faults and overlie faults, used to help date deposits or surfaces that have been deformed, or related to tectonics in other ways. They are also used in many environmental studies involving surface materials. Finally, paleosols in pre-Quaternary sedimentary rocks are important in understanding their environment of deposition, and in basin analysis.

Not covered here is the role of soils in forensic studies. Murray and Tedrow (1975) give examples of how the properties of soils on which a suspect walks or drives can be useful in solving crimes.

■ Use of Soils in Quaternary Stratigraphic Studies

Soils, whether at the surface or buried, are important to the subdivisions of Quaternary sediments. They are used primarily to aid in the subdivision of a local succession of deposits, to provide data on the lengths of time that separate periods of deposition, and to facilitate short- and long-range correlation. This rather special field of study is called soil stratigraphy, and its history and methods are reviewed quite thoroughly by Morrison (1968, 1978).

The use of soils for stratigraphic purposes demands that the investigator have a thorough background in pedology. It is important to be able to distinguish those features of the soil profile that are mainly geological in origin from those that are distinctly pedologic. A knowledge of processes responsible for profile development is one working tool of the soil stratigrapher, as is a good understanding of the influence of soil-forming factors on soil genesis. As an illustration, one of the first tasks of the investigator in soil-stratigraphic research is to reconstruct the landscape and delineate the factors that prevailed when the soil in question formed. In any stratigraphic study, soils should not be the only criteria used for correlation, to the exclusion of other field criteria. The best answers come from the integration of data from several disciplines. Thus, the soil evidence must be reconciled with evidence concerning geological events (volcanic ash stratigraphy, paleomagnetic stratigraphy), processes (postdepositional modification of the original landform), and knowledge of past climatic trends (global O-isotope stages and associated climates).

Stratigraphic studies of soils also are useful to pedologic studies, since a thorough knowledge of the soil in its stratigraphic setting is necessary to understand all the factors responsible for the development of a certain profile. I would go farther and say that soil chronosequences are crucial to the success of many soil process studies. Indeed, as more of the many factors in soil-profile development are seen to interact, it becomes more difficult to isolate the effects of any one factor. It is still possible in many places, however, to distinguish between the main soil-forming factors and those of relatively minor importance.

Soil Stratigraphy and Soil-Stratigraphic Units (Geosols and Pedoderms)

Stratigraphy is that branch of geology dealing with the sequence, age, and correlation of rock bodies. Soil stratigraphy involves the use of soils in defining a local stratigraphic succession, estimating the ages of units associated with the soils, and suggesting short- or long-range correlations. Soil-stratigraphic units can be formally or informally named, and examples of both follow.

Soils used formally in soil-stratigraphic studies are called pedostratigraphic units, according to the North American Commission on Stratigraphic Nomenclature (1983). Because the term "soil" has different connotations among geologists, engineers, and soil scientists, the fundamental pedostratigraphic unit is the geosol (Morrison, 1967). Geosols are recognized on pedologic criteria, are restricted to buried soils, must have a known stratigraphic position, and must be traceable laterally in the field. The top of the unit is placed at the top of the uppermost pedologic horizon, commonly an A horizon or a truncated B horizon, and the lower boundary is the base of the lowest pedologic horizon, commonly the Cox/Cu boundary. Names for formally named geosols are taken from the name of some nearby geographic place; an example is the Churchill Geosol of western Nevada. If one cares to keep the nomenclature informal, one can just use the term "soil" in combination with the underlying and overlying units. An example would be a post-Tahoe soil of the Sierra Nevada, the diagnostic soil formed on Tahoe till. Geosols and informal pedostratigraphic soils should be described well at the type locality, and morphological variation laterally should be indicated. It is hoped that some labora-

tory characterization will accompany the description. Although not specified in the code, laterally dissimilar morphologies of a geosol are called facies (Morrison, 1967; Morrison and Frye, 1965), as is done in geological studies (e.g., sedimentary facies). Thus, because of variation in pedogenic environments, a calcic facies of a particular geosol might grade laterally to a noncalcic facies and that to a gley facies characterized by properties induced by a reducing pedogenic environment.

The International Association for Quaternary Research (INQUA) had a Commission on Paleopedology that circulated an unpublished soil-stratigraphic guide (R.B. Parsons, written communication, 1981). Pedoderm is the formal pedostratigraphic unit they advocated at the time, and it was defined much as a geosol (should be mappable and have significant areal extent). One advantage pedoderm has over geosol is that relict and exhumed soils are included. Hence, soils associated with surface deposits of different ages (soils of postincisive chronosequences) can be given formal status. This is important because surface soils surely are more commonly used in stratigraphic studies than are the few known buried soils. Pedoderm has not been accorded formal status.

Although formal stratigraphic names are nice, they are not needed to do high-quality soil-stratigraphic work involving soils. We should not be formally naming all kinds of soils, thus contributing to a confusing proliferation of names. The main focus of such studies is to use the soil data as a guide to the age of a particular deposit, and this work can be done just as well if informal rather than formal soil names are used. Such informal names could be linked to the name of the deposit from which the soil formed. Thus, the strongly developed soil formed from the Rocky Flats alluvium in the Colorado Piedmont (Scott, 1963; Fig. 8.41) would be called the post-Rocky Flats soil. The post-Rocky Flats soil, even though it has regional significance, cannot be called a geosol because it is not buried. It could be called a pedoderm, such as the Louisville Pedoderm. Proliferation of names, however, is not always in the best interests of people trying to understand the stratigraphic succession in an area not too familiar to them.

Soil-Stratigraphic Units as an Aid in Defining Stratigraphic Units

The North American Commission on Stratigraphic Nomenclature (1983) allows soils to be used to help define allostratigraphic units, a new unit that is useful in Quaternary stratigraphic studies. Although lithostratigraphic units are the main mapping unit in stratigraphy, in many areas in which Quaternary deposits have been mapped, the deposits of all ages (e.g., till, alluvium, loess) commonly have identical lithologies, and there is no practical way to separate the deposits on age in the field by the properties of the deposits themselves. Thus, these deposits cannot be called formations. Allostratigraphic units are mappable stratified bodies of sediments whose boundaries are laterally traceable disconformities. The upper boundary can be a geomorphic surface and associated surface soil. Although not specified in the new code, I see no reason to exclude lichenometric or rock-weathering criteria in identifying such units (Birkeland and others, 1979). Indeed, the estimation of age and correlation are strengthened using more criteria than just soils. The lower boundary of the unit could be a buried soil. The fundamental unit is the alloformation. Thus, referring to the previously mentioned example, the Rocky Flats alluvium could be formally designated the Rocky Flats Alloformation. In the Sierra Nevada, tills could be defined as alloformations, such as the Tahoe Alloformation, and the latter would be distinguished on a multitude of relative-dating criteria, only one of which is a soil with a Bt horizon and included clasts weathered to grus (Burke and Birkeland, 1979).

A chronostratigraphic unit is a rock body whose boundaries depend on geological time; that is, the upper and lower boundaries are isochronous over large areas. Soils can be used to define such boundaries, as long as the boundaries of the soil-stratigraphic units are nearly the same age over large areas. The Illinois State Geological Survey uses these units in classifying the Quaternary geology of the state (Willman and Frye, 1970). Johnson and others (1997) argue that many boundaries in the midcontinent, including those of soil units, are diachronous.

■ Determining the Time of Soil Formation

General Concepts

The time of formation for a particular soil is determined by its stratigraphic position relative to adjacent deposits and soils (Morrison, 1965, 1967; Richmond, 1962). This time interval can be approximated for both buried and surface soils.

The time of soil formation is best dated with buried soils. In Fig. 11.1B, for example, buried soil "a" formed between the deposition of units I and II, and buried soil "b" between units II and III. Moreover, soil "a" is similar whether buried or at the surface, and so is soil "b"; hence, one interpretation is that the major properties of soils "a" and "b" were imparted to the soils during the time intervals between depositional episodes, whether the soil remained at the surface or was buried. Furthermore, because soil "a" is more strongly developed than soil "b," soil formation during soil-forming interval "b" had little effect on soil "a" where the latter remained at the surface. Figure 11.1C indicates that the strongest soil formed after the deposition of unit II; hence, soil "b" is recognizable as such only where buried. Although soil "b" was present on the landscape prior to the formation of soil "a," soil "a" is strong enough in development to mask most of the features of soil "b" in surface positions. Therefore, the major soil-forming interval here is that which formed soil "a."

It is more difficult to determine the timing of formation of surface soils where buried counterparts do not exist. In Fig. 11.1D, river terraces of various ages are depicted. In this case, soil-forming intervals cannot be proved. The maximum time available for the formation of each soil is from the time deposition ceased to the present. Thus, if soil formation proceeded at a relatively uniform rate since deposition, soil "a" may have had only about one-half its present amount of clay before soil "b" began to form, but soil "b" should have had most of its clay formed before soil "c" began to form. The soils could have formed in less time, but there is no evidence on the minimum time for formation.

Determining the ages of soils on slopes is even more difficult. Soil-stratigraphic relationships at the base of the slope may help put limits on the age of the slope and, thus, on the soils there. Hence, the entire slope catena has to be studied; examples in Chapter 9 provide data on what can be expected. Where possible, ^{14}C dates can help constrain the ages of the landforms.

Concept and Explanation of the Soil-Forming Interval

The concept of the soil-forming interval was one outgrowth of soil-stratigraphic studies in the western United States (Morrison, 1967, 1978). Where soil profiles are nearly identical in both buried and surface oc-

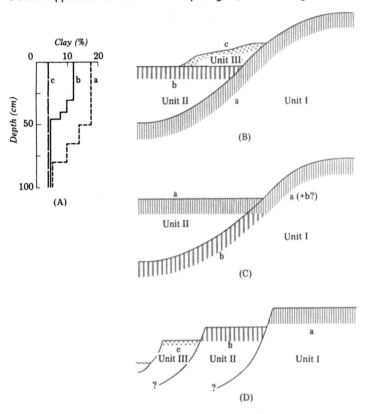

Figure 11.1 Hypothetical relations between soils and depositional units (I, old to III, young). Clay-content trends for soils in (B), (C), and (D) are shown in (A).

currences (Fig. 11.1B, soil a), the soil can be said to have formed during a discrete soil-forming interval bracketed by the ages of adjacent depositional units; the soil is younger than the unit from which it has formed, and it is older than the unit that buries it. A good example of a soil-forming interval is the Churchill Geosol of the Great Basin (Morrison, 1964). This soil was believed to have been on what was considered to be early-Wisconsin deposits and is now found either at the surface, where it has been exposed to soil formation since deposition ceased, or buried beneath late-Wisconsin deposits. Obviously, the buried soil had much less time to form relative to the surface soil, yet both are nearly identical in profile morphology and on laboratory analysis (Fig. 11.2). In addition, the clay mineralogy of both relict and surface occurrences of the soil is nearly identical, as is the degree of etching of the pyroxene grains (J.G. LaFleur, written report, 1972).

There were several ways to explain the fact that buried and surface occurrences of the same soil-stratigraphic units are nearly identical, although each has been exposed to soil-forming processes a different

length of time. Morrison (1964, 1967, 1978) and Morrison and Frye (1965) proposed that the soil-forming intervals were characterized by unique climatic conditions that resulted in accelerated soil formation over relatively short time spans; in contrast, the climates prevailing between the soil-forming intervals were not conducive to soil formation. In the Lake Lahontan area of Nevada, for example, the soil-forming intervals are thought to have occurred over short intervals of time during which both precipitation and temperature were above their present values. The time given for formation of the Churchill Geosol was 5 ky.

The above time resulted in an unusually rapid rate of soil formation for those conditions, and what we knew about rates and processes (Birkeland and Shroba, 1974), so other approaches were explored to explain the soils. From Chapter 8 we learned that soil development increases with precipitation, but also that low temperatures, such as those in the Colorado alpine where soils are frozen for about one-half the year, do not necessarily inhibit soil formation. This therefore raises questions of what happens to soils that remain at the surface during glacial climates, supposedly

Figure 11.2 Stratigraphic position and laboratory data on relict and buried occurrences of the Churchill soil (geosol where buried) in the Lake Lahontan area. *(Data from Morrison, 1964, Table II, and Birkeland, 1968, Fig. 3.)*

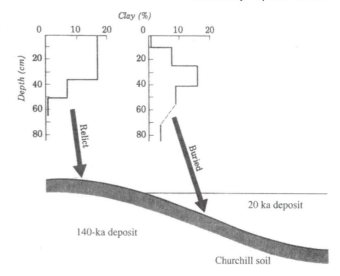

times when soil formation is inhibited. Surface erosion should also be considered as one means of explaining the similarity of a particular surface soil with a buried soil. As more work was done in the region, two other factors changed our ideas about such short intervals. First, the time available for formation of the Churchill Geosol increased about 20-fold (Morrison, 1991). A similar increase in time was suggested for a geosol of a similar age in the Lake Bonneville basin, where soil development rates played a distinct role in suggesting a much longer time span between deposits (Scott and others, 1983). So, sufficient time was available for soil formation. Second, episodic influx of dust from local dry-lake basins was identified as a source for the fines and carbonates in the soils (Chadwick and Davis, 1990). So, what started out as a unique climate to produce a soil in a short period of time, evolved into a climate-induced event (drying of lakes) that exposed dust sources to wind erosion, and the subsequent deposition of the dust into downwind soils over much longer periods of time. In places, the influx of dust could have taken place in less than the total time the surface was exposed to the atmosphere.

■ Using Soil-Stratigraphic Units for Subdivision and Correlation of Quaternary Deposits

Soils can be quite useful for subdivision and correlation of Quaternary successions in many areas (Leamy and others, 1973; Mulcahy and Churchward, 1973;

Morrison, 1968, 1978; Catt, 1986). In this section, examples of both short- and long-range correlation, using soils, will be presented. For the most part, correlations based solely on soils do not come easily. This is because soils change laterally in their properties, because the factors responsible for soil morphology commonly change laterally. Thus, soils of about the same age, or duration of formation, are compared in many different environmental settings. Spodosols are compared with Aridisols, thick soils with thin soils, and high-clay soils with low-clay soils, because these encompass some of the wide variety of soils and soil properties that can be expected at any one stratigraphic horizon.

Soils have long been an important stratigraphic tool in subdividing and correlating Quaternary deposits in the midcontinent; indeed, much of the early work was done there. Few students of Quaternary studies, for example, have not heard of the Sangamon soil (Follmer, 1978, 1983). In Illinois in particular, soils have had a major role in stratigraphic subdivision (Willman and Frye, 1970), and voluminous data have been published on the soils and sediments. Iowa is another area in which soils have played a key role in the development of the Quaternary stratigraphic succession (Ruhe, 1969; Hallberg, 1980).

U.S. Geological Survey and Soil Conservation Service Mapping Projects: Basic Data and Applied Maps

The U.S. Geological Survey has set a high standard in using soils as an aid in the mapping of Quaternary deposits, and these then can be used in land-use decisions (e.g., Costa and Baker, 1981). One landmark study was the La Sal Mountains, Utah, study of Richmond (1962). Combining the pedologic and geological record, Richmond recognized soils at many stratigraphic intervals (Fig. 11.3). Based partly on their development, it was demonstrated that some soils mark major interglacials, whereas others mark fairly short intervals of time during which deposition ceased. The area is one of rapid change in environmental factors over short distances. The mountains reach to about 4170 m, rising about 2100 m above the surrounding Colorado Plateau. Vegetation varies with altitude and climate. The highest summits are in the alpine zone, and, with decreasing altitude, one goes through a subalpine zone of spruce and fir; a montane zone of aspen; a foothills zone of scrub oak, mountain mahogany, ponderosa pine, juniper, and pinyon; and the lowest zone of sagebrush-grass. For each formation (alloformation under the new stratigraphic code), many depositional facies were recognized, and for each soil-stratigraphic unit, several pedologic facies were recognized. Soils formed on deposits of all ages vary in many of their properties, and this is mainly a function of the climate as related to altitude (Fig. 11.4; see also Fig. 10.19). Noncalcic soils formed at the higher altitudes and calcic soils at the lower. The problem of age assignment and correlation in the field, however, is confused by the fact that the altitude at which soils of a particular age change from noncalcic to calcic varies with age of the parent material (Fig. 11.5). Thus, in some places it is necessary to compare the development of the noncalcic facies of a young soil with that of the calcic facies of a nearby older soil to work out the stratigraphic succession.

Morrison (1964) was able to do some exceptionally detailed subdivisions of the deposits of Pleistocene Lake Lahontan, western United States, using soils. Soils are found both at the surface and buried and can be traced rather continuously. They are key markers to the mapping because in many places what are now allostratigraphic units are separable only on the basis of their position relative to the stratigraphically diagnostic soils.

In both of the above areas early work has been changed by later investigators, but this in no way diminishes the impact the early work had on subsequent workers and mappers.

Two more mapping examples can be given. A quadrangle map in the Colorado Piedmont by Ma-

chette (1977) includes a figure depicting the main soil features associated with each deposit, as well as some laboratory data. Alluvial deposits in the southwestern United States have been difficult to date, but irrespective of that, soils have been a major tool in differentiating the deposits (Christenson and Purcell, 1985). The most useful maps have tables of surface features, including soils, that aid the field mapper in reliably recognizing the various stratigraphic units (Table 11.1; Reheis and others, 1993).

Basic maps, such as the above, are essential to environmental assessment of an area, for whatever reason (later). An example is the Parker Quadrangle geologic map, at the south edge of Denver (Maberry and Lindvall, 1972). From that map, maps were generated that were essential to public officials in assessing future development of the area; one derivative map is of landslides and areas of potential landsliding (Mayberry, 1972a), and another delimits areas of the relative swelling potential of geologic materials (Mayberry, 1972b). The latter is important because some housing projects have been put on high-swelling clayey materials (smectite), causing widespread damage (Gill and others, 1996). In another quadrangle map, engineering properties of many of the units are tabulated (Shroba, 1980). Soil type and distribution were essential to the production of the above maps. In places, a soil mapper or soil-trained geologist is better able to delineate areas of problem materials (e.g., swelling clays) than are geologists not trained in soils.

The Soil Conservation Service also has included engineering characteristics of the mapped units in some soil surveys. They produced a basic soil map of the Malibu area, California, and their derivitative maps were important to planners (Kover, 1967). Maps included one on limitation for septic tank filter fields, and one on shrink–swell behavior (important to structures such as homes, roads, etc.). More recent reports tabulate data on the soil series as they relate to potential problem areas for various uses; examples are one for part of the Denver area (Price and Amen, 1984), and such distant areas as the northern Mariana Islands (Young, 1989).

Till and Moraine Correlations, and Comparison with Cosmogenic Isotope Dating

The Illinois State Geological Survey (ISGS) has been instrumental in using soils to decipher the complex

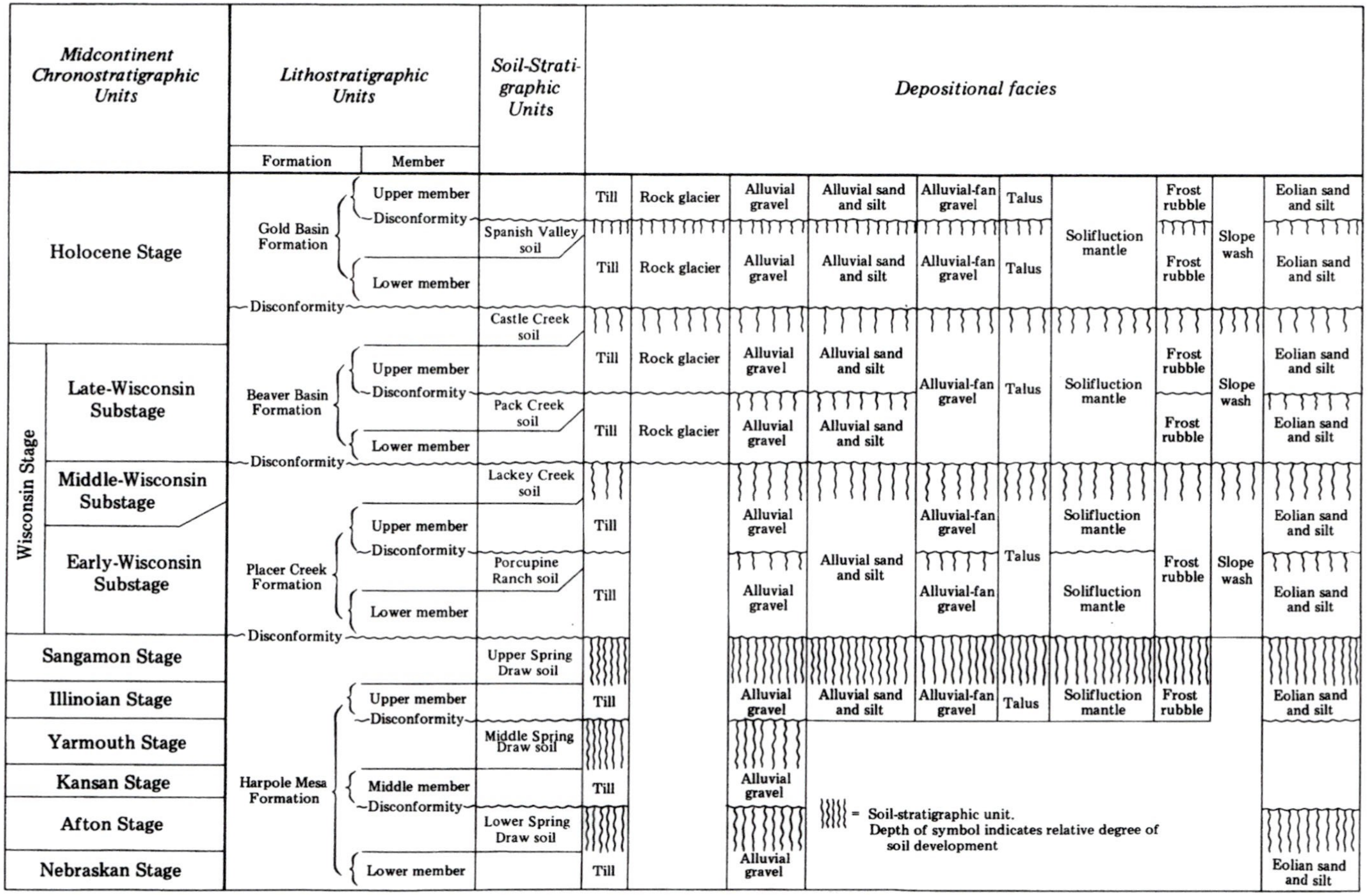

Table. Quaternary stratigraphic units in the La Sal Mountains, Utah.

Midcontinent Chronostratigraphic Units	Lithostratigraphic Units — Formation	Lithostratigraphic Units — Member	Soil-Stratigraphic Units	Depositional facies
Holocene Stage	Gold Basin Formation	Upper member / Disconformity / Lower member	Spanish Valley soil	Till, Rock glacier, Alluvial gravel, Alluvial sand and silt, Alluvial-fan gravel, Talus, Solifluction mantle, Frost rubble, Slope wash, Eolian sand and silt
Late-Wisconsin Substage	Beaver Basin Formation	Upper member / Disconformity / Lower member	Castle Creek soil; Pack Creek soil	Till, Rock glacier, Alluvial gravel, Alluvial sand and silt, Alluvial-fan gravel, Talus, Solifluction mantle, Frost rubble, Eolian sand and silt
Middle-Wisconsin Substage	(Disconformity)		Lackey Creek soil	Till, Rock glacier, Alluvial gravel, Alluvial sand and silt, Alluvial-fan gravel, Talus, Solifluction mantle, Frost rubble, Slope wash, Eolian sand and silt
Early-Wisconsin Substage	Placer Creek Formation	Upper member / Disconformity / Lower member	Porcupine Ranch soil	Till, Alluvial gravel, Alluvial sand and silt, Alluvial-fan gravel, Talus, Solifluction mantle, Frost rubble, Eolian sand and silt
Sangamon Stage	(Disconformity)		Upper Spring Draw soil	Alluvial gravel, Alluvial sand and silt, Alluvial-fan gravel, Solifluction mantle, Frost rubble, Eolian sand and silt
Illinoian Stage	Harpole Mesa Formation	Upper member / Disconformity	Middle Spring Draw soil	Till, Alluvial gravel
Yarmouth Stage				
Kansan Stage		Middle member / Disconformity	Lower Spring Draw soil	Till, Alluvial gravel
Afton Stage				
Nebraskan Stage		Lower member		Till, Alluvial gravel, Eolian sand and silt

≈≈≈ = Soil-stratigraphic unit. Depth of symbol indicates relative degree of soil development

Figure 11.3 Quaternary stratigraphic units in the La Sal Mountains, Utah, and tentative correlation with chronostratigraphic units of the midcontinent. Formations would be classified as alloformations under the new stratigraphic code, and many soils could be geosols. (*Redrawn from Richmond, 1962, Fig. 9 and Table 11.*)

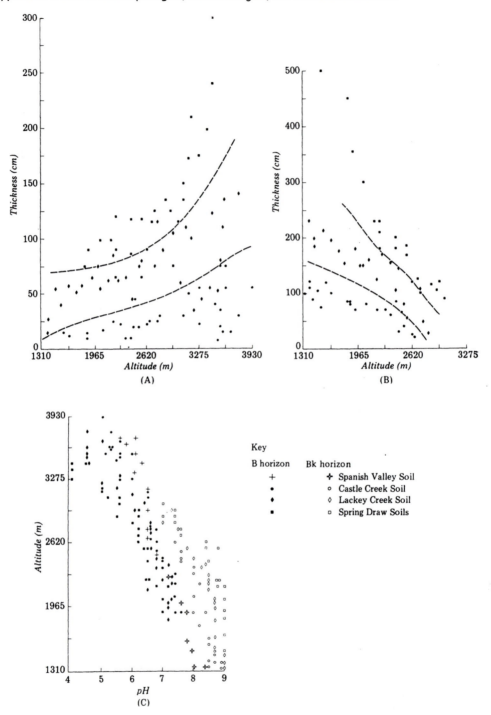

Figure 11.4 Variation with altitude and age in the properties of several soil-stratigraphic units, La Sal Mountains, Utah. Stratigraphic age assignment for the soils is given in Fig. 11.3. (A) Range in maximum thickness of B horizon, some of which at lower elevation contain CaCO₃; (B) range in maximum thickness of Bk horizon; (C) range of pH of B and Bk horizons. *(Redrawn from Richmond, 1962, Figs. 56, 60, and 61.)*

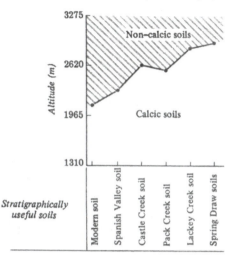

Figure 11.5 Average altitude of the boundary between non-calcic and calcic soils of different age, La Sal Mountains, Utah. *(Redrawn from Richmond, 1962, Fig. 19.)*

Quaternary stratigraphy of the state (Lineback, 1979). Indeed many of the principles concerning soils in stratigraphic studies were developed there, as well as in Iowa (Ruhe, 1969); Olson (1989) points out the important role that soils have played in understanding Quaternary stratigraphy and landscape evolution across the midcontinent. An early result was the publication of a book entitled "Pleistocene Stratigraphy of Illinois" (Willman and Frye, 1970), summarizing the depositional facies, soils, and geological ages. The detail of their subdivision can be seen in a chart (Fig. 11.6). An enormous amount of laboratory data have been published on both deposits and soils in numerous ISGS circulars, of which that in Fig. 11.7 is typical (Follmer and others, 1979). First, the combination of particle size, carbonate, and clay mineralogy can be used to help differentiate and correlate the parent material units as glaciers advanced over different geological terrains at different times depending on ice-flow patterns. These differences show up as texture and mineral differences in all depositional facies. The main soil features are the leaching of carbonates, and the dramatic increase in expandable clay minerals. The Roxana Silt and associated Farmdale Soil are quite thin here, as is the Sangamon Soil; erosion must account for some of this. Also the Farmdale Soil appears to extend into the Sangamon Soil, welding them together into a composite soil. Numerous sections have

been analyzed in detail across the state, giving workers a great deal of confidence in their mapping.

Soils and associated rock-weathering features have long been used to aid in moraine mapping in mountains in the western United States, correlations to type localities, and longer range correlations. Because of the so-called ambiguity of the age estimates using soils, many workers have switched to a more numerical method: cosmogenic isotope dating of surface boulders. A typical area is the Wind River Mountains of Wyoming. Richmond (1965) set the soil-stratigraphic framework, and this has been improved on and quantified by others. Mahaney (1993) has done the most extensive work in the alpine area, and recognizes tills of three Holocene ages. The mountain-front Pleistocene tills are the type localities for the Rocky Mountains, and their soil-stratigraphic relations and more recent subdivisions are being worked on by Hall and Shroba, (1993, 1995; see also discussion in Chapter 8).

Soils and associated surface and subsurface rock-weathering features were used to correlate moraines in several canyons over quite a distance along the eastern side of the Sierra Nevada (Burke and Birkeland, 1979). This was controversial because Burke and Birkeland (1979) used a conservative interpretation to assign the deposits to two age groups (Tioga = O-isotope stage 2, and perhaps a bit older; Tahoe = stage 6), far fewer than the number of groups derived from both previous and later studies. The argument was whether to lump the deposits together into a few groups or split them into more groups. As an aside, my experience has been that most of the succeeding stratigraphic studies in the same area commonly end up creating more stratigraphic units relative to those of the older studies, and the study of Burke and Birkeland created fewer. Most later studies in the Sierra Nevada recognized many more glacial deposits (Gillespie and Molnar, 1995). One cosmogenic isotope surface exposure dating study was undertaken in one of the canyons of interest (Phillips and others, 1990). This study recognized more glacial advances than Burke and Birkeland (1979), but the sequence of moraine deposition that they advocated is agreed on by few glacial geologists, including the original work of Sharp and Birman (1963), as well as later workers. The issues are still being worked on by more exposure dating studies, and at least in one study in a key canyon to the south (Gillespie and others, 1996), the latest subdivision supports much of the conservative

Table 11.1 Mapping Criteria and Age Controls for Alluvial-Fan Deposits[a]

Unit Name (Map Symbol)	Surface Form	Desert Varnish	Soil Development
Late alluvium of Marble Creek (Qfcl)	Unmodified bar-and-swale topography; well-preserved debris flows; sparse vegetation	None	From none to a few weak filaments of $CaCO_3$ on clast bottoms
Middle alluvium of Marble Creek (Qfcm)	Somewhat subdued bar-and-swale and debris-flow topography; moderately vegetated—cholla common	A few clasts with small spots in protected sites	Sandy vesicular A horizon, 2–3 cm thick; sometimes a weak Bw horizon; stage I $CaCO_3$—common weak filaments and spots on clast bottoms
Early alluvium of Marble Creek (Qfce)	Subdued bar-and-swale and debris-flow topography; incipient small areas of pavement—stones not interlocking; moderately vegetated—cholla common	Common, thin spots on many clasts	Vesicular A horizon, 3–5 cm thick—finer grained with moderate structure downwind of playas; Bw horizon; stage I–II $CaCO_3$—stronger in calcareous alluvium
Alluvium of Leidy Creek (Qfl)[a]	No depositional topography preserved; smooth areas of pavement surrounded by rougher pavement; smooth pavement has interlocking but unsorted clasts	Thin but continuous on most clasts; shiny and iridescent in protected sites	Continuous, loamy vesicular A horizon, 5–7 cm thick; Bw or weak argillic B horizon—common thin clay films; stage II $CaCO_3$—stronger in calcareous alluvium
Alluvium of Indian Creek (Qfi)	Commonly deeply dissected; well packed and sorted clasts in continuous pavement; prominent solifluction treads and risers	Continuous, thick varnish on most clasts—often shiny and iridescent	Continuous, silty vesicular A horizon, 5–10 cm thick; moderate argillic B horizon—abundant thin to moderately thick clay films; stage II–III $CaCO_3$—stronger in calcareous alluvium

From Reheis and others (1993).

[a]Age is early Holocene and late Pleistocene; younger deposits are Holocene, older are Pleistocene.

interpretation of Burke and Birkeland (1979). Other soil data support the conservative interpretation (Berry, 1994). The lesson learned here is that soil data do not provide a detailed subdivision of moraines, and they do point to extensive erosion of moraines (Chapter 9), something some cosmogenic daters have to consider more carefully as they evaluate their data. Basically, if people agree that the soil data indicate surface erosion, this same erosion has a high chance of impacting cosmogenic dating data (Hallet and Putkonen, 1994).

We can compare more soil versus cosmogenic dates at a famous outcrop for Sierra Nevada glacial geology, that where the Bishop Tuff (0.77 Ma) overlies the Sherwin Till (Sharp, 1968). A weathered zone exists at the top of the till, buried beneath the tuff, and the question is, how long was the till exposed before burial by the tuff? From our study of the weathered zone and knowledge of the regional soil-formation and rock weathering rates, we put the time at about 100 ky (Birkeland and others, 1980). Cosmogenic dating applied to the same soil came up with 53–67 ky (Nishi-

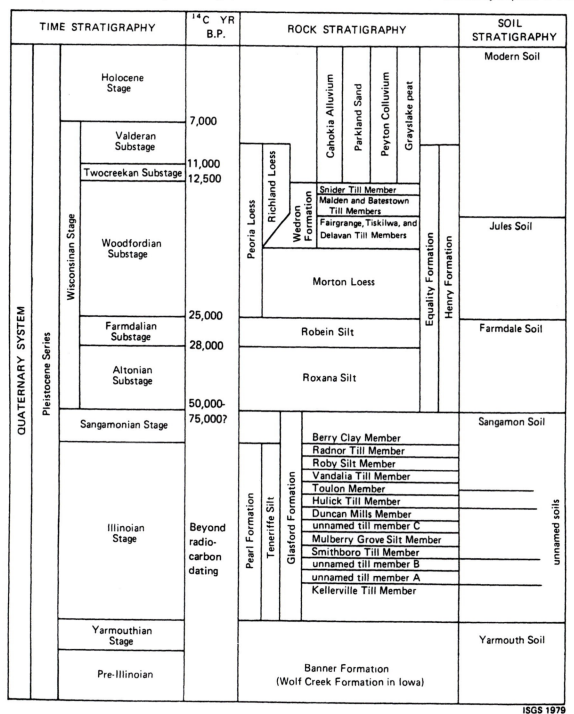

Figure 11.6 Time-stratigraphic, rock-stratigraphic, soil-stratigraphic, and numerical dating of the Quaternary deposits in the Peoria–Springfield area, Illinois. *(Reprinted from Lineback and others, 1979, Fig. 2.)*

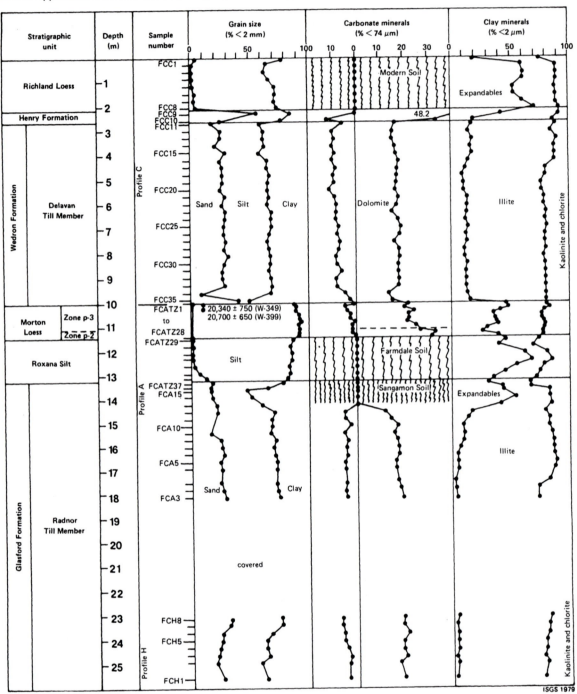

Figure 11.7 Particle-size data, and carbonate and clay minerals for soils and sediments at the Farm Creek Section, Illinois. *(Reprinted from Follmer and others, 1979, Fig. 5.)*

izumi and others, 1989). Thus, both methods, done without knowledge of the other study, came to similar conclusions.

Mahaney (1990) used soil stratigraphy extensively in his long-term study of the glacial history of Mt. Kenya in Africa. The morphology of both surface and buried soils and extensive laboratory analyses were used to aid both in the subdivision of units as well as in the interpretation of geologic history and paleoclimate. Mahaney used many other methods for dating (amino acid, ^{14}C, thermoluminescence, K–Ar, and paleomagnetism), and the importance of soils to these methods was that they provide important information as to the position and nature of the land surface for the times being dated. Only with this combined data set can we obtain the best interpretation of most numerical ages.

Europeans long have used soils in Quaternary stratigraphy and mapping. A recent example is the detailed map of Cremaschi (1987) of the central Po Plain of northern Italy. The main depositional facies are till, and glaciofluvial and loess deposits. Soils on all the units, both surface and buried, were used extensively in the subdivision of all facies and in basin-wide correlations. Billard (1993) has attempted more far-ranging correlations and made comparison of the soils present in these and other northern Italian deposits with those of the China loess plateau. One key to the correlation is the time of development of the well-developed red "ferretto soils" of northern Italy, some of which she puts at greater than the Brunhes-Matuyama paleomagnetic boundary (>0.78 Ma). A second key is the upsection change from dominantly red to less-red soils between loess sections in Italy, France, Hungary, and China.

Woodward and others (1994) used soils in a stratigraphic study of alluvial deposits in Greece, and their assessment of the usefulness of soils is true for the study of any stratigraphic facies (p. 281–282): "Despite recent major advances in the application and refinement of physical dating techniques such as ESR, luminescence and AMS ^{14}C, in the absence of datable materials, weathering profiles may provide the only available means of differentiating between otherwise similar units."

Using Soil Stratigraphy to Date Mountainous Landscapes and Assess Plant Nutrient Relations

Tonkin and Basher (1990) have used soil stratigraphy to understand landform evolution in mountainous

catchments in the South Island of New Zealand (Table 11.2). From a study of the soils they can determine how long particular slopes have been stable, and if their later history involved erosion or deposition. Furthermore, if more than one buried soil sequence is present, they can put them into a time framework, and by calculating the masses involved even obtain estimates on erosion rates. An example of the above comes from a catchment along the crest of the Southern Alps (Basher and others, 1988). Being at a compressional plate boundary, uplift is about 12 m/ky, and the exceptionally high MAP of 11 m erodes so rapidly (11 m/ky) that the landscape is in approximate balance between uplift and erosion. Although vegetation in many areas is used to recognize disturbance, the high precipitation results in such rapid revegetation and growth that vegetation patterns help recognize only very young (ca. 100 years) disturbances. Hence, soils are best for estimating land surface age, and from this it is possible to rank areas as to their relative contribution of eroded sediment. Soil removal is so rapid and widespread that few soils last more than 1 ky, and soil mean residence time is likely in the order of 100–200 years.

Soil fertility should be linked to soil-geomorphic patterns. In New Zealand there has been interest in damage to forests by browsing deer. In one study, forest types are associated with landform, landscape stability, and soils to such an extent that the deer prefer the habitat provided by young, fertile soils formed on relatively unstable landforms (Stewart and Harrison, 1987). In contrast, infertile, older soils on relatively stable landforms are a poor habitat for deer. Once these relations are established general conclusions can be drawn on the underlying soil by study of forest, no easy task in this area of high relief, steep slopes, and dense vegetation due to an MAP of 6–8 m. One of the co-authors told me we no longer have to dig holes to study soils; all we have to do is examine the vegetation and count the deer!

Using Soils to Date Mass Movements

A few studies use soils to date mass movements, and from these we can infer some applications. If the feature is a slump, the soil related to the block could be carried piggyback and not relate to the time of disturbance. If, however, movement was by flow (debris flow, etc.), the material and overlying soil are mixed up, soil formation begins at the time of stabilization,

Table 11.2 Derivation of Eroded, Composite, and Compound Soil-Profile Forms, Dry Acheron Stream and Ryton River, Eastern Basin-and-Range Subregion

Episodic Erosion Eroded Soils		Original Soil	Episodic Deposition Composite and Compound Soils		
					A
			A	A	C
	C	A	2Ab	2Ab	2Ab
		C	2Cb	2Cb	2Cb
					A
				A	2Ab
			A	C	2Cb
		A	2Ab	2Ab	3Ab
	Bw	Bwj or Bw	2Bwb	2Bwb	3Bwb
C	C	C	2Cb	2Cb	3Cb
					A
				A	2Ab
		O or A	A	C or Bwj	2Cb or Bwjb
	Bs	Bs or Ej	2Bsb	2Bsb	3Bsb
Bw	Bw	Bw	2Bwb	2Bwb	3Bwb
C	C	C	2Cb	2Cb	3Cb
					A
				A	2Ab
		O or A	A	C or Bwj	Bwjb
		E	2Eb	2Eb	3Eb
	Bs	Bs	2Bsb	2Bsb	3Bsb
Bw	Bw	Bw	2Bwb	2Bwb	3Bwb
C	C	C	2Cb	2Cb	3Cb

Increasing soil development →

←— Increasing erosion   Increasing deposition —→

From Tonkin and Basher (1990, Table 7) Soil-stratigraphic Techniques in the study of soil and landform evolution across the Southern Alps, New Zealand. Geomorphology 3, 547–575. © 1990, with kind permission from Elsevier Science, The Netherlands.

and so the soil could be used to estimate its age. With any landslide, buried soils at the toe of the feature give an estimate on the time the landscape was stable before burial, and ^{14}C dating the topmost part of the A horizon could put limits on the time of movement.

M.E. Berry (unpublished student report, 1981) shows how soils might be used to date a slump located on the terrace edge of a 600-ka alluvial deposit south of Boulder, Colorado (Fig. 11.8). After deposition of the alluvium, a strongly developed red Bt/stage III K profile formed (soil A); this required several 100 ky (see Fig. 8.27). Then the slump occurred, rotating the soil out of its original horizontal orientation. The slide occurred about 100 ka, as shown by the development of soil B formed in postlandslide colluvium and the buried soils in the colluvial wedge deposited in the basin between soil a and the buried landslide

slip planes. The methods used to date this feature are similar to those used to date faults (later).

Estimating ages of colluvium on slopes is also possible with soils. Many examples of this are given in Chapter 9, because it is necessary to understand the influence of slope position on soil properties to estimate ages of such soils on slopes. An example from the Allegheny Plateau of northern Pennsylvania (Waltman and others, 1990) is given here as the area lies just beyond the glaciated area, so the geomorphic history is one of alternating slope stability and instability. The lower part of the profile is a buried soil, characterized by a red (2.5YR), clay-rich (27–49% clay) Bt horizon with highly weathered clasts. In contrast, the overlying colluvium is 1–2 m thick , and the soil characterized by a brown (10-7.5YR) Bw horizon with 16–31% clay. Fe_d of the buried soil is about double that of the surface soil. In the context of the soil-geomorphic history of the region, the young colluvium seems to correlate mainly with the last glaciation (O-isotope stage 2), the older colluvium is pre-O-isotope stage 5, and the red soil is formed at least during O-isotope stage 5. It is interesting that the brown–red soil contrast is similar to that in many till sequences in the central and eastern United States that represent an age span similar to the one here.

Floodplains, River Terraces, and Flood Predictions

Soils have been used to estimate ages of a wide variety of fluvial deposits. Desert alluvial deposits have been notoriously difficult to date as so little datable material is present. Soils play a major role in estimating ages of the deposits as well as in past climate reconstruction, as shown by Bull (1991) (see also Chapters 8 and 10).

Soils can be used to subdivide parts of the floodplain, determine the limits of the next terrace above it that is beyond most flooding, as well as help guide sampling and interpret ^{14}C dating. Skully and Arnold (1981) do this for the Susquehanna River Basin of New York. The floodplain is subdivided

into three soil-stratigraphic units, with the younger having a C horizon alone, the next older having an A/C profile, and the oldest having a more complex profile: A/C/Ab/Cb. These are <1 ka. In any floodplain study it is necessary to know the frequency of flooding associated with each of these units. The first terrace above the floodplain encompasses the rest of the Holocene, it has no C horizon material at the surface but instead has an A/B profile, and there are as many as five buried A/B and A/C profiles at depth. In contrast, the second terrace (Pleistocene) has a less complicated A/B/C profile, either because environmental conditions were such that buried soils were never present or pedogenesis has altered and rearranged the sequence of buried soils into a single soil with a simpler morphology. In most areas, higher terraces of greatly older ages would have A/Bt/C profiles.

The Bureau of Reclamation has applied these concepts in evaluating the safety of Bradbury Dam, California (Ostenna and others, 1996). Their calculations for the probable maximum flood resulted in a calculated peak discharge of 460,900 cfs, and if this happens it would do structural damage to the dam, as the spillway capacity is only 160,000 cfs. The recurrence of such events is not well constrained, being between

Figure 11.8 Soils involved in a landslide near Boulder, Colorado. A buried soil formed at the former surface of 0.6-Ma alluvium and is a red Bt/stage III K (white material sloping down from left to right, center) profile that was rotated from the horizontal during slump movement. To the right is a zone of multiple slip surfaces (dip steeply to the left) at the head of the slump; some surfaces cut deposits thought to be Holocene in age. Colluvium fills the depression created by the slump, between the top of the buried soil and the slip surfaces. Development of surface soils and buried soils in the colluvium help to bracket the approximate time of landsliding at near the boundary between O-isotope stages 5 and 6. *(Data and photograph from M.E. Berry, 1981, unpublished student report; outcrop is 4.6 m high.)*

<100 years and 1 My. Ostenna and others decided to take a soil-stratigraphic approach to see if events of such magnitude happened. Terraces exist along the river at various levels, and the soils and stratigraphy of the deposits underlying the terraces are the key to whether inundation has occurred. Historic floodplain deposits have no soil, older floodplain deposits (0.5–1 ka) have historic alluvium overlying a buried A/C soil; terrace 1 (2.5–5 ka), where beyond the influence of the historic flood deposits, has soil profiles that range from A/C to A/B/C, terrace 2 (15–20 ka) has soil profiles that range from A/C to A/Bt/C, and terrace 3 (>60 ka) is characterized by A/Bt/C profiles. Regional soil correlation helped with age estimation of the pre-floodplain deposits. Terrace ages put bounds on the paleovalley dimensions for calculating paleodischarges, and the conclusion was that the chances of a flood equal to the spillway discharge capacity is less than 1 in 1 Ma. This suggests that the probable maximum flood has an exceedingly low chance of happening, and that alterations to the design of the dam do not need to be taken, thus saving taxpayers millions of dollars.

Long-Range Correlation, Including Loess and Deep-Sea Sediments

The methods of long-range correlation using field soil data were set forth by Richmond (1962) and Morrison (1964, 1967). The first task in any one area is to rank the soils in the local succession on the basis of their relative degree of profile development; some will be weakly developed, others moderately developed, and still others strongly developed. Ranking on development will not always be greater with age of the deposits, because some weakly developed soils can form and be preserved for a long time as buried soils. In the other area to which correlation is to be made, the soils are similarly ranked, again with respect to that local succession. Because the environment of each local succession will not be identical, the time-equivalent soils in one succession may bear little resemblance to those in another. As an example, a 12-cm-thick, weakly developed Aridisol in one environment may be correlated with a 50-cm-thick, moderately developed Alfisol in another environment. The soils and deposits in both areas are then matched. Commonly, the basic framework is provided by the youngest strongly developed soil in each area. If the timing of the periods of deposition and soil formation are ap-

proximately the same, and deposition occurred during the glaciations, the youngest strongly developed soil often is the soil that formed mainly during the last major interglacial in the midcontinent, the Sangamon. All other strongly developed soils of pre-Sangamon age also are matched assuming they too may mark major interglacials. The weakly and moderately developed soils that lie stratigraphically between the strongly developed ones can then be matched.

There are several problems with the above proposal for using soils in correlation. One is that the ranking on development is qualitative and relative within a sequence of soils in a particular environment; hence, one soil in one environment might be ranked as weakly developed, but the same soil in another environment could be strongly developed. The other is that laboratory data are not fitted into the scheme. Since this important work we know much more about how soil systems develop in different settings (Chapter 8), as well as pedogenic gradients, and we are in a much better position to make more accurate correlations.

Several methods can be used for more defendable long-range correlations using soils. One of the best successes in this regard is the chart of Machette (1985), showing the attainment of various carbonate morphological stages with time and environmental conditions (Fig. 8.27); in short, stage equates to age. One could also use the mass of carbonate that is associated with morphological stages for correlation, as demonstrated by Scott and others (1983) when estimating the hiatus between Quaternary lake deposits in the Lake Bonneville basin. Ponti (1985) used PDI and other soil indices to correlate deposits in the Mojave Desert with deposits in the San Joaquin Valley, for which there were numerical ages.

The need to assemble both geological and pedologic data for the most meaningful correlation cannot be overemphasized. It is necessary to have a thorough knowledge of the geological history in the area, including the kind and rate of geological processes that have been operative. Soils have to be studied in the field for their stratigraphic position and for their morphology and history, and they should be analyzed quantitatively for their critical properties. Only then can all the data be assembled to give the best reasonable correlations. One should avoid merely counting back in time and fitting all deposits and soils into a so-called known standard sequence, for we now know that few sections contain all possible deposits and soils, as erosion has removed some of them.

Thick sections of loesses and soils are favored sites for global change studies on land, long-range correlation, and correlation to the marine O-isotope record. An early example is the loess record of central Europe as discussed by Kukla (1970, 1977). Detailed study was made of loesses and their associated fossil content and buried soils. A marked glacial–interglacial cyclicity is apparent, and there are many similarities in the paleoenvironment deduced from both soils and the paleontological record. One cycle of deposition would be depicted by the diagram of Pye (Fig. 7.12), with the source areas being multiple glacial-outwash floodplains across that part of Europe. During interglacials, the loess deposition essentially stops, and soil formation ensues. These cycles were then shown to correlate well with those of the marine oxygen-isotope record to demonstrate a correlation between the marine and terrestrial records. Kukla argued that, of the land records, the loess–soil record is one of the best, since it contains the detail necessary for such correlation.

The loess of central China has been intensely studied (reviewed in Derbyshire, 1983; Pye, 1987; Liu, 1988). The loess forms a huge plateau southwest of Beijing. Deposition began about 2.4 Ma, when loess was deposited on top of the Red Clay, a regional marker that is a soil–sediment complex. As the depositing winds blow from the northwest and sweep across a huge desert, the sand component forms dune systems, and just downwind the silt forms loess, and the latter deposits both fine and thin to the southeast, much as depicted in Fig. 7.12. Thicknesses can exceed 300 m. Multiple soils occur within the loess (Fig. 11.9). The prevailing model is that loess is deposited during times of global glaciation, and that deposition is slow enough during the interglacials and interstadials that soils, or at least cumulic soils, formed. The climate during times of deposition was dominated by the dry, cold winter monsoon, and that during times of soil formation by the wet, warm summer monsoon (An and others, 1991). Also, the morphology of the soils northward in the sections can be used to infer the northward extent of the summer monsoon dominance for particular time intervals (see Porter and others, 1992). This loess/soil sequence was repeated many times throughout the Quaternary, and based on the paleomagnetic correlations, workers feel that the system is predictable and correlation is relatively easy as the same sequence is repeated from site to site.

The soils vary in development with position in the sections. The loesses are calcareous, and most soils are reddish, marked by carbonate leaching, and some have Bt horizons (Fig. 11.10; see also Bronger and Heinkele, 1989, Fig. 6). From the top down, the first most strongly developed soil is S5, formed from about 600 to 500 ka (Porter and others, 1992).

Many workers have noted that magnetic susceptibility varies between soils and loess in a very regular pattern, being highest during times of soil formation. Magnetic susceptibility is proportional to the amount of strongly magnetic minerals in either soil or loess. There are several reasons for the high magnetic susceptibility; much of the work on this is reviewed by Reynolds and King (1995). The conclusion is that the higher values in the soils are mainly due to the pedogenic formation of ultrafine ferrimagnetic grains, such as magnetite and/or maghemite (Banerjee, 1995). Hence, the magnetic susceptibility signal is one of pedogenesis following loess deposition, and one can speculate that the intensity of the signal may, in some way, be related to the intensity of weathering and duration of the soil-forming interval. Banerjee (1995) makes a case for paleorainfall estimations using these data.

Many workers have noted that river terraces underlie the loesses, and that the older terraces are mantled by more and older loess sheets (e.g., Kukla, 1977). Porter and others (1992) have used these relations to date terraces in China (Fig. 11.11), make terrace correlations along the rivers, and apply this to the dating of tectonic events.

Many studies have noted that the loess–soil record has a pattern that seems to match that of the marine O-isotope record, suggesting that both are linked to global glacial–interglacial cycles. Because the downwind portion of China loess is atmospheric dust transported across the north Pacific Ocean, Hovan and others (1989) estimated eolian fluxes to the marine environment. Their correlation indicated high flux during times of loess deposition in China, and low flux during times of soil formation, thus tying the land record to the marine record. This dust, as well as dust from other sources, blanketed Pacific Islands with an eolian mantle (Chapter 8).

A long-term loess–soil sequence that is stacked vertically is exposed in aridic eastern Washington, United States (Busacca, 1989). The thickest loess is 75 m, and the base may be 1.5 Ma or older. One 26-m-thick section, deposited over the last >0.8 Ma, has 19 or more paleosols. B horizons are mostly Bw, but a few are Bt. Calcic, petrocalcic, or duripan horizons are present at depth, or at the surface if the overly-

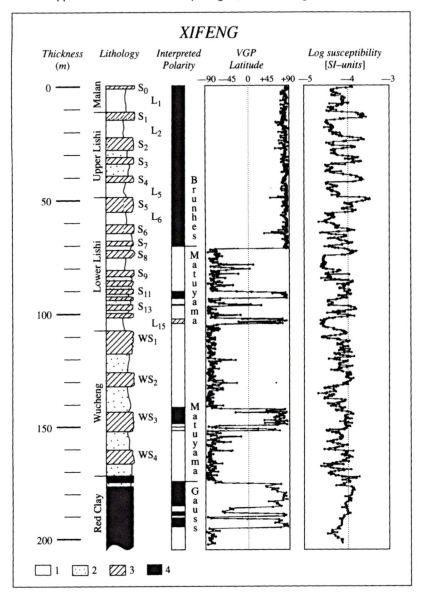

Figure 11.9 Data for the stratigraphic section at Xifeng. On the left side of the lithology column are the stratigraphic names, and on the right are the soils (S) and loess (L) units. They are numbered from top to bottom (S5 formed from L6, WS2 from WL2, etc.). Key to numbered boxes: 1, unweathered loess; 2, weathered loess; 3, soil or soil group (red clay should have this pattern); 4, normal polarity magnetozone. VGP is virtual geomagnetic pole position, with normals to the right and reversals to the left. Righthand column is log of the magnetic susceptibility value. *(Redrawn from Kukla, 1987, Fig. 4, Loess stratigraphy in central China. Quaternary Science Reviews 6, 191–219, © 1987, with kind permission from Elsevier Science, The Netherlands.)*

ing horizons have been stripped off by erosion. Inherited carbonate ranges from 1 to 5%, whereas pedogenic carbonate ranges from 7 to 40%. These sections have yet to be correlated on a global scale. Paleoclimate information can be gleaned from the present-day MAP gradient of 28 cm (west) to >70 cm (east), which results in A/Bw/Bk profiles in the west to A/Bw/E/Btx profiles in the east. Preliminary interpretations on paleoclimate suggest overall interglacial climates not too different from the present,

but some interglacials may have experienced a different climate where the above morphological packages shift laterally down section.

Loess has been studied in the Mississippi River basin for many decades, and soils have been instrumental in their subdivision (Follmer, 1996). The latest review is by Rodbell and others (1997; see also Fig. 11.6), and their data are rather typical for the region. At the top of the section is the modern soil formed in the Peoria Loess (O-isotope stage 2). They

Figure 11.10 Data for the stratigraphic section at Luochuan. (A) Depth below the top in meters. (B) Column symbols denote condition of loess and soils: a, weakly weathered loess; b, moderately weathered loess; c, strongly weathered loess; d, dark soil; e, weakly developed drab soil; f, drab soil; g, leached drab soil; h, strongly developed drab brown soil. (C) Loess and soils units as in Fig. 11.9. (D) Paleomagnetism. Righthand column is CaCO₃ percentage of soils (low values = leached) and loess (high values = parent material content). *(Reprinted from Kukla, 1987, Fig. 11, Loess stratigraphy in central China. Quaternary Science Reviews 6, 191–219, © 1987, with kind permission from Elsevier Science, The Netherlands.)*

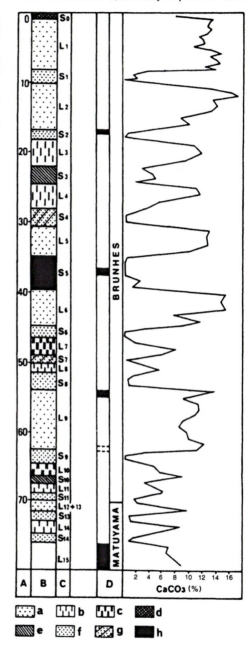

identify it as an A/Bw/C profile, but in some other areas it is A/Bt/C (Fig. 7.14). The next soil down is the buried Farmdale Soil, with either a Bt or Bw horizon, and this serves as the marker for the top of the Roxana Loess. Under the latter is the strongly developed Sangamon Soil with a thick Bt horizon, formed in Loveland Loess. The Sangamon Soil is a regional soil-stratigraphic marker, is present in many depositional facies, and has been the focus of many detailed studies (Ruhe and others, 1974; Follmer, 1978, 1983). John Frye, former Chief of the Illinois State Geological Survey, used to say that because it was so prominent he could walk the Sangamon Soil from the type locality south into Texas. Missing is the weakly developed Jules Soil (Fig. 11.6), which is stratigraphically above the Farmdale soil and not always present.

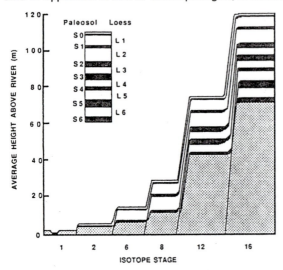

Figure 11.11 Alluvial terrace sequence for two river valleys near the southern margin of the Loess Plateau, China. The soil and loess units (numbering system as in Fig. 11.9) increase in number with age of terrace, and are the basis for assigning O-isotope stage ages to the terrace deposits. *(Reprinted from Porter and others, 1992, Fig. 5.)*

Correlation of still older loess using soils is difficult because many sections are separated from one another, and erosion has removed some units. Correlations based solely on soils and counting down the local succession have not always been verified when numerical dating and other methods have been used.

Rodbell and others (1997) point out that the young loess sheet has the expected exponential thinning away from the source (see Fig. 7.13), whereas older loess units do not. This difference is attributed to localized erosion of the older loesses. Likewise, Jacobs and others (1997) attribute erosion to the lack of some pre-O-isotope stage 6 loesses in the driftless Area of the Upper Mississippi Valley; others may be so weathered that they are difficult to recognize as loess. In eastern Colorado it is common to see the O-isotope stage 2 loess on bedrock. Although we would expect to see older soil–loess couplets downsection, this is seldom seen. This has important ramifications to the Quaternary stratigraphy; are we seeing evidence for nondeposition of loess, or are we seeing evidence of intense erosion over a large region?

It is common to have dune systems upwind of loess deposits, and soils in the dune sands can aid in their subdivision and age estimation (e.g., Jorgenson, 1992; Muhs and others, 1996; see also methods in Wells and others, 1990a). Because the dunes are prone to erosion, the in-

terpretation of the soils is not always easily made. Young soils are A/C profiles and older ones are A/Bt/C profiles, and soil development correlates with the duration of landscape stability (e.g., Fig. 2.5). The soils are times of landscape stability, and the underlying deposits can be either (1) a dune sand whose age immediately precedes that of the soil or (2) a dune sand much older than the time of initiation of the soil. In the latter case, wind erosion could have truncated an old dune sand (and any soils), and the soil formed on still older sediment once erosion ceased. A further complication in an eroding system is that downward erosion can be halted by a well-developed B horizon. Thus it is difficult to correlate the dune sand–soil sections throughout a dune field. Finally, because of the parent-material differences, correlation with the loess–soil sequences downwind is not easy.

■ Using Soils to Date Tectonic Activity

Dating Faults

Faults cut Quaternary deposits in many areas of active faulting, and in places, soils help to decipher the age of faulting. The best success is with normal faults, for there the sediments and soils are vertically displaced. The impetus for many studies was to determine the seismic hazard of known faults near nuclear power plant sites (e.g., Shlemon, 1978), and, in later years, the proposed high-level nuclear waste repository at Yucca Mountain, Nevada (later). The reason for using soils to help with, or verify, dating is that material for radiocarbon dating either is not present or is not present in the most useful stratigraphic position, or the dated materials might be contaminated. Studies have been conducted in many environments and tectonic settings (Kirkham, 1977; Borchardt and others, 1980a,b; Douglas, 1980; Swan and others, 1980; McCalpin, 1982, 1996; Gerson and others, 1993; Mueller and Rockwell, 1995; Amit and others, 1995, 1996). One by Crone and others (1992) is unusual in that ferricrete is involved in the faulting. The U.S. Bureau of Reclamation has undertaken a study of the long-term seismicity near the location of their dams in the western United States, and soils play an integral role in identifying the stratigraphic units involved and deciphering the trench stratigraphy (e.g., Piety and others, 1992).

Commonly, the work is done in artificial exposures, such as backhoe trenches. Detailed maps are made of the trench walls showing the relationships of the deposits and soils to the fault(s). Sediments and soils overlying the fault, but not displaced by it, are younger than the fault, whereas those cut by the fault predate one or more episodes of faulting. There are a variety of sediments deposited on the footwall block, whose source is the fault scarp. Nelson (1992) provides a convenient description of lithofacies assemblages, common to normal fault terrain, that can be used in trench-wall mapping.

Using soils in these studies is not easy. First, soils have to be recognized as such, and those on both blocks studied, as well as those on the scarp. Many of them on the downthrown block are buried, many have complex parent materials, and the relations of soils on the downthrown block and on the scarp can be complex. Understanding these relations is essential to unraveling the total history of the site (McCalpin and Berry, 1996). McCalpin and Berry recognize four soil catena phenomena on multiple-event fault scarps (Fig. 11.12). (1) There is truncation of the summit soil on the upthrown block. Parts of the eroded soil become inherited soil properties in the downslope colluvium, and these inherited soil properties are taken into account in assigning slope soil ages. (2) Soil beyond the reach of the colluvial wedge represents all of the time since deposition of the associated prefaulting deposit, but its development decreases toward the fault scarp where it is buried under the colluvial wedge. The decreased development of these buried soils corresponds to the decreased time available for soil formation. (3) Soils (all buried except for the surface soil)

are developed on individual colluvial wedge sediments, and these merge downslope to form a cumulative soil at the toeslope of the scarp catena. (4) Soils along the surface, and those buried and related to individual colluvial units, generally thicken and increase in development downslope. Amit and others (1995) warn us that all soil–colluvium packages might not be fault related, as some could relate to other slope processes. Perhaps climatic events could trigger erosion and colluvial deposition in places. One should be able to sort this out with detailed work and unambiguous exposures.

The downslope changes in soil development along the modified fault scarp surface, and the reasons for these changes are the same as those for catenas (Chapter 9). Berry (1990) demonstrates the methods using Sierra Nevada fault-scarp catenas. In one place she sees no buried soils in thick colluvium, probably due to the combination of slow regional soil-formation rates and short time intervals between faulting events.

In areas with no age control, fault-scarp catenas can be placed into a time frame by study of a chronosequence of nearby river-terrace-scarp catenas. One problem, however, is that the two different catenas are not always comparable. Two studies compare fluvial-terrace catenas with nearby multievent fault-scarp catenas, using PDI and laboratory data, one from Utah (McCalpin and Berry, 1996) and one from southern Israel (Amit and others, 1996). In Utah, there are the usual increases in values downslope in both catenas, and the footslope properties for the fault scarp are

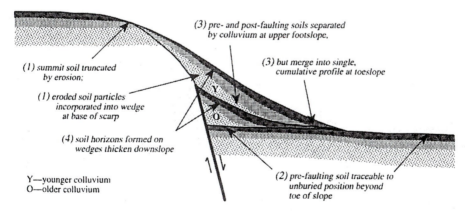

Figure 11.12 Schematic diagram showing four soil–catena phenomena on a fault scarp with several episodes of fault movement. Patterns depict generalized vertical arrangement of soil horizons, with thicker soils being either older or those at the lower end of catenas, and thinner soils being either younger, at the upper end of catenas, or eroded. *(Reprinted from McCalpin and Berry, 1996, Fig. 5. Soil catenas to estimate ages of movements on normal fault scarps, with an example from the Wasatch fault zone, Utah. Catena 27, 265–286, © 1996, with kind permission from Elsevier Science, The Netherlands.)*

greater than those for the terrace scarp; this latter trend is explained by having properties inherited by the colluvium from the eroded soils in the upthrown block (clay being a common one). One useful parameter they define is the footslope correction factor (F), which is the ratio of the sum of soil development index values (PDI or clay) of the colluvial-wedge soils at the footslope to that of the summit soil minus that of any prefaulting soil buried at the footslope. The ratio is used to estimate development times of individual soils, and thereby paleoseismic events. In Israel, for contrast, the terrace-scarp soil increases in PDI at the footslope, whereas PDI remains at a low level downslope in the fault-scarp catena. The reason for the latter is that renewed faulting exposes mainly unaltered material, and it was transported downslope, keeping the soils young. One important factor is the thickness of soils exposed

in the upthrown block relative to the scarp height. In more humid environments the soils are richer in clay and thicker, so more of this soil material in the upthrown block has the opportunity to be incorporated into the downslope colluvium. By comparison, thin aridic soils contribute less pedogenic materials to the colluvium, and if pedogenic gypsum is one of the materials, it might be leached from the soil in time.

A model study of Quaternary faulting, using calcic soils west of Albuquerque, New Mexico, indicates how one can calculate recurrent fault movement over the past 0.5 Ma (Machette, 1978). Analyses were made both on the percentage and amount (grams) of carbonate for each soil profile. Because of recurrent normal down-faulting, several buried soils are present just east of the fault (Fig. 11.13, locations c, d, and e); materials shed from the fault scarp buried each soil soon

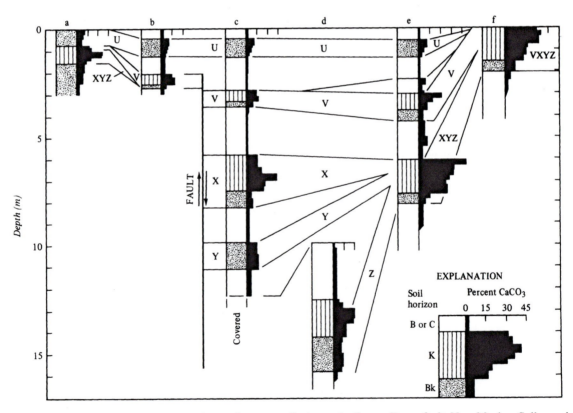

Figure 11.13 Position of soils with depth along a line perpendicular to the County Dump fault, New Mexico. Soils are denoted by u, v, x, y, and z, from younger to older. Combinations of letters imply that individual soils have merged laterally to form one composite soil (e.g., xyz) with a carbonate content that approximately equals the cumulative amount of the included soils. Letters (a, b, c, d, e, f) at the top of the figure denote position of the soils in the field, shown to scale in the lower diagram of Fig. 11.14. *(Redrawn and modified from Machette, 1978, Fig. 6.)*

Figure 11.14 Schematic cross sections showing the sequence of events for the last 500 ka in the vicinity of the County Dump fault, New Mexico. Depositional units are I (old) to V (young), soils are the same as in Fig. 11.13, and horizontal spacing of the hachure pattern is proportional to the duration of soil formation. Letters a, b, c, d, e across the top of the lowest diagram match those across the top of Fig. 11.13. *(Redrawn from Machette, 1978, Fig. 9.)*

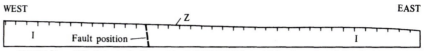

a. 400,000 Years B.P. Prior to fault event 1. Soil Z (100,000 yr old) has formed, and then fault event 1 took place at the indicated position

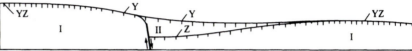

b. 310,000 Years B.P. Prior to fault event 2. Soil Z buried by unit II following fault event 1, surface restabilized, and soil Y (90,000 yr old) has formed

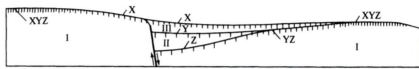

c. 120,000 Years B.P. Prior to fault event 3. Soil Y buried by unit III following fault event 2, surface restabilized, and soil X (190,000 yr old) has formed

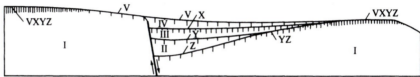

d. 20,000 Years B.P. Prior to fault event 4. Soil X buried by unit IV following fault event 3, surface restabilized, and soil V (100,000 yr old) has formed

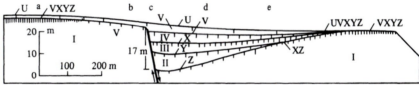

e. Present Day Soil V buried by unit V following fault event 4, surface restabilized, and soil U (20,000 yr old) has formed

after movement. Farther east, the buried soils merge to form one surface soil (location f); west of the fault, the soils also merge to form a single composite profile (location a). Because some of the carbonate is primary, it was subtracted from the total carbonate to calculate the pedogenic carbonate in each profile. Geological relationships suggest that the geomorphic surface cut by the fault is about 0.5 Ma. Dividing the total pedogenic carbonate, as represented by the sum of pedogenic carbonate of the buried soils, by 0.5 My gives a long-term average rate of accumulation of pedogenic carbonate. Knowing the amount of pedogenic carbonate in each buried soil, one can estimate the duration of soil formation between recurrent episodes of fault movement. Four faulting events are recognized

over the past 0.5 Ma, the most recent of which probably took place about 20 ka (Fig. 11.14). Individual fault displacements ranged from about 2 to 8 m, and the recurrence intervals range from 90 to 190 ka. Two fairly reasonable assumptions were used in this study. One is that the rate of carbonate accumulation has been relatively constant. The other is that the amount of time encompassed by erosion of the fault scarp and burial of the soil on the down-thrown side is negligible relative to the amount of time necessary for surface stability and soil formation. Similar studies can be done in other parts of the western United States where calcic soils exist. Age estimation could be done knowing the grams of $CaCO_3/cm^2$/profile for the soils of interest, and either knowing or estimating the regional rate of accumulation of pedogenic carbonate (Fig. 8.27).

In another New Mexico study, Machette (1988) used masses of both pedogenic clay and $CaCO_3$ to estimate episodes of faulting. Key trenches were studied in each of the fault segments (Fig. 11.15) to demonstrate that each segment has its own, unique history.

Episodic thrust faulting along the coast of northern California was dated with the help of soils (Fig. 11.16; Burke and Carver, 1989, in Carver and McCalpin, 1996, p. 207–208). A thrust fault at shallow depth is expressed at the surface as an overturned fold involving a Pleistocene marine terrace. Episodic movement on the thrust produced coseismic growth of the fold. Material was shed off the steepened toe of the thrust and buried the preexisting soil downslope. This happened six times. Two ^{14}C dates constrained the age of one of the movements, and soils were used to date the

rest. PDI was used to show that periods of quiescence varied between about 3 and 6 ky. Parent material for the PDI calculation was difficult to assess, so for each soil many properties of the next underlying soil were used as that soil would have been exposed on the scarp immediately after movement, and its erosion would form the downslope colluvial material.

Soils can also be used to help date other stratigraphic features associated with earthquakes. During the large earthquakes in the New Madrid (Missouri) seismic zone in the early 1800s, saturated near-surface sand layers liquified and moved upward toward or to the surface, crosscutting the overlying stratigraphic units and soils. Where they surface they form a circular sand body called a sand blow. In studying the soil stratigraphy of the blows and parent terrace deposits, Rodbell and Schweig (1993) concluded that all were formed during the historic earthquakes, and that

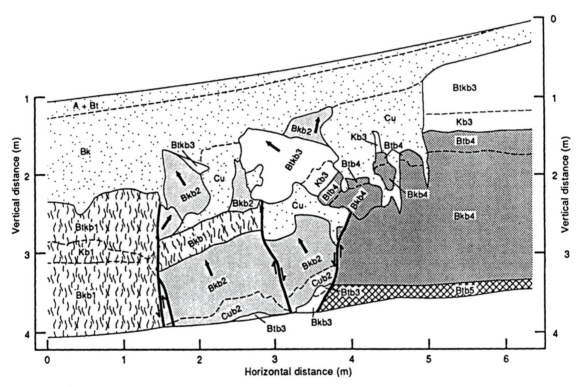

Figure 11.15 Map of stratigraphic units and soils in a trench across the La Jencia fault, New Mexico. The oldest unit is to the right and consists of a sequence of alluvial-fan deposits and buried soils. Fault at 3.5 m horizontal distance displaced the above units below the base of the trench to the left of the fault. Intact blocks of soil rotated to the left (arrow is original up direction). Because a free face is necessary for rotation to occur, the base of rotated blocks gives a minimum measure of the height of the fault-scarp free face following movement. Younger faults between the 1.5 and 3.5 m horizontal distance displace younger units and soils, and some blocks of them are also rotated. *(From Machette, 1988; redrawn by Birkeland and others, 1991, Fig. 6-14.)*

Figure 11.16 (A) Map of stratigraphic units and soils in a trench across one trace of the Mad River fault, near Humboldt Bay, northern California. Blind thrust at depth shows up at the surface as an overturned fold of the marine terrace platform (heavy line), and overlying sand and gravel. Colluvial units each have a stone line at the base (stones shed from toe of thrust following each movement). During quiescence, soils form and are denoted from 6 (oldest) to M (modern). Left side of older colluvial units and soils are also deformed. (B) Total and incremental (between times of soil development) slip history derived for the site using ^{14}C dates and soil PDIs. The long-term slip rate is derived from the approximately 10 m of slip in 30 ka. *(Redrawn from Carver and McCalpin, 1996, Fig. 5.13, © 1996, with permission of Academic Press Inc., Orlando, Florida.)*

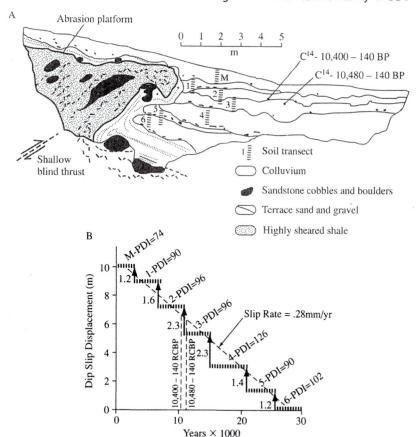

there were no older liquifaction features. This suggested that comparable earthquakes had not occurred in the region for at least 10 ka.

Roy Shlemon is one of the few consulting geologists relying heavily on soils in geologic hazard assessment. I cite one of his later reports (1986) as an example of the interpretations that can be made from field examination. He was called in to help assess Quaternary movement on a bedrock fault near the proposed site of the Two Forks dam, near Denver. Soil-stratigraphic relations of mainstream and fan deposits, and surface and buried soils in trenchs and roadcuts suggested ages that span 200 ka. As the units are not cut by the fault, this is the minimum time for tectonic quiescence.

Dating Folds

Soils also can be used to date geomorphic surfaces that have been folded. River terraces serve as a good example of this. Terraces might converge (e.g., Sug-

gate, 1965; Milne, 1973) or diverge downstream (Reheis, 1985) for a variety of reasons, only one of which might be tectonic. If one can be assured that the pattern is tectonic, broad age ranges can be determined to bracket the time of deformation. Ages can be estimated from the soil formed in the uppermost terrace materials. Or if loess or ash deposition is interspersed with soil formation on the top of the terraces, the degree of complexity of the soils and deposits may increase in the older units as long as erosion has not eradicated the record (Milne, 1973). Or it might be that the soils on top of the terrace are eroded and that the soils and deposits that form the colluvial wedge on the terrace scarp may display a complexity with time that relates to the ages of the associated surfaces.

Terraces in Beaver Basin, Utah, dated by soils, are an example of bracketing the ages of deformation (Fig. 11.17). The profile for the Greenville alluvial surface parallels that of Indian Creek, so little or no deformation has occurred in the last 120–140 ka. The In-

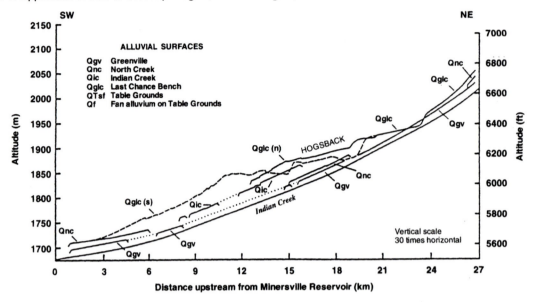

Figure 11.17 Topographic profile of Indian Creek and alluvial terraces, Beaver Basin, Utah. Lower-case letters in parentheses denote profiles both north (n) and south (s) of the creek. *(Data of M.N. Machette, reprinted from Birkeland and others, 1991, Fig. 6–6A.)*

dian Creek alluvial surface diverges upstream, placing tilting between its age (350–400 ka) and 120–140 ka. In contrast, the Last Chance Bench alluvial surface (500 ka) is warped into an anticline in the same area where the Indian Creek alluvial surface is divergent; anticlinal warping, therefore, was dominant between about 500 and 400 ka. Rockwell (1988) shows somewhat similar relations related to a series of faults in southern California.

Colman and others (1988) used soils to help date deposits and hiatuses in deformed basin-fill deposits in southeastern Utah. These data helped date tectonics related to flowage of a bedrock salt unit. Salt beds in the region were being studied as possible disposal sites of nuclear wastes, and this study, although not at a proposed site, helped provide information on the more recent tectonics of salt-cored anticlines in the area.

Soils and Safety of the Yucca Mountain Repository

Soils have played a large role in assessing the geological hazards associated with the Yucca Mountain site, Nevada. The U.S. Department of Energy is studying the site as a possible repository for high-level nu-

clear waste. The repository will be 320 m below the ground surface, with the water table another 200–400 m deeper. The intent is to keep the repository isolated from the water table, or from water moving up fissures from the water table, for at least 10 ky, the life span of the radioactive elements of interest. The soil studies address several issues.

It was important to know past climate change in the area, for many geologists assert that the past is the key to the future. If the area has experienced a wetter past climate, might that happen in the future, and either downward percolating water or the rising water table reach the repository? Taylor (1986) studied an alluvial chronosequence of soils along one of the major drainages in the area (MAP <20; MAT 15). Her calculation of the glacial climate produces these estimates: MAP 20, MAT <10. The soils are characterized by the pedogenic accumulation of carbonate and clay from atmospheric sources, and pedogenic opal silica from weathering of the parent rhyolitic ash-flow tuffs. All trends are typical chronosequence trends (see Birkeland and others, 1991, Fig. 4.6). This is the study that helped define the morphology stages of pedogenic silica accumulation (Appendix 1, Table A1.8). Because soils of about 100–250 ka have accumulation of all three above components in the top 3 m, it is doubtful if there has been very deep soil leaching over the past 0.5 Ma.

A major geological hazard in the area is Quaternary faulting. Intensive regional studies have been undertaken, and soils have been used to identify Quaternary mapping units and estimate fault ages. Quaternary faults have been identified and characterized within 100 km of Yucca Mountain (Piety, 1996), and soils help decipher the fault rupture histories (Whitney and Taylor, in press). This study also encouraged the use of numerous dating techniques, and these have been used to estimate ages of different components of the Quaternary soils. The conclusions are that the fault intersecting the repository has no Quaternary offset, and bounding faults to the east and west of the repository block have low frequency, small (<0.5 m total) Pleistocene offset, with some evidence for small (<0.2 m) offset on some of the larger faults.

The fault exposed in trench 14 has drawn national attention (Taylor and Huckins, 1995). The fault zone has several fractures, and they are filled with $CaCO_3$ and opaline silica. The question was whether these latter components (1) were pedogenic and therefore precipitated from vertically descending soil water or (2) were spring deposits indicating a past condition of vertically ascending ground water. If the latter is the case, the same waters could reach the repository. Extremely detailed work on field morphology, combined with chemical, mineralogical, and thin-section data, strongly supported the pedogenic origin (see also Monger and Adams, 1996). Quade and Cerling (1990) came to the same conclusion in a stable isotope study of the carbonates in the trench, after comparing their signature with those of regional pedogenic and spring carbonates.

A final geological hazard is the threat of nearby volcanism, as a young volcanic cone and associated flows lie about 20 km from the site. Young Quaternary flows are sometimes difficult to date; in addition, flows and cinder cones do not necessarily have to be the same age. A basaltic ash, the same age as the cone, is important to the tectonic and volcanic history of the area; for example, the ash occurs as a dilute component of the fills of some fault-created fissures in the area. Young Quaternary volcanic deposits are difficult to data. $^{40}Ar/^{39}Ar$ ages on the ash provide an age estimate of 75–80 ka (Whitney and Taylor, in press). Wells and others (1990b) studied soils on the youngest ash deposit of the cone and

found them to be no better developed than regional Holocene soils. Combining the latter with geomorphic data led them to estimate the age of the youngest volcanism at <20 ka. If the soil dates turn out to be too young, it could be that the processes used to estimate the ages proceed at rates slower than originally thought.

■ Using Soils in Archeological Studies

Soils can be of use to archeologists in the interpretation of archeological sites (reviewed in Holliday, 1989, 1990, 1992, 1994). Many articles on archeology include soils in the stratigraphic sections, and they are shown to be useful in many ways (e.g., Lasca and Donahue, 1990; Bettis, 1995; Collins and others, 1995). They can be used to indicate the ages of certain layers, past climates or local moisture conditions, and perhaps past vegetation, and to reconstruct the impact of landscape evolution on human occupation and vice versa. In addition, archeological excavations, because of their extreme detail, yield much information that is useful to soils studies.

The initial major problem with stratified archeological sites is to differentiate between geological and pedologic materials, as well as materials altered by humans. Throughout this book, there are examples by which one can identify pedologic materials, whether at the surface or buried (see also Stein and Farrand, 1985). Farrand (1975) has been active in sediment analysis of archeological sites and how they might be used to infer paleoenvironment. Useful analytical techniques are particle size, shape, and sorting, bulk density, thin section, mineralogy of all size fractions, and the usual soil-laboratory analyses (Catt and Weit, 1976). These must be used in the context of the field setting for the best geological-pedologic interpretation. Thus, it is important that the person responsible for the pedologic interpretation study the site rather than just be handed bags of materials for analysis ("cigarbox geoarcheology" of Thorson and Holliday, 1990).

One major use of soils at archeological sites is to identify depositional hiatuses, for if materials are not being deposited, soils can form. C.V. Haynes, Jr., has been a leader in these studies, and his regional review of the western United States (1968), although it was later altered (Hawley and others, 1976), is an example of how to incorporate soils into the study of archeologically important, late-Quaternary alluvial deposits. However, few detailed soil analyses accompanied these early studies. Reider (1990) solved this by expanding the data set for Colorado and Wyoming, and including soil laboratory data. Specifically, he shows the variations in morphology for key soils used in correlation, and groups them into three periods: late Pleistocene, Altithermal, and post-Altithermal; it is particularly important to recognize the Altithermal soil to put sediments into the right stratigraphic context, and it can be used to help with paleoenvironmental reconstruction. This article also is useful for the many high-quality photographs of soils at archeological sites.

Holliday (1985a,b,c, 1995) has set the standard for soil-archeological studies in his work at the Lubbock Lake State and National Archaeological Landmark Site, Texas, in which field relationships of deposits and soils are combined with quantitative soil data. The geology of the area is complex, since the following contrasting depositional facies are compressed into a small area: lacustrine, fluvial under both poorly and well-drained conditions, hill-slope colluvial, and eolian. Soils have formed from deposits of the above facies. The work is based on the description of over 150 back-hoe trenches and over 100 radiocarbon ages.

The soils vary in their parent materials, in the duration of time they formed, and in pedogenic environment, and so contrast markedly (Figs. 1.15C and 11.18). Specifically, there commonly is a great difference between the poorly drained valley-axis soils versus those formed in better drained, and perhaps sandier, soils of the valley margins. The Firstview Soil is an Entisol formed in about 1.8 ky; those profiles along the former valley axis have high clay content and are gleyed due to reducing conditions that accompanied a high water table. In contrast, the valley-margin soil facies is a sandy, calcareous A/C profile formed in eolian sand. Near the middle Holocene, the weakly developed Yellowhouse Soil formed over a period of about 0.5 ky. This soil is formed in the same two depositional facies as above, and Entisols, Inceptisols, and Mollisols are represented. The Lubbock Lake Soil, the most distinctive one at the site, formed mainly in eolian sediments; the duration of soil formation varies from about 3.5 ky, where buried, to 4.5 ky where it occurs at the surface. The A horizon is so distinctive that it was given a geological designation

and used as a marker bed in the early studies at the site. This emphasizes the necessity to be certain that pedologic units are distinguished from geological ones. Where best developed, the Lubbock Lake Soil is an A/Bt/Bk profile with stage I to II carbonate morphology. Because this soil occurs in several topographic settings, catena relationships (Chapter 9) are commonly well expressed. Mottling is present in the valley-axis sites. Inceptisols, Mollisols, and Alfisols are present. The Apache Soil represents from 250 to 450 years of pedogenesis and has formed mainly in eolian materials. Both Bw and Bt horizons are present, and stage I carbonate morphology in Bk horizons is common. Orders represented are the same as for the Lubbock Lake Soil. Finally, the Singer Soil has formed in the Historic period and usually is an A/weak Bw or C/Bk profile with weak stage I carbonate morphology. Extensive laboratory and thin-section data are used to characterize these soils.

The above sediment and soils are intimately associated with archeological remains (Fig. 11.19). These relations are very important in the dating of remains and in the interpretation of numerical dates. For example, if a soil underlies a date, and the archeological site can be shown to be older than the soil, the age of the soil has to be added to the numerical age to obtain the best minimum age for the site. Figure 11.19 shows some of the complexities, for although the Lubbock Lake Soil interveens between the hearth site (b), and the human debris in (d), the soil was forming while the debris was being deposited. If an unweathered sediment separated the debris from the top of the soil, then we could add the soil age to the hearth age to date the debris.

Many well-known archeological sites are present in the Southern High Plains, United States, and although many numerical dates are available, soil-stratigraphic relations identified at Lubbock Lake persist throughout a large region and are essential to the best regional correlations (Holliday, 1989, Fig. 1; Holliday, 1995).

Holliday (1988) shows how the above soils can be arranged into a chronosequence, with age defined as the duration of pedogenesis for each soil. Important conclusions are as follows: (1) Mollic epipedons form within 100 years, most of the gains in organic carbon occur in <1 ka, and gains slow considerably after that; (2) Minimal argillic horizons form in 450 years; (3) there is thin-section evidence for clay translocation into calcareous horizons before they are decalcified;

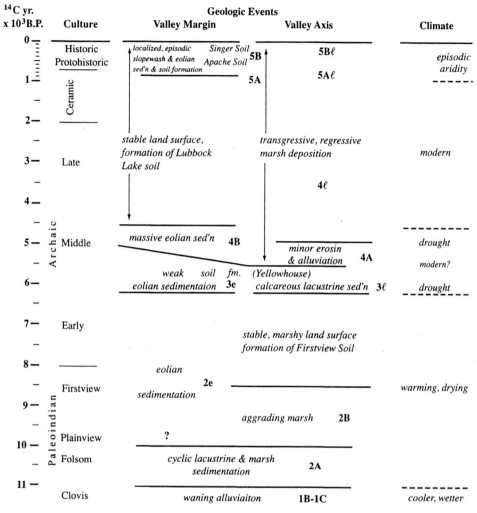

Figure 11.18 Summary of geologic events, soil formation, and cultural succession at the Lubbock Lake Site. *(Redrawn from Holliday, 1985a, Fig. 7 from the Geological Society of America Bulletin 96, 1483–1492, © 1985, with permission from the Geological Society of America, Boulder, Colorado.)*

(4) calcic horizons form within 200 years; and (5) the change in order is

Entisol $\Rightarrow$ Inceptisol, Mollisol (100 yr) $\Rightarrow$
   Inceptisol (with calcic), Mollisol (with calcic),
                          Alfisol (450 years)

Because the soils occur both at the surface and buried, Holliday is able to estimate the time it takes to remove most of the organic carbon from a buried soils: approximately 800 years.

Applications of soils to archeology are well illustrated throughout Holliday (1992). Ferring (1992) discusses alluvial plain settings, whether the associated soil is cumulative or noncumulative, and the eventual vertical distribution of artifacts. He concludes that cumulative soils tend to preserve the artifacts within a narrow depth range, whereas pedogenic processes (turbation) tend to widen the range in which they can occur in noncumulative soils. One also should be aware of what pedoturbation of soil does to the final resting place of artifacts (Johnson and Watson-Stegner, 1990).

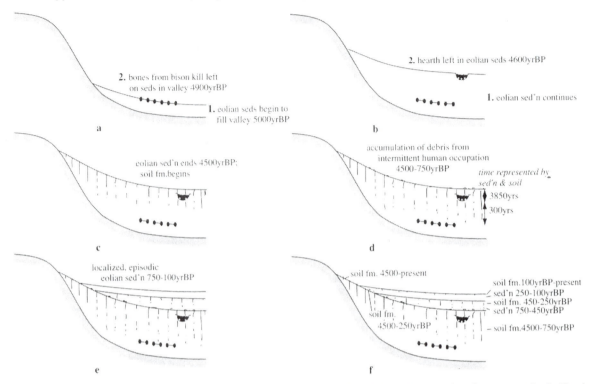

Figure 11.19 Generalized sequence of late-Holocene sedimentological, pedological, and cultural events at the Lubbock Lake site. Numbers in (a) and (b) indicate sequence of events in each diagram. Diagram (f) denotes the present-day relationships. Seds, sediments; sed'n, sedimentation; fm, formation. *(Redrawn from Holliday, 1990, Fig. 5 from the Geological Society of America Centennial Special Vol. 4, 525–540, © 1990, with permission from the Geological Society of America, Boulder, Colorado.)*

Mandel (1992) uses a correlation chart for soils and sediments at archeological sites in the Central Plains to show that the match from site to site is complex, and that deposition in the entire region does not stop and undergo soil formation all at once. From this he was able to predict the potential for finding buried materials of all of the Holocene cultural periods in terraces and fan deposits in large versus small valleys in two river basins.

Bettis (1992) carries on with the theme of helping the archeologist predict ahead of time whether a surface is worth excavating. Surface soils are used to put the landforms into a time sequence, and from this a strategy for exploration can be developed.

Goldberg (1992) is one of the few investigators to use soil micromorphology in archeological investigations. He shows that even the best laboratory methods cannot always separate features attributed to sedimentation from those attributed to pedogenesis or human activities, and that thin-section analysis has an important role to play in this regard (see also Courty and others, 1989).

Sandor (1992) has been active in the study of ancient agricultural practices on soils. New Mexico soils were sensitive to such practices and remain degraded after 900 years of abandonment, whereas Peruvian soils are fertile after 1.5 ky of cultivation. The European experience of agriculture and erosion from the present back to the archaeological records is well treated in Bell and Boardman (1992). In short, the eroded soils of the hill country supply sediment that might end up at footslope and toeslope sites. Various aspects of the soil profiles (truncated soils grade downvalley to cumulic soils) are the key to understanding these relations.

Benedict (1981, 1985) is known for his detailed geological–archeological work in alpine areas of the Colorado Front Range, and although soils are not the basic tool used in age assignment, soil horizons are commonly related to the geological and cultural materials. Most of the cultural materials are found at shallow depth in the upper soil horizons, in contrast to arroyo sites in which the deposits can be thick and the soils and cultural materials widely separated vertically. Till, alluvium, slopewash, colluvium, and loess, generally modified to some extent by frost-sorting processes, form the complex parent materials; a loess or mixed loess unit commonly is the uppermost unit throughout the area. Soils are formed through all materials and, in places, buried soils are present. Hearths usually predate the youngest loess, and the latter commonly is the parent material for the A horizon and, in places, part of the B horizon. The detailed archeological work points out that the uppermost loess, although thin, contains artifacts and was deposited at different times. These studies helped soil stratigraphers better understand the soils related to Holocene alpine tills.

Soils have also been used either to help date or to evaluate the paleoenvironment of much older archeological sites around the world (reviewed in Holliday, 1989). Although not essential in dating the hominid remains and archeological materials, calcic soils occur within the deposits at Olduvai Gorge, Africa and are used to help date some intervals of deposition and to verify the fact that a semiarid climate existed throughout the Pleistocene (Hay, 1976). The Calico archeological site in the Mojave Desert is controversial because workers are not unanimous as to whether the materials are human artifacts, although many archaeologists think thay are. At any rate, the soil formed in the uppermost part of the deposit containing "artifacts" is a strongly developed A/Bt/Bk profile with stage II carbonate morphology (Shlemon and Budinger, 1990). Comparison with other calcic soils in the region suggests an age of at least 80 ka, and this is compatible with uranium-series dates of around 200 ka on groundwater carbonate near the base of the deposit.

The above examples should alert archeologists to the potential use of soils in their research. They also can help in the search for early human sites. For example, early humans are considered to be no older than latest Pleistocene in the United States, and soil morphology should indicate that deposits beneath a particular soil are too old to yield meaningful results.

Finally, it should be pointed out that some chemical tests might be useful in determining the impact of humans on present or buried land surfaces (Eidt, 1985). Phosphorus determinations have been found to be useful, because human remains and debris are high in this element (Bethell and Mate, 1989). Some of these interpretations might be difficult to make, however, and workers should be aware of the pedologically distinctive extracts and trends (Chapters 5 and 8).

■ Using Soils in Environmental Studies

Soils have a role to play in a wide variety of environmental studies: wetland, geochemical, global change, acid rain, and resource and engineering projects. Singer and Warkentin (1996) discuss the contributions of soils to this wide field.

The recognition, mapping, and preservation of wetlands is a concern of many countries (Hughes and Heathwaite, 1995; Mulamoottil and others, 1996). Wetlands are areas in which the water table is at or near the surface, or land is covered by shallow water. The definition of a wetland includes three factors (Environmental Laboratory, Dept. of the Army, 1987): (1) vegetation species are adapted to anaerobic soil conditions; (2) soils have properties associated with reducing conditions; and (3) either the area is inundated with water permanently or periodically, or the soil is saturated to the surface at some time during the growing season. Most Histosols form in wetlands. The aquic soil moisture regime defines wetlands, and these soils will have redoximorphic features.

There are many other consulting opportunities using redoximorphic features. For example, the siting of septic drain fields requires that the site is well drained, and a trained soils person can make that judgment rather quickly, by looking for the presence or absence of redoximorphic features.

There are many texts on geochemistry and environmental sciences that have a strong soils component (e.g., Alloway, 1990; Fergusson, 1990; McBride, 1994; Pierzynski and others, 1994; Langmuir, 1997). The processes that are important to these studies are also important to soil genesis (Chapter 5), as these dictate the mobility and immobility of various ions of interest in aqueous environments. Furthermore, many of the ions are adsorbed onto soil particles, so knowledge of clay minerals and associated cation-exchange capacities is important. Finally, many properties that are the key to environmental assessment are part of the soil profile, and these properties vary with depth. Hence it is important to recognize horizons in the context of the soil profile. In some instances one might be able to reduce the number of samples taken and analyzed by sampling by horizon rather than by some arbitrary incremental depth.

Two examples of the above can be given. Meriwether and others (1995) were interested in the distribution of naturally occurring radionuclides in parts of Louisiana. One of the co-authors once remarked to me that once the results were in, they made little sense. Questioning the people who did the sampling revealed that incremental depth sampling was done without regard for the soil profile, and possible surface disturbance, such as erosion of the surface or treethrow. Examining the sites in the field with the soil environment in mind led to these conclusions: (1) In well-drained soils with no redoximorphic features, the highest concentration will be in the best developed B horizon (Bt, Bs, Bw, depending on the soil order). The radionuclides get leached downward and sequestered in the B horizon. (2) Because some soils do not leach radionuclides downward, a representative sample can be obtained from any depth. These include soils with redoximorphic features (little vertical leaching), Vertisols (vertical mixing inhibits concentration at depth), and Entisols (too young for leaching). (3) If treethrow is involved, the pattern can be complex, with the highest concentrations now at the surface. Another point is that in dealing with concentrations across a landscape, sampling using the catena concepts (position, process, etc.) could give the best results (Burns and others, 1991).

The Rocky Flats Plant, near Denver, Colorado, produced plutonium for nuclear weapons until 1989. Some radioactive materials were released into the environment and dispersed eastward by wind. Litaor and others (1994) and Litaor and Ibrahim (1996) characterized the present distribution of ^{234}Pu, ^{240}Pu, and ^{241}Am by incremental sampling keyed to the soil horizons and soil components. Concentrations are highest at the surface and decrease with depth; downward transport mechanisms in the soils seem to be macropore-associated decayed root channels and earthworm activities. Most of the Pu is associated with organic C and sesquioxides. Under the present average geochemical soil conditions, little Pu is available for transport, but one wet spring seems to have created conditions conducive to partial sesquioxide dissolution, and desorption of Pu and its increased mobility. Combining soils and geochemistry provided a better understanding of the system than would have been possible without this combination.

Soils have an important role to play in the neutralization of acid rain (Kennedy, 1992). Johnson (1984) points out that the weathering of the common rock-forming minerals takes up the acidity at varying rates. Cation exchange is another way in which soils neutralize the incoming acidity; the incoming acid is exchanged for the common cations. One reaction that neutralizes acid rain most rapidly is the weathering of calcite. Hence the more obvious relations are that calcic soils of the western United States are high in buffering capacity, whereas sandy, acidic Spodosols of the eastern United States are low. The mountains in the western United States are at high enough altitude that the soils are not calcic, so they are particularly vulnerable to acid deposition from upwind point sources. Many alpine basins have soils in addition to expanses of bedrock, and detailed analysis of the soils indicates that they have a high capacity to buffer acid deposition (Litaor and Thurman, 1988; Clayton et al., 1991a,b). How effective the soils are on a regional basis depends on their properties and distribution within the basins. One property that helps with the buffering is the fine-grained eolian mantle of many of the alpine soils (higher CEC); another is calcite present in some contemporary dust (Litaor, 1987; Bockheim and Koerner, 1997). The source of the calcite is difficult to pinpoint, but an obvious source is calcic soils in dry intermontane basins upwind of the mountains of interest.

A person schooled in soils and geology should be able to predict the vulnerability of specific watersheds to acid rain. April and others (1986), for example, compare two watersheds in the Adirondack Mountains of New York. In one watershed with thick till, acid deposition seems to have been neutralized by a combination of mineral weathering and cation exchange, and the lake of interest has a mean pH of 6.2. The other watershed has much thinner till, on average, less capacity to neutralize the acid rain, and the lake has a mean pH of 4.7. Some data suggest that because of the acid loading, the weathering rate in the first watershed has increased threefold.

Yet another aspect of soils in environmental studies is to anticipate how soils will change in an earth predicted to be warmer in the near future (Scharpenseel and others, 1990). Much of the focus is on carbon and nitrogen cycles and gaseous emissions (Lal and others, 1995). The results are crude, but will be improved as global change models improve. Most of the changes are predictable from what we know of the soil-forming factors, and alterations in those properties that are least resistant to change would take place the most rapidly (e.g., pH and CEC, see Table 1.5). Major changes include the following: (1) Changes will occur in soil moisture, with wetter ones experiencing greater leaching, and drier ones accumulating salts. (2) Organic matter will equilibrate at new levels as a function of changing climate–vegetation patterns. (3) Greater weathering will release more nutrients, which could influence biomass production, and the real impact of this will vary from place to place, and not always increase agricultural productivity where it is needed the most. Some soils will become more reducing, such as along coasts with rising sea level, or in polar areas where the melting (partial or total) of permafrost could produce poorly drained soils. In addition, with population pressure on agricultural lands, soil erosion could be a major problem.

There are several ways in which soils relate to the resources and engineering projects. One is the identification of gravel resources. For the Colorado Front Range, geologists characterized the gravel by landform, and whether they have fines, decomposed clasts, and $CaCO_3$ (Schwochow and others, 1974). Older terrace landforms have more developed soils, and therefore detrimental properties such as high clay and perhaps $CaCO_3$ contents, and the more weatherable clasts are decomposed; these are low-quality resources. Road construction in granitic terrain can involve blasting where the rock is hard, but bulldozing where grus is present. Hence in submitting bids for a project, knowledge of the distribution of grus can result in a bid that is more competitive. Maps such as those of Pierce and Schmidt (1975) would be valuable in this regard.

Finally, one of the more bizarre soil research projects has to do with geophagy, the practise of eating soils. Mahaney and others (1996) has studied this for many animals, and in this study they report on chimpanzees in Tanzania. Chemical and clay-mineral analyses suggest that ingestion of termite mound material (reworked soils) can serve as a dietary supplement, because of its mineralogical similarity to Kaopectate™; taken by some humans to soothe gastric attacks, it also might serve the same purpose for the chimpanzees.

■ Paleosols

A rapidly growing subdiscipline of soil science is the study of paleosols. One photograph that caught my attention early in my career was on the June 1976 cover of Geology; the man used for scale was upside-down. What the authors were trying to show was that a possible Middle Jurassic calcic paleosol subsequently was overturned. Retallack (1990) is probably publishing the most work in this field, and his book chronicles paleosols back to Precambrian times on Earth, and discusses those on Mars and Venus. These are important data for sedimentologists as in some sections the only paleoenvironmental data available are from paleosols (see review by Dahms and Holliday, in press). There are disagreements, however, about how easy these are to recognize, how far can we carry the interpretations, and how best to classify them.

Definition of the Term Paleosol

The definition of the term paleosol is not straightforward. Many of the early workers define a paleosol as a soil that formed on a landscape of the past (Ruhe, 1965; Yaalon, 1971). It can be at the surface as a relict soil, but the soil-forming environment has changed. For a soil at the surface, the argument is how long should a soil be at the surface before it is termed a paleosol? Should it be there long enough for a well developed soil to form? Ruhe (1965) gives the example of the red Ruston Soil of the southeastern United States that is present only beyond O-isotope stages 2 and 6 loesses of the Mississippi Valley, and in places passes under the loesses; obviously it formed its main properties some time in the distant past. Another question is how much of an environmental change has to occur for this to be a paleosol? Some areas experienced environmental change during the mid-Holocene, so are all older soils paleosols? Nobody doubts that drastic climatic change has occurred during the Quaternary, and this can show up in dramatic ways such as periglacial features impressed on paleoargillic horizons along the Canadian border (Karlstrom, 1990). Yet many other soils are on old surfaces that have gone through climatic change, but have few properties that reflect the change (soils on coarse-grained tills of the western United States, a parent material and environment relatively immune to having properties that reflect environmental change); in contrast, loess could be one of the better substrates for detecting such change. Still other highly developed soils at low latitudes could have formed under relatively uniform conditions for hundreds of thousands of years. Cremaschi (1987) discusses this latter concept for soils in northern Italy, and coins the term "vetusol." These are very old soils that have formed generally under the same set of pedological conditions and processes since their inception.

The other kind of paleosol is the buried soil (Chapter 1). Nobody would doubt that these are associated with a landscape of the past, but was the past environment always different; some were and some were not.

Catt (1990) discusses much of the above as arguments against using the term paleosol in lieu of soil, as the definition leaves much to be desired. He concludes that few would argue against use of paleosol for soils buried before the late Pleistocene, so this seems to be the best compromise. Many books with paleosol in the title have a pre-Quaternary emphasis, and these soils will be the focus here.

I have difficulty accepting some aspects of paleosols, especially that all published paleosols were once soils. The problem could be that my experience has been mainly with soils associated with coarse-grained deposits at the Earth's surface. Other workers will have a much more extensive background with a larger variety of parent materials. Regardless of that, and in fairness to the topic, I will follow this outline: review their recognition and classification, bring out problems in the latter, including diagenesis; assume that all paleosols were once surface soils, and present a few examples of how they are interpreted.

Recognizing Paleosols

It takes no special experience to work with and describe paleosols. One should use the same techniques used to describe surface soils and recognize buried soils (Chapter 1 and Appendix 1).

Retallack (1988, 1990) lists three main field features for differentiating paleosols from the enclosing sediments. First is root traces. If they are present, they indicate (1) the position of the ancient land surface, (2) that it was vegetated, and (3) by their pattern,

whether the vegetation was dominated by trees, and perhaps the kind of trees, shrubs, or grasses. They do not occur in soils older than the Devonian, for no large woody plants existed then. Apparently the root traces appear during the burial diagenesis of the soil as they are not common in contemporary surface soils. Trace fossils related to surfaces also can help with paleosol identification and, in one case, with paleo-ground-water conditions (Hasiotis and others, 1993). Second is the presence of soil horizons, but Retallack states that the uppermost part of the soil usually is truncated by erosion. Hence, one may not expect to find the entire A horizon. Horizons are expected to have the same association with adjacent horizons as do surface soils (e.g., a paleosol with stage III carbonate morphology should have an overlying red Bt horizon). The upper boundary of a well-expressed horizon usually has a sharper contact than the lower boundary. All horizons are described the same as surface soils. If the soils are calcic and one wants to classify horizons on the stage of morphological development, use the one in use by soil morphologists (Appendix 1), rather than define yet another scheme. Third is the presence of structures, and associated features such as clay films. If structures are bounded by slickensides, that is important information to note. Nodules, and their compositions, can help with environmental interpretations: Fe/Mn ones in poorly drained soils, calcareous ones in well-drained aridic soils, etc.

Retallack (1988, Fig. 11) offers a set of symbols he has found useful in depicting various pedogenic and geologic features in paleosols.

Laboratory Analyses

Two sets of laboratory analyses are usually done (Retallack, 1983). One of the more common ones is thin-section study, as this seems to be one of the more unambiguous ways of settling the argument as to whether the body is a soil or a sediment. One could attempt particle-size analysis, but if the materials are difficult to disperse, one can estimate particle-size distribution by point counting thin sections, as well as determine nonclay mineralogy. The second one is total chemical analysis. Because down-profile trends can be either inherited or pedogenic, one should first discount parent material trends (Chapter 7). Few investigators determine abundances using extractive chemical techniques (Fe_d, Fe_o, etc.).

Classification of Paleosols

There are two approaches to the classification of paleosols in the United States. Retallack (1990) classifies paleosols in the subgroup level of Soil Taxonomy. The problems with this are many. First, however, recall that Soil Taxonomy was designed for use in soil surveys, not for geological interpretation. The obvious problem in using it with paleosols is that epipedon properties are critical to the classification of many orders, as is base saturation with depth. All of these properties are among the most easily altered on burial (Table 1.5). With burial, for example, organic carbon content is rapidly lost, leading to problems in the identification of the thickness of the paleo-A horizon, and base saturation cannot only increase throughout the profile, but the composition of bases change, as vertically descending waters introduce bases from the overlying sediments. The record of soil-moisture regime generally is lost. Depending on the geochemical conditions, carbonate could be leached or introduced after burial. Retallack (1993, 1994) uses proxy data to defend his use of Soil Taxonomy. For example, thin-section data are used to aid in the identification of diagnostic horizons, and molar ratios from to-

tal chemical analysis used to indicate the degree of leaching, a crude measure of what the base saturation might have been. He finds, for example, that a value of 2 or higher in the alumina/bases molar ratio separates Ultisols form Alfisols.

A simple classification alternative is offered by Mack and others (1993). Some terms are new, and some are borrowed from Soil Taxonomy. First, one has to evaluate the relative importance of six pedologic features or processes: organic-matter content, horizonation, redox conditions, in situ mineral alteration, illuviation of insoluble compounds, and accumulation of soluble compounds. The more dominant of the above puts the soil into one of nine orders (Fig. 11.20). It is fairly obvious from the figure what the main property/process are for each order. One enters the classification in any box, and continues in the direction of the arrows until the paleosol properties fit the definition. They also list modifiers for the orders, so that a Calcisol with an argillic horizon is an argillic Calcisol. One that does not seem to fit too well is their example of an argillic, gleyed, calcic Spodosol. Because most Spodosols form mainly in an environment low in clay and carbonate, this one might be misclassified or have some diagenetic overprinting.

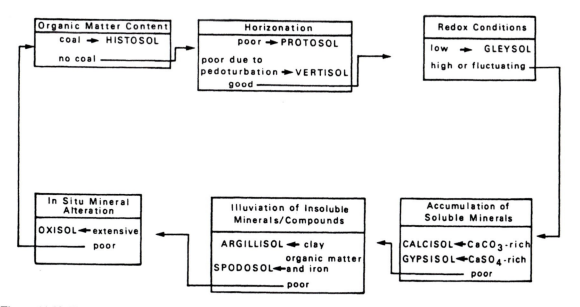

Figure 11.20 Flow chart of the nine paleosol orders. The chart can be entered at any position and the arrows followed until the paleosol being classified fits the definition. (*Reprinted from Mack and others, 1993, Fig. 1 from the Geological Society of America Bulletin 105, 129–136, © 1993, with permission from the Geological Society of America, Boulder, Colorado.)*

Criticism of Paleosol Identification and Classification

Few articles on paleosols differentiate those properties due to pedogenesis from those due to diagenesis (physical and chemical changes of soils and sediments after burial). Some of these materials have been buried 10s to 100s of My, and waters with chemical compositions much different from those in the soil-forming environment have been in contact and moving through them. Surely these must have had an impact on the paleosols. Here we will stay with temperatures lower than those that induce metamorphism.

Retallack (1990, 1991) lists several diagenetic changes that paleosols undergo. Among them are compaction, cementation, mineral recrystallization (common one involves calcite), authigenesis (formation of new minerals in situ with euhedral shapes), replacement of an initial grain by another, dissolution of grains or cements, dehydration [anhydrite converts to gyp-sum, or goethite to hematite, but this is a complex re-action involving more than dehydration according to Schwertmann and Taylor (1989, Fig. 8.8)], reduction [a problem in saturated floodplains with continual burial, soil formation, and a water table that rises into the accumulating sediments; see Pimentel and others (1996) for diagenic versus pedogenic criteria], and base exchange (which may convert smectite into illite, common in thick clay-rich sections; see Eberl, 1984).

Diagenesis can alter the original materials to the extent that diagenetic features mimic pedogenic ones (e.g., see discussion/reply by Patterson and others, 1990, and Lehman, 1990). One example can be given. Red color is common in many paleosols, and the question is how much of the redness is pedogenic and how much is diagenetic. Patterson (1990) tried to separate one origin from the other in her study of varicolored floodplain Eocene sediments of Wyoming. The red-colored beds have attributes that could classify them as paleosols. It was found, however, that the red color

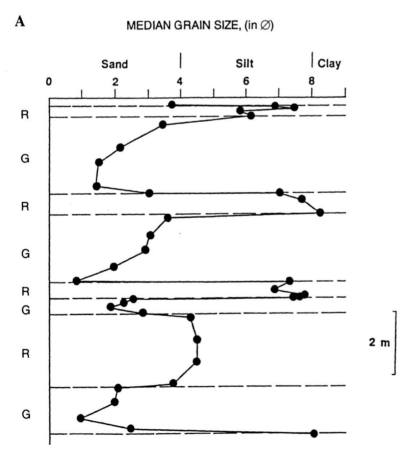

A

MEDIAN GRAIN SIZE, (in ∅)

Figure 11.21 Section through the Wind River Formation near Dubois, Wyoming, showing median grain size (A) and iron content (B) with depth. R, reddish layers; G, greenish layers. *(Reprinted from Patterson, 1990, Figs. 14 and 126.)*

is highly texture dependent: fine-grained beds are red, coarse grained beds are greenish, and all contacts are sharp (Fig. 11.21). I have looked at similar beds at other localities with experts in diagenesis, some of which are described as paleosols, and many color patterns are similarly texture dependent. In places, if the fine-grained beds are traced laterally to a coarse-grained channel deposit, the red color does not continue across the latter. This is strange because most coarse-grained Quaternary deposits have well-expressed soils. Furthermore, beneath some coarse-grained channel deposits, the fine-grained beds are locally leached of the red pigment. In this case, and in many others involving red beds (Walker, 1975), dia-

genesis initially imparted a red pigment on coarse and fine beds alike, and subsequent leaching selectively removed the hematitic pigment from the more permeable coarse-grained beds, leaving them greenish, in contrast to neighboring reddish fine-grained beds. This also might help explain why these red layers seem to all have a rather similar Munsell color notation.

Other properties of paleosols should also be evaluated for diagenetic alteration. In sandstones, for example, Walker and others (1978) document in detail widespread diagenetic alteration involving many of the changes mentioned above (formation of clay, alteration of inherited clay, weathering of Fe-bearing minerals, production of red pigment, precipitation of

Figure 11.21 (continued). **B**

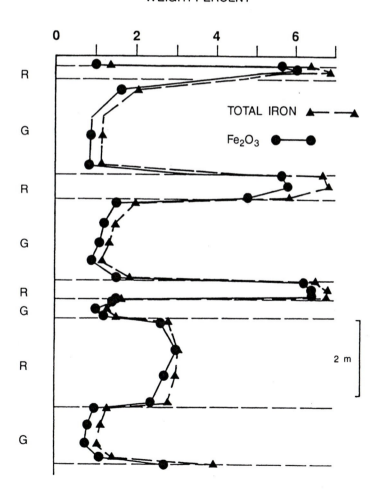

calcite, change in bulk chemistry, etc.). The question to be asked is, if they occur in these beds, why not in any soil associated with them? For example, underlying late Paleozoic red beds in the Front Range of Colorado is an oft-cited paleosol formed in granodiorite (Wahlstrom, 1948). If the red bed sediments have undergone extensive diagenesis, it is fair to ask which properties of the paleosol are also the product of diagenesis; Nesbitt (1992), for example, believes the clay minerals have undegone diagenetic change.

Thus, I have a problem believing that paleosol properties such as structure, clay mineralogy, and some attributes of total chemistry will survive unchanged. Several workers are addressing the diagenetic overprints in possible paleosols, using different approaches (Retallack, 1991; Nesbitt, 1992; Blodgett and others, 1993). Patterson and others (1990, p. 845) concluded by saying that "future work should focus on ways of looking through the veil of diagenesis to detect those chemical and physical attributes that could be related to pedogenesis." This can be depicted in a diagram that shows the veil, assuming greater im-

portance of diagenesis the longer that diagenetic reactions have been going on (Fig. 11.22). In short, diagenesis has little impact on young paleosols, but in old paleosols the challenge is to separate diagenesis from pedogenesis. My goal in this discussion is to add a bit of caution in paleosol recognition and thereby question how far one can carry their interpretation. This should be addressed in paleosol research.

Walkers's (1967) strategy for the study of red bed origins would be useful for the study of diagenetic alteration of paleosols. He followed the development of red beds and all the accompanying diagenetic stages by a study of the initial development in young sediments to mature development in old sediments. With paleosols, one could study the present-day soil in detail (usual laboratory properties, thin-section analysis, total chemistry, etc.), then do the same for paleosols with progressively greater durations of burial, all in the same general environment. This would be analogous to going vertically down in Fig. 11.22.

Finally, the above leads me also to be cautious with paleosol classification. Hence, I prefer a classification that does not rely on properties that can be so readily altered that it is necessary to interpret what was there prior to burial. Hence, the classification of Mack and others (1993), used with appropriate modifiers, could be less subject to error.

Examples of Use of Paleosols

There are many examples of the use of paleosols in geology (e.g., Wright, 1986; Reinhardt and Sigleo, 1988; Retallack, 1990; Martini and Chesworth, 1992), and only a few will be given here. Here I will assume the published features are indeed paleosols, even though they might not meet the diagenetic versus pedogenic tests mentioned above.

Retallack (1990) reviewed the presence of soils throughout the history of the Earth and listed the time span of the soil orders (Fig. 11.23). The oldest paleosols predate 3000 Ma. Inceptisol and Entisols seem to have existed throughout most geologic time, as little pedogenic rearrangement is required for their formation, and they form in all environments. Vertisols, Aridisols, and Oxisols also have a long history: Vertisols, because all that is needed is clayey parent materials and wet-dry cycles; Aridisols, because all that is needed is an aridic soil-moisture regime, accompanied by aridic soil features; and Oxisols, because they formed whenever landscapes were stable for long pe-

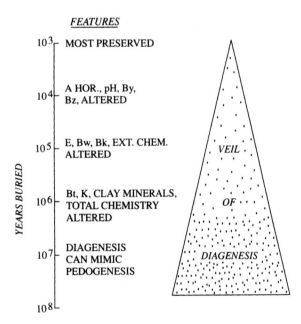

Figure 11.22 General diagram of the approximate burial time required for alteration of various soil features, most of which were ranked by Yaalon (1971) (see Table 1.5). The veil of diagenesis is depicted as widening with time, with width approximately equal to the importance of diagenesis in altering soil features.

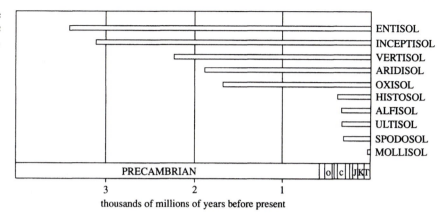

Figure 11.23 Geological range of soil orders. *(Redrawn from Retallack, 1990, part of Fig. 14.1.)*

riods of time under warm and humid environments. The other orders had to await the evolution of various ecosystems on the land surface. Land plants date to the Silurian, so thereafter in the Devonian soils are seen that are closely linked with land vegetation (Histosols, Alfisols, Ultisols, and Spodosols). In contrast, Mollisols had to await the appearance of grasslands in the Tertiary.

Paleosols are good markers of unconformities. Sigleo and Reinhardt (1988) report on Cretaceous paleosols across a large part of the southeastern United States. Red soils, with Bt horizons and kaolinite, formed irrespective of the parent materials. Similarities to present-day soils suggest a tropical climate during their formation, and this is in accord with continental and paleoclimate reconstructions and the global pattern of similar paleosols.

In the European Alps, Barrientos and Silverstone (1987) report on a Paleozoic schist related to a regional unconformity, formed under temperatures of about 550°C at 35 km depth, that might be a metamorphosed paleosol. The material of interest is 5 m thick, and Fe and Al rich. They use vertical variation in bulk chemistry, show that the trends are similar to those of some paleosols, including that of Wahlstrom (1948), and what is expected from pedogenesis, to argue that it is a highly altered paleosol.

Soil formation on alluvial plains has been common throughout the history of the Earth. Allen (1989) and Wright (1992) reviewed these, and Allen has presented a useful summary diagram (his Fig. 3.2, highly modified as Fig. 11.24). Essentially, the best preserved soils are located in sites with low rates of both erosion and deposition. The most mature of these soils

have undergone sufficient duration of formation that horizonation and profile form are readily distinguished. Soil formation (weak, moderate, strong) is related to both duration of time and stability of the landscape. Finally, floodplain soils are best preserved in aggrading systems and least preserved in degrading systems. These paleosols are important to the interpretation of basin tectonics and paleoclimate.

Eocene floodplain sediments of northwestern Wyoming are about 700 m thick and contain from 500 to 1200 paleosols of varying development (Bown and Kraus, 1981, 1987, 1993; Kraus and Bown, 1986). First, paleosol maturity is proportional to the time between depositional events; the longer these depositional hiatuses are, and the greater the pedogenic imprint, the farther one is from the high sedimentation rates close to the paleochannel. The lateral differences in paleosol morphological development due to differing rates of sedimentaion are termed pedofacies (Kraus and Bown, 1988). As the channel migrates with time (up section), so do the pedofacies. The paleosol development was transcribed to a paleosol maturation stage (stage 0 to stage 6, with 6 being the most mature) based on solum thickness. Each stage is given a temporal weighting, and one can assign these to any paleosols in the sequence. The overall trend is that paleosols are more mature at the base of the sediment section and become less mature upsection. In concert with this, accumulation rate is low at the base of the section (time is needed to form mature soils) and increases upsection. Put another way, soil formation takes up more of the total time at the base of the section and progressively less upsection, where more time is recorded as rock deposition. While this is happen-

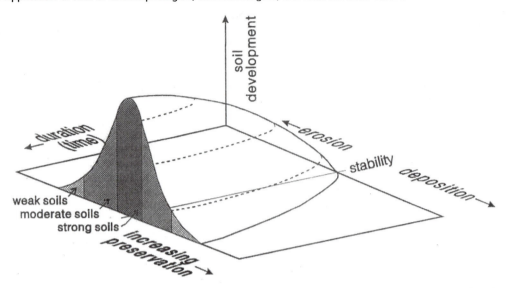

Figure 11.24 Diagram depicting relative development of soils in an alluvial plain setting, as a function of landscape stability (erosion vs deposition) and time. *(Diagram idea from Allen, 1989, Fig. 3.2, modified after consultation with Emmett Evanoff, 1997, and diagram courtesy of Evanoff.)*

ing, the soil morphology suggests that climate becomes drier upsection, perhaps due to the progressive uplift of the mountains surrounding the basin.

One of the thinnest soils on Earth is a very unique paleosol at the Tertiary/Cretaceous boundary in eastern Montana (Fastovsky and others, 1989). Common at many localities across a large area, a two-part iridium-bearing claystone marks the boundary. Both layers were considered by some to result from an asteroid impact. The claystone ranges in thickness from 0.5 to 1.5 cm, with the upper part the thinner of the two (0.25 cm). There are major differences in the two layers: (1) The upper one is dark brown, has aligned stringers of organic matter that are not disrupted, and is smectitic; total chemical analysis shows it to be relatively enriched in immobile elements. (2) The lower layer is pink, is dominated by halloysite and/or kaolinite, is more leached of immobile elements, and in thin section has pedogenic features such as voids, iron ox-

ide and clay coatings, and root traces. Although the latter has undergone pedogenesis, the underlying siltstone has not been altered. Their interpretation is that a depositional layer (1) overlies a soil (2), and that the hiatus represented by the soil is less than 6 ky.

Stable isotopes can be used to identify the environment of paleosols (Chapter 10). For example, the sediments that were deposited along the southern flank of the Himalaya, as the latter was uplifted, contain many paleosols (1 every 10 m on average in one section) (Quade and others, 1995). Sections studied span 11 Ma. Soils near the base are red Bt/Bk profiles, the middle ones are gleyed, and the upper ones contain yellow Bt horizons. The carbon isotope component shows a shift at 7–8 Ma from an older isotopic signal that probably signifies a semideciduous forest to a younger signal that signifies grasslands. Of interest to problems of diagenesis in paleosols, they applied an isotopic test to detect diagenesis, and found little.

Describing Soil Properties

■ Horizon Nomenclature

It should be emphasized that horizon nomenclature depends not only on the properties of the particular horizon, but also on those of both the overlying and underlying horizons and the parent material (or the presumed parent material). Hence, it is not uncommon to change horizon designations during the soil description process, because the nomenclature chosen for horizons may change after more is learned of the adjacent horizons and the parent material. Because some of the older soil literature in the United States uses the old soil-horizon nomenclature, Table A1.1 is provided to convert the old symbols to the new ones.

There are certain rules to follow when more than one lower-case letter is used with the master horizon designation (Soil Survey Division Staff, 1993, with some modification). In most cases, more than one lower-case letter can be used. If more than one such letter is used, these letters are always written first: a, e, h, i, r, s, t, v (when used for vesicular), and w. Further, none of these letters is used in combination, except for Bhs and Crt. If more than one lower-case letter is used and the horizon is not buried, the following, if used, are commonly written last: c, f, g, m, v (when used for plinthite), and x. For buried soils, b commonly is last, unless field or laboratory data indicate horizon features that formed after burial. For example, if carbonate accumulates in a Bt horizon after burial, the designation is Btbk. For B-horizon designation, t has precedence over w, s, and h, and the latter symbols are not used in combination with t. If other lower-case letters are used with t, the t comes first (e.g., Btg).

■ Digging and Photographing Soils

Several steps should be taken in preparing the soil for description:

1. Study the geomorphic surface carefully before selecting the spot that is best for the purpose of the study. If a roadcut is selected, dig back a few 10s of centimeters. One problem in dry areas is that salts might have been collecting on and near the face of the roadcut. Be sure you clean the cut back far enough to be beyond this influence of salt concentration. If a hole is to be dug by hand, use a long-handle shovel and pick, if necessary. Cut the turf with the shovel and place it aside so it can be replaced later. Decide which face of the hole will be photographed. This face should be vertical and should be in direct sunlight at the time the hole is completed. Do not disturb the vegetation a the top of the face to be photographed.

2. Clean off the face to be photographed with a stiff knife and perhaps a hand broom. Let the structure be expressed. Avoid multiple knife marks on the face to be photographed. Place a scale neatly to one side. When taking a light meter reading, be sure to have the meter in the hole directed to the face of interest. Include the surface vegetation in the photograph, and keep packs, knives, etc. out of view.

3. Once ready for description of a deep hand-dug hole, consider describing the deeper horizons first, for if you do not, the soil dislodged as one describes the upper horizons might cover the deeper horizons.

4. Channel sample each horizon. Dislodging the soil with the knife, obtain a representative sample of the entire thickness of the horizon, but stay away from

Table A1.1 Soil-Horizon Symbol Conversions

Master Horizons		Subordinate Departures		
Old	New	Old	New	Description
O	O	—	a	Highly decomposed organic matter
O1	Oi, Oe	b	b	Buried soil horizon
O2	Oa, Oe	cn	c	Concretions or nodules
A	A	—	e	Intermediately decomposed organic matter
A1	A	f	f	Frozen soil
A2	E	g	g	Strong gleying
A3	AB or EB	h	h	Illuvial accumulation of organic matter
AB	—	—	i	Slightly decomposed organic matter
A&B	E/B	ca	k	Accumulation of carbonates
AC	AC	m	m	Strong cementation
B	B	sa	n	Accumulation of sodium
B1	BA or BE	—	o	Residual accumulation of sesquioxides
B&A	B/E	p	p	Plowing or other disturbance
B2	B or Bw	si	q	Accumulation of silica
B3	BC or CB	r	r	Weathered or soft bedrock
C	C	ir	s	Illuvial accumulation of sesquioxides
R	R	t	t	Accumulation of clay
		—	v	Plinthite
		—	w	Color or structural B
		x	x	Fragipan character
		cs	y	Accumulation of gypsum
		sa	z	Accumulation of salts

From Guthrie and Witty (1982).

the boundary areas. If the gravel is not wanted, screen the sample in the field with inexpensive 3-mm door screen, followed by a 2-mm screening in the laboratory. To determine the gravel weight, screen the sample in the field and weigh with a field scale.

■ Soil Properties

The descriptive terminology for horizons developed by soil scientists is used to describe soils (Soil Survey Division Staff, 1993). An example of a completed soil-profile description is given in Table A1.2. A field recording sheet that we have found useful is given in Table A1.3; because describing soils can be quite messy with dirt on the hands, all one has to do is cir-

cle the abbreviation (n,f,abk, etc.) of the correct property. In the text that follows, all of the abbreviations in Tables A1.2 and A1.3 are defined. Because these data are used to calculate the PDI, for uniformity one person should collect all of the data for all of the soils of a particular project.

Field site characteristics should be noted at the top of Table A1.3. The configuration of the landform associated with the soil should be described, as concave parts of the landscape can receive sediment or additional water from upslope. In contrast, convex parts of the landscape might be undergoing erosion. Ruhe (1975) has devised a classification useful for this purpose (Fig. A1.1).

The key properties of soils (Soil Survey Division Staff, 1993) are described in the following section.

Depth

The top of the uppermost mineral horizon (A or E) is taken as zero depth. The O-horizon thickness is measured up from that point (i.e., 2–0 cm), and all other horizons down from that point (i.e., 0–8 cm).

Percentage Estimate

In the description of soils, one often has to estimate the percentage by volume of various soil features (e.g., gravel, carbonate stage, mottles). Many charts are available for estimating percentages, many color charts include them, and Fig. A1.2 is a useful one.

Color

List dominant color, both moist and dry, if feasible. Take them in direct sunlight, without the aid of sunglasses. For many studies, to be consistent, dry colors of sieved soil samples could give the most consistent results, all taken on the same day outdoors in direct sunlight. Use the Munsell Soil Color Charts (1954) or other suitable charts that use the Munsell color notation. List the moisture state when taken. If a color lies between two hue pages or color chips, this can be denoted, for example, as 7.5-10YR or 10YR 6/3.5, respectively. If the soil color name is to be used, be aware that the names used in the Japanese and American charts differs; the latter names can be found in Soil Survey Staff (1975).

Mottling refers to repetitive color changes not related to parent material compositional differences. Wetting and drying commonly produce redoximorphic features of several colors that are repeated throughout the soil mass. A color pattern related to a soil feature, such as a ped, is not mottling. For mottled soils, record the following:

1. Where two or more colors occur in an intricate pattern and they occur in equal amounts, this should be stated.

Table A1.2 Field Description of a Soil Formed at the Summit (SU) Position on a Bull Lake moraine, Idaho[a]

Horizon	Depth (cm)	Horizon Boundary	Munsell Color (dry)	>2 mm Fraction (%)	Soil Texture
O	2–0	—	—	—	—
A	0–4	a,s	10YR 4.5/2	27	SL
AB	4–8	a,s	10YR 5/3	18	SL
Bw	8–26	a,s	10YR 6/3	53	SL
2Bt1	26–48	g,w	10YR 6/3	62	SL
2Bt2	48–84	g,w	10YR 6/4	64	SL
2Btj	84–125+	—	10-7.5YR 6/4	56	LS

Horizon	Consistence Wet	Consistence Moist	Structure	Clay Films	Parent Material[b]
O	—	—	—	—	—
A	so,ps	lo	1f,gr	o	Eo/till
AB	so,ps	lo	1f,sbk	o	Eo/till
Bw	so,ps	lo	1f,sbk	o	Eo/till
2Bt1	ss,ps	fi	2m,sbk	2d br, 2d pf	Till
2Bt2	ss,ps	vfi	2m,abk	3d br, 2d pf	Till
2Btj	ss,po	fi	1m,sbk	1f br, 1f pf	Till

From Berry (1987, Table 2).

[a] Abbreviations are given in the text.

[b] Eo/till indicates a mixed parent material of eolian material and till.

Table A1.3 Work Sheet for Recording Soil Properties in the Field

Soil Description: Location _____

Site No. _____   Date _____   Time _____   Vegetation _____

Elevation _____   Slope _____   Aspect _____   Geomorphic Surface _____

Parent Material(s) _____   Described by _____

Depth (cm)	Horizon	Color moist / dry	Structure	Gravel %	Wet	Moist	Dry	Texture	pH	Clay Films	Boundaries	Notes[a]
		....	m vf gr sg f pl 1 m pr 2 c cpr 3 vc abk sbk	0 50 <10 75 10 >75 25	so po ss ps s p vs vp	lo vfr fr fi vfi efi	lo so sh h vh eh	S SiCL LS SiL SL Si SCL SiC L C CL SC		vl f pf 1 po 2 d br 3 co p cobr	a s c w g i d b	
		....	m vf gr sg f pl 1 m pr 2 c cpr 3 vc abk sbk	0 50 <10 75 10 >75 25	so po ss ps s p vs vp	lo vfr fr fi vfi efi	lo so sh h vh eh	S SiCL LS SiL SL Si SCL SiC L C CL SC		vl f pf 1 po 2 d br 3 co p cobr	a s c w g i d b	
		....	m vf gr sg f pl 1 m pr 2 c cpr 3 vc abk sbk	0 50 <10 75 10 >75 25	so po ss ps s p vs vp	lo vfr fr fi vfi efi	lo so sh h vh eh	S SiCL LS SiL SL Si SCL SiC L C CL SC		vl f pf 1 po 2 d br 3 co p cobr	a s c w g i d b	
		....	m vf gr sg f pl 1 m pr 2 c cpr 3 vc abk sbk	0 50 <10 75 10 >75 25	so po ss ps s p vs vp	lo vfr fr fi vfi efi	lo so sh h vh eh	S SiCL LS SiL SL Si SCL SiC L C CL SC		vl f pf 1 po 2 d br 3 co p cobr	a s c w g i d b	
		....	m vf gr sg f pl 1 m pr 2 c cpr 3 vc abk sbk	0 50 <10 75 10 >75 25	so po ss ps s p vs vp	lo vfr fr fi vfi efi	lo so sh h vh eh	S SiCL LS SiL SL Si SCL SiC L C CL SC		vl f pf 1 po 2 d br 3 co p cobr	a s c w g i ·d b	
		....	m vf gr sg f pl 1 m pr 2 c cpr 3 vc abk sbk	0 50 <10 75 10 >75 25	so po ss ps s p vs vp	lo vfr fr fi vfi efi	lo so sh h vh eh	S SiCL LS SiL SL Si SCL SiC L C CL SC		vl f pf 1 po 2 d br 3 co p cobr	a s c w g i d b	
		....	m vf gr sg f pl 1 m pr 2 c cpr 3 vc abk sbk	0 50 <10 75 10 >75 25	so po ss ps s p vs vp	lo vfr fr fi vfi efi	lo so sh h vh eh	S SiCL LS SiL SL Si SCL SiC L C CL SC		vl f pf 1 po 2 d br 3 co p cobr	a s c w g i d b	

Table A1.3 (continued)

Depth (cm)	Horizon	Color moist / dry	Structure	Gravel %	Consistence Wet	Moist	Dry	Texture	pH	Clay Films	Boundaries	Notes[a]
		....	m vf gr sg f pl 1 m pr 2 c cpr 3 vc abk sbk	0 50 <10 75 10 >75 25	so po ss ps s p vs vp	lo vfr fr fi vfi efi	lo so sh h vh eh	S SiCL LS SiL SL Si SCL SiC L C CL SC		vl f pf 1 po 2 d br 3 co p cobr	a s c w g i d b	
		....	m vf gr sg f pl 1 m pr 2 c cpr 3 vc abk sbk	0 50 <10 75 10 >75 25	so po ss ps s p vs vp	lo vfr fr fi vfi efi	lo so sh h vh eh	S SiCL LS SiL SL Si SCL SiC L C CL SC		vl f pf 1 po 2 d br 3 co p cobr	a s c w g i d b	
		....	m vf gr sg f pl 1 m pr 2 c cpr 3 vc abk sbk	0 50 <10 75 10 >75 25	so po ss ps s p vs vp	lo vfr fr fi vfi efi	lo so sh h vh eh	S SiCL LS SiL SL Si SCL SiC L C CL SC		vl f pf 1 po 2 d br 3 co p cobr	a s c w g i d b	

Courtesy of D. Jorgenson, written communication (1989).

[a]The notes column can be used to record properties not universal to all soils.

2. Where one color is the continuous phase or matrix with mottles contained in it, designate the former as the dominant color and specify the quantity, size, contrast, and color of the mottles as indicated below.

Quantity:

f—few (mottles <2% of surface area).
c—common (mottles 2–20% of surface area).
m—many (mottles >20% of surface area).

Size:

1—fine (<5 mm in diameter).
2—medium (5–15 mm in diameter).
3—large (>15 mm in diameter).

Contrast:

f—faint. Evident only on close examination. Faint mottles commonly have the same hue as the matrix color to which they are compared and differ by no more than 1 unit of chroma or 2 units of value.

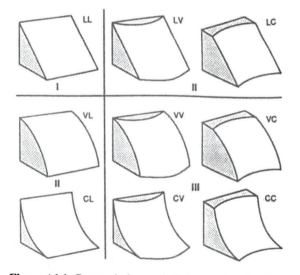

Figure A1.1 Geometric forms of hillslopes. Slope length is down the form; slope width is across the form: L, linear; V, convex; and C, concave. The simplest form (group I) is colinear (LL); group III forms, the most complex, are doubly curved; group II forms are linear in one dimension and curved in the other. (*Redrawn from Ruhe, 1975.*)

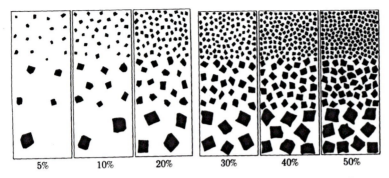

5%   10%   20%   30%   40%   50%

Figure A1.2 Charts for estimating percent, by volume, of various soil features. *(Reprinted from Yaalon, 1966, Fig. 1, ©1966, The Williams & Wilkins Co., Baltimore.)*

Some faint mottles of similar but low chroma and value differ by 2.5 units (one page) of hue.

d—distinct. Readily seen but contrast only moderately with the matrix color to which they are compared. Distinct mottles commonly have the same hue as the color to which they are compared but differ by 2 to 4 units of chroma and 3 to 4 units of value; or differ from the chroma to which they are compared by 2.5 units (one page) of hue but no more than 1 unit of chroma or 2 units of value.

p—prominent. Contrast strongly with the matrix color to which they are compared. Prominent mottles are commonly the most obvious color feature of the section described. Prominent mottles that have medium chroma and value commonly differ from the color to which they are compared by at least 5 units (two pages) of hue if chroma and value are the same, at least 4 units of value or chroma if the hue is the same, or at least 1 unit of chroma or 2 units of value if hue differs by 2.5 units (one page). One can also describe the shape of the mottles, their location, and the boundary (sharp-no color gradation; clear-color grades over 2 mm; diffuse-color grades over >2 mm).

In line with the redoximorphic features described by Vepraskas (1994; see Chapter 5), one should describe both concentrations and depletions. Concentrations are Fe, Mn, or Fe/Mn, and they can occur on ped surfaces, along pores, or within peds as soft masses, nodules (cemented mass), or concretions (cemented and layered mass). Depletions are either Fe or clay, and they occur on ped surfaces.

Structure

Describe type, grade, and structure size. If the structure is not apparent, take a spade full of the soil and tap it horizontally on the ground and look for repeating patterns.

Type of Structure: Use Table 1.3 to define the type of soil structure.

Grade:

m—massive. Enough aggregation to maintain a vertical face but no formation of structure type (structureless).

sg—single grain. No aggregation (structureless). Loose grains of a sand dune are a good example.

1—weak. Peds barely observable in place, and, when disturbed, few entire peds are observed; much of the material is unaggregated.

2—moderate. Peds easily observable in place. When disturbed, there is a mixture of whole peds, broken peds, and some material not organized into peds.

3—strong. Peds are distinctly visible in place, and, when disturbed, nearly the entire mass consists of whole peds.

Size: Size differs with the kind of structure as shown in Table A1.4. Smaller structural units may be held together in such a way as to form larger units. For example, small subangular blocky units may combine in such a way to form larger prismatic units. The dominant structure is the primary structure when calculating PDI values, and the subordinate structure is the secondary structure.

Gravel Content

Estimate volume percentage occupied by gravel (>2 mm). Weight percentage can be determined in the field with a screen (one can use an inexpensive 3-mm door screen) and a hand-held portable scale. Be watchful for shape and lithologic changes during the screening process, as they may indicate parent materials of more than one origin.

Consistence

Consistence is a measure of the adherence of the soil particles to the fingers, the cohesion of soil particles to one another, and the resistance of the soil mass to deformation. Soil Survey Division Staff (1993) has changed some of the terms, but the older terms are kept here as PDI values are based on them. Because this property varies with moisture content, it is taken when the soil is dry, moist, and wet. The wet consistence (natural or artificial wetness) is useful in determining texture classes in the field.

Dry Consistence (naturally dry in exposure):

lo—loose. Noncoherent, such as grains of a sand dune.
so—soft. Easily fails to powder or single grain, with very slight force between thumb and forefinger.
sh—slightly hard. Easily fails under slight force between thumb and forefinger.
h—hard. Fails in the hands without difficulty; requires strong force to fail between thumb and forefinger.
vh—very hard. Fails in hands with difficulty, but not between thumb and forefinger.
eh—extremely hard. Cannot be failed in hands.

Moist Consistence (usual moisture when one digs back into exposure):

lo—loose. Noncoherent.
vfr—very friable. Easily fails to powder or single grain, with very slight force between thumb and forefinger.
fr—friable. Fails under slight force between thumb and forefinger.
fi—firm. Fails under moderate force between thumb and forefinger.
vfi—very firm. Fails under strong force between thumb and forefinger.
efi—extremely firm. Fails under very strong force between hands but cannot be crushed between thumb and forefinger.

Wet Consistence (usually wetted artificially, but not so much the mass flows):
Stickiness is measured by pressing the wet soil between the thumb and forefinger and noting its adherence.

so—nonsticky. Practically no adherence to thumb and forefinger when pressure released.
ss—slightly sticky. After release of pressure, soil adheres to both thumb and forefinger but comes off one or the other rather cleanly. Does not appreciably stretch.
s—sticky. After release of pressure, soil adheres to both thumb and forefinger and tends to stretch somewhat before pulling apart from either digit.
vs—very sticky. After release of pressure, soil adheres strongly to both digits and is markedly stretched when they are separated.

Table A1.4 Classes of Soil Structure

	0	20	40	
		Scale (in mm)		

Size Class	Diameter of Granules (mm)	Thickness of Plates (mm)	Diameter of Blocks (mm)	Diameter of Prisms (mm)
vf—very fine	<1	<1	<5	<10
f—fine	1–2	1–2	5–10	10–20
m—medium	2–5	2–5	10–20	20–50
c—coarse	5—10	5–10	20–50	50–100
vc—very coarse	>10	>10	>50	>100

Plasticity is measured by rolling the wet soil between the thumb and forefinger and observing whether a roll can be formed and maintained.

po—nonplastic. No roll can be formed.

ps—slightly plastic. A roll 4 cm long and 6 mm thick can be formed and, if held on end, will support its own weight. A 4-mm-thick roll will not support its own weight. The roll is easily deformed and broken.

p—plastic. A roll 4 cm long and 4 mm thick can be formed and support its own weight. A 2-mm-thick roll will not support its own weight.

vp—very plastic. A roll 4 cm long and 2 mm thick can be formed and support its own weight. The roll is readily bent into a half or full circle.

Texture

Use established names from the textural triangle (Fig. A1.3). Screen out gravels and determine the textural class of the <2-mm fraction by noting the grittiness and wet consistence as shown in Fig. A1.4 (see also useful table of properties in Foss and others, 1975). Broad guidelines are given in the figure, but for more accuracy one should calibrate one's fingers by texturing samples with known particle-size distribution.

pH

Record the specific pH value using a field kit.

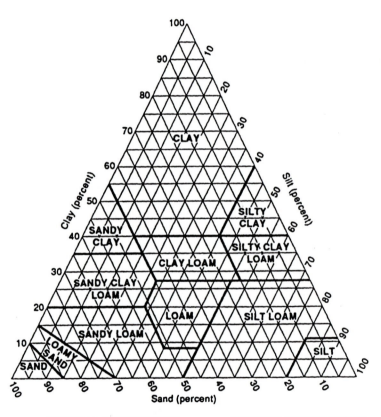

Figure A1.3 Textural names and abbreviations of names versus sand–silt–clay contents. *(Redrawn from Soil Survey Division Staff, 1993, Fig. 3.16.)*

	TEXTURAL ABBREVIATIONS:			MODIFIER ABBREVIATIONS:	
C	Clay	SCL	Sandy Clay Loam	vf	very fine
CL	Clay Loam	SL	Sandy Loam	f	fine
L	Loam	Si	Silt	co	coarse
LS	Loamy Sand	SiC	Silty Clay	vco	very coarse
S	Sand	SiCL	Silty Clay Loam	g	gravelly
SC	Sandy Clay	SiL	Silt Loam		

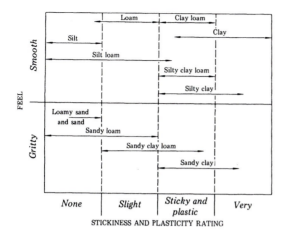

Figure A1.4 Approximate relations between texture class, grittiness, and wet consistence.

Clay Films

Clay films are thin layers of oriented clay and are described by recording their amount, distinctness, and locations. Study the peds with a hand lens in the field, or with a binocular microscope in the laboratory.

Amount:

v1—very few. Occupies less than 5% of the total area of the kind of surface described.
1—few. Occupies 5–25% of the total area of the kind of surface described.
2—common. Occupies 25–50% of the total area of the kind of surface described.
3—many. Occupies more than 50% of the total area of the kind of surface described.

The same classes are used to describe the amount of bridges connecting particles of structureless soil bodies. The amount is judged on the basis of the percentage of particles of the size designated that are joined to adjacent particles of similar size by bridges at contact points.

Distinctness: Distinctness refers to the ease and degree of certainty with which a surface feature can be identified. Distinctness is related to thickness, color contrast with the adjacent material, and other properties, but is not itself a measure of any one of them. Some thick films, for example, are faint, whereas some thin ones are prominent. The distinctness of some surface features changes markedly as the amount of mois-

ture changes; therefore, the soil-water state might be specified. Clay films are difficult to recognize in wet soils. If classifying films on ped faces, compare features on a ped face with those on a nonstructural face broken across the ped. Three distinctness classes are used.

f—faint. Evident only on close examination with 10× magnification and cannot be identified positively in all places without greater magnification. The contrast with the adjacent material in color, texture, and other properties is small.

d—distinct. Can be detected without magnification, although magnification or tests may be needed for positive identification. The feature contrasts enough with the adjacent material that a difference in color, texture, or other properties is evident.

p—prominent. Conspicuous without magnification when compared with a surface broken through the soil. Color, texture, or some other property or combination of properties contrasts sharply with properties of the adjacent material, or the feature is thick enough to be conspicuous.

Location of Clay Films: Oriented clay is present as films on peds, inside of pores, or as bridges between grains and coats on grains. If films are preferential to some orientation (horizontal vs vertical), this should be noted.

pf—clay films occur on ped faces. Where the structure grade is weak or the soil is structureless, ped faces are indistinct or absent. It is probable that only when the structure grade is moderate or strong are the clay films on ped faces discernible.
po—clay films line tubular or interstitial pores.
br—oriented clay occurs as bridges holding mineral grains together. This is probably an initial step that occurs before clay films coat grains and is best observed in coarse-textured soils.
co—colloid coats mineral grains.
cobr—coats and bridges are present. This is probably more common than coats or bridges alone.

In describing clay films, care must be exercised not to confuse pressure faces with clay films. The former are common in soils with high clay content (Vertisols; shrink–swell clay such as smectite is best), and seasonal wetting and drying. Pressure faces arise when swelling pushes structural aggregates together and makes their sides smooth and, in places, reflective. At times these are difficult to differentiate from clay films, but some clay films can also be partly pressure faces. Slickensides are produced in the same manner, but are better developed, being polished and striated, and usually at >50 cm depth. Where slickensides are prominent, they are extensive and oriented at 20–30° from the horizontal to form wedges (Ahmad, 1983). If the shrinking and swelling that produce slickensides are extensive enough, wide and deep ground cracks will form during the dry season.

Examples of Clay-Film Descriptions:

3d po—many distinct clay films in pores.
2f pf and po—common faint clay films on peds and in pores.
3p pf, 2f po—many prominent clay films on ped faces, common faint clay films in pores.

It is important to record clay films because their presence is strong evidence for pedogenically illuviated clay. However, be warned that in places clay films can be original depositional (parent material) features. Waters charged with fine sediment that infiltrate a flood plain can produce clay films at depth (Walker and others, 1978), as can similar waters infiltrating till at the base of a glacier. If these latter parent-material films are present below the main soil-forming zone, their color will be closer to that of the parent material than to that of the soil.

Horizon Boundaries

Describe the lower boundary of each horizon, indicating distinctness and general topography.

Distinctness:

a—abrupt. Transition is less than 2 cm.
c—clear. Transition is 2–5 cm thick.
g—gradual. Transition is 5–15 cm thick.
d—diffuse. Transition is more than 15 cm thick.

Topography: Topography refers to the nature of the surface that separates the horizons. The modifiers sl (slightly) and v (very) may be used in combination with the following abbreviations.

s—smooth. Boundary is planar or parallel to the geomorphic surface.
w—wavy. Undulating surface with pockets wider than they are deep.

i—irregular. If pockets are deeper than their width.
b—broken. If one or both of the horizons separated by the boundary are discontinuous, so that boundary is interrupted.

Stages of Carbonate Morphology

Describe the stage of morphology (Fig. A1.5, Tables A1.5 and A1.6). In some places, there may not be stage II morphology in a sequence of nongravelly soils; rather, filaments of stage I become so common that the horizon meets the approximate percentage requirements for stage II. Holliday (1982) suggests that these latter occurrences be termed IIf to indicate their filamentous morphology.

I want to inject a word of caution on the recognition of carbonate morphological stages. In places, carbonate can be deposited on vertical faces by laterally seeping waters and thereby mask the pedogenic carbonate morphology (Lattman, 1973). In addition, M.N. Machette and R.E. Anderson (personal communication, 1991) have observed strong lateral (on contour) variations in carbonate morphology and accumulation along natural arroyos in arid parts of the eastern Great Basin. Hence, to study the morphology of pedogenic carbonate and avoid surficial cementation, one may have to dig back a meter or more.

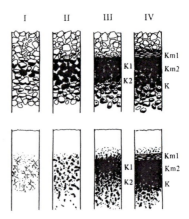

Figure A1.5 Sketch of carbonate buildup stages (I, II, III, IV) for gravelley (top) and nongravelley (bottom) parent materials. Machette (1985) added two more stages beyond stage IV (Table A1.5). In general, the stage morphologies merge to a common form at about stage III. *(Redrawn and modified from Gile and others, 1966, © 1966, The Williams & Wilkins Co., Baltimore.)*

Table A1.5 Stages of Carbonate Morphology

Stage	Gravelly Parent Material	Nongravelly Parent Material
I	Thin discontinuous clast coatings; some filaments; matrix can be calcareous next to stones; about 4% $CaCO_3$	Few filaments or coatings on sand grains; <10% $CaCO_3$
I+	Many or all clast coatings are thin and continuous	Filaments are common
II	Continuous clast coatings; local cementation of few to several clasts; matrix is loose and calcareous enough to give somewhat whitened appearance	Few to common nodules; matrix between nodules is slightly whitened by carbonate (15–50% by area), and the latter occurs in veinlets and as filaments; some matrix can be noncalcareous; about 10–15% $CaCO_3$ in whole sample, 15–75% in nodules
II+	Same as stage II, except carbonate in matrix is more pervasive	Common nodules; 50–90% of matrix is whitened; about 15% $CaCO_3$ in whole sample
Continuity of fabric high in carbonate		
III	Horizon has 50–90% K fabric with carbonate forming an essentially continuous medium; color mostly white; carbonate-rich layers more common in upper part; about 20–25% $CaCO_3$	Many nodules, and carbonate coats so many grains that over 90% of horizon is white; carbonate-rich layers more common in upper part; about 20% $CaCO_3$
III+	Most clasts have thick carbonate coats; matrix particles continuously coated with carbonate or pores plugged by carbonate; cementation more or less continuous; >40% $CaCO_3$	Most grains coated with carbonate; most pores plugged; >40% $CaCO_3$
Partly or entirely cemented		
IV	Upper part of K horizon is nearly pure cemented carbonate (75–90% $CaCO_3$) and has a weak platy structure due to the weakly expressed laminar depositional layers of carbonate; the rest of the horizon is plugged with carbonate (50–75% $CaCO_3$)	
V	Laminar layer and platy structure are strongly expressed; incipient brecciation and pisolith (thin, multiple layers of carbonate surrounding particles) formation	
VI	Brecciation and recementation, as well as pisoliths, are common	

Taken from Gile and others (1981) and Machette (1985), with further modification by R.R. Shroba (written communication, 1982).

Carbonate Effervescence

If dilute HCl (use a 1:10 ratio of concentrated HCl:water) is added to a soil containing $CaCO_3$, it will effervesce. The classes of effervescence are generally related to the amount of carbonate as well as to particle size (more rapid with smaller size) and mineralogy (slight with dolomite). Four classes are recognized:

Very slightly effervescent—few bubbles seen.
Slightly effervescent—bubbles readily seen.
Strongly effervescent—bubbles form low foam.
Violently effervescent—thick foam forms quickly.

For most geomorphic purposes, carbonate morphology stage is more useful than the classification of effervescence.

Salts and Silica Development

Pedogenic gypsum and silica have developmental stages that are similar to the stages of carbonate morphology (Table A1.7). One could devise a similar scheme for halite or any other accumulation of interest.

Cementation

Cementation refers to the brittle, hard consistence caused by some cementing agent, such as silica or $CaCO_3$, which, unlike clay, does not deform under pressure.

cw—weakly cemented. Mass is brittle and hard, but can be broken in hands.

Table A1.6 Further Subdivision of Carbonate Stages I and II in Gravelly Material on the Basis of Coating Thickness and Coverage

Stage of Carbonate Development	Characteristics		
	Bottom coverage (%)	Thickness (mm)	Remarks
IA	5–30	<0.5	Thin coats.
IB	30–70	≤0.5	
IC	70–100	0.5–1.0	
IIA	90–100	1–2	Thick coat; 0.5–1.0 mm pendants
IIB	100	2–5	Thick plates; 1–3 mm pendants

From Forman and Miller (1984).

Table A1.7 Carbonate, Gypsum, and Silica Stages in Gravelly Soil

Stage	Precipitate	Morphology
I	Carbonate	Thin, discontinuous under clasts
	Gypsum	Thin, discontinuous under clasts
	Silica	Scale-like coatings under clasts
II	Carbonate	Continuous around clasts; loose matrix
	Gypsum	Abundant pendants under clasts, crystals, nodules in matrix
	Silica	Stalactitic, pendants below undercoatings
III	Carbonate	Continuous around clasts, through matrix; interpebble fillings
	Gypsum	Continuous thru matrix; large pendants under clasts
	Silica	Cemented matrix
IV	Carbonate	Platy, weak laminar caps
	Gypsum	Continuous, plugged, wedging clasts
	Silica	Laminar, indurated platelets

From Harden and others, 1991, Table 1.1.

Table A1.8 Stages of Silt-Cap Development and Their Characteristics

Stage of Silt-Cap Development	Characteristics		
	Thickness (mm)	Coverage (%)	Remarks
1 (cap)	<1	5–50	
2 (cap)	1–4	50–90	
3 (cap)	4–7	75–90	
4 (cap)	5–10	90–100	
5 (cap and bridge)	10–20	100	Caps interconnected to form bridges
6 (encapsulate)	10–30	100	Clast completely encapsulated by infiltrated silt

From Forman and Miller (1984).

cs—strongly cemented. Mass is brittle, cannot be broken in hands, easily broken with hammer.

ci—indurated. Very strongly cemented, brittle, does not soften under prolonged wetting; breaks only with a sharp blow with a hammer, rings to hammer blow.

Cementation may be continuous or discontinuous and this feature should be described.

Silt Caps

In places, one can detect translocation of silt and clay by the presence of these materials on the tops of clasts in the soil (Table A1.8). The caps become thicker and cover a greater proportion of the clast surface with time (stages 1–4), form bridges between clasts (stage 5), and eventually encapsulate all clasts (stage 6). Stages 5 and 6 characterize the B*l* horizon. Some silt caps can be parent-material features; these will have colors closer to those of the parent material than to those of the soil.

Not all U.S. Department of Agriculture soil properties are given above, and some of the above (e.g., cementation) is no longer used. Consult Soil Survey Division Staff (1993) for the description of additional properties (e.g., roots, pores, concretions and nodules, and plinthite).

Calculation of
Profile-Development Index

There are nine steps in calculating the PDI (Fig. A2.1) as described by Harden (1982) and Harden and Taylor (1983), with some modification.

1. Describe the soil profile. In any study, the same individual should describe all the soils, so that uniform descriptions are obtained. Reheis and others (1989) reported on the variations in PDI that can occur (1) when several individuals describe the same soil and (2) when the number of profiles/geomorphic surface is increased.

2. Assess the parent material or parent materials, and describe them in the same way you would a soil horizon (color, texture, clay films, etc.). In many places, the Cu horizon is the parent material. In other places, weathering may extend so deep that unaltered parent material is not reached in the exposure of the soil; therefore, you must find a reasonable substitute (surrogate) for the parent material. For example, modern river alluvium can be used for the parent material of a river terrace, or "Little Ice Age" (latest Holocene) till at the front of glaciers might be used for the parent material for tills farther downvalley. Contemporary floodplain silt (overbank alluvium) might be representative of the parent material for loess that is floodplain derived. If, however, the loess is derived from the reworking of an older loess, one may use the properties of the latter. Parent-material properties for colluvium are difficult to ascertain; in places one could use the properties of ground-up rock as a parent-material surrogate. Finally, in the unusual case in which an older soil is the parent material for a younger soil

(e.g., if properties formed before and after a significant climatic change, respectively), then the properties of the older soil are valid for the parent-material properties of the younger soil.

3. Quantify the degree to which the soil-horizon properties differ from the parent-material properties. Ten-point intervals are used in the quantification, and five-point values can be used for properties with intermediate values (Table A2.1, Part A). It should be emphasized that these point assignments for various degrees of development of a property are somewhat arbitrary and a soil property that is assigned 20 points does not necessarily mean the property is twice as developed as one that is assigned 10 points. The number of properties is unlimited, with 13 given in Table A2.1. In each case, the direction of point increase is the direction of the progression of that property with time. Several rules apply, however. If rubification proceeds with time, and color paling does not occur within the soil sequence of interest, the latter should not be included as a zero property but should be excluded from the index. The same goes for the melanization-color lightening pair. However, if clay films occur in older soils but not in younger soils, they should be entered as zero properties in the younger soils.

4. The properties are then normalized to what is called the "current maximum" (see Table A2.1, part B). This results in a number between 0 (no development) and 1 (maximum development) for that property. However, in places the horizon value might exceed that of the current maximum (i.e., 1.25), and rather than redefine the current maximum, we proceed

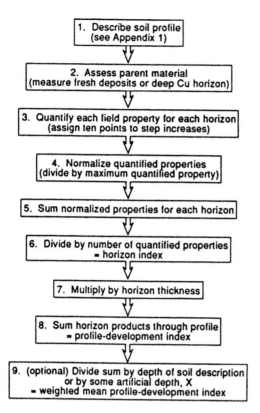

1. Describe soil profile
(see Appendix 1)

2. Assess parent material
(measure fresh deposits or deep Cu horizon)

3. Quantify each field property for each horizon
(assign ten points to step increases)

4. Normalize quantified properties
(divide by maximum quantified property)

5. Sum normalized properties for each horizon

6. Divide by number of quantified properties
= horizon index

7. Multiply by horizon thickness

8. Sum horizon products through profile
= profile-development index

9. (optional) Divide sum by depth of soil description
or by some artificial depth, X
= weighted mean profile-development index

Figure A2.1 Flow diagram for deriving the profile-development index. *(Redrawn and modified from Harden and Taylor, 1983 Fig. 1.)*

with the calculation, using the normalized value of 1.25. The system is very flexible and allows for alterations as given in footnotes to Table A2.1, as well as the introduction of new properties.

5. Sum the normalized values for properties for each horizon.

6. Divide the sum (in 5) by the total number of properties for that horizon. This value is called the horizon index, which varies between 0 and 1 unless the soil horizon being describing exceeds the current maxima. In the case in which a horizon has a zero property, it should be included in the total number of properties. As in the example mentioned earlier, young soils may not have any clay films yet older soils do. The property value therefore is zero in the young soils, and it is included in the number of total properties investigated.

7. Multiply by horizon thickness in centimeters to weight the horizon properties for the proportion of the total profile that the horizon represents.

8. Sum the horizon products for the profile. This is the profile-development index (PDI) of most workers. The value is open ended at the high end.

9. Divide sum (in 8) by depth of the described soil or by some prescribed (artificial, in places) depth. This produces a weighted mean PDI, with values between 0 (no development) and 1 (maximum development). The depth used in this calculation is important. Most soils are described to variable depths, usually to the parent material; in particular, young soils are described to shallow depths and old soils to great depths. However, if the Cu horizon begins at 20 cm in a young soil and at 120 cm in an old soil, the best comparison between the two is to dig both to 120 cm, but this is seldom done. The resulting PDIs for the young soil can differ significantly depending on the total soil depth considered. If the PDI for the young soil is weighted to the same depth as the old soil, the inclusion of 100 cm of zero properties (Cu horizon) for the younger soil will result in a smaller PDI for the young soil than if its properties are weighted to the maximum depth of the hole. The result is a greater difference in PDIs between young and old soils. It may be preferable to compare soils to an equivalent apparent described thickness (termed artificial in Fig. A2.1) by increasing the thickness of the lowest horizon in all but the thickest soil in the local sequence. The problem with this manipulation is the assumption that the properties of the lowest horizon in all profiles continue to that uniform depth, and this may not be true for all soils. J.W. Harden (written communication, 1990) does not endorse these manipulations and prefers these be termed optional. Table A2.1 gives examples of the calculations mentioned above, Table A2.2 is useful for making the calculations, and Table A2.3 is a PDI calculation for an ~75-ka aridic soil in southern New Mexico described by the author (unpublished).

■ Plotting and Comparing Soil-Index Data

There are many ways to plot either the horizon or PDI as illustrated in the following examples.

1. Plot individual properties with depth to depict changes that take place over the same time range in different environments (Fig. A2.2).
2. Plot profile-property data versus age for soils in different environments to show the variation in development with environment (Fig. A2.3).

Table A2.1 Quantification and Normalization of Soil Properties for Calculating PDI

	Part A — Quantification of Soil Properties	Example — Horizon Property	Example — Parent Material Property	Part B — Normalization of Quantified Properties — Calculation	Example
	Soil Property				
Rubification	10 pts/increase in *hue* redness 5Y ⇒ 2.5Y ⇒ 10YR ⇒ 7.5YR ⇒ 5YR ⇒ PM HP 2.5YR ⇒ 10R ⇒ 5R 10 pts/increase in *chroma* HP(d) 0 ⇒ 1 ⇒ 2 ⇒ 3 ⇒ 4 ⇒ 5 ⇒ 6 ⇒ 7 ⇒ 8 PM HP(m) PM Note: For multiple colors, record and calculate each[a]	7.5YR 5/4 (d) 7.5YR 4/5 (m)	10YR 7/2 (d) 10YR 6/2 (m)	Divide by current maximum (190)	
	$X_r = 10[(\text{hue } \Delta X_0 + (\text{chroma } \Delta X_0)]_{dry} + 10[(\text{hue } \Delta X_0) + (\text{chroma } \Delta X_0)]_{moist}$	$X_r = 10(1 + 2)_{dry} + 10(1 + 3)_{moist}$ $X_r = 70$		$X_m = X_r \div 190$	$X_m = 70 \div 190$ $X_m = 0.37$
Color paling	10 pts/decrease in *hue* redness 5Y ⇐ 2.5Y ⇐ 10YR ⇐ 7.5YR ⇐ 5YR ⇐ 2.5YR ⇐ 10R ⇐ 5R HP PM HP(d) PM (d) 10 pts/decrease in *chroma* HP(d) PM (d) 0 ⇐ 1 ⇐ 2 ⇐ 3 ⇐ 4 ⇐ 5 ⇐ 6 ⇐ 7 ⇐ 8 HP PM(m)				
	$X_{cp} = 10[(\text{hue } \Delta X_0) + (\text{chroma } \Delta X_0)]_{dry} + 10[(\text{hue } \Delta X_0) + (\text{chroma } \Delta X_0)]_{moist}$	2.5Y 6/2 (d) 2.5Y 6/4 (m) $X_{cp} = 10(1 + 2)_{dry} + 10(1 + 1)_{moist}$ $X_{cp} = 50$	10YR 4/4 (d) 10YR 4/5 (m)	Divide by current maximum (60) $X_{cpn} = X_{cp} \div 60$	$X_{cpn} = 50 \div 60$ $X_{cpn} = 0.83$
Melanization	For horizons with upper boundary at ≤100 cm: 10 pts/decrease in *value* HP(d) PM(d) 0 ⇐ 1 ⇐ 2 ⇐ 3 ⇐ 4 ⇐ 5 ⇐ 6 ⇐ 7 ⇐ 8 ⇐ 9 ⇐ 10 HP(m) PM(m)				
	$X_m = 10(\text{value } \Delta X_0)_{dry} + 10(\text{value } \Delta X_0)_{moist}$	7.5YR 5/4(d) 7.5YR 4/5 (m) $X_m = 10(2)_{dry} + 10(2)_{moist}$ $X_m = 40$	10YR 7/2(d) 10YR 6/2 (m)	Divide by current maximum (85) $X_{mn} = X_m \div 85$	$X_{mn} = 40 \div 85$ $X_{mn} = 0.47$
Color lightening	10 pts/increase in *value* 0 ⇒ 1 ⇒ 2 ⇒ 3 ⇒ 4 ⇒ 5 ⇒ 6 ⇒ 7 ⇒ 8 ⇒ 9 ⇒ 10 PM HP				
	$X_{cl} = 10(\text{value } \Delta X_0)_{dry} + 10(\text{value } \Delta X_0)_{moist}$	10YR 8/2(d) 10YR 8/3(m) $X_{cl} = 10(2)_{dry} + 10(2)_{moist}$ $X_{cl} = 40$	10YR 6/3(d) 10YR 6/4(m)	Divide by current maximum (80) $X_{cln} = X_{cl} \div 80$	$X_{cln} = 40 \div 80$ $X_{cln} = 0.50$
Total texture	*Texture* 10 pts/line crossing toward clay on texture triangle 10 pts/increase in *stickiness* so ⇒ ss ⇒ s ⇒ vs PM HP 10 pts/increase in *plasticity* po ⇒ ps ⇒ p ⇒ vp PM HP	SCL ps s	SL so po	Divide by current maximum (90)	
	$X_t = 10[(\text{textural } \Delta X_0) + (\text{stickiness } \Delta X_0) + (\text{plasticity } \Delta X_0)]$	$X_t = 10(1 + 2 + 1)$ $X_t = 40$		$X_{tn} = X_t \div 90$	$X_{tn} = 40 \div 90$ $X_{tn} = 0.44$
Dry consistence	10 pts/increase in *hardness* lo ⇒ so ⇒ sh ⇒ h ⇒ vh ⇒ ch PM HP	vh $X_{dc} = 10(4)$ $X_{dc} = 40$	lo	Divide by 2 × current maximum $(2 \times 50 = 100)$	$X_{dcn} = 40 \div 100$ $X_{dcn} = 0.40$
	$X_{dc} = 10(\text{dry consistence } \Delta X_0)$			$X_{dcn} = X_{dc} \div 100$	

Property	Scale / definition	Example values	Calculation	Example calc	Normalization	Normalized
Moist consistence[b]	10 pts/increase in *firmness* $lo \Rightarrow vfr \Rightarrow fr \Rightarrow fi \Rightarrow vfi \Rightarrow efi$ PM .... HP	fi $X_{mc} = 10(3)$ $X_{mc} = 30$			Divide by 2 × current maximum $(2 \times 50 = 100)$ $X_{mcn} = X_{mc} \div 100$	$X_{mcn} = 30 \div 100$ $X_{mcn} = 0.30$
Clay films[b]	points 10 20 30 40 | 1 2 3 amount: vl 1 1 | f d p distinctness | po br location | co pf Note: If classes are equal, choose 1° class with greatest abundance. $X_{cf} = [(\text{abundance} + \text{distinctness} + \text{location}) \text{ of } 1° \text{ class} + 1/2 (\text{abundance of } 2° \text{ class})]$	3fbr 1ppf, 2dbr $X_{cf} = (40 + 10 + 20) + 1/2 (20 + 30)$ $X_{cf} = 95$	No films		Subtract 20 pts from all $X_{cf} > 0^c$ Divide by current maximum (130) $X_{cfn} = (X_{cf} - 20) \div 130 \geq 0$	$X_{cfn} = (95 - 20) \div 130$ $X_{cfn} = 0.58$
Structure[d]	points 5 10 20 30 | 1 2 3 grade | pl gr pr col type | sbk abk $X_s = [(\text{grade} + \text{type}) \text{ of } 1°] + 1/2 [(\text{grade} + \text{type}) \text{ of } 2°]$	1co sbk $X_s = 10(1) + 10(1)$ $X_s = 20$	Structureless		Divide by maximum primary possible (60) $X_{sn} = X_s \div 60$	$X_{sn} = 20 \div 60$ $X_{sn} = 0.33$
pH[e]	Difference between pH of horizon and parent material $X_{pH} = pH \, \Delta X_0$	7.2 $X_{pH} = 7.9 - 7.2$ $X_{pH} = 0.7$	7.9		Divide by decrease of 3.5 or increase of 1.5 $X_{pHn} = X_{pH} + 3.5 \geq 0$	$X_{pHn} = 0.7 \div 3.5$ $X_{pHn} = 0.20$
Carbonate morphology[f]	Use carbonate stage as multiplier (disseminated, 0.5; stage I to VI, 1 to 6). $X_{cm} = (X_{cp} + X_{cl})(\text{carbonate stage})$	Using above color paling and color lightening, and stage II carbonate morphology $X_{cm} = (50 + 40) \times 2$ $X_{cm} = 180$			Divide by current maximum (240) $X_{cmn} = X_{cm} \div 240$	$X_{cmn} = 180 \div 240$ $X_{cmn} = 0.75$
Color mottling[g]	10 pts/increase in *abundance* none ⇒ few ⇒ common ⇒ many PM .... HP 10 pts/increase in *degree of contrast* none ⇒ faint ⇒ distinct ⇒ prominent PM .... HP $X_{mo} = \text{abundance } \Delta[pch]_0 + \text{contrast } \Delta X_0$	Common 2.5YR 4/6(m) $X_{mo} = 10(2) + 10(3)$ $X_{mo} = 50$	None 2.5Y 7/3(m)		Divide by current maximum (60) $X_{mon} = X_{mo} \div 60$	$X_{mon} = 50 \div 60$ $X_{mon} = 0.83$
Clast weathering[h]	10 pts/increase in *weathering stage* none ⇒ slight ⇒ moderate ⇒ high ⇒ extreme PM .... HP $X_{cw} = 10(\text{weathering stage } \Delta X_0)$	Moderate $X_{cw} = 10(2)$ $X_{cw} = 20$	None		Divide by current maximum (40) $X_{cw} = X_{cw} \div 40$	$X_{cwn} = 20 \div 40$ $XX_{cwn} = 0.50$

Modified from procedures outlined by Harden (1982), Harden and Taylor (1983), and Taylor (1988). See Taylor (1988) for current values (Part B).

[a] See Taylor (1988) for details of calculating multiple colors.

[b] For this property and structure property, 1° denotes primary and 2° secondary. The new terms amount and distinctness replace frequency and thickness. Workers might consider a change in location points: br = 10; co, cobr, po = 20; pf = 30. We believe this is a more common progression with time.

[c] Harden (1982) subtracts 20 pts; we do not, and instead compare the horizon to the parent material, and record only positive values.

[d] Workers might consider a change in type points: gr, sbk = 10; abk, pr, pl = 20; col = 30.

[e] The original intent here was to assign points for a lowering of pH. However, in some places the pH will increase with time as pedogenic carbonate accumulates. If this is done, use a maximum pH increase of 1.5 (maximum pH of 8.5).

[f] Stages from Machette (1985). Modified from Sowers and others (1988) and Taylor (1988).

[g] From Knuepfer (1988).

[h] From Knuepfer (1988). The stages are slightly weathered (minor spalling and cracking), moderately weathered (easily broken by hammer), highly weathered (disintegrated by hand), and extremely weathered (so completely weathered that only outline of clast remains). Workers can alter this for different kinds of rocks and progressive weathering stages.

Table A2.2 Work Sheet for Making PDI Calculations

Horizon					
Horizon thickness					
Values					
Rubification					
Paling					
Melanization					
Lightening					
Texture					
Dry consistence					
Moist consistence					
Clay films					
Structure					
pH					
Carbonate morphology					
Color mottling					
Clast weathering					
Normalized Values					
Rubification					
Paling					
Melanization					
Lightening					
Texture					
Dry consistence					
Moist consistence					
Clay films					
Structure					
pH					
Carbonate morphology					
Color mottling					
Clast weathering					
1. Sum of normalized values					
2. Divide by number of properties					
3. Multiply by horizon thickness					
4. Sum of (3)					
5. Divide (4) by profile thickness					

Table A2.3 PDI Calculation for ~75-ka Aridic Soil in Southern New Mexico[a]

	PM	Av	Bk	2Kmj	2Btk
Horizon Properties					
Horizon thickness		7	25	48	31
Color—dry	10 YR 6/2	7.5YR 6/3	7.5YR 6/3	7.5YR 8/2	7.5YR 7.5/2
Color—moist	10 YR 5/2	7.5YR 4/3	7.5YR 4.5/3	7.5YR 7.5/3	7.5YR 6/3.5
Structure	sg	3, pl	m-1, sbk	n/a	m-sg
Wet consistence	so, po	s, ps-p	s, ps-p	n/a	s, p
Dry Consistence	lo	sh	so	h-vh	lo-so
Texture	S	SL	SL	n/a	SCL
Clay films	none	3, f, cobr	3, f, cobr	none	none
Carbonate stage	none	0	I	III.5	II
Quantified Soil Field Properties					
Rubification		40	40	30	35
Melanization		10	5	0	0
Lightening		0	0	45	25
Total texture		55	55	0	70
Dry consistence		20	10	35	5
Structure		50	15	0	0
Clay films		70	70	0	0
Carbonate stage		0	0	157.5	50
Normalized Values for Properties					
Rubification		0.21	0.21	0.16	0.18
Melanization		0.12	0.06	0.00	0.00
Lightening		0.00	0.00	0.56	0.31
Total texture		0.61	0.61	0.00	0.78
Dry consistence		0.40	0.20	0.70	0.10
Structure		0.83	0.25	0.00	0.00
Clay films		0.54	0.54	0.00	0.00
Carbonate stage		0.00	0.00	0.66	0.21
Index Results					
$\sum$ of normalized values		2.71	1.87	2.08	1.58
$\div$ by number of properties (8)	HI	0.34	0.23	0.26	0.20
$\times$ by horizon thickness		2.37	5.84	12.48	6.12
$\sum$ of horizon properties	PDI	26.82			
$\div$ by described thickness (111 cm)	WM PDI	0.24			

[a]Data from unpublished work of author, and calculations and table courtesy of M.A. Franseen.

3. Plot the horizon-development-index data with depth to portray changes in soils with time (Fig. A2.4).
4. Plot PDI versus estimated age (Fig. A2.5) to portray soil development as a function of deposit age. The data are best portrayed as semilog or log–log plots because of the great range in soil age that is typically encountered in soil chronosequence studies (i.e., several thousand years to a million years), plus the fact that many ages are estimates.

5. Finally, one can compare PDI with time for different environments (Fig. A2.6). Various manipulations can be made; Harden and Taylor (1983) preferred using the four "best" properties in each environment (not the same properties in each environment) rather than all eight of the properties for which they collected data. When the four best properties were used, the data from soils formed in greatly different environments plotted within a tighter envelope (curve) than when

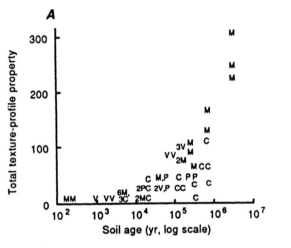

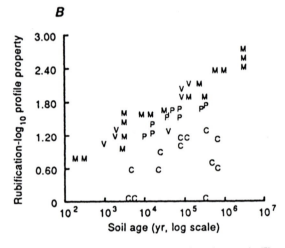

Figure A2.2 Depth plots of two field properties of soils that are approximately 40–120 ka. (A) Normalized total texture plotted against basal horizon depth. (B) Normalized rubification plotted against basal horizon depth. Locations are V, Ventura, California; M, Merced, California; C, Las Cruces, New Mexico; P, Pennsylvania. *(Redrawn from Harden and Taylor, 1983 Fig. 3.)*

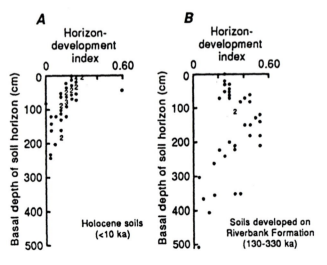

Figure A2.3 Profile calculations versus age for soils of four chronosequences. Letters represent same locations as in Fig. A2.2, small numbers indicate multiple data points. (A) Total texture. (B) Rubification. *(Redrawn from Harden and Taylor, 1983 Fig. 4.)*

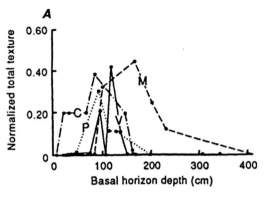

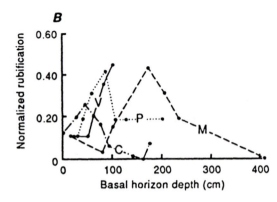

Figure A2.4 Horizon-development index versus depth to base of horizon for two different age soils near Merced, California. Small numbers indicate multiple data points. (A) Holocene soils. (B) Soils on Riverbank Formation, estimated at 130–330 ka. *(Redrawn from Harden, 1983 Fig. 7.)*

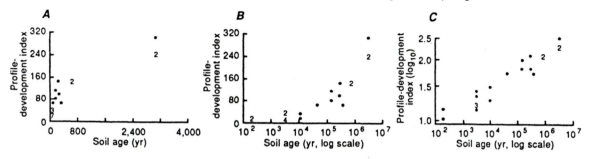

Figure A2.5 Profile-development index (PDI) versus age for soils near Merced, California. Small numbers indicate multiple data points. The PDI versus soil age are in (A) years, (B) log$_{10}$ years; (C) PDI (log$_{10}$) versus soil age in log$_{10}$ years. *(Redrawn from Harden and Taylor, 1982, Fig. 6.)*

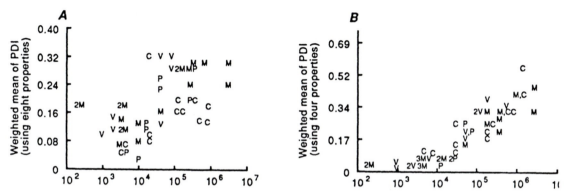

Figure A2.6 PDI calculations for soils of four chronosequences in different environments versus age. Letters represent same locations as in Fig. A2.2. (A) PDI calculated using eight soil properties. (B) PDI calculated using four soil properties. *(Redrawn from Harden and Taylor, 1983 Fig. 5.)*

all eight properties were used. I usually include all soil properties.

Many applications can be made of the PDI data. Data commonly are plotted against estimated age.

However, the applications are unlimited. Data can be plotted against landscape position, or some quantified portion of the parent material (e.g., SiO$_2$%). Examples will be presented throughout the book.

References

■ Chapter 1

Alexander, E.B. 1980. Bulk densities of California soils in relation to other soil properties. Soil Sci. Soc. Am. J. 44, 689–692.

Amundson, R., and Yaalon, D.H. 1995. E.W. Hilgard and J.W. Powell: efforts for a joint agricultural and geological survey. Soil Sci. Soc. Am. J. 59, 4–13.

Arkley, R.J. 1963. Calculations of carbonate and water movement in soil from climatic data. Soil Sci. 96, 239–248.

Arkley, R.J. 1964. Soil survey of the eastern Stanislaus area, California. U.S. Dept. Agric., Soil Surv. Series 1957, No. 20. 160 p.

Aylor, D.E., and Parlange, J. 1973. Vertical infiltration into a layered soil. Soil Sci. Soc. Am. Proc. 37, 673–676.

Baver, L.D. 1956. Soil physics. John Wiley & Sons, New York, 489 p.

Bigham, J.M., and Ciolkosz, E.J., eds. 1993. Soil color. Soil Sci. Soc. Am. Spec. Publ. No. 31, 159 p.

Bilzi, A.F., and Ciolkosz, E.J. 1977. A field morphology rating scale for evaluating pedological development. Soil Sci. 124, 45–48.

Birkeland, P.W. 1984. Holocene soil chronofunctions, Southern Alps, New Zealand. Geoderma 34, 115–134.

Black, C.A. 1957. Soil-plant relationships. John Wiley & Sons, New York, 332 p.

Blodget, R.H., Crabaugh, J.P., and McBride, E.F. 1993. The color of red beds—a geologic perspective. Soil Sci. Soc. Am. Spec. Publ. No. 31, 127–159.

Bockheim, J.G. 1990. Soil development rates in the Transantarctic Mountains. Geoderma 47, 59–77.

Bodman, G.B., and Mabmud, A.J. 1932. The use of the moisture equivalent in the textural classification of soils. Soil Sci. 33, 363–374.

Boggs, S., Jr. 1992. Petrology of sedimentary rocks. Macmillan, New York, 707 p.

Bowen, D.Q. 1978. Quaternary geology. Pergamon Press, Oxford, England, 221 p.

Bradley, R.S. 1985. Quaternary paleoclimatology. Allen & Unwin Inc., Winchester, MA, 472 p.

Brewer, R. 1964. Fabric and mineral analysis of soils. John Wiley & Sons, New York, 407 p.

Brook, G.A., Folkoff, M.E., and Box, E.O. 1983. A world model of soil carbon dioxide. Earth Surface Proc. Landforms 8, 79–88.

Buckman, H.O., and Brady, N.C. 1969. The nature and properties of soils. Macmillan, Toronto, 653 p.

Bull, W.B. 1991. Geomorphic responses to climatic change. Oxford University Press, New York, 326 p.

Bullock, P., and Murphy, C.P. 1979. Evolution of a paleo-argillic brown earth (Paleudalf) from Oxfordshire, England. Geoderma 22, 225–252.

Bullock, P., and Murphy, C.P., eds. 1983. Soil micromorphology, v. 1 and 2. A B Academic Publ., Berkhamsted, UK.

Bullock, P., Federoff, N., Jongerius, A., Stoops, G., and Babel, U. 1985. Handbook for soil thin section description. Waine Research Publ., Wolverhampton, England, 152 p.

Bunting, B.T. 1977. The occurrence of vesicular structures in arctic and subarctic soils. Zeitschrift Geomorphol. 21, 87–95.

Buntley, G.J., and Westin, F.C. 1965. A comparative study of developmental color in a Chestnut-Chernozem-Brunizem soil climosequence. Soil Sci. Soc. Am. Proc. 29, 579–582.

Buol, S.W., Hole, F.D., McCracken, R.J., and Southard, R.J. 1997. Soil genesis and classification. Iowa State Univ. Press, Ames, 544 p.

Carroll, D. 1962. Rainwater as a chemical agent of geologic processes: A review. U.S. Geol. Surv. Water-Supply Pap. 1535-G, 18 p.

Catt, J.A. 1986. Soils and Quaternary geology. Oxford University Press, New York, 267 p.

Catt, J.A.,ed. 1990. Paleopedology manual. Quaternary Int. 6, 1–95.

Chiang, S.C., West, L.T., and Radcliffe, D.E. 1994. Morphological properties of surface seals in Georgia soils. Soil Sci. Soc. Am. J. 58, 901–910.

Clayden, B., and Hewitt, A.E. 1993. A new system of horizon notation for New Zealand soils. Catena 20, 405–410.

Clayton, J.L., and Jensen, C.E. 1973. Water retention of granitic soils in the Idaho batholith. U.S.D.A. Forest Service Res. Pap. INT-143, 20 p.

Clothier, B.E., Scotter, D.R., and Kerr, J.P. 1977. Water retention in soil underlain by a coarse-textured layer. Theory and field application. Soil Sci. 123, 392–399.

Daniels, R.B. 1988. Pedology, a field or laboratory science? Soil Sci. Soc. Am. J. 52, 1518–1519.

Evenari, M., Yaalon, D.H., and Gutterman, Y. 1974. Note on soils with vesicular structure in deserts. Zeitschrift Geomorphol. 18, 162–172.

Fairbridge, R.W., and Finkl, C.W., Jr. 1984. Tropical stone lines and podzolized sand plains as paleoclimatic indicators for weathered cratons. Quaternary Sci. Res. 3, 41–72.

FitzPatrick, E.A. 1993. Soil microscopy and micromorphology. John Wiley & Sons, Ltd., England, 304 p.

Flint, R.F. 1971. Glacial and Quaternary geology. John Wiley & Sons, New York, 892 p.

Follmer, L.R., McKay, E.D., Lineback, J.A., and Gross, D.L. 1979. Wisconsinan, Sangamonian, and Illinoian stratigraphy in central Illinois. Illinois State Geol. Surv. Guidebook 13, 139 p.

Forman, S.L., and Miller, G.H. 1984. Time dependent soil morphologies and pedogenic processes on raised beaches, Bröggerhalvöya, Spitzbergen, Svaldbard Archipelago. Arc. Alp. Res. 16, 381–394.

Gile, L.H., Peterson, F.F., and Grossman, R.B. 1965. The K horizon. A master soil horizon of carbonate accumulation. Soil Sci. 99, 74–82.

Gile, L.H., Peterson, F.F., and Grossman, R.B. 1966. Morphological and genetic sequences of carbonate accumulation in desert soils. Soil Sci. 101, 347–360.

Gile, L.H., Hawley, J.W., and Grossman, R.B. 1981. Soils and geomorphology in the Basin and Range area of southern New Mexico—Guide-

book to the Desert Project. New Mexico Bur. Mines and Mineral Resources Mem. 39, 222 p.

Graham, R.C., Guertal, W.R., and Tice, K.R. 1994. The pedologic nature of weathered rock. Soil Sci. Soc. Am. Spec. Publ. No. 34, 21–40.

Harden, J.W. 1982. A quantitative index of soil development from field descriptions. Examples from a chronosequence in central California. Geoderma 28, 1–28.

Harden J.W. 1988. Measurements of water penetration and volume percentage water-holding capacity for undisturbed, coarse-textured soils in southwestern California. Soil Sci. 146, 374–383.

Harden, J.W., and Taylor, E.M. 1983. A quantitative comparison of soil development in four climatic regimes. Quaternary Res. 20, 342–359.

Harden, J.W., Taylor, E.M., McFadden, L.D., and Reheis, M.C. 1991. Calcic, gypsic, and siliceous soil chronosequences in arid and semiarid environments. Soil Sci. Soc. Am. Spec Publ. No. 26, 1–16.

Harrison, J.B.J. 1982. Soil periodicity in a formerly glaciated drainage basin, Ryton Valley, Craigieburn Range, Canterbury, New Zealand. Unpublished M.A.S. thesis, Lincoln College, Canterbury, New Zealand, 189 p.

Hodson, J.M. 1976, compiler and ed. Soil survey field handbook. Soil Surv. Tech. Mono. No. 5, Rothamsted Exp. Sta., Harpenden, Herts., England, 99 p.

Holliday, V.T. 1988. Genesis of the late-Holocene soil chronosequence at the Lubbock Lake archaeological site, Texas Ann. Assoc. Am. Geogr. 78, 594–610.

Hurst, V.J. 1977. Visual estimation of iron in saprolite. Geol. Soc. Am. Bull. 88, 174–176.

Jackson, M.L. 1973. Soil chemical analysis, advanced course. Dept. Soil Sci., Univ. of Wisconsin, Madison, WI, 894 p.

Janitzky, P. 1986. Determination of soil pH. U.S. Geol. Surv. Bull. 1648, 19–21.

Jenny, H., and Raychaudhuri, S.P. 1960. Effect of climate and cultivation on nitrogen and organic matter reserves in Indian soils. Indian Council of Agricultural Research, New Delhi, 126 p.

Joffe, J.S. 1949. Pedology. Pedology Publ., New Brunswick, NJ, 662 p.

Johnson, D.L. 1990. Biomantle evolution and the redistribution of earth materials and artifacts. Soil Sci. 149, 84–102.

Johnson, D.L., and Watson-Stegner, D. 1990. The soil-evolution model as a framework for evaluating pedoturbation in archaeological site formation. In N.P. Lasca and J. Donahue, eds., Archaeological geology of North America, p. 541–560. Geol. Soc. Am. Centennial Spec., v. 4. Geological Society of America, Boulder, CO.

Kemp, R.A. 1985. Soil micromorphology and the Quaternary. Quaternary Res. Assoc. Tech. Guide No. 2, Cambridge, England, 80 p.

Klute, A., ed. 1986. Methods of soil analysis. Pt. 1. Physical and mineralogical methods. Soil Sci. Soc. Am., No. 9 in Agron. Ser., 1188 p.

Knuepfer, P.L.K., and McFadden, L.D., eds. 1990. Soils and landscape evolution. Geomorphology 3, nos. 3/4.

Kubiena, W.L. 1970. Micromorphological features of soil geography. Rutgers Univ. Press, New Brunswick, NJ, 254 p.

Levine, E.L., and Ciolkosz, E.J. 1983. Soil development in till of various ages in northeastern Pennsylvania. Quaternary Res. 19, 85–99.

Machette, M.N. 1985. Calcic soils of the southwestern United States. Geol. Soc. Am. Spec. Pap. 203, 1–21.

Mausbach, M.J., Wingard, R.C., Jr., and Gamble, E.E. 1982. Modification of buried soils by postburial pedogenesis, southern Indiana. Soil Sci. Soc. Am. J. 36, 364–369.

McFadden, L.D. 1988. Climatic influences on rates and processes of soil development in Quaternary deposits of southern California. Geol. Soc. Am. Spec. Pap. 216, 153–177.

McFadden, L.D., and Knuepfer, P.L.K. 1990. Soil geomorphology: The linkage of pedology and surficial processes. Geomorphology 3, 197–205.

McFadden, L.D., and Tinsley, J.C. 1985. The rate and depth of accumulation of pedogenic carbonate in soils: formulation and testing of a compartment model. Geol. Soc. Am. Spec. Pap. 203, 23–41.

McFadden, L.D., Wells, S.G., and Jercinovic, M.J. 1987. Influences of eolian and pedogenic processes on the origin and evolution of desert pavements. Geology 15, 504–508.

Meixner, R.E., and Singer, M.J. 1981. Use of a field morphology rating system to evaluate soil formation and discontinuities. Soil Sci. 131, 114–123.

Miller, C.D., and Birkeland, P.W. 1974. Probable pre-Neoglacial age of the type Temple Lake moraine, Wyoming: Discussion and additional relative-age data. Arc. Alp. Res. 6, 301–306.

Morrison, R.B. 1964. Lake Lahontan. Geology of the southern Carson desert. U.S. Geol. Surv. Prof. Pap. 401, 156 p.

Morrison, R.B. 1978. Quaternary soil stratigraphy concepts, methods and problems. In W.C. Mahaney, ed., Quaternary soils, p. 77–108. Geo Abstracts Ltd., Univ. of East Anglia, Norwich, England.

Mualem, Y., Assouline, S., and Rohdenburg, H. 1990. Rainfall induced soil seal. (A) A critical review of observations and models. Catena, 17, 185–203.

Nielsen, D.R., and Shaw, R.H. 1958. Estimation of the 15-atmosphere moisture percentage from hydrometer data. Soil Sci. 86, 103–105.

Page, A.L., ed. 1982. Methods of soil analysis. Pt. 2. Chemical and microbiological properties. Soil Sci. Soc. Am., No. 9 in Agron. Ser.

Reheis, M.C. 1987. Gypsic soils on the Kane alluvial fans, Big Horn County, Wyoming. U.S. Geol. Surv. Bull. 1590-C, 39 p.

Reheis, M.C., Harden, J.W., McFadden, L.D., and Shroba, R.R. 1989. Development rates of late Quaternary soils, Silver Lake playa, California. Soil Sci. Soc. Am. J. 53, 1127–1140.

Retallack, G.J. 1990. Soils of the past: an introduction to paleopedology. Unwin Hyman, Boston, 520 p.

Richmond, G.M. 1962. Quaternary stratigraphy of the La Sal Mountains, Utah. U.S. Geol Surv. Prof. Pap. 324, 135 p.

Rodbell, D.T. 1990. Soil-age relationships on late Quaternary moraines, Arrowsmith Range, Southern Alps, New Zealand. Arctic Alp. Res. 22, 355–365.

Rode, A.A. 1962. Soil science. Israel Program for Scientific Translations, Jerusalem, 517 p.

Ruhe, R.V. 1956. Geomorphic surfaces and the nature of soils. Soil Sci. 82, 441–455.

Ruhe, R.V. 1959. Stone lines in soils. Soil Sci. 87, 223–231.

Ruhe, R.V. 1965. Quaternary paleopedology. In H.E. Wright, Jr. and D.G. Frey, eds., The Quaternary of the United States, p. 755–764. Princeton Univ. Press, Princeton, 922 p.

Ruhe, R.V. 1968. Identification of paleosols in loess deposits in the United States. In C.B. Schultz and J.C. Frye, eds., Loess and related eolian deposits of the world, p. 49–65. Int. Assoc. Quaternary Res., VII Cong., Proc. v. 12.

Ruhe, R.V. 1969. Quaternary landscapes in Iowa. Iowa State Univ. Press, Ames, 255 p.

Ruhe, R.V., and Daniels, R.B. 1958. Soils, paleosols, and soil-horizon nomenclature. Soil Sci. Soc. Am. Proc. 22, 66–69.

Ruhe, R.V., and Olson, C.G. 1980. Soil welding. Soil Sci. 130, 132–139.

Salter, P.J., and Williams, J.B. 1965. The influence of texture on the moisture characteristics of soils: I. A critical comparison of techniques for determining the available-water capacity and moisture characteristic curve of a soil: J. Soil Sci. 16, 1–15.

Salter, P.J., and Williams, J.B. 1967. The influence of texture on the moisture characteristics of soils: IV. A method of estimating the available-water capacities of profiles in the field. J. Soil Sci. 18, 174–181.

Schaetzl, R.J., and Sorenson, C.J. 1987. The concept of "buried" versus "isolated" paleosols: Examples from northeastern Kansas. Soil Sci. 143, 426–435.

Schaetzl, R.J., Mokma, D.L., and Delbert, L. 1988. A numerical index of podzol and podzolic soil development. Phys. Geogr. 9, 232–246.

Schulze, D.G., Nagel, J.L., Van Scoyoc, G.E., Henderson, T.L., Baumgardner, M.F., and Stott, D.E. 1993. Significance of organic matter in determining soil colors. Soil Sci. Soc. Am. Spec. Publ. No. 31, 71–90.

Schwertmann, U. 1993. Relations between iron oxides, soil color, and soil formation. Soil Sci. Soc. Am. Spec. Publ. No. 31, 51–69.

Singer, M.J., and Janitsky, P. 1986. Field and laboratory procedures used in a soil chronosequence study. U.S. Geol. Surv. Bull. 1648, 49 p.

Singer, M.J., and Munn, D.N. 1987. Soils. Macmillan, New York, 492 p.

Slate, J.L., Bull, W.B., Ku, T-L., Shafiqullah, M., Lynch, D.J., and Huang, Y-P. 1991. Soil-carbonate genesis in the Pinacate volcanic field, northwestern Sonora, Mexico. Quaternary Res. 35, 400–416.

Soil Survey Division Staff. 1993. Soil survey manual. U.S. Dept. Agri. Handbook No. 18, 437 p.

Soil Survey Staff. 1960. Soil classification, a comprehensive system (7th approximation). U.S. Dept. Agri., Soil Cons. Service, 265 p.

Soil Survey Staff. 1975. Soil taxonomy. U.S. Dept. Agri. Handbook No. 436, 754 p.

Springer, M.E. 1958. Desert pavement and vesicular layer of some soils of the desert of the Lahontan Basin, Nevada. Soil Sci. Soc. Am. Proc. 22, 63–66.

Stevenson, F.J. 1982. Humus chemistry. John Wiley & Sons, New York, 443 p.

Stolt, M.H., and Baker, J.C. 1994. Strategies for studying saprolite and saprolite genesis. Soil Sci. Soc. Am. Spec. Publ. No. 34, 1–19.

Stuart, D.M., and Dixon, R.M. 1973. Water movement and caliche formation in layered arid and semi-arid soils. Soil Sci. Soc. Am. Proc. 37, 323–324.

Tandarich, J.P., Follmer, L.R., and Darmody, R.G. 1994. The pedo-weathering profile: A paradigm for whole-regolith pedology from the glaciated midcontinental United States of America. Soil Sci. Soc. Am. Spec. Publ. No. 34, 97–117.

Torrent, J. Schwertmann, U., and Schulze, D.G. 1980. Iron oxide mineralogy of two river terrace sequences in Spain. Geoderma 23, 191–208.

Torrent, J., Schwertmann, U., Fechter, H., and Alferez, F. 1983. Quantitative relationship between soil color and hematite content. Soil Sci. 136, 354–358.

Ugolini, F.C., and Sletten, R.S. 1991. The role of proton donors in pedogenesis as revealed by soil solution studies. Soil Sci. 151, 59–75.

Valentine, K.W.G., and Dalrymple, J.B. 1976. Quaternary buried paleosols. A critical review. Quaternary Res. 6, 209–222.

Vincent, K.R., and Chadwick, O.A. 1994. Synthesizing bulk density for soils with abundant rock fragments. Soil Sci. Soc. Am. J. 58, 455–464.

Walker, P.H., and Green, P. 1976. Soil trends in two valley fill sequences. Aust. J. Soil Res. 14, 291–303.

Walker, T.R. 1967. Formation of red beds in ancient and modern deserts. Geol. Soc. Am. Bull. 78, 353–368.

Wells, N.A., Andriamihaja, B., and Rakotovololona, H.F.S. 1990. Stonelines and landscape development on the laterized craton of Madagasgar. Geol. Soc. Am. Bull. 102, 615–627.

White, E.M. 1966. Subsoil structure genesis. Theoretical consideration. Soil Sci. 101, 135–141.

Wright, H.E., Jr. 1984. Sensitivity and response time of natural systems to climatic change in the late Quaternary. Quaternary Sci. Rev. 3, 91–131.

Wright, H.E., Jr., Kutzbach, J.E., Webb, III, T., Ruddiman, W.F., Street-Perrott, F.A., and Bartlein, P.J., eds. 1993. Global climates since the last glacial maximum. Univ. Minnesota Press, Minneapolis.

Yaalon, D.H. (chairman). 1971a. Criteria for the recognition and classification of Paleosols. In D.H. Yaalon, ed., Paleopedology, p. 153–158. Israel Univ. Press, Jerusalem, 350 p.

Yaalon, D.H. 1971b. Soil-forming processes in time and space. In D.H. Yaalon, ed., Paleopedology, p. 29–39. Israel Univ. Press, Jerusalem, 350 p.

■ Chapter 2

Amen, A.E., and Foster, J.W. 1987. Soil landscape analysis project (SLAP) methods in soil surveys. U.S. Bur. Land management Tech. Note 379, 42 p.

Arkley, R.J. 1962. Soil survey of Merced area, California. U.S. Dept. Agri., Soil Survey Series 1950, No. 7, 131 p.

Arkley, R.J. 1964. Soil survey of the eastern Stanislaus area, California. U.S. Dept. Agri., Soil Survey Series 1957, No. 20, 160 p.

Bartelli, L.J., Klingebiel, A.A., Baird, J.V., and Heddleson, M.R., eds. 1966. Soil surveys and land use. Soil Sci. Soc. Am., 196 p.

Blackstock, D.A. 1979. Soil survey of Lubbock County, Texas. Soil Cons. Service, U.S. Dept. Agri., 105 p.

Bockheim, J.G. 1990. Properties and classification of cold desert soils from Antarctica. Soil Sci. Soc. Am. J. 61, 224–231.

Bockheim, J.G., Ping, C.L., Moore, J.P., and Kimble, J.M. 1994. Gelisols: A new proposed order for permafrost-affected soils. In J.M. Kimble and R. Ahrens, eds., Proceedings of a meeting on classification, correlation, and management of permafrost-affected soils, July 18–30, 1993, Alaska, U.S. and Yukon and Northwest Territories, Canada, p. 25–45. USDA-SCS, Washington, D.C.

Bockheim, J.G., Tarnocai, C., Kimble, J.M., and Smith, C.A.S. 1997. The concept of gelic materials in the new Gelisol order for permafrost-affected soils. Soil Sci. 162, 927–939.

Bockheim, J.G., and Tarnocai, C., 1998. Recognition of cryoturbation for classifying permafrost-affected soils. Geoderma 81, 281–293.

Buol, S.W., Hole, F.D., McCracken, R.J., and Southard, R.J. 1997. Soil genesis and classification. Iowa State Univ. Press, Ames, Iowa, 544 p.

Christenson, G.E., and Purcell, C. 1985. Correlation and age of Quaternary alluvial-fan sequences, Basin and Range province, southwestern United States. Geol. Soc. Am. Spec. Pap. 203, 115–122.

Crabb, J.A. 1980. Soil survey of Weld County, Colorado, southern part. Soil Conservation Service, U.S. Dept. Agri., 135 p.

FAO/UNESCO. 1972. Soil map of the world. FAO, Paris.

Hole, F.D. 1978. An approach to landscape analysis with emphasis on soils. Geoderma 21, 1–23.

Hole, F.D., and Campbell, J.B. 1985. Soil landscape analysis. Rowman & Allanheld, Totowa, NJ, 196 p.

Holliday, V.T. 1995. Stratigraphy and paleoenvironments of late Quaternary valley fills on the Southern High Plains. Geol. Soc. Am. Memoir 186, 136 p.

Holliday, V.T., McFadden, L.D., Bettis, A., and Birkeland, P.W. In press. Soil survey and geomorphology.

Hubble, G.D., Isbell, R.F., and Northcote, K.H. 1983. Features of Australian soils. In Division of Soils, CSIRO, Soils: An Australian viewpoint, p. 17–47. Academic Press, Melbourne.

Hudson, B.D. 1992. The soil survey as paradigm-based science. Soil Sci. Soc. Am. J. 56, 836–841.

Jenny, H. 1946. Arrangement of soil series and types according to functions of soil-forming factors. Soil Sci. 61, 375–391.

Jorgenson, D.W. 1992. Use of soils to differentiate dune age and to document spatial variation in eolian activity, northeast Colorado, U.S.A. J. Arid Environ. 23, 19–34.

Lof, P. (compiler), 1987, Soils of the world (chart). Elsevier, Amsterdam, The Netherlands.

Maat, P.B. 1992. Eolian stratigraphy, soils and geomorphology in the Hudson dune field. Unpublished M.S. thesis, Univ. Colorado, Boulder.

Mausbach, M.J., and Wilding, L.P. 1991. Spatial variabilities of soils and landforms. Soil Soc. Sci. Am. Spec. Publ. No. 28, 270 p.

Olson, G.W. 1984. Field guide to soils and the environment. Chapman and Hall, New York, 219 p.

Oschwald, W.R., Riecken, F.F., Dideriksen, R.I., Scholtes, W.H., and Schaller, F.W. 1965. Principal soils of Iowa—their formation and properties. Iowa State Univ. Coop. Ext. Serv. Spec. Rep. 42, 76 p.

Ponti, D.J. 1985. The Quaternary alluvial sequence of the Antelope Valley, California. Geol. Soc. Am. Spec. Pap. 203, 79–96.

Reybold, W.U., and Peterson, G.W. 1987. Soil survey techniques. Soil Sci. Soc. Am. Spec. Publ. No. 20, 98 p.

Soil Conservation Service. 1967. Soils of the Malibu area, California. U.S. Dept. Agri., 89 p.

Soil Conservation Service. 1994. Keys to soil taxonomy. U.S. Dept. Agri., 306 p.

Soil Survey Division Staff. 1993. Soil survey manual. U.S. Dept. Agri. Handbook No. 18, 437 p.

Soil Survey Staff. 1975. Soil taxonomy. U.S. Dept. Agri. Handbook No. 436, 754 p.

Smith, G.D. 1986. The Guy Smith interviews: Rationale for concepts in soil taxonomy. U.S. Dept. Agri., SMSS Tech. Monograph No. 11, 259 p.

Swanson, D.K. 1990a. Soil landform units for soil survey. Soil Surv. Horizons 31(1), 17–21.

Swanson, D.K. 1990b. Landscape classes: Higher-level map units for soil survey. Soil Surv. Horizons 31(2), 52–54.

Ulrich, R., and Stromberg, L.K. 1962. Soil survey of the Madera area, California. U.S. Dept. Agri., Soil Survey Series 1951, No. 11, 155 p.

Weir, W.W. 1952. Soils of San Joaquin County, California. Univ. Calif. Agri. Exp. Sta., Soil Survey No. 9, 137 p.

Woodruff, G.A., McCoy, W.J., and Sheldon, W.B. 1970. Soil survey of the Antelope Valley area, California. U.S. Dept. Agri., Soil Cons. Service, 187 p.

Yaalon, D.H. 1995. The soils we classify: Essay review of recent publications on soil taxonomy. Catena 24, 233–241.

Yaalon, D.H., and Kalmar, D. 1978. Dynamics of cracking and swelling clay soils: Displacement of skeletal grains, optimum depth of slickensides, and rate of intra-pedonic turbation. Earth Surf. Proc. Landforms 3, 31–42.

■ Chapter 3

Abrahams, A.D., and Parsons, A.J., eds. 1994. Geomorphology of desert environments. Chapman & Hall, London, 674 p.

Allison, R.J., and Goudie, A.S. 1994. The effects of fire on rock weathering: An experimental study. In D.A. Robinson and R.B.G. Williams, eds., Rock weathering and landform evolution, p. 41–56. John Wiley & Sons Ltd., Chichester, England, 519 p.

Amit, R., Gerson, R., and Yaalon, D.H. 1993. Stages and rate of the gravel shattering process by salts in desert Reg soils. Geoderma 57, 295–324.

Arkley, R.J. 1963. Calculation of carbonate and water movement in soil from climatic data. Soil Sci. 96, 239–248.

Barman, A.K., Varadachari, C., and Ghosh, K. 1992. Weathering of silicate minerals by organic acids. I. Nature of cation solubilization. Geoderma 53, 45–63.

Barshad, I. 1964. Chemistry of soil development. In F.E. Bear, ed., Chemistry of the soil, Reinhold, New York, 515 p.

Berner, E.K., and Berner, R.A. 1987. The global water cycle. Prentice-Hall, Englewood Cliffs, NJ, 397 p.

Berner, R.A. 1995. Chemical weathering and its effect on atmospheric CO_2 and climate. Rev. Mineral. 31, 565–583.

Berner, R.A., and Holden, G.R., Jr. 1977. Mechanism of feldspar weathering: Some observational evidence. Geology 5, 369–372.

Berthelin, J. 1988. Microbial weathering processes in natural environments. In A. Lerman and M. Meybeck, eds., Physical and chemical weathering in geochemical cycles, p. 33–59. Kluwer Academic Publ., Dordrecht, The Netherlands, 375 p.

Bierman, P., and Gillespie, A. 1991. Range fires: A significant factor in exposure-age determination and geomorphic surface evolution. Geology 19, 641–644.

Birkeland, P.W. 1973. Use of relative age-dating methods in a stratigraphic study of rock glacier deposits, Mt. Sopris, Colorado. Arctic Alp. Res. 5, 401–416.

Birkeland, P.W., and Noller, J.S. 1998. Rock and mineral weathering. In J.M. Sowers, J.S. Noller, and W.R. Lettis, eds., Dating and earthquakes: Review of Quaternary geochronology and its application to paleoseismology, p. 2-467–2-496. U.S. Nuclear Regulatory Commission, NUREG/CR-5562.

Birkeland, P.W., Burke, R.M., and Shroba, R.R. 1987. Holocene alpine soils in gneissic cirque deposits, Colorado Front Range. U.S. Geol. Surv. Bull. 1590-E, 21 p.

Blackwelder, E.B. 1927. Fire as an agent in rock weathering. J. Geol. 35, 134–140.

Blackwelder, E.B. 1933. The insolation hypothesis of rock weathering. Am. J. Sci. 226, 97–113.

Bloom, A.L. 1998. Geomorphology. Prentice-Hall, Upper Saddle River, NJ, 482 p.

Boettinger, J.L., and Southard, R.J. 1991. Silica and carbonate sources for Aridisols on a granitic pediment, western Mojave Desert. Soil Sci. Soc. Am. J. 55, 1057–1067.

Bohn, H.I., McNeal, B.L., and O'Connor, G.A. 1979. Soil chemistry: John Wiley & Sons, New York, 329 p.

Bradley, W.C. 1963. Large-scale exfoliation in massive sandstones of the Colorado Plateau. Geol. Soc. Am. Bull. 74, 519–528.

Bradley, W.C., Hutton, J.T., and Twidale, C.R. 1978. Role of salts in development of granitic tafoni, South Australia. J. Geol. 86, 647–654.

Bradley, W.C., Hutton, J.T., and Twidale, C.R. 1980. Role of salts in development of granitic tafoni, South Australia: A reply. J. Geol. 88, 121–122.

Brimhall, G.H., Chadwick, O.A., Lewis, C.J., Compston,

W., Williams, I.S., Danti, K.J., Dietrich, W.E., Power, M.E., Hendricks, D., and Bratt, J. 1991. Deformational mass transport and invasive processes in soil evolution. Science, 255, 695–702.

Brook, G.A., Folkoff, M.E., and Box, E.O. 1983. A world model of soil carbon dioxide. Earth Surface Proc. Landforms, 8, 79–88.

Bull, W.B. 1991. Geomorphic responses to climatic change. Oxford Univ. Press, New York, 326 p.

Burke, R.M., and Birkeland, P.W. 1979. Reevaluation of multiparameter relative dating techniques and their application to the glacial sequence along the eastern escarpment of the Sierra Nevada, California. Quaternary Res. 11, 21–51.

Caine, N. 1979. Rock weathering rates at the soil surface in an alpine environment. Catena 6, 131–144.

Campbell, I.B., and Claridge, G.G.C. 1987. Antarctica: Soils, weathering processes and environment. Elsevier, Amsterdam, 368 p.

Chesworth, W. 1992. Weathering systems. In I.P. Martini and W. Chesworth, eds., Weathering, soils & paleosols, p. 19–40. Elsevier, Amsterdam, 618 p.

Clayton, J.C. 1979. Nutrient supply to soil by rock weathering. In Proceedings, Impact of intensive harvesting on forest nutrient cycling, p. 75–96. State Univ. of New York, Syracuse, NY.

Clayton, J.L. 1983. A rational basis for estimating elemental supply rate from weathering. In E.L. Stone, ed., Forest soils and treatment impacts, p. 405–419. The Univ. of Tennessee, Knoxville.

Clayton, J.L. 1986. An estimate of plagioclase weathering rate in the Idaho Batholith based upon geochemical transport rates. In S.M. Colman and D.P. Dethier, eds., Rates of chemical weathering of rocks and minerals, p. 453–466. Academic Press, Orlando, FL, 603 p.

Clayton, J.L., and Megahan, W.F. 1986. Erosional and chemical denudation rates in the southwestern Idaho batholith. Earth Surface Proc. Landforms 11, 389–400.

Clayton, J.L., Megahan, W.F., and Hampton, D. 1979. Soil and bedrock properties: Weathering and alteration products and processes in the Idaho batholith. U.S.D.A. Forest Service Res. Pap. INT-237, 35 p.

Cleaves, E.T., Godfrey, A.E., and Bricker, O.P. 1970. Geochemical balance of a small watershed and its geomorphic implications. Geol. Soc. Am. Bull. 81, 3015–3032.

Colman, S.M. 1982. Chemical weathering of basalts and andesites: Evidence from weathering rinds. U.S. Geol. Surv. Prof. Pap. 1246, 51 p.

Colman, S.M., and Dethier, D.P., eds. 1986. Rates of chemical weathering of rocks and minerals. Academic Press, Orlando, FL, 603 p.

Colman, S.M., Pierce, K.L., and Birkeland, P.W. 1987. Suggested terminology for Quaternary dating methods: Quaternary Res. 28, 314–319.

Cooke, R.U., Warren, A., and Goudie, A. 1993. Desert geomorphology. UCL Press, London, 526 p.

Curtis, C.D. 1976. Chemistry of rock weathering: Fundamental reactions and controls. In E. Derbyshire, ed., Geomorphology and climate, p. 25–57. John Wiley & Co., New York.

Dethier, D.P. 1986. Weathering rates and the chemical flux from catchments in the Pacific Northwest, U.S.A. In S.M. Colman and D.P. Dethier, eds., Rates of chemical weathering of rocks and minerals, p. 503–530. Academic Press, Orlando, FL, 603 p.

Dixon, J.C., and Young, R.W. 1981. Character and origin of deep arenaceous weathering mantles on the Bega batholith, southeastern Australia. Catena 8, 97–109.

Drever, J.I. 1982. The geochemistry of natural waters. Prentice-Hall, Englewood Cliffs, NJ, 388 p.

Drever, J.I., ed., 1985. The chemistry of weathering. D. Reidel, Dordrecht, Holland, 324 p.

Drever, J.I. 1997. Weathering processes. In O.M. Saether, and P. de Caritat, eds., Geochemical processes, weathering and groundwater recharge in catchments, p. 3–19. A.A. Balkema, Rotterdam, 400 p.

Drever, J.I., and Clow, D.W. 1995. Weathering rates in catchments. Rev. Mineral. 31, 463–483.

Easterbrook, D.J. 1993. Surface processes and landforms. Macmillan, New York, 520 p.

Eggler, D.H., Larson, E.E., and Bradley, W.C. 1969. Granites, grusses, and the Sherman erosion surface, southern Laramie Range, Wyoming. Am. J. Sci. 267, 510–522.

Feller, M.C. 1977. Nutrient movement through western hemlock-western redcedar ecosystems in southwestern British Columbia. Ecology 58, 1269–1282.

Fry, E.J. 1927. The mechanical action of crustaceous lichens on substrata of shale, schist, gneiss, limestone, and obsidian. Ann. Bot. XLI, 437–463.

Gardner, L.R. 1992. Long-term isovolumetric leaching of aluminum from rocks during weathering: Implications for the genesis of saprolite. Catena 19, 521–537.

Garrels, R.M., and MacKenzie, F.T. 1971. Evolution of sedimentary rocks. W.W. Norton, New York, 397 p.

Gerrard, J. 1994. Weathering of granitic rocks: environment and clay mineral formation. In D.A. Robinson and R.B.G. Williams, eds., Rock weathering and landform evolution, p. 3–20. John Wiley & Sons Ltd., Chichester, England, 519 p.

Grant, W.H. 1969. Abrasion pH, an index of weathering. Clays Clay Minerals 17, 151–155.

Griggs, D.T. 1936. The factor of fatigue in rock exfoliation. J. Geol. 44, 781–796.

Harden, J.W. 1988. Genetic interpretations of elemental and chemical differences in a soil chronosequence, California. Geoderma 43, 179–193.

Hendricks, D.M., and Whittig, L.D. 1968. Andesite weathering II. Geochemical changes from andesite to saprolite. J. Soil Sci. 19, 147–153.

Hendricks, D.M., Whittig, L.D., and Jackson, M.L. 1967. Clay mineralogy of andesite saprolite. Clays Clay Minerals, Proc. 15th Conf., 395–407.

Hudson, B.D., 1995, Reassessment of Polynov's ion mobility series: Soil Sci. Soc. Am. J. 59, 1101–1103.

Isherwood, D., and Street, A. 1976. Biotite-induced grussification of the Boulder Creek Granodiorite, Boulder County, Colorado. Geol. Soc. Am. Bull. 87, 366–370.

Jackson, M.L. 1973. Soil chemical analysis advanced course. Dept. Soil Sci., Univ. of Wisconsin, Madison, Wisc., 894 p.

Jackson, T.A., and Keller, W.D. 1970. A comparative study of the role of lichens and "inorganic" processes in the chemical weathering of recent Hawaiian lava flows. Am. J. Sci. 269, 446–466.

Jenny, H. 1941. Factors of soil formation. McGraw-Hill, New York, 281 p.

Jenny, H. 1950. Origin of soils. In P.D. Trask, ed., Applied sedimentation, p. 41–61. John Wiley & Sons, New York.

Knuepfer, P.L.K. 1988. Estimating ages of late Quaternary stream terraces from analysis of weathering rinds and soils. Geol. Soc. Am. Bull. 100, 1224–1236.

Krauskopf, K.B. 1979. Introduction to geochemistry, 2nd ed. McGraw-Hill, New York, 617 p.

Lehman, D.S. 1963. Some principles of chelation chemistry. Soil Sci. Soc. Amer. Proc. 27, 167–170.

Li, Y-H. 1976. Denudation of Taiwan Island since the Pliocene Epoch. Geology 4, 105–107.

Likens, G.E., Bormann, F.H., Pierce, R.S., Eaton, J.S., and Johnson, N.M. 1977. Biogeochemistry of a forested ecosystem. Springer-Verlag, New York, 146 p.

Likens, G.E. and ten co-authors. 1998. The biogeochemistry of calcium at Hubbard Brook. Biogeochem. 41, 89–173.

Litaor, M.I. 1988. Review of soil solution samplers. Water Resources Res. 24, 727–733.

Loughnan, F.C. 1969. Chemical weathering of the silicate minerals. American Elsevier, New York, 154 p.

Mather, J. 1997. Relationship between rock, soil and groundwater compositions. In O.M. Saether and P. de Caritat, eds., Geochemical processes, weathering and groundwater recharge in catchments, p. 305–328. A.A. Balkema, Rotterdam, 400 p.

McFadden, L.D., Wells, S.G., and Ritter, J.B. 1989. Use of multiparameter relative-age methods for age determination and correlation of alluvial-fan surfaces on a desert piedmont, eastern Mojave Desert, California. Quaternary Res. 32, 276–290.

Meierding, T.C. 1981. Marble tombstone weathering rates: A transect of the United States. Phys. Geogr. 2, 1–18.

Meierding, T.C. 1993. Marble tombstone weathering and air pollution in North America. Ann. Assoc. Am. Geogr. 83, 568–588.

Merritts, D.J., Chadwick, O.A., Hendricks, D.M., Brimhall, G.H., and Lewis, C.J. 1992. The mass balance of soil evolution on late Quaternary marine terraces, northern California. Geol. Soc. Am. Bull. 104, 1456–1470.

Miller, D.C., Birkeland, P.W., and Rodbell, D.T. 1993. Evidence for Holocene stability of steep slopes, northern Peruvian Andes, based on soils and radiocarbon dates. Catena 20, 1–12.

Muir, J.W., and Logan, J. 1982. Eluvial/illuvial coefficients of major elements and the corresponding loses and gains in three soil profiles. J. Soil Sci. 33, 295–308.

Nahon, D.B. 1991. Introduction to the petrology of soils and chemical weathering. John Wiley & Sons, New York, 313 p.

Netoff, D.I., Cooper, B.J., and Shroba, R.R. 1995. Giant sandstone weathering pits near Cookie Jar Butte, southeastern Utah. In C. van Riper, III, ed., Proceedings of the second biennial conference on research in Colorado Plateau National Parks, p. 25–53. National Park Service Trans. and Proc. Series NPS/NRNAU/NRTP-95/11, 305 p.

Nettleton, W.D., Flach, K.W., and Nelson, R.E. 1970. Pedogenic weathering of tonalite in southern California. Geoderma 4, 387–402.

Nordstrom, D.K., and Munoz, J.L. 1994. Geochemical thermodynamics. Blackwell Sci. Publ., Boston, 493 p.

Ollier, C.D. 1969. Weathering. Oliver and Boyd, Edinburgh, 304 p.

Parker, A. 1970. An index of weathering for silicate rocks. Geol. Mag. 107, 501–504.

Pavich, M.J. 1986. Processes and rates of saprolite production and erosion on a foliated granitic rock of the Virginia Piedmont. In S.M. Colman and D.P. Dethier, eds., Rates of chemical weathering of rocks and minerals, p. 522–590. Academic Press, Orlando, FL, 603 p.

Peel, R.F. 1974. Insolation weathering: Some measurements of diurnal temperature changes in exposed rocks in the Tibesti region, central Sahara. Zeit. Geomorph. Supple. 21, 19–28.

Pierce, K.L., and Schmidt, P.W. 1975. Reconnaissance map showing relative amounts of soil and bedrock in the mountainous part of the Ralston Buttes Quadrangle and adjoining areas to the east and west in Jefferson County, Colorado. U.S. Geol. Surv. Misc. Field Studies Map MF-689.

Pinet, P., and Souriau, M. 1988. Continental erosion and large-scale relief. Tectonics 7, 563–582.

Ponomareva, V.V. 1969. Theory of podzolization. Israel Program for Scientific Translations, Jerusalem, 309 p.

Pope, G.A., Dorn, R.I., and Dixon, J.C. 1995. A new conceptual model for understanding geographical variations in weathering. Ann. Assoc. Am. Geogr. 85, 38–64.

Raymo, M.E., and Ruddiman, W.F. 1992. Tectonic forcing of late Cenozoic climate. Science 359, 117–122.

Reheis, M.C. 1988. Pedogenic replacement of aluminosilicate grains by $CaCO_3$ in Ustollic Haplargids, south-central Montana, U.S.A. Geoderma 41, 243–261.

Reheis, M.C. 1990. Influence of climate and eolian dust on the major-element chemistry and clay mineralogy of soils in the northern Bighorn Basin, U.S.A. Catena 17, 219–248.

Reiche, P. 1943. Graphical representation of chemical weathering. J. Sed. Petrol. 13, 58–68.

Rice, A. 1976. Insolation warmed over. Geology 4, 61–62.

Ritter, D.F., Kochel, R.C., and Miller, J.R. 1995. Process geomorphology. Wm. C. Brown, Dubuque, Iowa, 538 p.

Robinson, D.A., and Williams, R.B.G., eds. 1994. Rock weathering and landform evolution. John Wiley & Sons Ltd., Chichester, England, 519 p.

Rodbell, D.T. 1993. Subdivision of late-Pleistocene moraines in the Cordillera Blanca, Peru, based on rock-weathering features, soils, and radiocarbon dates. Quaternary Res. 39, 133–143.

Schalascha, E.B., Appelt, H., and Schatz, A. 1967. Chelation as a weathering mechanism I. Effect of complexing agents on the solubilization of iron from minerals and granodiorite. Geochim. Cosmochim. Acta 31, 587–596.

Schatz, A. 1963. Chelation in nutrition, soil microrganisms and soil chelation. The pedogenic action of lichens and lichen acids. J. Agri. Food Chem. 11, 112–118.

Schott, J., and Berner, R.A. 1985. Dissolution mechanisms of pyroxenes and olivines during weathering. In J.I. Drever, ed., The chemistry of weathering, p. 35–53. D. Reidel, Dordrecht, Holland, 324 p.

Selby, M.J. 1985. Earth's changing surface. Clarendon Press, Oxford, 607 p.

Siever, R., and Woodward, N. 1979. Dissolution kinetics and the weathering of mafic minerals. Geochim. Cosmochim. Acta 43, 717–724.

Smith, B.J. 1977. Rock temperature measurements from the northwest Sahara and their implications for rock weathering. Catena 4, 41–63.

Smith, B.J. 1994. Weathering processes and forms. In A.D. Abrahams and A.J. Parsons, eds., Geomorphology of desert environments, p. 39–63. Chapman & Hall, London, 674 p.

Sowers, J.M., Noller J.S., and Lettis, W.R. 1998. Dating and earthquakes: Review of Quaternary geochronology and its application to paleoseismology U.S. Nuclear Regulatory Commission, NUREG/CR-5562.

Stallard, R.F. 1985. River chemistry, geology, geomorphology, and soils in the Amazon and Orinoco basins. In J.I. Drever, ed., The chemistry of weathering, p. 293–316. D. Reidel, Dordrecht, Holland, 324 p.

Stallard, R.F. 1992. Tectonic processes, continental freeboard, and the rate-controlling step for continental denudation. In S.S. Butcher, R.J. Charlson, G.H. Orians, and G.V. Wolfe, eds., Global biogeochemical cycles, p. 93–121. Academic Press, London.

Stallard, R.F. 1995. Relating chemical and physical erosion. Rev. Mineral. 31, 543–564.

Stevens, R.E., and Carron, M.K. 1948. Simple field test for distinguishing minerals by abrasion pH. Am. Mineral. 33, 31–49.

Summerfield, M.A. 1991. Global geomorphology. Longman Scientific & Technical, Essex, England, 357 p.

Tan, K.H. 1993. Principles of soil chemistry. Marcel-Dekker, New York, 362 p.

Taylor, A.B., and Velbel, M.A. 1991. Geochemical mass balances and weathering rates in forested watersheds of the southern Blue Ridge II. Effects of botanical uptake terms. Geoderma 51, 29–50.

Thomas, M.F. 1994. Geomorphology in the tropics. John Wiley & Sons Ltd., Chichester, England, 460 p.

Thorn, C.E. 1979. Bedrock freeze-thaw weathering

regime in an alpine environment, Colorado Front Range: Earth Surf. Proc. Landforms 4, 211–228.

Trudgill, S.T. 1976. Rock weathering and climate: quantitative and experimental aspects. In E. Derbyshire, ed., Geomorphology and climate, p. 59–99. John Wiley & Co., New York.

Ugolini, F.C., and Sletten, R.S. 1991. The role of proton donors in pedogenesis as revealed by soil solution studies. Soil Sci. 151, 59–75.

Ugolini, F.C., Dahlgren, R., LaManna, J., Nuhn, W., and Zachara, J. 1991. Mineralogy and weathering processes in recent and Holocene tephra deposits of the Pacific Northwest, USA. Geoderma 51, 277–299.

Vidic, N.J. 1994. Pedogenesis and soil-age relationships of soils on glacial outwash terraces in the Ljubljana Basin. Ph.D. thesis, Univ. of Colorado, Boulder, 229 p.

Wahrhaftig, C. 1965. Stepped topography of the southern Sierra Nevada, California. Geol. Soc. Am. Bull. 76, 1165–1190.

Wakatsuki, T., and Rasyidin, A. 1992. Rates of weathering and soil formation. Geoderma 52, 251–263.

Wakatsuki, T., Furukawa, H., and Kyuma, K. 1977. Geochemical study of the redistribution of elements in soil—1. Evaluation of degree of weathering of transported soil materials by distribution of major elements among the particle size fractions and soil extract. Geochim. Cosmochim. Acta 41, 891–902.

Walder, J., and Hallet, B. 1985. A theoretical model of the fracture of rock during freezing. Geol. Soc. Am. Bull. 96, 336–346.

Walling, D.E. 1987. Rainfall, runoff and erosion of the land: A global view. In K.J. Gregory, ed., Energetics of physical environment, p. 89–117. John Wiley & Sons Ltd., Chichester, England.

Warke, P.A., and Smith, B.J. 1994. Short-term temperature fluctuations under simulated hot desert conditions: some preliminary data. In D.A. Robinson and R.B.G. Williams, eds., Rock weathering and landform evolution, p. 57–70. John Wiley & Sons Ltd., Chichester, England, 519 p.

Watts, S.H. 1983. Weathering processes and products under arid arctic conditions. Geograf. Ann. 65A, 85–98.

White, A.F., and Brantley, S.L., eds. 1995. Chemical weathering rates of silicate minerals. Rev. Mineral. 31, 583 p.

White, S.E. 1976. Is frost action really only hydration shattering? A review. Arc. Alp. Res. 8, 1–6.

White, W.B. 1988. Geomorphology and hydrology of karst terrains. Oxford Univ. Press, New York, 464 p.

Whitehouse, I.E. 1988. Geomorphology of the central Southern Alps, New Zealand: The interaction of plate collision and atmospheric circulation. Zeit. Geomorphol. Suppl. 69, 105–116.

Winkler, E.M. 1965. Weathering rates as exemplified by Cleopatra's Needle in New York City. J. Geol. Educ. 13, 50–52.

Winkler, E.M. 1980. Role of salts in development of granitic tafoni, South Australia: A discussion. J. Geol. 88, 119–120.

Winkler, E.M. 1975. Stone: properties, durability in man's environment. Springer-Verlag, New York, 230 p.

Wolff, R.G. 1967. Weathering of Woodstock granite near Baltimore, Maryland. Am. J. Sci. 265, 106–117.

Wollast, H. 1967. Kinetics of the alteration of K-feldspar in buffered solutions at low temperature. Geochim. Cosmochim. Acta 31, 635–648.

Wray, R.A.L. 1997. A global review of solutional weathering forms on quartz sandstones. Earth-Sci. Rev. 42, 137–160.

Young, A.R.M. 1987. Salt as an agent in the development of cavernous weathering. Geology 15, 962–966.

Yuretich, R.F., and Batchelder, G.L. 1988. Hydrogeochemical cycling and chemical denudation in the Fort River watershed, central Massachusetts: An appraisal of mass-balance studies. Water Resources Res. 24, 105–114.

Zimmerman, S.G., Evenson, E.B., Gosse, J.C., and Erskine, C.P. 1994. Extensive boulder erosion resulting from a range fire on the type-Pinedale moraines, Fremont Lake, Wyoming. Quaternary Res. 42, 255–265.

■ Chapter 4

Allen, B.L., and Fanning, D.S. 1983. Compositon and soil genesis. In L.P. Wilding, N.E. Smeck, and G.F. Hall, eds., Pedogenesis and soil taxonomy 1. Concepts and interactions, p. 141–192. Elsevier Sci. Publ. Co., Inc., New York.

Allen, B.L., and Hajek, B.F. 1989. Mineral occurrence in soil environments. In J.B. Dixon and S.B. Weed, eds., Minerals in soil environments, p. 199–278. Soil Sci. Soc. Am. Book Series No. 1, 1244 p.

Altschuler, Z.S., Dwornik, E.J., and Kramer, H. 1963. Transformation of montmorillonite to kaolinite during weathering. Science 141, 148–152.

Amonette, J.E., and Zelazny, L.W. 1994. Quantitative methods in soil mineralogy. Soil Sci. Soc. Am. Mscl. Publ., 462 p.

Bachman, G.O., and Machette, M.N. 1977. Calcic soils and calcretes in the southwestern United States. U.S. Geological Survey Open-File Report 77-794, 163 p.

Bailey, S.W. 1980. Structures of layer silicates. Rev. Mineral. 19, 1–8.

Baldar, N.A., and Whittig, L.D. 1968. Occurrence and synthesis of soil zeolites. Soil Sci. Soc. Am. Proc. 32, 235–238.

Barshad, I. 1964. Chemistry of soil development. In F.E. Bear, ed., Chemistry of the soil, p. 1–70. Reinhold, New York, 515 p.

Birkeland, P.W. 1984. Holocene soil chronofunctions, Southern Alps, New Zealand. Geoderma 34, 115–134.

Black, A.B. 1967. Applications: electrokinetic characteristics of hydrous oxides of aluminum and iron. In S.D. Faust and J.V. Hunter, eds. Principles and applications of water chemistry, p. 247–300. John Wiley & Sons, New York, 643 p.

Blume, H.P., and Schwertmann, U. 1969. Genetic evaluation of profile distribution of aluminum, iron, and manganese oxides. Soil Sci. Soc. Am. Proc. 33, 438–444.

Brindley, G.W., and Brown, G., eds. 1980. Crystal structures of clay minerals and their X-ray identification. Mineralogical Soc. (London) Monograph No. 5, 495 p.

Brindley, G.W., and Gillery, F.H. 1954. A mixed-layer kaolin-chlorite structure. Proc., 2nd National Conf., Clays and Clay Minerals (Columbia, MO, Oct. 15–17, 1953), Publ. 327 of the committee on clay minerals of the National Acad. Sci., National Res. Council, Washington, D.C., 349–353.

Carroll, D. 1959. Ion exchange in clays and other minerals. Geol. Soc. Am. Bull. 70, 749–780.

Chesworth, W. 1990. Weathering systems. In I.P. Martini and W. Chesworth, eds., Weathering, soils & paleosols, p. 19–40. Elsevier, Amsterdam, 618 p.

Chesworth, W. 1991. Geochemistry of micronutrients. Soil Sci. Soc. Am. Book Series, No. 4, 1–30.

Cleaves, E.T., Fisher, D.W., and Bricker, O.P. 1974. Chemical weathering of serpentinite in the eastern Piedmont of Maryland. Geol. Soc. Am. Bull. 85, 437–444.

Colman, S.M. 1982. Clay mineralogy of weathering rinds and possible implications concerning the sources of clay minerals in soils. Geology 10, 370–375.

Dahlgren, R.A. 1994. Quantification of allophane and imogolite. In J.E. Amonette and L.W. Zelazny, eds., Quantitative methods in soil mineralogy, p. 430–451. Soil Sci. Soc. Am. Misc. Publ., 462 p.

Dahlgren, R., Shoji, S., and Nanzyo, M. 1993. Mineralogical characteristics of volcanic ash soils. In S. Shoji, M. Nanzyo, and R. Dahlgren, eds., Volcanic ash soils: Genesis, properties and utilization, p. 101–143. Elsevier, Amsterdam, 288 p.

Davis, S.N. 1964. Silica in streams and ground water. Am. J. Sci. 262, 870–891.

Dixon, J.B., and Weed, S.B. 1989. Minerals in soil environments. Soil Sci. Soc. Am. Book Series No. 1, 1244 p.

Drever, J.I. 1982. The geochemistry of natural waters. Prentice-Hall, Englewood Cliffs, NJ, 388 p.

Duchaufour, P. 1982. Pedology. George Allen & Unwin, London, 448 p.

Dymond, J., Biscaye, P.E., and Rex, R.W. 1974. Eolian origin of mica in Hawaiian soils. Geol. Soc. Am. Bull. 85, 37–40.

Eberl, D.D. 1984. Clay mineral formation and transformation in rocks and soils. Phil. Trans. Royal Soc. London A 311, 241–257.

Feth, J.H., Roberson, C.E., and Polzer, W.L. 1964. Sources of mineral constituents in water from granitic rocks, Sierra Nevada, California and Nevada: U.S. Geol. Surv. Water-Supply Pap. 1535-I, 70 p.

Fieldes, M., and Perrott, K.W. 1966. The nature of allophane in soils. Part 3. Rapid field and laboratory test for allophane. New Zealand J. Sci. 9, 623–629.

Foster, M.D. 1962. Interpretation of the composition and a classification of the chlorites. U.S. Geol. Surv. Prof. Pap. 414-A, 33 p.

Garrells, R.M., and Mackenzie, F.T. 1971. Evolution of sedimentary rocks. W.W. Norton, New York, 397 p.

Gjems, O. 1960. Some notes on clay minerals in podzol profiles in Fennoscandia. Clay Minerals Bull. 4, 208–211.

Glenn, R.C., Jackson, M.L., Hole, F.D., and Lee, G.B. 1960. Chemical weathering of layer silicate clays in loess-derived Tama silt loam of southwestern Wisconsin. Clays Clay Minerals 8, 63–83.

Gradusov, B.P. 1974. A tentative study of clay mineral distribution in soils of the world. Geoderma 12, 49–55.

Grim, R.E. 1968. Clay mineralogy. McGraw-Hill, New York, 596 p.

Hay, R.L. 1963. Zeolitic weathering in Olduvai Gorge, Tanganyika. Geol. Soc. Am. Bull. 74, 1281–1286.

Hay, R.L. 1964. Phillipsite of saline lakes and soils. Am. Mineral. 49, 1366–1387.

Helgeson, H.C., Brown, T.H., and Leeper, R.H. 1969. Handbook of theoretical activity diagrams depicting chemical equilibria in geologic systems involving an aqueous phase at one atm and 0-300C. Freeman, Cooper, San Francisco, 253 p.

Hendricks, D.M., Whittig, L.D., and Jackson. M.L. 1967. Clay mineralogy of andesite saprolite. Clays Clay Minerals, Proc. 15th Conf., 395–407.

Hess, P.C. 1966. Phase equilibria of some minerals in the K_2O-Na_2O- Al_2O_3-SiO_2-H_2O system at 25C and 1 atmosphere. Am. J. Sci. 264, 289–309.

Jackson, M.L. 1964. Chemical composition of soils. In F.E. Bear, ed., Chemistry of the soil, p. 71–141. Reinhold, New York, 515 p.

Jackson, M.L., Tyler, S.A., Willis, A.L., Bourbeau, G.A., and Pennington, R.P. 1948. Weathering sequence of clay-size minerals in soils and sediments-I. Fundamental generalizations. J. Phys. Colloid Chem. 52, 1237–1260.

Krauskopf, K.B. 1967. Introduction to geochemistry. McGraw-Hill, New York, 721 p.

Langmuir, D. 1997. Aqueous environmental geochemistry. Prentice-Hall, Englewood Cliffs, NJ, 600 p.

McKeague, J.A., Ross, G.J., and Gamble, D.S. 1978. Properties, criteria of classification, and concepts of genesis of podzolic soils in Canada. In W.C. Mahaney, ed., Quaternary soils, p. 27–60. Geo Abstracts Ltd., University of East Anglia, Norwich, England.

Monger, H.C., and Daugherty, L.A. 1991. Neoformation of palygorskite in a southern New Mexico Aridisol. Soil Sci. Soc. Am. J. 55, 1646–1650.

Moore, D.M., and Reynolds, R.C., Jr. 1989. X-ray diffraction and the identification and analysis of clay minerals. Oxford Univ. Press, New York, 332 p.

Neall, V.E., and Paintin, I.K. 1986. Rates of weathering of ^{14}C-dated late Quaternary volcaniclastic deposits in the western United States. In S.M. Colman and D.P. Dethier, eds., Rates of chemical weathering of rocks and minerals, p. 331–350. Academic Press, Orlando, FL, 603 p.

Nettleton, W.D., and Brasher, B.R. 1983. Correlation of clay minerals and properties of soils in the western United States. Soil Sci. Soc. Am. J. 47, 1032–1036.

Newman, A.C.D. 1987. Chemistry of clays and clay minerals. Mineralogical Soc. Mon. No. 6, John Wiley & Sons, 480 p.

Newman, A.C.D., and Brown, G. 1987. The chemical constitution of clays. In A.C.D. Brown, ed., Chemistry of clays and clay minerals, p. 1–128. Mineralogical Soc. Mon. No. 6, John Wiley & Sons, New York, 480 p.

Parfitt, R.L. 1980. Chemical properties of variable charge soils. In B.K.G. Theng, ed., Soils with variable charge, p. 167–194. New Zealand Soc. Soil Sci., Lower Hutt, 448 p.

Parfitt, R.L., and Childs, C.W. 1988. Estimation of forms of Fe and Al: A review, and analysis of contrasting soils by dissolution and Moesbauer methods. Aust. J. Soil Res. 26, 121–144.

Pavich, M.J. 1986. Processes and rates of saprolite production and erosion on a foliated granitic rock of the Virginia Piedmont. In S.M. Colman and D.P. Dethier, eds., Rates of chemical weathering of rocks and minerals, p. 522–590. Academic Press, Orlando, FL.

Pedro, G., Jamagne, M., and Begon, J.C. 1969. Mineral interactions and transformations in relation to pedogenesis during the Quaternary. Soil Sci. 107, 462–469.

Pierce, J.W., and Siegel, F.R. 1969. Quantification in clay mineral studies of sediments and sedimentary rocks. J. Sed. Petrol. 39, 187–193.

Rodbell, D.T. 1990. Soil-age relationships on late Quaternary moraines, Arrowsmith Range, Southern Alps, New Zeland. Arctic Alp. Res. 22, 355–365.

Ross, G.J. 1980. The mineralogy of Spodosols. In B.K.G. Theng, ed., Soils with variable charge, p. 127–143. New Zealand Society of Soil Science, Lower Hutt.

Ruhe, R.V., and Olson, C.G. 1979. Estimate of clay-mineral content: Additions of proportions of soil clay to constant standard. Clays Clay Minerals 27, 322–326.

Schultz, L.G. 1964. Quantitative interpretation of mineralogical composition from X-ray and chemical data for the Pierre Shale. U.S. Geol. Surv. Prof. Pap. 391-C, 31 p.

Schultz, L.G. 1978. Mixed-layer clay in the Pierre Shale and equivalent rocks, northern Great Plains region: U.S. Geol. Surv. Prof. Pap. 1064-A, 28 p.

Schulze, D.G. 1994. Differential X-ray diffraction analysis of soil minerals. In J.E. Amonette and L.W. Zelazny, eds., Quantitative methods in soil mineralogy, p. 412–419. Soil Sci. Soc. Am. Mscl. Publ., 462 p.

Schwertmann, U. 1993. Relations between iron oxides, soil color, and soil formation. Soil Sci. Soc. Am. Spec. Publ. No. 31, 51–69.

Schwertmann, U., and Taylor, R.M. 1989. Iron oxides. In J.B. Dixon and S.B. Weed, eds., Minerals in soil environments, p. 379–437. Soil Sci. Soc. Am. Book Series No. 1, 1244 p.

Siffert, B. 1967. Some reactions of silica in solution: Formation of clay. Israel Program for Scientific Translations, Jerusalem, 100 p.

Simonson, R.W. 1995. Airborne dust and its significance to soils. Geoderma 65, 1–43.

Singer, A. 1988. Illite in aridic soils, desert dusts and desert loess. Sediment. Geol. 59, 251–259.

Singer, A. 1989a. Palygorskite and sepiolite group minerals. In J.B. Dixon and S.B. Weed, eds., Minerals in soil environments, p. 829–872. Soil Sci. Soc. Am. Book Series No. 1, 1244 p.

Singer, A. 1989b. Illite in the hot-aridic soil environment. Soil Sci. 147, 126–133.

Singer, A., and Norrish, K. 1974. Pedogenic palygorskite occurrences in Australia. Am. Mineral. 59, 508–517.

Singer, M.J., and Munns, D.N. 1987. Soils, an introduction. Macmillan, New York, 492 p.

Soil Survey Staff. 1975. Soil taxonomy. U.S. Dept. Agri., Agri. Handbook 436, 754 p.

Swindale, L.D., and Uehara, G. 1966. Ionic relationships in the pedogenesis of Hawaiian soils. Soil Sci. Soc. Am. Proc. 30, 726–730.

Tamura, T., and Jackson, M.L. 1953. Structural and energy relationships in the formation of iron and aluminum oxides, hydroxides, and silicates. Science 117, 381–383.

Tardy, Y. 1971. Characterization of the principal weathering types by the geochemistry of waters from some European and African crystalline massifs. Chem. Geol. 7, 253–271.

Tardy, Y., Bocquier, G., Paquet, H., and Millot, G. 1973. Formation of clay from granite and its distribution in relation to climate and topography. Geoderma 10, 271–284.

Velde, B. 1992. Introduction to clay minerals. Chapman & Hall, London, 198 p.

Wada, K. 1989. Allophane and imogolite. In J.B. Dixon and S.B. Weed, eds., Minerals in soil environments, p. 1051–1087. Soil Sci. Soc. Am. Book Series No. 1, 1244 p.

Walker, A.L. 1983. The effects of magnetite on oxalate- and dithionite-extractable iron. Soil Sci. Soc. Am. J. 47, 1022–1026.

Weaver, C.E., and Pollard, L.D. 1973. The chemistry of clay minerals. Elsevier Scientific Publishing Co., New York.

White, A.F. 1995. Chemical weathering rates of silicate minerals in soils. Rev. Mineral. 31, 407–461.

Wiklander, L. 1964. Cation and anion exchange phenomena. In F.E. Bear, ed., Chemistry of the soil, p. 163–205. Reinhold, New York, 515 p.

■ Chapter 5

Alexander, E.B., and Nettleton, W.D. 1977. Post-Mazama Natrargids in Dixie Valley, Nevada. Soil Sci. Soc. Am. J. 41, 1210–1212.

Alexander, L.T., and Cady, J.G. 1962. Genesis and hardening of laterite in soils. U.S. Dept. Agri. Tech. Bull. 1282, 90 p.

Amit, R., and Harrison, J.B.J. 1995. Biogenic calcic horizon development under extremely arid conditions, Nizzana sand dunes, Israel. Adv. GeoEcol. 28, 65–88.

Amundson, R.G., Chadwick, O.A., Sowers, J.M., and Doner, H.E. 1989. The stable isotope chemistry of pedogenic carbonates at Kyle Canyon, Nevada. Soil Sci. Soc. Am. J. 53, 201–210.

Amundson, R., Wang, Y., Chadwick, O., Trumbore, S., McFadden, L., McDonald, E., Wells, S., and DeNiro, M. 1994. Factors and processes governing the ^{14}C content of carbonate in desert soils. Earth Planetary Sci. Lett. 125, 385–405.

Anderson, H.A., Berrow, M.L., Farmer, V.C., Hepburn, A., Russell, J.D., and Walker, A.D. 1982. A reassessment of podzol formation processes. J. Soil Sci. 33, 125–136.

Anderson, J.L. 1984. Part II. Soil mottling, an indicator of saturation. Soil Horizons, Winter issue, 13–16.

Aristarian, L.F. 1970. Chemical analyses of caliche profiles from the High Plains, New Mexico. J. Geol. 78, 201–212.

Arkley, R.J. 1963. Calculation of carbonate and water movement in soil from climatic data. Soil Sci. 96, 239–248.

Asumadu, K., Churchward, H.M., and Gilkes, R.J. 1991. The origins of surficial sands within highly weathered terrain in southwest Australia. Aust. J. Earth Sci. 38, 45–54.

Bachman, G.O., and Machette, M.N. 1977. Calcic soils and calcretes in the southwestern United States. U.S. Geol. Surv. Open-File Report 77-794, 163 p.

Bárdossy, G., and Aleva, G.J.J. 1990. Lateritic bauxites. Elsevier, Amsterdam, 624 p.

Barshad, I. 1964. Chemistry of soil development. In F.E. Bear, ed., Chemistry of the soil, p. 1–70. Reinhold, New York, 515 p.

Benedict, J.B. 1966. Radiocarbon dates from a stone-banked terrace in the Colorado Rocky Mountains. Geogr. Ann. 48A, 24–31.

Berg, R.C. 1984. The origin and early genesis of clay bands in youthful sandy soils along Lake Michigan, U.S.A. Geoderma 32, 45–62.

Berner, E.K., and Berner, R.A. 1996. Global environment: Prentice-Hall, Englewood Cliffs, NJ, 376 p.

Birkeland, P.W. 1978. Soil development as an indication of relative age of Quaternary deposits, Baffin Island, N.W.T., Canada. Arc. Alp. Res. 10, 733–747.

Blume, H.P., and Schwertmann, U. 1969. Genetic evaluation of profile distribution of aluminum, iron, and manganese oxides. Soil Sci. Soc. Am. Proc. 33, 438–444.

Bockheim, J.G. 1979. Properties and relative age of soils of southwestern Cumberland Peninsula, Baffin Island, N.W.T., Canada. Arc. Alp. Res. 11, 289–306.

Bockheim, J.G. 1980. Properties and classification of some desert soils in coarse-grained glacial drift in the Arctic and Antarctic. Geoderma 24, 45–69.

Bond, W.J. 1986. Illuvial clay band formation in a column study of sand. Soil Sci. Soc. Am. J. 50, 265–267.

Bouma, J. 1983. Hydrology and soil genesis of soils with aquic moisture regimes. In L.P. Wilding, N.E. Smeck, and G.F. Hall, eds., Pedogenesis and soil taxonomy, I, Concepts and interactions, p. 253–281. Elsevier, Amsterdam, 303 p.

Bowler, J.M., and Polach, H.A. 1971. Radiocarbon analyses of soil carbonates: An evaluation from paleosols in southeastern Australia. In D.H. Yaalon, ed., Paleopedology, p. 97–108. Israel Univ. Press, Jerusalem, 350 p.

Brewer, H., and Haldane, A.D. 1957. Preliminary experiments in the development of clay orientation in soils. Soil Sci. 84, 301–309.

Brewer, R. 1955. Mineralogical examination of a yellow podzolic soil formed on granodiorite. Commonwealth Scientific and Industrial Res. Organ. (Australia), Soil Publ. No. 5, 28 p.

Brewer, R. 1964. Fabric and mineral analysis of soils. John Wiley & Sons, New York, 470 p.

Brewer, R. 1968. Clay illuviation as a factor in particle-size differentiation in soil profiles. 1968 Int. Cong. Soil Sci. Trans. 4, 489–499.

Brimhall, G.H., Lewis, C.J., Ague, J.J., Dietrich, W.E., Hampel, J., Teague, T., and Rix, P. 1988. Metal enrichment in bauxites by deposition of chemically mature aeolian dust. Nature (London) 333, 819–824.

Brinkman, H. 1970. Ferrolysis, a hydromorphic soil forming process. Geoderma 3, 199–206.

Broecker, W.S., Kulp, J.L., and Tucek, C.S. 1956. Lamont natural radiocarbon measurements III. Science 124, 154–165.

Brook, G.A., Folkoff, M.E., and Box, E.O. 1983. A world model of soil carbon dioxide. Earth Surface Proc. and Landforms 8, 79–88.

Bull, W.B. 1991. Geomorphic responses to climatic change. Oxford Univ. Press., New York, 326 p.

Bullock, P., and Murphy, C.P., eds. 1983. Soil micromorphology. A B Academic Publ., Berkhamsted, UK, 705 p.

Bullock, P., and Thompson, M.L. 1985. Micromorphology of Alfisols. In L.A. Douglas and M.L. Thompson, eds., Soil micromorphology and soil classification, p. 17–47. Soil Sci. Soc. Am. Spec. Publ. No. 15, 216 p.

Bullock, P., Milford, M.H., and Cline, M.G. 1974. Degradation of argillic horizons in Udalf soils of New York State. Soil Sci. Soc. Am. Proc. 38, 621–628.

Buol, S.W., 1990, Spodosols in the tropics. In J.M. Kimble and R.D. Yeck, eds., Proceedings of the fifth international soil correlation meeting (ISCOM): Correlation, classification, and utilization of Spodosols, p. 41–45. U.S. Dept. Agri., Soil Conservation Service, Lincoln, NE, 438 p.

Buol, S.W., and Hole, F.D. 1959. Some characteristics of clay skins on peds in the B horizon of a Gray-Brown Podzolic soil. Soil Sci. Soc. Am. Proc. 23, 239–241.

Buol, S.W, and Hole, F.D. 1961. Clay skin genesis in Wisconsin soils. Soil Sci. Soc. Am. Proc. 25, 377–379.

Buol, S.W., Hole, F.D., McCracken, R.J., and Southard, R.J. 1997. Soil genesis and classification. Iowa State Univ. Press, Ames, 544 p.

Burns, S.F. 1990. Alpine Spodosols: Cryaquods, Cryohumods, Cryorthods, and Placaquods above treeline. In J.M. Kimble and R.D. Yeck, eds., Proceedings of the fifth international soil correlation meeting (ISCOM): Correlation, classification, and utilization of Spodosols, p. 46–62. U.S. Dept. Agri., Soil Conservation Service, Lincoln, NE, 438 p.

Buurman, P., ed. 1984. Podzols: Van Nostrand Reinhold, New York, 450 p.

Campbell, C.A., Paul, E.A., Rennie, D.A., and McCallum, K.J. 1967a. Factors affecting the accuracy of the carbon-dating method in soil humus studies. Soil Sci. 104, 81–85.

Campbell, C.A., Paul, E.A., Rennie, D.A., and McCallum, K.J. 1967b. Applicability of the carbon-dating method of analysis to soil humus studies. Soil Sci. 104, 217–224.

Campbell, I.B., and Claridge, G.G.C. 1987. Antarctic soils, weathering processes and environment. Elsevier, Amsterdam, 366 p.

Chartres, C.J., and Walker, P.H. 1988. The effect of aeolian accessions on soil development on granitic rocks in south-eastern Australia. III. Micromorphological and geochemical evidence of weathering and soil development. Aust. J. Soil Res. 26, 33–53.

Chartres, C.J., Chivas, A.R., and Walker, P.H. 1988. The effect of aeolian accessions on soil development on granitic rocks. II. Oxygen-isotope, mineralogical and geochemical evidence for aeolian deposition. Aust. J. Soil Res. 26, 17–31.

Chen, X.Y. 1997. Pedogenic gypcrete formation in arid central Australia. Geoderma 77, 39–61.

Chesworth, W. 1973. The parent material effect in the genesis of soil. Geoderma 10, 215–225.

Chesworth, W., and Macias-Vasquez, F. 1985. pe, pH, and podzolization. Am. J. Sci. 285, 128–146.

Childs, C.W., Parfitt, R.L., and Lee, H. 1983. Movement of aluminum as an inorganic complex in some podzolized soils, New Zealand. Geoderma 29, 139–155.

Chittleborough, D.J., and Oades, J.M. 1979. The development of a red-brown earth. I. A reinterpretation of published data. Aust. J. Soil Res. 17, 371–381.

Chittleborough, D.J., and Oades, J.M. 1980a. The development of a red-brown earth. II. Uniformity of the parent material. Aust. J. Soil Res. 18, 375–382.

Chittleborough, D.J., and Oades, J.M. 1980b. The development of a red-brown earth. III. The degree of weathering and translocation of clay. Aust. J. Soil Res. 18, 383–393.

Ciolkosz, E.J., Waltman, W.J., Simpson, T.W., and Dobos, R.R. 1989. Distribution and genesis of soils of the northeastern United States. Geomorphology 2, 285–302.

Clark, F.W. 1924. The data of geochemistry. U.S. Geol. Surv. Bull. 770, 841 p.

Clayden, B., and Hewitt, A.E. 1989. Horizon notation for New Zealand soils. Division of Land and Soil Sciences (New Zealand) Scientific Report 1, 30 p.

Collins, J.F., and Buol, S.W. 1981. Effects of fluctuations in the Eh-pH environment on iron and/or manganese equilibria. Soil Sci. 110, 111–118.

Conacher, A.J. 1991. The laterite profile, ferricrete and unconformity, a discussion. Catena 18, 585–588.

Cutler, E.J.B. 1983. Soil classification in New Zealand. Occasional Rept. No. 2, Soil Sci. Dept., Lincoln College, New Zealand, 148 p.

Dahms, D.E. 1993. Mineralogical evidence for eolian contribution to soils of late Quaternary moraines, Wind River Mountains, Wyoming, USA. Geoderma 59, 175–196.

Daly, B.K. 1982. Identification of podzols and podzolized soils in New Zealand by relative absorbance of oxalate extracts of A and B horizons. Geoderma 28, 29–38.

Dan, J., and Yaalon, D.H. 1982. Automorphic saline soils in Israel. Catena, Supple. 1, 103–115.

Dan, J., Gerson, R., Koyumdjisky, H., and Yaalon, D.H. 1981. Aridic soils of Israel. Special Publication No. 190, Agricultural Research Organization, The Volcani Center, Israel, 353 p.

Daniels, R.B., Perkins, H.F., Hajek, B.F., and Gamble, E.E. 1978. Morphology of discontinuous phase plinthite and criteria for its field identification in the southern United States. Soil Sci. Soc. Am. J. 42, 944–949.

Dawson, J.J., Ugolini, F.C., Hrutfiord, B.F., and Zachara, J. 1978. Role of soluble organics in the soil processes of a podzol, central Cascades, Washington. Soil Sci. 126, 290–296.

Dent, D. 1986. Acid sulphate soils: A baseline for research and development. Int. Inst. Land Reclamation and Improvement, Publ. 39, Wageningen, The Netherlands, 204 p.

Dijkerman, J.C., Cline, M.G., and Olson, G.W. 1967. Properties and genesis of textural subsoil lamellae. Soil Sci. 104, 7–16.

Division of Soils, CSIRO. 1983. Soils, an Australian viewpoint. CSIRO, Melbourne/Academic Press, London, 928 p.

Dixon, J.C. 1994. Duricrusts. In A.D. Abrahams and A.J. Parsons, eds., Geomorphology of desert environments, p. 82–105. Chapman & Hall, London.

Drees, L.R., Wilding, L.D., Smeck, N.E., and Senkayi, A.L. 1989. Silica in soils: Quartz and disorder silica polymorphs. In J.B.Dixon and S.B. Weed, eds., Minerals in soil environments, p. 913–974. Soil Sci. Soc. Am. Book Series, No. 1.

Drever, J.I. 1988. The geochemistry of natural waters. Prentice-Hall, Englewood Cliffs, NJ, 437 p.

Duchaufour, P. 1982. (English ed.). Pedology. George Allen & Unwin, London, 448 p.

Ericksen, G.E. 1981. Geology and origin of the Chilean nitrate deposits. U.S. Geol.Surv. Prof. Pap. 1188, 37 p.

Eswaran, H., and Tavernier, R. 1980. Classification and genesis of Oxisols. In B.K.G. Theng, ed., Soils with variable charge, p. 427–442. New Zealand Soc. Soil Sci., Lower Hutt.

Eswaran, H., and Zi-Tong, G. 1991. Properties, genesis, classification, and distribution of soils with gypsum. Soil Sci. Soc. Am. Spec. Publ. No. 26, 89–119.

Evans, C.V., and Franzmeier, D.P. 1986. Saturation, aeration, and color patterns in a toposequence of soils in north-central Indiana. Soil Sci. Soc. Am. J. 50, 975–980.

Fanning, D.S., and Fanning, M.C.B. 1989. Soil, morphology, genesis, and classification. John Wiley & Sons, New York, 395 p.

Fanning, D.S., Rabenhorst, M.C., and Bigham, J.M. 1993. Colors of acid sulphate soils. Soil Sci. Soc. Am. Spec. Publ. No. 31, 91–108.

Farmer, V.C., and Fraser, A.R. 1982. Chemical and colloidal stability of sols in the Al_2O_3-Fe_2O_3-SiO_2-H_2O system: Their role in podzolization. J. Soil Sci. 33, 737–742.

Farmer, V.C., Russell, J.D., and Berrow, M.L. 1980. Imogolite and protoimogolite allophane in spodic horizons—evidence for a mobile aluminum silicate complex in podzol formation. J. Soil Sci. 31, 673–684.

Federoff, N. 1997. Clay illuviation in Red Mediterranean soils. Catena 28, 171–189.

Flach, K.W., Nettleton, W.D., Gile, L.H., and Cady, J.G. 1969. Pedocementation: Induration by silica, carbonates, and sesquioxides in the Quaternary. Soil Sci. 107, 442–453.

Flach, K.W., Holzhey, C.S., De Coninck, F., and Bartlett, R.J. 1980. Genesis and classification of Andepts and Spodosols. In B.K.G.Theng, ed., Soils with variable charge, p. 411–426. New Zealand Soc. Soil Sci., Lower Hutt.

Forman, S.L., and Miller, G.H. 1984. Time-dependent soil morphologies and pedogenic processes on raised beaches, Bröggerhalvöya, Spitzbergen, Svalbard Archipeligo. Arc. Alp. Res. 16, 381–394.

Frenot, Y., Van Vliet-Lanoë, B., and Gloaguen, J. 1995. Particle translocation and initial soil development on a glacier foreland, Kerguelen Islands, Subantarctica. Arc. Alp. Res. 27, 107–115.

Fyfe, W.S., and Kronberg, B.I. 1980. Nutrient conservation, the key to agricultural strategy. Mazingira 13, 4, 64–69.

Gardner, L.R. 1972. Origin of the Mormon Mesa caliche. Geol. Soc. Am. Bull. 83, 143–156.

Gile, L.H. 1979. Holocene soils in eolian sediments of Bailey County, Texas. Soil Sci. Soc. Am. Proc. 43, 994–1003.

Gile, L.H., and Grossman R.B. 1979. The desert soil monograph. Soil Conservation Service, U.S. Department of Agriculture, 984 p.

Gile, L.H., and Grossman, R.B. 1968. Morphology of the argillic horizon in desert soils of southern New Mexico. Soil Sci. 106, 6–15.

Gile, L.H., Peterson, F.F., and Grossman, R.B. 1965. The K horizon: A master soil horizon of carbonate accumulation. Soil Sci. 99, 74–82.

Gile, L.H., Peterson, F.F., and Grossman, R.B. 1966. Morphological and genetic sequences of carbonate accumulation in desert soils. Soil Sci. 101, 347–360.

Gile, L.H., Hawley, J.W., and Grossman, R.B. 1970. Distribution and genesis of soils and geomorphic surfaces in a desert region of southern New Mexico. Soil Sci. Soc. Am. Guidebook, soil-geomorphology field conferences, Aug. 21–22, 29–30, 1970, 156 p.

Gile, L.H., Hawley, J.W., and Grossman, R.B. 1981. Soils and geomorphology in the Basin and Range area of southern New Mexico—Guidebook to the Desert Project. New Mexico Bureau of Mines and Mineral Resources, Mem. 39, 222 p.

Goh, K.M., and Pullar, W.A. 1977. Radiocarbon dating techniques for tephras in central North Island, New Zealand. Geoderma 18, 265–278.

Goh, K.M., Molloy, B.P.J., and Rafter, TA. 1977. Radiocarbon dating of Quaternary loess deposits, Banks Peninsula, Canterbury, New Zealand. Quaternary Res. 7, 177–196.

Goh, K.M., Tonkin, P.J., and Rafter, T.A. 1978. Implications of improved radiocarbon dates of Timaru peats on Quaternary loess stratigraphy. New Zealand J. Geol. Geophys. 21, 463–466.

Goss, D.W., Smith, S.J., and Stewart, B.A. 1973. Movement of added clay through calcareous materials. Geoderma 9, 97–103.

Goudie, A. 1973. Duricrusts in tropical and subtropical landscapes. Oxford Univ. Press, London, 174 p.

Goudie, A.S. 1983. Calcrete. In A.S. Goudie and K. Pye, eds., Chemical sediments and geomorphology: Precipitates and residua in the near-surface environment, p. 93–131. Academic Press, London, 439 p.

Gustafsson, J.P., Bhattacharya, P., Bain, D.C., Fraser, A.R., and McHardy, W.J. 1995. Podzolisation mechanisms and the synthesis of imogolite in northern Scandinavia. Geoderma 66, 167–184.

Hallsworth, E.G. 1963. An examination of some factors affecting the movement of clay in an artificial soil. J. Soil Sci. 14, 360–371.

Hammond, A.P., Goh, K.M., and Tonkin, P.J. 1991. Chemical pretreatments for improving the radiocarbon dates of peats and organic silts in a gley podzol environment: Grahams Terrace, North Westland. New Zealand J. Geol. Geophys. 34, 191–194.

Harden, J.W., Taylor, E.M., McFadden, L.D., and Reheis, M.C. 1991. Calcic, gypsic, and siliceous soil chronosequences in arid and semiarid environments. Soil Sci. Soc. Am. Spec. Publ. No. 26, 1–16.

Hay, R.L., and Reeder, R.J. 1978. Calcretes of Olduvai Gorge and the Ndolanya Beds of northern Tanzania. Sedimentology 25, 649–673.

Hay, R.L., and Wiggins, B. 1980. Pellets, ooids, sepiolite and silica in three calcretes of the southwestern United States. Sedimentology 27, 559–576.

Holliday, V.T. 1988. Genesis of a late-Holocene soil chronosequence at the Lubbock Lake Archaeological Site, Texas. Ann. Assoc. Am. Geogr. 78, 594–610.

Hubble, G.D., Isbell, R.F., and Northcote, K.H. 1983. Features of Australian soils. In Division of soils, CSIRO, Soils: An Australian viewpoint, p. 17–47. CSIRO, Melbourne/Academic Press, London, 928 p.

Hutton, J.T., and Dixon, J.C. 1981. The chemistry and mineralogy of some South Australian calcretes and associated soft carbonates and their dolomitisation. J. Geol. Soc. Aust. 28, 71–79.

Isbell, R.F., Reeve, R., and Hutton, J.T. 1983. Salt and sodicity. In Division of Soils, CSIRO, Soils: An Australian viewpoint, p. 107–117. CSIRO, Melbourne/Academic Press, London, 928 p.

Jersak, J., Amundson, R., and Brimhall, G., Jr. 1995. A mass balance analysis of podzolization: Examples from the northeastern United States. Geoderma 66, 15–42.

Johnson, D.L., and Balek, C.L. 1991. The genesis of Quaternary landscapes with stone-lines. Phys. Geogr. 12, 385–395.

Johnson, D.L., and Watson-Stegner, D. 1990. The soil-evolution model as a framework for evaluating pedoturbation in archaeological site information. In N.P. Lasca and J. Donahue, eds., Archaeological geology of North America, p. 541–560. Geol. Soc. Am., Centennial Spec. Vol. 4.

Jordan, C.F. 1982. The nutrient balance of an Amazonian rain forest. Ecology 63, 647–654.

Junge, C.E., and Werby, R.T. 1958. The concentration of chloride, sodium, potassium, calcium, and sulphate in rain water over the United States. J. Meteorol. 15, 417–425.

Kemp, R.A. 1985. Soil micromorphology and the Quaternary. Quaternary Res. Assoc. Tech. Guide No. 2, Cambridge, 80 p.

Kimble, J.M., and Yeck, Y.D., eds. 1990. Proceedings of the fifth international soil correlation meeting (IS-COM IV): Characterization, classification, and utilization of Spodosols. U.S. Dept. Agri., Soil Conservation Service.

King, R.H., and Brewster, G.R. 1976. Characteristics and genesis of some subalpine Podzols (Spodosols), Banff National Park, Alberta. Arc. Alp. Res. 8, 91–104.

Kittrick, J.A., Fanning, D.S., and Hossner, L.R. 1982. Acid sulphate weathering. Soil Sci. Soc. Am. Spec. Publ. No. 10, 234 p.

Kononova, M.M. 1961. Soil organic matter. Pergamon Press, New York, 450 p.

Krauskopf, K.B., and Bird, D.K. 1995. Introduction to geochemistry. McGraw-Hill, New Yoek, 647 p.

Kronberg, B.I., and Fyfe, W.S. 1989. Tectonics, weathering and environment. In K.S. Balasubramanian and 11 others, advisory ed. board, Weathering; its products and deposits, Vol. I. Processes, p. 3–13. Theophrastus Publ., S.A., Athens, 462 p.

Kronberg, B.I., Fyfe, W.S., McKinnon, B.J., Couston, J.F., Stilianidi, F.B., and Nash, R.A. 1982. Model for bauxite formation: Paragominas (Brazil). Chem. Geol. 35, 311–320.

Ku, T., Bull, W.B., Freeman, S.T., and Knauss, K.G. 1979. Th232/U234 dating of pedogenic carbonates in gravelly desert soils of Vidal Valley, southeastern California. Geol. Soc. Am. Bull. 90, 1063–1073.

Lattman, L.H., and Lauffenburger, S.K. 1974. Proposed role of gypsum in the formation of caliche. Zeit. Geomorphol. N. F., Suppl. Bd. 20, 140–149.

Lee, R., ed. 1980. Soil groups of New Zealand, Part 5: Podzols and gley podzols. New Zealand Society of Soil Science, Lower Hutt, 452 p.

Litaor, M. 1996. The influence of pocket gophers on the status of nutrients in alpine soils. Geoderma 70, 37–48.

Locke, W.W. 1986. Fine particle translocation in soils developed on glacial deposits, southern Baffin Island, N.W.T., Canada. Arct. Alp. Res. 18, 33–43.

Machette, M.N. 1985. Calcic soils and calcretes of the southwestern United States. Geol. Soc. Am. Spec. Pap. 203, 1–21.

Macias, F., and Chesworth, W. 1992. Weathering in humid regions, with emphasis on igneous rocks and their metamorphic equivalents. In I.P. Martini and W. Chesworth, eds., Weathering, soils & paleosols, p. 283–306. Elsevier, Amsterdam, 618 p.

Magaritz, M., and Amiel, A.J. 1980. Calcium carbonate in a calcareous soil from the Jordan Valley, Israel: Its origin as revealed by the stable carbon isotope method. Soil Sci. Soc. Am. J. 44, 1059–1062.

Malde, H.E. 1955. Surficial geology of the Louisville Quadrangle, Colorado. U.S. Geol. Surv. Bull. 996-E, 217–259.

Marshall, C.E. 1977. The physical chemistry and mineralogy of soils. Vol.II: Soils in place. John Wiley & Sons, New York, 313 p.

Martin, C.W., and Johnson, W.C. 1995. Variation in radiocarbon ages of soil organic matter fractions from late Quaternary buried soils. Quaternary Res. 43, 232–237.

Matthews, J.A. 1985. Radiocarbon dating of surface and buried soils: Principles, problems and prospects. In K.S. Richards, R.R. Arnett, and S. Ellis, eds., Geomorphology and soils, p. 269–288. George Allen & Unwin, London, 441 p.

Mayer, L., McFadden, L.D., and Harden, J.W. 1988. Distribution of calcium carbonate in desert soils: A model. Geology 16, 303–306.

McDonald, E.V., Pierson, F.B., Flerchinger, G.N., and McFadden, L.D. 1996. Application of a soil-water balance model to evaluate the influence of Holocene climate change on calcic soils, Mojave Desert, California, U.S.A. Geoderma 74, 167–192.

McFadden, L.D. 1988. Climatic influences on rates and processes of soil development in Quaternary deposits of southern California. Geol. Soc. Am. Spec. Pap. 216, 153–181.

McFadden, L.D., Amundson, R.G., and Chadwick, O.A. 1991. Numerical modeling, chemical, and isotopic studies of carbonate accumulation in soils of arid regions. Soil Sci. Soc. Am. Spec. Publ. No. 26, 17–35.

McFarlane, M.J. 1976. Laterite and landscape. Academic Press, London, 151 p.

McFarlane, M.J. 1983. Laterites. In A.S. Goudie and K. Pye, eds., Chemical sediments and geomorphology: Precipitates and residua in the near-surface environment, p. 7–58. Academic Press, London, 439 p.

McFarlane, M.J., ed. 1987. Laterites, some aspects of current research. Zeit. Geomorphol. Suppl. 64, 180 p.

McKeague, J.A., and Day, J.H. 1966. Dithiomite and oxalate-extractable Fe and Al as aids in differentiating various classes of soils. Can. J. Soi Sci. 46, 13–22.

McKeague, J.A., and St. Arnaud, R.J. 1969. Pedotranslocation: Eluviation-illuviation in soils during the Quaternary. Soil Sci. 107, 428–434.

McKeague, J.A., Ross, G.J., and Gamble, D.S. 1978. Properties, criteria of classification, and concepts of genesis of podzolic soils in Canada. In W.C. Mahaney, ed., Quaternary soils, p. 27–60. Geo Abstracts, Norwich, England.

McKeague, J.A., DeConinck, F., and Franzmeier, D.P. 1983. Spodosols. In L.P. Wilding, N.E. Smeck, and G.F. Hall, eds., Pedogenesis and soil taxonomy, II, The soil orders, p. 217–252. Elsevier, Amsterdam, 410 p.

Meixner, R.E., and Singer, M.J. 1985. Phosphorus fractions from a chronosequence of alluvial soils, San Joaquin Valley, California. Soil Sci. 139, 37–46.

Miles, R.J. and Franzmeier, D.P. 1981. A lithosequence of soils formed in dune sand. Soil Sci. Soc. Am. J. 45, 362–367.

Milnes, A.R., Wright, M.J., and Thiry, M. 1991. Silica accumulations in saprolites and soils in South Australia. Soil Sci. Soc. Am. Spec. Publ. No. 26, 121–149.

Mokma, D.L., and Sprecher, S.W. 1994a. Water table depths and color patterns in Spodosols of two hydrosequences in northern Michigan, USA. Catena 22, 275–286.

Mokma, D.L., and Sprecher, S.W. 1994b. Water table depths and color patterns in soils developed from red parent materials in Michigan, USA. Catena 22, 287–298.

Monger, H.C., Daugherty, L.A., and Gile, L.H. 1991a. A microscopic examination of pedogenic calcite in an aridisol of southern New Mexico. Soil Sci. Soc. Am. Spec. Publ. No. 26, 37–60.

Monger, H.C., Daugherty, L.A., Lindemann, W.C., and Liddell, C.M. 1991b. Microbial precipitation of pedogenic carbonate. Geology 19, 997–1000.

Muhs, D.R., Bush, C.A., Stewart, K.C., Rowland, T.R., and Crittendon, R.C. 1990. Geochemical evidence for Saharan dust parent material for soils developed on Quaternary limestones of Caribbean and western Atlantic islands. Quaternary Res. 33, 157–177.

Nahon, D.B. 1991. Introduction to the petrology of soils and chemical weathering. John Wiley & Sons, New York, 313 p.

NADP/NTN. 1984. NADP/NTN annual data summary, precipitation chemistry in the United States, 1984. National Atmospheric Deposition Program/National Trends Network Coordinator's Office, Colorado State Univ., Ft. Collins, 240 p.

Naidu, R., Sumner, M.E., and Rengasamy, P., eds. 1995. Australian sodic soils—distribution, properties and management. CSIRO Publ., Melbourne, 351 p.

Nettleton, W.D., Flach, K.W., and Brasher, B.R. 1969. Argillic horizons without clay skins. Soil Sci. Soc. Am. Proc. 33, 121–125.

Nettleton, W.D., Witty, J.E., Nelson, R.E., and Hawley, J.E. 1975. Genesis of argillic horizons in soils of desert areas of the southwestern United States. Soil Sci. Soc. Am. Proc. 39, 919–926.

Nettleton, W.D., Nelson, R.E., Brasher, D.R., and Derr, P.H. 1982. Gypsiferous soils in the western United States. In J.A. Kittrick, D.S. Fanning, and L.R. Hossner, eds., Acid sulphate weathering, p. 147–168. Soil Science Society of America, Madison, WI.

Noller, J.S. 1993. Late Cenozoic stratigraphy and soil geomorphology of the Peruvian desert, 3-18 degrees

S: A long-term record of hyperaridity and El Nino. Ph.D. dissertation, Univ. Colorado, Boulder, 279 p.

Norrish, K., and Pickering, J.G. 1983. Clay minerals. In Division of Soils, CSIRO, Soils: An Australian viewpoint, p. 281–308. CSIRO, Melbourne and Academic Press, London.

Oertal, A.C. 1968. Some observations incompatible with clay illuviation. 1968 Internat. Cong. Soil Sci. Trans. 4, 481–488.

Olphen, H. van. 1963. An introduction to clay colloid chemistry. John Wiley & Sons, New York, 301 p.

Parakshin, Y.P. 1982. Some regularities in the formation and development of Solonetz soils, Kokchetav Upland, Kazakh SSR. Pochvovedeniye No. 11, 8–16.

Paramananthan, S., and Eswaran, H. 1980. Morphological properties of Oxisols. In B.K.G. Theng, ed., Soils with variable charge, p. 35–43. New Zealand Soc. Soil Sci., Lower Hutt.

Parfitt, R.L. 1980. Chemical properties of variable charge soils. In B.K.G. Theng, ed., Soils with variable charge, p. 167–194. New Zealand Soc. Soil Sci., Lower Hutt, 448 p.

Parfitt, R.L., and Childs, C.W. 1988. Estimation of forms of Fe and Al: A review, and analysis of contrasting soils by dissolution and Moessbauer methods. Aust. J. Soil Res. 26, 121–144.

Parizek, E.J., and Woodruff, J.F. 1957. Description and origin of stone layers in soils of the southeastern states. J. Geol. 65, 24–34.

Paton, T.R., Humphreys, G.S., and Mitchell, P.B. 1995. Soils, a new global view. Yale Univ. Press, New Haven, 213 p.

Paul, E.A. 1969. Characterization and turnover rate of soil humic constituents. In S. Pawluk, ed., Pedology and Quaternary research, p. 63–76. Univ. of Alberta Printing Dept., Edmonton, Alberta, 218 p.

Pedro, G., Jamagne, M., and Begon, J.C. 1978. Two routes in genesis of strongly differentiated acid soils under humid, cool-temperate conditions. Geoderma 20, 173–189.

Petersen, L. 1976. Podzols and podzolization. DSR Forlag, Copenhagen, 293 p.

Peterson, F.F. 1980. Holocene desert soil formation under sodium salt influence in a playa-margin environment. Quaternary Res. 13, 172–186.

Ponomareva, V.V. 1969. Theory of podzolization. Israel Program for Scientific Translations, Jerusalem, 309 p.

Rabenhorst, M.C., West, L.T., and Wilding, L.P. 1991. Genesis of calcic and petrocalcic horizons in soils over carbonate rocks. Soil Sci. Soc. Am. Spec. Publ. No. 26, 61–74.

Reeves, C.C., Jr. 1976. Caliche-origin, classification, morphology and uses. Estacado Books, Lubbock, Texas, 233 p.

Reheis, M.C. 1988. Pedogenic replacement of aluminosilicate grains by $CaCO_3$ in Ustollic Haplargids, south-central Montana, U.S.A. Geoderma 41, 243–261.

Reheis, M.C., Goodmacher, J.C., Harden, J.W., McFadden, L.D., Rockwell, T.K., Shroba, R.R., Sowers, J.M., and Taylor, E.M. 1995. Quaternary soils and dust deposition in southern Nevada and California. Geol. Soc. Am. Bull. 107, 1003–1022.

Reheis, M.C., Sowers, J.M., Taylor, E.M., McFadden, L.D., and Harden, J.W. 1992. Morphology and genesis of carbonate soils on the Kyle Canyon fan, Nevada, U.S.A. Geoderma 52, 303–342.

Reider, R.G. 1985. Soil formation at the McKean archaeological site, northeastern Wyoming. Current research, occasional papers on Wyoming archaeology, No. 4, Wyoming Recreation Commission, Cheyenne, 51–61.

Richardson, J.L., and Daniels, R.B. 1993. Stratigraphic and hydraulic influences on soil color development. Soil Sci. Soc. Am. Spec. Publ. No. 31, 109–125.

Ross, C.W., Mew, G., and Searle, P.L. 1977. Soil sequences on two terrace sequences in the North Westland area, New Zealand. New Zealand J. Sci. 20, 231–244.

Rowell, D.L. 1981. Oxidation and reduction. In D. J. Greenland and M.H.B. Hayes, eds., The chemistry of soil processes, p. 401–461. John Wiley & Sons, New York.

Ruhe, R.V. 1967. Geomorphic surfaces and surficial deposits in southern New Mexico. New Mex. Bur. Mines and Mineral Resources, Memoir 18, 66 p.

Ruhe, R.V. 1959. Stone lines in soils. Soil Sci. 87, 223–231.

Salomons, W., and Mook, W.G. 1976. Isotope geochemistry of carbonate dissolution and reprecipitation in soils. Soil Sci. 122, 15–24.

Salomons, W., Goudie, A., and Mook, W.G. 1978. Isotopic composition of calcrete deposits from Europe, Africa and India. Earth Surface Processes 3, 43–57.

Schaetzl, R.J. 1992. Texture, mineralogy, and lamellae development in sandy soils in Michigan. Soil Sci. Soc. Am. J. 56, 1538–1545.

Schaetzl, R.J. 1996. Spodosol-Alfisol intergrades: Bisequal soils in NE Michigan, USA. Geoderma 74, 23–47.

Schellmann, W. 1994. Geochemical differentiation in laterite and bauxite formation. Catena 21, 131–143.

Schlesinger, W.H. 1985. The formation of caliche in soils of the Mojave Desert, California. Geochim. Cosmochim. Acta 49, 57–66.

Schnitzer, M. 1969. Reactions between fulvic acid, a soil humic compound and inorganic soil constituents. Soil Sci. Soc. Am. Proc. 33, 75–81.

Schwarz, T. 1994. Ferricrete formation and relief inversion: An example from central Sudan. Catena 21, 257–268.

Schwarz, T., and Germann, K., eds. 1994. Laterization processes and supergene ore formation. Catena 21, Nos. 2,3.

Sharpenseel, H.W. 1971. Radiocarbon dating of soils—problems, troubles, hopes. In D.H. Yaalon, ed., Paleopedology, p. 77–88. Israel Univ. Press, Jerusalem, 350 p.

Shlemon, R.J., and Hamilton, P. 1978. Late Quaternary rates of sedimentation and soil formation, Camp Pendleton-San Onofre State Beach coastal area, Southern California, U.S.A. Tenth Int. Cong. Sedimentol. (Jerusalem) (Abstracts), 603–604.

Simonson, R.W. 1978. A multiple-process model of soil genesis. In W. C. Mahaney, ed., Quaternary soils, p. 1–25. Geo Abstracts, University of East Anglia, Norwich, England.

Singer, A. 1994. The chemistry of precipitation in Israel. Israel J. Chem. 34, 315–326.

Singer, M., Ugolini, F.C., and Zachara, J. 1978. In situ study of podzolization on tephra and bedrock. Soil Sci. Soc. Am. J. 42, 105–111.

Singer, M.J., and Munns, D.N. 1987. Soils, an introduction. Macmillan, New York, 492 p.

Sivarajasingham, S., Alexander, L.T., Cady, J.G., and Cline, M.G. 1962. Laterite. Adv. Agron. 14, 1–60.

Slate, J.L., Bull, W.B., Teh-lung, K., Shafiqullah, M., Lynch, D.J., and Yi-Pu, H. 1991. Soil-carbonate genesis in the Pinacate volcanic field, northwestern Mexico. Quaternary Res. 35, 400–416.

Smeck, N.E., and Ciolcosz, E.J. eds. 1989. Fragipans: Their occurrence, classification, and genesis. Soil Sci. Soc. Am. Spec. Publ. No. 24, 153 p.

Smeck, N.E., Runge, E.C.A., and Mackintosh, E.E. 1983. Dynamics and genetic modelling of soil systems. In L.P. Wilding, N.E. Smeck, and G.F. Hall, eds., Pedogenesis and soil taxonomy, I. Concepts and interactions, p. 51–81. Elsevier, Amsterdam, 303 p.

Soil Survey Division Staff. 1993. Soil survey manual. U.S. Dept. Agri. Handbook No. 18, 437 p.

Soil Survey Staff. 1960. Soil classification, a comprehensive system, 7th approximation. U.S. Dept. Argi. Soil Conservation Service, 503 p.

Soil Survey Staff. 1975. Soil Taxonomy. U.S. Dept. Agri. Handbook 436, 754 p.

Soil Survey Staff. 1994. Keys to soil taxonomy (6th ed.). U.S. Dept. Argi., Soil Conservation Service, Washington, D.C., 306 p.

Sowers, J.M., Szabo, B., Jull, T., Ku, T.L., Reheis, M.C., Chadwick, O.A., Robinson, S.W., and Amundson, R.G. 1988. Age data for Kyle Canyon soils and deposits. In D.L. Weide and M.L. Faber, eds., This extended land, geological journeys in the southern basin and range, p. 141–142. Geol. Soc. Am., Cordilleran Section, Field Trip Guidebook, 330 p.

Stobbe, P.C., and Wright, J.R. 1959. Modern concepts of the genesis of podzols. Soil Sci. Soc. Am. Proc. 23, 161–164.

Stolt, M.H., Ogg, C.M., and Baker, J.C. 1994. Strongly contrasting redoximorphic patterns in Virginia Valley and Ridge paleosols. Soil Sci. Soc. Am. J. 58, 477–484.

Stoops, G., and Eswaran, H. eds. 1986. Soil Micromorphology. Van Nostrand Reinhold, New York, 345 p.

Szabolcs, I. 1989. Salt-affected soils. CRC Press, Inc., Boca Raton, FL, 274 p.

Tardy, Y. 1992. Diversity and terminology of lateritic profiles. In I.P. Martini and W. Chesworth, eds., Weathering, soils and paleosols, p. 379–405. Elsevier, Amsterdam, 618 p.

Tardy, Y., and Roquin, C. 1992. Geochemistry and evolution of lateritic landscapes. In I.P. Martini and W. Chesworth, eds., Weathering, soils & paleosols, p. 407–443. Elsevier, Amsterdam, 618 p.

Tedrow J.C.F. 1977. Soils of the polar landscapes. Rutgers Univ. Press, New Brunswick, NJ, 638 p.

Timpson, M.E., Lee, S.Y., Ammons, J.T., and Foss, J.E. 1996. Mineralogical investigation of soils formed in calcareous gravelly alluvium, eastern Crete, Greece. Soil Sci. Soc. Am. J. 60, 299–308.

Ugolini, F.C., and Mann, D.H. 1979. Biopedological origin of peatlands in southeast Alaska. Nature (London) 281, 366–368.

Ugolini, F.C., Zachara, J.M., and Reanier, R.E. 1982. Dynamics of soil- forming processes in the Arctic. In Proc. 4th Canadian Permafrost Conf., p. 103–115. National Research Council of Canada, Ottawa.

Ugolini, F.C., Dahlren, R., and Vogt, K. 1990. The genesis of Spodosols and the role of vegetation in the Cascade Range of Washington, U.S.A. In J. Kimble and R. Yeck, eds., Proceedings of the fifth international soil correlation meeting (IS-COM IV), characterization, classification, and utilization of Spodosols, p. 370–380. Soil Conservation Service, U.S. Dept. Agri., Lincoln, Nebraska, 438 p.

Valeton, I. 1994. Element concentration and formation of ore deposits by weathering. Catena 21, 99–129.

Valeton, I., and Wilke, F. 1993. Tertiary bauxites in subsidence areas and associated laterite-derived sediments in northwestern India: Contributions to sedimentology, 18. E. Schweizerbart'sche Verlagsbuchlhandlung (Nägele u Obermiller), Stuttgart, 104 p.

Van Vliet-Lanoë, B. 1985. Frost effects in soils. In J. Boardman, ed., Soils and Quaternary landscape evolution, p. 117–158. John Wiley & Sons Ltd., Chichester, 391 p.

Vepraskas, M.J. 1994. Redoximorphic features for identifying aquic conditions. North Carolina Agri. Res. Service, Tech. Bull. 301, 33 p.

Walker, P.H., and Chittleborough, D.J. 1986. Development of particle-size distributions in some Alfisols of southeastern Australia. Soil Sci. Soc. Am. J. 50, 394–400.

Walker, P.H., and Hutka, J. 1979. Size characteristics of soils and sediments with special reference to clay fractions. Aust. J. Soil Res. 17, 383–404.

Walker, P.H., Chartres, C.J., and Hutka, J. 1988. The effect of aeolian accessions on soil development on granitic rocks. I. Soil morphology and particle-size distributions. Aust. J. Soil Res. 26, 1–16.

Walker, T.R., Waugh, B., and Crone, A.J. 1978. Diagenesis in first-cycle desert alluvium of Cenozoic age, southwestern United States and northwestern Mexico. Geol. Soc. Am. Bull. 89, 19–32.

Walker, T.W. 1964. The significance of phosphorus in pedogenesis. In E.G. Hallsworth and D.V. Crawford, eds., Experimental pedology, p. 295–315. Butterworths, London.

Walker, T.W., and Syers, J.K. 1976. The fate of phosphorus during pedogenesis. Geoderma 15, 1–19.

Wang, C., McKeague, J.A., and Kodama, H. 1986. Pedogenic imogolite and soil environments: Case study of Spodosols in Quebec, Canada. Soil Sci. Soc. Am. J. 50, 711–718.

Wang, Y., Nahon, D., and Merino, E. 1994. Dynamic model of the genesis of calcretes replacing silicate grains in semiarid regions. Geochim. Cosmochim. Acta 58, 5131–5145.

Wang, Y., Amundson, R., and Trumbore, S. 1996. Radiocarbon dating of soil organic matter. Quaternary Res. 45, 282–288.

Washburn, A.L. 1980. Geocryology: A survey of periglacial processes and environments. Halsted Press, John Wiley & Sons, New York, 406 p.

Watson, A. 1979. Gypsum crusts in deserts. J. Arid Environ. 2, 3–20.

Watson, A. 1983. Gypsum crusts. p. 133–161 in A.S. Goudie and K. Pye, eds., Chemical sediments and geomorphology: Precipitates and residua in the near-surface environment. Academic Press, London, 439 p.

Watson, A. 1985. Structure, chemistry and origins of gypsum crusts in southern Tunisia and the central Namib Desert. Sedimentology 32, 855–875.

Watts, N.L. 1978. Displacive calcite: Evidence from recent and ancient calcretes. Geology 6, 699–703.

Wells, N.A., Andriamihaja, B., and Rakotovololona, H.F.S. 1990. Stonelines and landscape development on the laterized craton of Madagasgar. Geol. Soc. Am. Bull. 102, 615–627.

Williams, G.E., and Polach, H.A. 1971. Radiocarbon dating of arid-zone calcareous paleosols. Geol. Soc. Am. Bull. 82, 3069–3086.

Wright, J.R., and Schnitzer, M. 1963. Metallo-organic interactions associated with podzolization. Soil Sci. Soc. Am. Proc. 27, 171–176.

Wright, P.V., and Tucker, M.E. 1991. Calcretes: An introduction. In V.P. Wright and M.E. Tucker, eds., Calcretes, p. 1–22. Reprint series vol. 2, Internat. Assoc. Sedimentologists, Blackwell Scientific, Oxford, 352 p.

Wright, W.R., and Foss, J.E. 1968. Movement of silt-sized particles in sand columns. Soil Sci. Soc. Am. Proc. 32, 446–448.

Yaalon, D.H. 1963. On the origin and accumulation of salts in groundwater and in soils of Israel. Bull. Res. Council Israel 11G, 105–131.

Yaalon, D.H., ed. 1971. Paleopedology. Israel Univ. Press, Jerusalem, 350 p.

Yaalon, D.H. 1997. Soils in the Mediterranean region: What makes them different? Catena 28, 157–169.

Yaalon, D.H., and Ganor, E. 1973. The influence of dust on soils during the Quaternary. Soil Sci. 116, 146–155.

■ Chapter 6

Amundson, R., Harden, J., and Singer, M. 1994. Factors of soil formation: A fiftieth anniversary retrospective. Soil Sci. Soc. Am. Spec. Publ. No. 33, 160 p.

Barshad, I. 1964. Chemistry of soil development. In F.E. Bear, ed., Chemistry of the soil, p. 1–70. Reinhold, New York, 515 p.

Bockheim, J.G. 1980. Solution and use of chronofunctions in studying soil development. Geoderma 24, 71–85.

Brinkman, R. 1970. Ferrolysis, a hydromorphic soil forming process. Geoderma 3, 199–206.

Bunting, B.T. 1965. The geography of soil. Aldine Publ. Co., Chicago, 213 p.

Chesworth, W. 1973. The parent rock effect on the genesis of soil. Geoderma 10, 215–225.

Ciolkosz, E.J., Petersen, G.W., Cunningham, R.L., and Matelski, R.P. 1979. Soils developed from colluvium in the ridge and valley area of Pennsylvania. Soil Sci. 128, 153–162.

Ciolkosz, E.J., Waltman, W.J., Simonson, T.W., and Dobos, R.R. 1989. Distribution and genesis of soils of the northeastern United States. Geomorphology 2, 285–302.

Colman, S.M. 1981. Rock-weathering rates as functions of time. Quaternary Res. 15, 250–264.

Cremeens, D.L., Brown, R.B., and Huddleston, J.H. 1994. Whole regolith pedology. Soil Sci. Soc. Am. Spec. Publ. No. 34, 136 p.

Crocker, R.L. 1952. Soil genesis and the pedogenic factors. Quat. Rev. Biol. 27, 139–168.

Easterbrook, D.J. 1993. Surface processes and landforms. Macmillan, New York, 520 p.

Graham, R.C., and Buol, S.W. 1990. Soil-geomorphic relations on the Blue Ridge Front: II. Soil characteristics and pedogenesis. Soil Sci. Soc. Am. J. 54, 1367–1377.

Graham, R.C., Daniels, R.B., and Buol, S.W. 1990. Soil-geomorphic relations on the Blue Ridge Front: I. Regolith types and slope processes. Soil Sci. Soc. Am. J. 54, 1362–1367.

Harden, J.W. 1982. A quantitative index of soil development from field descriptions: Examples from a chronosequence in central California. Geoderma, 28, 1–28.

Harden, J.W., and Taylor, E.M. 1983. A quantitative comparison of soil development in four climatic regimes. Quaternary Res. 20, 342–359.

Harradine, F., and Jenny, H. 1958. Influence of parent material and climate on texture and nitrogen and carbon contents of virgin California soils: 1: Texture and nitrogen contents of soils. Soil Sci. 85, 235–243.

Holliday, V.T. 1992. Soil formation, time and archaeology. In V.T. Holliday, ed., Soils in archaeology, p. 101–117. Smithsonian Inst. Press, Washington, D.C., 254 p.

Jenny, H. 1941. Factors of soil formation. McGraw-Hill, New York, 281 p.

Jenny, H. 1946. Arrangement of soil series and types according to functions of soil-forming factors. Soil Sci. 61, 375–391.

Jenny, H. 1958. Role of the plant factor in the pedogenic functions. Ecology 39, 5–16.

Jenny, H. 1961. Derivation of state factor equations of soils and ecosystems. Soil Sci. Soc. Am. Proc. 25, 385–388.

Jenny, H. 1980. The soil resource—origin and behavior. Springer-Verlag, New York, 377 p.

Johnson, D.L., and Watson-Stegner, D. 1987. Evolution model of pedogenesis. Soil Sci. 143, 349–366.

Johnson, D.L., Keller, E.A., and Rockwell, T.K. 1990. Dynamic pedogenesis: New views on some key soil concepts, and a model for interpreting Quaternary soils. Quaternary Res. 33, 306–319.

Lavkulich, L.M. 1969. Soil dynamics in the interpretation of paleosols. In S. Pawluk, ed., Pedology and Quaternary research, p. 25–37. Univ. of Alberta Printing Dept., Edmonton, Alberta, 218 p.

Major, J. 1951. A functional, factorial approach to plant ecology. Ecology 32, 392–412.

Markewich, H.W., Pavich, M.J., Mausbach, M.J., Hall, R.L., Johnson, R.G., and Hearn, P.P. 1987. Age relations between soils and geology in the Coastal Plain of Georgia and Virginia. U.S. Geol. Surv. Bull. 1589-A, 34 p.

McFadden, L.D., and Tinsley, J.C. 1985. Rate and depth of pedogenic-carbonate accumulation in soils: Formation and testing of a compartment model. Geol. Soc. Am. Spec. Pap. 203, 23–41.

Morrison, R.B. 1964. Lake Lahontan: Geology of the southern Carson Desert, Nevada. U.S. Geol. Surv. Prof. Pap. 401, 156 p.

Pedro, G., Jamagne, M., and Begon, J.C. 1978. Two routes in genesis of strongly differentiated acid soils under humid, cool-temperate conditions. Geoderma 20, 173–189.

Peterson, G.M., Webb III, T., Kutzbach, J.E., van der Hammen, T., Wijmstra, T.A., and Street, F.A. 1979. The continental record of environmental conditions at 18,000 yr B.P.: An initial evaluation. Quaternary Res. 12, 47–82.

Phillips, J.D. 1993. Progressive and regressive pedogenesis and complex soil evolution. Quaternary Res. 40, 169–176.

Retallack, G.J. 1994. The environmental factor approach to the interpretation of paleosols. Soil Sci. Soc. Am. Spec. Publ. No. 33, 31–64.

Richmond, G.M. 1962. Quaternary stratigraphy of the La Sal Mountains, Utah. U.S. Geol. Surv. Prof. Pap. 324, 135 p.

Ruhe, R.V. 1975. Geomorphology. Houghton Mifflin, Boston, 246 p.

Runge, E.C.A. 1973. Soil development sequences and energy models. Soil Sci. 115, 183–193.

Smeck, N.E., Runge, E.C.A., and Mackintosh, E.E. 1983. Dynamics and genetic modelling of soil systems. In L.P. Wilding, N.E. Smeck, and G.F. Hall, eds., Pedogenesis and soil taxonomy I. Concepts and interactions, p. 51–81. Elsevier, Amsterdam, 303 p.

Tandarich, J.P., and Specher, S.W. 1994. The intellectual background for the factors of soil formation. Soil Sci. Soc. Am. Spec. Publ. No. 33, 1–13.

Tonkin, P.J., and Basher, L.R. 1990. Soil-stratigraphic techniques in the study of soil and landform evolution across the southern Alps, New Zealand. Geomorphology 3, 547–575.

Yaalon, D.H. 1975. Conceptual models in pedogenesis: Can soil-forming functions be solved? Geoderma 14, 189–205.

Yaalon, D.H. 1983. Climate, time and soil development. In L.P. Wilding, N.E. Smeck, and G.F. Hall, eds., Pedogenesis and soil taxonomy. I. Concepts and interactions, p. 233–251. Elsevier Science Publ. B.V., Amsterdam, 303 p.

■ Chapter 7

Alexander, E.B., Ping, C.L., and Krosse, 1994. Podzolization in ultramafic materials in southeast Alaska. Soil Sci. 157, 46–52.

Allen, B.L., and Hajek, B.F. 1989. Mineral occurrence in soil environments. In J.B. Dixon and S.B. Weed, eds., Minerals in soil environments, p. 199–278. Soil Sci. Soc. Am. Book Series No. 1, 1244 p.

Barnhisel, R.I., and Rich, C.I. 1967. Clay mineral formation in different rock types of a weathering boulder conglomerate. Soil Sci. Soc. Am. Proc. 31, 627–631.

Barshad, 1. 1958. Factors affecting clay formation. 6th Natl. Conf. Clays Clay Minerals Proc. 110–132.

Barshad, I. 1964. Chemistry of soil development. In F.E. Bear, ed., Chemistry of the soil, Reinhold, New York, 515 p.

Barshad, I. 1966. The effect of a variation in precipitation on the nature of clay mineral formation in soils from acid and basic igneous rocks. 1966 Internat. Clay Conf. (Jerusalem), Proc. V. 1, 167–173.

Bates, T.F. 1962. Halloysite and gibbsite formation in Hawaii. 9th Natl. Conf. Clays Clay Minerals Proc. 307–314.

Benedict, J.B. 1981. The Fourth of July Valley: Glacial geology and archeology of the timberline ecotone. Center for Mountain Archeology, Res. Rept. No. 2, Ward, Colorado, 139 p.

Berry, M.E. 1994. Soil-geomorphic analysis of late-Pleistocene glacial sequences in the McGee, Pine, and Bishop creek drainages, east-central Sierra Nevada, California. Quaternary Res. 41, 160–175.

Best, M.G. 1982. Igneous and metamorphic petrology. W.H. Freeman, San Francisco, 630 p.

Birkeland, P.W. 1969. Quaternary paleoclimatic implications of soil clay mineral distribution in a Sierra Nevada-Great Basin transect. J. Geol. 77, 289–302.

Birkeland, P.W. 1973. Use of relative age dating methods in a stratigraphic study of rock glacier deposits, Mt. Sopris, Colorado. Arct. Alp. Res. 5, 401–416.

Birkeland, P.W. 1994. Variation in soil-catena characteristics of moraines with time and climate, South Island, New Zealand. Quaternary Res. 42, 49–59.

Birkeland, P.W., and Janda, R.J. 1971. Clay mineralogy of soils developed from Quaternary deposits of the eastern Sierra Nevada, California. Geol. Soc. Am. Bull. 82, 2495–2514.

Birkeland, P.W., Burke, R.M., and Walker, A.L. 1980. Soils and subsurface rock-weathering features of Sherwin and pre-Sherwin glacial deposits, eastern Sierra Nevada, California. Geol. Soc. Am. Bull. 91, pt. 1, 238–244.

Boggs, S., Jr. 1991. Petrology of sedimentary rocks. Macmillan, New York, 707 p.

Boulding, B.H., and Boulding, J.R. 1981. Genesis of silty and clayey material in some alpine soils in the Teton Mountains, Wyoming and Idaho. Indiana Acad. Sci. Proc. 91, 552–562.

Bouma, J., Hoeks, J., van der Plas, L., and van Scherrenburg, B. 1969. Genesis and morphology of some alpine podzol profiles. J. Soil Sci. 20, 384–398.

Brewer, R. 1976. Fabric and mineral analysis of soils. Kriger, New York.

Brimhall, G.H., Lewis, C.J., Ague, J.J., Dietrich, W.E., Hampel, J., Teague, T., and Rix, P. 1988. Metal enrichment in bauxites by deposition of chemically mature aeolian dust. Nature (London) 33, 819–824.

Burke, R.M., and Birkeland, P.W. 1979. Reevaluation of multiparameter relative dating techniques and their application to the glacial sequence along the eastern escarpment of the Sierra Nevada, California. Quaternary Res. 11, 21–51.

Cady, J.G. 1960. Mineral occurrence in relation to soil profile differentiation. 7th Internat. Cong. Soil Sci. Trans. 4, 418–424.

Chesworth, W. 1973. The parent material effect and the genesis of soils. Geoderma 10, 215–225.

Cline, M.G. 1953. Major kinds of profiles and their relationships in New York. Soil Sci. Soc. Am. Proc. 17, 123–127.

Colman, S.M., and Pierce, K.L. 1981. Weathering rinds on andesitic and basaltic stones as a Quaternary age indicator, western United States. U.S. Geol. Surv. Prof. Pap. 1210, 56 p.

Colman, S.M., and Pierce, K.L. 1986. Glacial sequence near McCall, Idaho: Weathering rinds, soil development, morphology, and other relative-age criteria. Quaternary Res. 25, 25–42.

Curtis, C.D. 1976a. Chemistry of rock weathering: fundamental reactions and controls. In E. Derbyshire, ed., Geomorphology and climate, p. 25–57. John Wiley & Sons, London.

Curtis, C.D. 1976b. Stability of minerals in surface weathering reactions: A general thermochemical approach. Earth Surf. Proc. Landforms 1, 63–70.

Dahms, D.E. 1992. Comment on "Origin of silt-enriched mantles in Indian Basin, Wyoming". Soil Sci. Soc. Am. J. 56, 991–992.

Dahms, D.E. 1993. Mineralogical evidence for eolian contribution to soils of late Quaternary moraines, Wind River Mountains, Wyoming, USA. Geoderma 59, 175–196.

Dan, J. 1990. The effect of dust deposition on the soils of the land of Israel. Quaternary Int. 5, 107–113.

Dan, J., and Yaalon, D.H. 1966. Trends of soil development with time in the Mediterranean environments of Israel. In Trans. Conf. Mediterranean Soils (Madrid) 139–145.

Dan, J., Gerson, R., Koyumdjisky, H., and Yaalon, D.H. 1981. Aridic soils of Israel-properties, genesis and management. Agric. Res. Organ. Spec. Publ. No. 190, The Volcani Center, Bet Dagan, Israel.

Drees, L.R., Wilding, L.P., Smeck, N.E., and Senkayi, A.L. 1989. Silica in soils: Quartz and disordered silica polymorphs. In J.B. Dixon and S.B. Weed, eds., Minerals in soil environments, p. 913–974. Soil Sci. Soc. Am. Book Series No. 1, 1244 p.

Drever, J.I., and Clow, D.W. 1995. Weathering rates in catchments. Rev. Mineral. 31, 463–483.

Dymond, J., Biscaye, P.E., and Rex, R.W. 1974. Eolian origin of mica in Hawaiian soils. Geol. Soc. Am. Bull. 85, 37–40.

Farrand, W.R. 1975. Sediment analysis of a prehistoric rockshelter: The Abri Pataud. Quaternary Res. 5, 1–26.

Fieldes, M., and Perrott, K.W. 1966. The nature of allophane in soils. Part 3. Rapid field and laboratory test for allophane. New Zealand J. Sci. 9, 623–629.

Fine, P., Singer, M.J., and Verosub, K.L. 1992. Use of magnetic-susceptibility measurements in assessing soil uniformity in chronosequence studies. Soil Sci. Soc. Am. J. 56, 1195–1199.

Flach, K.W., Holzhey, C.S., De Coninck, F., and Bartlett, R.J. 1980. Genesis and classification of Andepts and Spodosols. In B.K.G. Theng, ed., Soils with variable charge, p. 411–426. New Zealand Soc. Soil Sci., Lower Hutt.

Folk, R.L. 1974. Petrology of sedimentary rocks. Hemphill, Austin, Texas.

Garrels, R.M., and Mackenzie, F.T. 1971. Evolution of the sedimentary rocks. W.W. Norton, New York.

Gerson, R., Amit, R., and Grossman, S. 1985. Dust availability in desert terrains—a study in the desert of Israel and the Sinai. Phys. Geogr., Inst. Earth Sci., The Hebrew Univ., Jerusalem.

Goldich, S.S. 1938. A Study in rock-weathering. J. Geol. 46, 17–58.

Graham, R.C., Daniels, R.B., and Buol, S.W. 1990. Soil-geomorphic relations on the Blue Ridge front: I. Regolith types and slope processes. Soil Sci. Soc. Am. J. 54, 1362–1367.

Harradine, F., and Jenny, H. 1958. Influence of parent material and climate on texture and nitrogen and carbon contents of virgin California soils. I. Texture and nitrogen contents of soils. Soil Sci. 85, 235–243.

Hay, R.L. 1959. Origin and weathering of late Pleistocene ash deposits on St. Vincent, B.W.I. J. Geol. 67, 65–87.

Hembree, C.H., and Rainwater, F.H. 1961. Chemical degradation of opposite flanks of the Wind River Range, Wyoming. U.S. Geol. Surv. Water Supply Pap. 1535-E, 9 p.

Hoover, M.T., and Ciolkosz, E.J. 1988. Colluvial soil parent material relationships in the Ridge and Valley physiographic province of Pennsylvania. Soil Sci. 145, 163–172.

Hovan, S.A., Rea, D.K., Pisias, N.G., and Shackleton, N.J. 1989. A direct link between the China loess and marine O-18 records: eolian flux to the North Pacific. Nature (London) 340, 296–298.

Hutton, C.E. 1951. Studies of the chemical and physical characteristics of a chrono-litho-sequence of loess-derived prairie soils of southwestern Iowa. Soil Sci. Soc. Am. Proc. 15, 318–324.

Ito, T., Shoji, S., Shirato, Y., and Ono, E. 1991. Differentiation of a spodic horizon from a buried A horizon. Soil Sci, Soc. Am. J. 55, 438–442.

Jackson. M.L., Levelt, T.W.M., Syers, J.K., Rex, R.W., Clayton, R.N., Sherman, G.D., and Uehara, G.

1971. Geomorphological relationships of tropospherically derived quartz in the soils of the Hawaiian Islands. Soil Sci. Soc. Am. Proc. 35, 515–525.

Jenny, H. 1941. Calcium in the soil: III. Pedologic relations. Soil Sci. Soc. Am. Proc. 6, 27–35.

Jenny, H. 1962. Model of a rising nitrogen profile in Nile Valley alluvium, and its agronomic and pedogenic implications. Soil Sci. Soc. Am. Proc. 26, 588–591.

Johnson, D.L., and Stegner-Watson, D. 1990. The soil-evolution model as a framework for evaluating pedoturbation in archaeological site formation. In N.P. Lasca and J. Donahue, eds., Archaeological geology of North America, p. 541–560. Geological Soc. Am. Centennial Spec. Vol. 4, 633 p.

Leamy, M.L., Smith, G.D., Colmet-Daage, F., and Otowa, M. 1980. The morphological characteristics of Andisols. In B.K.G. Theng, ed., Soils with variable charge, p. 17–34. New Zealand Soc. Soil Sci., Lower Hutt.

Litaor, M.I. 1987. The influence of dust on the genesis of alpine soils in the Front Range, Colorado. Soil Sci. Soc. Am. J. 51, 142–147.

Litaor, M.I., Mancinelli, R., and Penny, J.C. 1996. The influence of pocket gophers on the status of nutrients in alpine soils. Geoderma 70, 37–48.

Loughnan, F.C. 1969. Chemical weathering of the silicate minerals. American Elsevier, New York, 154 p.

Lowe, D.L. 1986. Controls on the rates of weathering and clay mineral genesis in airfall tephras: A review and New Zealand case study. In S.M. Colman and D.P. Dethier, eds., Rates of chemical weathering of rocks and minerals, p. 265–330. Academic Press, Orlando, FL, 603 p.

Macleod, D.A. 1980. The origin of the red mediterranean soils in Epirus, Greece. J. Soil Sci. 31, 125–136.

Madole, R.F., and Shroba, R.R. 1979. Till sequence and soil development in the North St. Vrain drainage basin, east slope, Front Range. In F.G. Ethridge, ed., Field guide, northern Front Range and northwest Denver Basin, Colorado, p. 123–178. Dept. Earth Resources, Colorado State Univ., Ft. Collins, 209 p.

Marchand, D.E. 1970. Soil contamination in the White Mountains, eastern California. Geol. Soc. Am. Bull. 81, 2497–2506.

Marsan, F.A., Bain, D.C., and Duthie, D.M.L. 1988. Parent material uniformity and degree of weathering in a soil chronosequence, northwestern Italy. Catena 15, 507–517.

Marshall, C.E. 1977. The physical chemistry and mineralogy of soils, Vol. II: Soils in place. John Wiley & Sons, New York, 313 p.

McCalpin, J.P. 1982. Quaternary geology and neotectonics of the west flank of the northern Sangre de Cristo Mountains, south-central Colorado. Colorado School Mines Quart. 7, No. 3, 97 p.

McFadden, L.D. 1991. A model of pedogenesis in Quaternary siliclastic fan deposits in an arid climate: Implications for interpreting pre-Quaternary paleosols. Geol. Soc. Am. Abstr. with Programs 23, No. 5, A63.

McFadden, L.D., Wells, S.G., and Dohrenwend, J.C. 1986. Influences of Quaternary climatic changes on soil development on desert loess deposits of the Cima volcanic field, California. Catena 13, 361–389.

McFadden, L.D., Wells, S.G., and Jercinovich, M.J. 1987. Influences of eolian and pedogenic processes on the origin and evolution of desert pavements. Geology 15, 504–508.

Mew, G., and Lee, R. 1981. Investigation of the properties and genesis of West Coast wet land soils, South Island, New Zealand. 1. Type localities, profile morphology, and soil chemistry. New Zealand J. Sci. 24, 1–24.

Mew, G., Whitton, J.S., Robertson, S.M., and Lee, R. 1983. Investigation of the properties and genesis of West Coast wet land soils, South Island, New Zealand. 3. Soil Mineralogy. New Zealand J. Sci. 26, 85–97.

Mew, G., and Lee, R. 1988. The genesis and classification of soils on wet terraces and moraines in New Zealand. J. Soil Sci. 39, 125–138.

Miles, R.J., and Franzmeier, D.P. 1981. A lithochronosequence of soils formed in dune sand. Soil Sci. Soc. Am. J. 45, 362–367.

Mizota, C., Kusakabe, M., and Noto, M. 1988. Eolian contribution to soil development on cretaceous limestones in Greece by oxygen isotope composition of quartz. Geochem. J. 22, 41–46.

Mizota, C., Izuhara, H., and Noto, M. 1992. Eolian influence on oxygen isotope abundance and clay minerals in soils of Hokkaido, northern Japan. Geoderma 52, 161–171.

Mokma, D.L., Syers, J.K., Jackson, M.L., Clayton, R.N., and Rex, R.W. 1972. Aeolian additions to soils and sediments in the South Pacific area. J. Soil Sci. 23, 147–162.

Muhs, D.R. 1983. Airborne dust fall on the California Channel Islands, U.S.A. J. Arid Environ. 6, 222–238.

Muhs, D.R., Crittenden, R.C., Rosholt, J.N., Bush, C.A., and Stewart, K.C. 1987. Genesis of marine terrace soils, Barbados, West Indies: Evidence from mineralogy and geochemistry. Earth Surf. Proc. Landforms 12, 605–618.

Muhs, D.R., Bush, C.A., Stewart, K.C., Rowland, T.R., and Crittenden, R.C. 1990. Geochemical evidence of Saharan dust parent material for soils developed on Quaternary limestones of Caribbean and western Atlantic islands. Quaternary Res. 33, 157–177.

Muhs, D.R. in preparation. Origin of soils and paleosols on carbonate islands and implications for stratigraphy and sea level history.

Munn, L.C. 1992. Response to "Comments on 'Origin of silt-enriched surface mantles in Indian Basin, Wyoming' ". Soil Sci. Soc. Am. J. 56, 992–993.

Munn, L.C., and Spackman, L.K. 1990. Origin of silt-enriched alpine surface mantles in Indian Basin, Wyoming. Soil Sci. Soc. Am. J. 54, 1670–1677.

Nahon, D.B. 1991. Introduction to the petrology of soils and chemical weathering. John Wiley & Sons, New York, 313 p.

Neall V.E., and Paintin, I.K. 1986. Rates of weathering of ^{14}C-dated late Quaternary volcaniclastic deposits in the western United States. In S.M. Colman and D.P. Dethier, eds., Rates of chemical weathering of rocks and minerals, p. 331–350. Academic Press, Orlando, FL, 603 p.

Nettleton, W.D., Flach, K.W., and Nelson, R.E. 1970. Pedogenic weathering of tonalite in southern California. Geoderma 4, 387–402.

Nikiforoff, C.C. 1949. Weathering and soil evolution. Soil Sci. 67, 219–223.

Olson, C.G., Ruhe, R.V., and Mausbach, M.J. 1980. The terra rossa limestone contact phenomena in karst, southern Indiana. Soil Sci. Am. J. 44, 1075–1079.

Pettijohn, F.J. 1941. Persistence of heavy minerals and geologic age. J. Geol. 49, 610–625.

Pettijohn, F.J., Potter, P.E., and Siever, R. 1987. Sand and sandstone. Springer-Verlag, New York, 553 p.

Porter, S.C. 1975. Weathering rinds as a relative-age criterion. Geology 3, 101–104.

Pye, K. 1987. Aeolian dust and dust deposits. Academic Press, London, 334 p.

Raeside, J.D. 1959. Stability of index minerals in soils with particular reference to quartz, zircon, and garnet. J. Sed. Petrol. 29, 493–502.

Rapp, A., and Nihlén, T. 1991. Desert dust storms and loess deposits in north Africa and south Europe. Catena Suppl. 20, 43–55.

Reheis, M.C. 1980. Loess sources and loessial soil changes on a downwind transect, Boulder-Lafayette area, Colorado. Mountain Geologist 17, 7–12.

Reheis, M.C. 1990. Influence of climate and eolian dust on the major-element chemistry and clay mineralogy of soils in the northern Bighorn Basin, U.S.A. Catena 17, 219–248.

Reheis, M.C., and Kihl, R. 1995. Dust deposition in southern Nevada and California, 1984–1989: Relations to climate, source area, and source lithology. J. Geophys. Res. 100, No. D5, 8893–8918.

Reheis, M.C., Goodmacher, J.C., Harden, J.W., McFadden, L.D., Rockwell, T.K., Shroba, R.R., Sowers, J.M., and Taylor, E.M. 1995. Quaternary soils and dust deposition in southern Nevada. Geol. Soc. Am. Bull. 107, 1003–1022.

Rex, R.W., Syers, J.K., Jackson, M.L., and Clayton, R.N. 1969. Eolian origin of quartz in soils of Hawaiian Islands and in Pacific pelagic sediments. Science 163, 277–279.

Rieger, S., and Juve, R.L. 1961. Soil development in Recent loess in the Matanuska Valley, Alaska. Soil Sci. Soc. Am. Proc. 25, 243–248.

Robertson, S.M., and Mew, G. 1982. The presence of volcanic glass in soils on the west coast, South Island, New Zealand. New Zealand J. Geol. Geophys. 25, 503–507.

Rodbell, D.T. 1993. Subdivision of late Pleistocene moraines in the Cordillera Blanca, Peru, based on rock-weathering features, soils, and radiocarbon dates. Quaternary Res. 39, 133–143.

Ross, C.W., Mew, G., and Searle, P.L. 1977. Soil sequences on two terrace systems in the north Westland area, New Zealand. New Zealand J. Sci. 20, 231–244.

Ruhe, R.V. 1969a. Quaternary landscapes in Iowa. Iowa State Univ. Press, Ames, 255p.

Ruhe, R.V. 1969b. Application of pedology to Quaternary Research. In S. Pawluk, ed., Pedology and Quaternary research, p. 1–23. Univ. of Alberta Printing Dept., Edmonton, Alberta, 218 p.

Ruhe, R.V. 1973. Backgrounds of model for loess-derived soils in the upper Mississippi River Basin. Soil Sci. 115, 250–253.

Ruhe, R.V. 1984. Clay-mineral regions in Peoria loess, Mississippi River basin. J. Geol. 92, 339–343.

Ruhe, R.V., Cady, J.G., and Gomez, R.S. 1961. Paleosols of Bermuda: Geol. Soc. Am. Bull. 72, 1121–1142.

Schumacher, B.A., Lewis, G.C., Miller, B.J., and Day, W.J. 1988. Basal mixing zones in loesses of Louisiana and Idaho: I. Identification and characterization. Soil Sci. Soc. Am. J. 52, 753–758.

Sherman, G.D., and Uehara, G. 1956. The weathering of olivine basalt in Hawaii and its pedogenic significance. Soil Sci. Soc. Proc. 20, 337–340.

Shoji, S., Nanzyo, M., and Dahlgren, R. 1993. Volcanic ash soils: Genesis, properties, and utilization. Elsevier, Amsterdam, 288 p.

Shroba, R.R., and Birkeland, P.W. 1983. Trends in late-Quaternary soil development in the Rocky Mountains and Sierra Nevada of the western United States. In S.C. Porter, ed., Late-Quaternary environments of the United States. Vol. 1. The late-Pleistocene, p. 145–156. University of Minnesota Press, Minneapolis.

Simonson, R.W. 1995. Airborne dust and its significance to soils. Geoderma 65, 1–43.

Skully, R.W., and Arnold, R.W. 1981. Holocene alluvial stratigraphy in the Upper Susquehanna River Basin, New York. Quaternary Res. 15, 327–344.

Smith, G.D. 1942. Illinois loess-variations in its properties and distribution: A pedologic interpretation. Univ. Illinois Agri. Exp. Sta. Bull. 490, 137–184.

Smith, W.W. 1962. Weathering of some Scottish basic igneous rocks with reference to soil formation. J. Soil Sci. 13, 202–215.

Soil Survey Staff. 1994. Keys to Soil Taxonomy. U.S. Dept. Agri., 306 p.

Tardy, Y., Bocquier, G., Paquet, H., and Millot, G. 1973. Formation of clay from granite and its distribution in relation to climate and topography. Geoderma 10, 271–284.

Tejan-Kella, M.S., Fitzpatrick, R.W., and Chittleborough, D.J. 1991. Scanning electron microscope study of zircons and rutiles from a podzol chronosequence at Cooloola, Queensland, Australia. Catena 18, 11–30.

Thomas, R.F., and Lee, R. 1983. Investigation of the properties and genesis of West Coast wet land soils, South Island, New Zealand. 2. Particle size distribution. New Zealand J. Sci. 26, 73–83.

Velde, B. 1992. Introduction to clay minerals. Chapman & Hall, London, 198 p.

Wada, K. 1980. Mineralogical characteristics of Andisols. In B.K.G. Theng, ed., Soils with variable charge, p. 89–107. New Zealand Soc. Soil Sci., Lower Hutt.

Wada, K. 1989. Allophane and imogolite. In J.B. Dixon and S.B. Weed, eds., Minerals in soil environments, p. 1051–1087. Soil Sci. Soc. Am. Book Series No. 1, 1244 p.

Whitehouse, I.E., McSaveney, M.J., Knuepfer, P.L.K., and Chinn, T.J.H. 1986. Growth of weathering rinds on Torlesse Sandstone, southern Alps, New Zealand. In S.M. Colman and D.P. Dethier, eds., Rates of chemical weathering of rocks and minerals, p. 419–435. Academic Press, Orlando, FL, 603 p.

Willman, H.B., Glass, H.D., and Frye, J.C. 1966. Mineralogy of glacial tills and their weathering profile in Illinois II. Weathering profiles. Illinois State Geol. Surv. Circ. 400, 76 p.

Wood, W.R., and Johnson, D.L. 1978. A survey of disturbance processes in archaeological site formation. In M.B. Schiffer, ed., Advances in archaeological method and theory, vol. 1, p. 315–381. Academic Press, New York.

Yaalon, D.H. 1997. Soils in the Mediterranean region: What makes them different? Catena 28, 157–169.

Yaalon, D.H., and Dan, J. 1974. Accumulation and distribution of loess derived deposits in the semi-desert and desert fringe areas of Israel. Zeitscht. Geomorphol. Suppl. Bd. 20, 91–105.

Yaalon, D.H., and Ganor, E. 1973. The influence of dust on soils during the Quaternary. Soil Sci. 116, 146–155.

Yaalon, D.H., and Ganor, E. 1979. East Mediterranean trajectories of dust-carrying storms from the Sahara and Sinai. In C. Morales, ed., Saharan dust-mobilization, transport, deposition, p. 187–193. John Wiley & Sons, New York.

Yerkes, R.F., and Wentworth, C.M. 1965. Structure, Quaternary history, and general geology of the Corral Canyon area, Los Angeles County, California. U.S. Geol. Surv. open-file report, 215 p.

■ Chapter 8

Amit, R., and Gerson, R. 1986. The evolution of Holocene reg (gravelly) soils in deserts—an example from the Dead Sea area. Catena 13, 59–79.

Amit, R., Gerson, R., and Yaalon, D.H. 1993. Stages and rate of the gravel shattering process by salts in desert reg soils. Geoderma 57, 295–324.

Aniku, J.R.F., and Singer, M.J. 1990. Pedogenic iron oxide trends in a marine terrace chronosequence. Soil Sci. Soc. Am. J. 54, 147–152.

Arkley, R.J. 1963. Calculation of carbonate and water movement in soil from climatic data. Soil Sci. 96, 239–248.

Bachman, G.O., and Machette, M.N. 1977. Calcic soils and calcretes in the southwestern United States. U.S. Geol. Surv. Open-File Report 77-794, 163 p.

Barton, D.C. 1916. The disintegration of granite in Egypt. J. Geol. 24, 382–393.

Benedict, J.B. 1985. Apapaho Pass: Glacial geology and archeology at the crest of the Colorado Front Range. Center for Mountain Archeology, Res. Report No. 3, Ward, Colorado 197 p.

Berry, M.E. 1994. Soil-geomorphic analysis of late-Pleistocene glacial sequences in the McGee, Pine, and Bishop Creek drainages, east-central Sierra Nevada, California. Quaternary Res. 41, 160–175.

Bilzi, A.F., and Ciolkosz, E.J. 1977. A field morphology rating scale for evaluating pedological development. Soil Sci. 124, 45–48.

Birkeland, P.W. 1964. Pleistocene glaciation of the northern Sierra Nevada north of Lake Tahoe, California. J. Geol. 72, 810–825.

Birkeland, P.W. 1968. Correlation of Quaternary stratigraphy of the Sierra Nevada with that of the Lake Lahontan area. In R.B. Morrison and H.E. Wright, Jr., eds., Means of correlation of Quaternary successions, p. 469–500. Internat. Assoc. Quaternary Res., VII Cong., Proc. vol. 8, 631 p.

Birkeland, P.W. 1969. Quaternary paleoclimatic implications of soil clay mineral distribution in a Sierra Nevada-Great Basin transect. J. Geol. 77, 289–302.

Birkeland, P.W. 1978. Soil development as an indication of relative age of Quaternary deposits, Baffin Island, N.W.T., Canada. Arct. Alp. Res. 10, 733–747.

Birkeland, P.W. 1982. Subdivision of Holocene glacial deposits, Ben Ohau Range, New Zealand, using relative-dating methods. Geol. Soc. Am. Bull. 93, 433–449.

Birkeland, P.W. 1984. Holocene soil chronofunctions, Southern Alps, New Zealand. Geoderma 34, 115–134.

Birkeland, P.W. 1990. Soil-geomorphic research—a selective review. Geomorphology 3, 207–224.

Birkeland, P.W., and Janda, R.J. 1971. Clay mineralogy of soils developed from Quaternary deposits of the eastern Sierra Nevada, California. Geol. Soc. Am. Bull. 82, 2495–2514.

Birkeland, P.W., and Noller, J.S. 1998. Rock and mineral weathering. In J.M. Sowers, J.S. Noller, and W.R. Lettis, eds., Dating and earthquakes: Review of Quaternary geochronology and its application to paleoseismology, p. 2-467–2-496. U.S. Nuclear Regulatory Commission, NUREG/CR-5562.

Birkeland, P.W., Burke, R.M., and Walker, A.L. 1980. Soils and subsurface rock-weathering features of Sherwin and pre-Sherwin glacial deposits, eastern Sierra Nevada, California. Geol. Soc. Am. Bull. 91 (Part 1), 238–244.

Birkeland, P.W., Burke, R.M., and Shroba, R.R. 1987. Holocene alpine soils in gneissic cirque deposits, Colorado Front Range. U.S. Geol. Surv. Bull. 1590, 21 p.

Birkeland, P.W., Burke, R.M., and Benedict, J.B. 1989. Pedogenic gradients of iron and aluminum accumulation and phosphorus depletion on arctic and alpine soils as a function of time and climate. Quaternary Res. 32, 193–204.

Birkeland, P.W., Miller, D.C., Patterson, E., Price, A.B., and Shroba, R.R. 1996. Soil-geomorphic relationships near Rocky Flats, Boulder and Golden, with a stop at the pre-Fountain Formation paleosol of Wahlstrom (1948). Field trip 27 in R.A. Thompson, M.R. Hudson, and C.L. Pillmore, eds., Geologic excursions to the Rocky Mountains and beyond, Field trip guidebook for the 1996 annual meeting Geological Society of America. Colorado Geological Survey Special Publ. 44.

Birman, J.H. 1964. Glacial geology across the crest of the Sierra Nevada, California. Geol. Soc. Am. Spec. Pap. 75, 80 p.

Blackwelder, E. 1931. Pleistocene glaciation in the Sierra Nevada and Basin Ranges. Geol. Soc. Am. Bull. 42, 865–922.

Bloom, A.L. 1980. Late Quaternary sea level change on South Pacific coasts: A study in tectonic diversity. In Mörner, Nils-Axel, ed., Earth rheology, isostacy and eustacy, p. 505–516. John Wiley & Sons, New York, 599 p.

Boardman, J. 1985. Comparison of soils in midwestern United States and western Europe with the interglacial record. Quaternary Res. 23, 62–75.

Bockheim, J.G. 1979. Properties and relative age of soils of southwestern Cumberland Peninsula, Baffin Island, N.W.T., Canada. Arct. Alp. Res. 11, 289–306.

Bockheim, J.G. 1980a. Solution and use of chronofunctions in studying soil development. Geoderma 24, 71–85.

Bockheim, J.G. 1980b. Properties and classification of some desert soils in coarse-textured glacial drift in the Arctic and Antarctic. Geoderma 24, 45–69.

Bockheim, J.G. 1982. Properties of a chronosequence of ultraxerous soils in the Trans-Antarctic Mountains. Geoderma 28, 239–255.

Bockheim, J.G. 1990. Soil development rates in the Transantarctic Mountains. Geoderma 47, 59–77.

Bockheim, J.G. 1997. Properties and classification of cold soils from Antarctica. Soil Sci. Soc. Am. J. 61, 224–231.

Bockheim, J.G., and Ugolini, F.C. 1990. A review of pedogenic zonation in well-drained soils of the southern circumpolar region. Quaternary Res., 34, 47–66.

Bockheim, J.G., Wilson, S.C., and Leide, J.E. 1990. Soil development in the Beardmore Glacier region, Antarctica. Soil Sci. 149, 144–157.

Bockheim, J.G., Marshall, J.G., and Kelsey, H.M. 1996. Soil-forming processes and rates on uplifted marine terraces in southwestern Oregon, USA. Geoderma 73, 39–62.

Bradley, R.S. 1985. Quaternary paleoclimatology. Allen & Unwin, Boston, 472 p.

Brimhall, G.H., Lewis, C.J., Ague, J.J., Dietrich, W.E., Hampel, J., Teague, T., and Rix, P. 1988. Metal enrichment in bauxites by deposition of chemically mature aeolian dust. Nature (London) 333, 819–824.

Brook, G.A., Folkoff, M.E., and Box, E.O. 1983. A world model of soil carbon dioxide. Earth Surface Proc. Landforms 8, 79–88.

Brophy, J.A. 1959. Heavy mineral ratios of Sangamon weathering profiles in Illinois. Illinois State Geol. Surv. Circ. 273, 22 p.

Bruce, J.G. 1973. Loessial deposits of the South Island, with a definition of the Stewarts Claim Formation. New Zealand J. Geol. Geophys. 16, 533–548.

Bull, W.B. 1991. Geomorphic responses to climatic change. Oxford Univ. Press, New York, 326 p.

Burke, R.M. 1979. Multiparameter relative dating (RD) techniques applied to morainal sequences along the eastern Sierra Nevada, California and Wallowa Lake area, Oregon. Ph.D. thesis, Univ. of Colorado, Boulder, 166 p.

Burke, R.M., and Birkeland, P.W. 1979. Reevaluation of multiparameter relative dating techniques and their application to the glacial sequence along the eastern escarpment of the Sierra Nevada, California. Quaternary Res. 11, 21–51.

Burke, R.M., and Carver, G.A., coordinators. 1992. A look at the southern end of the Cascadia Subduction Zone and the Mendicino Triple Junction. Pacific Cell, Friends of the Pleistocene, Guidebook for the field trip to northern coastal California, 266 p.

Burns, S.F. 1979. The northern pocket gopher (Thomomys talpoides): A major geomorphic agent in the alpine tundra. J. Colorado-Wyoming Acad. Sci. 11, 86.

Busacca, A.J. 1987. Pedogenesis of a chronosequence in the Sacramento Valley, California, California, U.S.A. I. Application of soil development index. Geoderma 41, 123–148.

Campbell, A.S. 1975. Chemical and mineralogical properties of a sequence of terrace soils near Reefton, New Zealand. Ph.D. thesis, Lincoln College, Canterbury, New Zealand, 477 p.

Campbell, I.B., and Claridge, G.G.C. 1987. Antarctica: Soils, weathering processes and environment. Elsevier, Amsterdam, 368 p.

Campbell, I.B., and Claridge, G.G.C. 1992. Soils of cold climate regions. In I.P. Martini and W. Chesworth, eds., Weathering, soils & paleosols, p. 183–201. Elsevier, Amsterdam, 618 p.

Chadwick, O.A., and Davis, J.O. 1990. Soil-forming intervals caused by eolian sediment impulses in the Lahontan basin, northwestern Nevada: Geology 18, p. 243–246.

Claridge, G.G.C., and Campbell, I.B. 1982. A comparison between hot and cold desert soils and soil processes. Catena Suppl. 1, 1–28.

Colman, S.M., and Pierce, K.L. 1981. Weathering rinds on andesitic and basaltic stones as a Quaternary age indicator, western United States. U.S. Geol. Surv. Prof. Pap. 1210, 56 p.

Colman, S.M., and Pierce, K.L. 1986. Glacial sequence near McCall, Idaho: Weathering rinds, soil development, morphology, and other relative-age criteria. Quaternary Res. 25, 25–42.

Colman, S.M., and Pierce, K.L. 1992. Varied records of early Wisconsinan alpine glaciation in the western United States derived from weathering-rind thickness. Geol. Soc. Am. Spec. Pap. 270, 269–278.

Colman, S.M., Pierce, K.L., and Birkeland, P.W. 1987. Suggested terminology for Quaternary dating methods. Quaternary Res. 28, 314–319.

Cremeens, D.L. 1995. Pedogenesis of Cotiga Mound, a 2100-year-old woodland mound in southwest West Virginia. Soil Sci. Soc. Am. J. 59, 1377–1388.

Crocker, R.L., and Major, J. 1955, Soil development in relation to vegetation and surface age at Glacier Bay, Alaska. J. Ecol. 43, 427–448.

Dahms, D.E., Shroba, R.R., Gosse, J.C., Hall, R.D., Sorenson, C.J., and Reheis, M.C. 1997. Relation between soil age and silicate weathering rates determined from the chemical evolution of a glacial chronosequence: Comment. Geology 25, 381–382.

Dan, J., and Yaalon, D.H. 1982. Automorphic saline soils in Israel. Catena Suppl. 1, 103–115.

Dan, J., Yaalon, D.H., Moshe, R., and Nissim, S. 1982. Evolution of Reg soils in southern Israel and Sinai. Geoderma 28, 173–202.

Daniels, R.B., and Hammer, R.D. 1992. Soil geomorphology. John Wiley & Sons, New York, 236 p.

Daniels, R.B., Gamble, E.E., and Cady, J.C. 1971. The relation between geomorphology and soil morphology and genesis. Adv. Agron. 23, 51–88.

Danin, A., and Ganor, E. 1991. Trapping of airborne dust by mosses in the Negev Desert, Israel. Earth Surface Process. Landforms 16, 153–162.

Dawson, B.S.W., Fergusson, J.E., Campbell, A.S., and Cutler, E.J.B. 1991. Depletion of first-row transition metals in a chronosequence of soils in the Reefton area of New Zealand. Geoderma 48, 271–296.

DeBano, L.F. 1980. Water-repellant soil: A state-of-the-art. U.S. Dept. Agri., Forest Serv., General Technical Report PSW-46, Pacific Southwest Forest and Range Experiment Sta., Berkeley, CA., 21 p.

Dethier, D.P. 1987. The soil chronosequence along the Cowlitz River, Washington. U.S. Geological Survey Bull., 1590-F, 47 p.

Division of Soils, CSIRO. 1983. Soils: An Australian viewpoint. CSIRO, Melbourne and Academic Press, London, 928 p.

Dixon, J.C. 1986. Solute movement on hillslopes in the alpine environment of the Colorado Front Range. In A.D. Abrahams, ed., Hillslope processes, p. 139–159. Allen & Unwin, Boston, 416 p.

Dixon, J.C. 1991. Alpine and subalpine soil properties as paleoenvironomental indicators. Phys. Geogr. 12, 370–384.

Doerr, S.H., Shakesby, R.A., and Walsh, R.P.D. 1996. Soil hydrophobicity variations with depth and particle size fraction in burned and unburned Eucalyptus globulus and Pinus pinaster forest terrain in the Águeda Basin, Portugal. Catena 27, 25–47.

Dorronsoro, C., and Alonso, P. 1994. Chronosequence of Almar River fluvial-terrace soil. Soil Sci. Soc. Am. J. 58, 910–925.

Evans, L. J., and Cameron, B.H. 1979. A chronosequence of soils developed from granitic morainal material, Baffin Island, N.W.T. Can. J. Soil Sci. 59, 203–210.

Follmer, L.R., McKay, E.D., Lineback, J.A., and Gross, D.L. 1979. Wisconsinian, Sangamonian, and Illinoian stratigraphy in central Illinois. Illinois State Geol. Surv. Guidebook 13, 139 p.

Foos, A.M. 1991. Aluminous lateritic soils, Eleuthera, Bahamas: A modern analog to carbonate paleosols. J. Sed. Petrol. 61, 340–348.

Forman, S.L., and Miller, G.H. 1984. Time-dependent soil morphologies and pedogenic processes on raised beaches, Bröggerhalvöya, Spitsbergen, Svalbard Archipelago. Arctic Alp. Res. 16, 381–394.

Forsyth, J.L. 1965. Age of the buried soil in the Sidney, Ohio, area. Am. J. Sci. 263, 571–597.

Frye, J.C., Willman, H.B., and Glass, H.D. 1960. Gumbotil, accretion-gley, and the weathering profile. Illinois State Geol. Surv. Circ. 295, 39 p.

Frye, J.C., Glass, H.D., and Willman, H.B. 1968. Mineral zonation of Woodfordian loesses of Illinois. Illinois State Geol. Surv. Circ. 427, 44 p.

Frye, J.C., Glass, H.D., Kempton, J.P., and Willman, H.B. 1969. Glacial tills of northwestern Illinois. Illinois State Geol. Surv. Circ. 437, 47 p.

Gerson, R., and Amit, R. 1987. Rates and modes of dust accretion and deposition in an arid region—the Negev, Israel. In L. Frostick and I. Reid, eds., Desert sediments: Ancient and modern, p. 157–169. Geol. Soc. Spec. Publ. 35, 401 p.

Gerson, R., Amit, R., and Grossman, S. 1985. Dust availability in desert terrains—a study in the desert in Israel and the Sinai. Phys. Geogr., Inst. Earth Sci., The Hebrew University, Jerusalem.

Gile, L.H. 1975. Holocene soils and soil-geomorphic relations in an arid region of southern New Mexico. Quaternary Res. 5, 321–360.

Gile, L.H. 1979. Holocene soils in eolian sediments of Bailey County, Texas. Soil Sci. Soc. Am. J. 43, 994–1003.

Gile, L.H., and Grossman, R.B. 1979. The desert project soil monograph. Soil Conservation Service, U.S. Dept. Agri., 984 p.

Gile, L.H., Hawley, J.W., and Grossman, R.B. 1981. Soils and geomorphology in the Basin and Range area of southern New Mexico—Guidebook to the Desert Project. New Mexico Bur. Mines and Mineral Resources Memoir 39, 222 p.

Gosse, J.C., Klein, J., Evenson, E.B., Lawn, B., and Middleton, R. 1995. Beryllium-10 dating of the duration and retreat of the last Pinedale glacial sequence. Science 268, 1329–1333.

Graham, R.C., and Wood, H.B. 1991. Morphologic development and clay redistribution in lysimeter soils under chaparral and pine. Soil Sci. Soc. Am. J. 55, 1638–1646.

Guccione, M.J. 1982. Stratigraphy, soil development and mineral weathering of Quaternary deposits, mid-continent, USA. Ph.D. thesis, Univ. of Colorado, Boulder.

Guccione, M.J. 1985. Quantitative estimates of clay-mineral alteration in a soil chronosequence in Missouri, U.S.A. Catena Suppl. 6, 137–150.

Hall, R.D., and Martin, R.E. 1986. The etching of hornblende grains in the matrix of alpine tills and periglacial deposits. In S.M. Colman and D.P. Dethier, eds., Rates of chemical weathering of rocks and minerals, p. 101–128. Academic Press, Orlando, FL.

Hall, R.D., and Shroba, R.R. 1993. Soils developed in the glacial deposits of the type areas of the Pinedale and Bull Lake glaciations, Wind River Range, Wyoming, U.S.A. Arc. Alp. Res. 25, 368–373.

Hall, R.D., and Shroba, R.R. 1995. Soil evidence for a glaciation intermediate between the Bull Lake and Pinedale glaciations at Fremont Lake, Wind River Range, Wyoming, U.S.A. Arc. Alp. Res. 27, 89–98.

Hallberg, G.R., Wollenhaupt, N.C., and Miller, G.A. 1978. A century of soil development in spoil derived from loess in Iowa. Soil Sci. Soc. Am. J. 42, 339–343.

Harden, D.R., Biggar, N.E., and Gillam, M.L. 1985. Quaternary deposits and soils in and around Spanish Valley, Utah. Geol. Soc. Am. Spec. Pap. 203, 43–64.

Harden, J.W. 1987. Soils developed in granitic alluvium near Merced, California. U.S. Geol. Surv. Bull. 1590-A, 65 p.

Harden, J.W. 1988. Genetic interpretations of elemental and chemical differences in a soil chronosequence, California. Geoderma 43, 179–193.

Harden, J.W. 1990. Soil development on stable landforms and implications for landscape studies. Geomorphology 3, 391–398.

Harden, J.W., and Taylor, E.M. 1983. A quantitative comparison of soil development in four climatic regimes. Quaternary Res. 20, 342–359.

Harden, J.W., Sarna-Wojcicki, A.M., and Dembroff, G.R. 1986. Soils developed on coastal and fluvial terraces near Ventura, California. U.S. Geol. Surv. Bull. 1590-B, 34 p.

Harden, J.W., Taylor, E.M., Hill, C., Mark, R.K., McFadden, L.D., Reheis, M.C., Sowers, J.M., and Wells, S.G. 1991a. Rates of soil development from four chronosequences in the southern great Basin. Quaternary Res. 35, 383–399.

Harden, J.W., Taylor, E.M., McFadden, L.D., and Reheis, M.C. 1991b. Calcic, gypsic, and siliceous soil chronosequences in arid and semiarid environments. Soil Sci. Soc. Am. Spec. Publ. No. 26, 1–16.

Harden, J.W., Sundquist, E.T., Stallard, R.F., and Mark, R.K. 1992. Dynamics of soil carbon during deglaciation of the Laurentide Ice Sheet. Science 258, 1921–1924.

Harrison, J.B.J., McFadden, L.D., and Weldon III, R.J. 1990. Spatial soil variability in the Cajon Pass chronosequence: Implications for the use of soils as a geochronological tool. Geomorphology 3, 399–416.

Harrison, R., Swift, R.S., Campbell, A.S., and Tonkin, J. 1990. A study of two soil development sequences located in a montane area of Canterbury, New Zealand. I. Clay mineralogy and cation exchange properties. Geoderma 47, 261–282.

Harrison, R., Swift, R.S., and Tonkin, P.J. 1991. A study of two soil development sequences located in a montane area of Canterbury, New Zealand. II. Soil development indices based on major element analyses. Geoderma 48, 163–177.

Herwitz, S.R., Muhs, D.R., Prospero, J.M., Mahan, S., and Vaughn, B. 1996. Origin of Bermuda's clay-rich Quaternary paleosols and their paleoclimatic significance. J. Geophys. Res. 101, No. D18, 23,389–23,400.

Holliday, V.T. 1985a. Early and middle Holocene soils at the Lubbock Lake archeological site, Texas. Catena 12, 61–78.

Holliday, V.T. 1985b. Morphology of late Holocene soils at the Lubbock Lake archeological site, Texas. Soil Sci. Soc. Am. J. 49, 938–946.

Holliday, V.T. 1988. Genesis of the late-Holocene soil chronosequence at the Lubbock Lake archaeological site, Texas Ann. Assoc. Am. Geogr. 78, 594–610.

Howard, J.L., Amos, D.F., and Daniels, W.L. 1993. Alluvial soil chronosequence in the inner Coastal Plain, central Virginia. Quaternary Res. 39, 201–213.

Howard, J.L., Amos, D.F., and Daniels, W.L. 1995. Micromorphology and dissolution of quartz sand in some exceptionally ancient soils. Sed. Geol. 105, 51–62.

Hunt, C.B., and Sokoloff, V.P. 1950. Pre-Wisconsin soil in the Rocky Mountain region, a progress report. U.S. Geol. Surv. Prof. Pap. 221-G.

Isherwood, D.J. 1975. Soil geochemistry and rock weathering in an arctic environment. Ph.D. thesis, Univ. Colorado, Boulder, 173 p.

Jackson, M.L. 1964. Chemical composition of the soil. In F.E. Bear, ed., Chemistry of the soil, p. 71–141. Reinhold, New York, 515 p.

Jackson, M.L., Tyler, S.A., Willis, A.L., Bourbeau, G.A., and Pennington, R.P. 1948. Weathering sequence of clay-size minerals in soils and sediments. I. Fundamental generalizations. J. Phys. Colloid. Chem. 52, 1237–1260.

Johnson, D.L., and Watson-Stegner, D. 1987. Evolution model of pedogenesis. Soil Sci. 143, 349–366.

Johnson, D.J., and Watson-Stegner, D. 1990. The soil-evolution model as a framework for evaluating pedoturbation in archaeological site formation. In N.P. Lasca and J. Donahue, eds., Archaeological geology of North America. Centennial Special Volume 4, p. 541–560. Geological Society of America, Boulder, CO.

Karlstrom, E.T. 1988. Rates of soil formation on Black Mesa, northeast Arizona. A chronosequence in late Quaternary alluvium. Phys. Geogr. 9, 301–327.

Knuepfer, P.L.K. 1988. Estimating ages of late Quaternary stream terraces from analysis of weathering rinds and soils. Geol. Soc. Am. Bull. 100, 1224–1236.

Kutiel, P., Lavee, H., Segev, M., and Benyamini, Y. 1995. The effect of fire-induced surface heterogeneity on rainfall-runoff-erosion relationships in an eastern Mediterranean ecosystem, Israel. Catena 25, 77–87.

Latham, M., and Mercky, P. 1983. Etude des sols des Iles Loyaute. ORSTOM, Paris, 45 p.

Leigh, D.S. 1996. Soil Chronosequence of Brasstown Creek, Blue Ridge Mountains, USA. Catena 26, 99–114.

Lessig, H.D. 1961. The soils developed on Wisconsin and Illinoian-age glacial outwash terraces along Little Beaver Creek and the adjoining upper Ohio Valley, Columbiana County, Ohio. Ohio J. Sci. 6, 286–294.

Levine, E.L., and Ciolkosz, E.J. 1983. Soil development in till of various ages in northeastern Pennsylvania. Quaternary Res. 19, 85–99.

Litaor, M.I., Mancinelli, R., and Penny, J.C. 1996. The influence of pocket gophers on the status of nutrients in alpine soils. Geoderma 70, 37–48.

Locke, W.W., III, 1979. Etching of hornblende grains in arctic soils: An indicator of relative age and paleoclimate. Quaternary Res. 11, 197–212.

Locke, W.W. 1986. The etching of hornblende grains on glacial deposits, Baffin Island, Canada. In S.M. Colman and D.P. Dethier, eds., Rates of chemical weathering of rocks and minerals, p. 129–145. Academic Press, Orlando, FL.

Lowe, D.J. 1986. Controls on the rates of weathering and clay mineral genesis in airfall tephras: A review and New Zealand case study. In S.M. Colman and D.P. Dethier, eds., Rates of chemical weathering of rocks and minerals, p. 265–330. Academic Press, Orlando, FL.

Lowe, D.J., and Percival, H.J. 1993. Clay mineralogy of tephras and associated paleosols and soils, and hydrothermal deposits, North Island. Guidebook for New Zealand preconference field trip F1, 10th Internat. Clay Conf., Adelaide, Australia, 110 p.

Machette, M.N. 1982. Guidebook to the late Cenozoic geology of the Beaver Basin, south-central Utah. U.S. Geol. Surv. Open-File Rept. 82–850, 42 p.

Machette, M.N. 1985. Calcic soils of the southwestern United States. Geol. Soc. Am. Spec. Pap. 203, 1–21.

Machette, M.N., Birkeland, P.W., Markos, G., and Guccione, M.J. 1976a. Soil development in Quaternary deposits in the Golden-Boulder portion of the Colorado Piedmont. In R.C. Epis and R.J. Weimer, eds., Studies in Colorado field geology, p. 339–357. Prof. Contributions of the Colorado School of Mines, No. 8, 552 p.

Machette, M.N., Birkeland, P.W., Burke, R.M., Guccione, M.J., Kihl, R., and Markos, G. 1976b. Field descriptions and laboratory data for a Quaternary soil sequence in the Golden-Boulder portion of the Colorado Piedmont. U.S. Geol. Surv. Open-File Report 76–804.

Mahaney, W.C., and Halvorson, D.L. 1986. Rates of mineral weathering in the Wind River Mountains, western Wyoming. In S.M. Colman and D.P. Dethier, eds., Rates of chemical weathering of rocks and minerals, p. 147–167. Academic Press, Orlando, FL.

Markewich, H.W., and Pavich, M.J. 1991. Soil chronosequence studies in temperate to subtropical, low-latitude, low-relief terrain with data from the eastern United States. Geoderma 51, 213–239.

Markewich, H.W., Pavich, M.J., Mausbach, M.J., Stuckey, B.N., Johnson, R.G., and Gonzalez V. 1986. Soil development and its relation to the ages of morphostratigraphic units in Horry County, South Carolina. U.S. Geol. Surv. Bull. 1589-B, 61 p.

Markewich, H.W., Pavich, M.J., Mausbach, M.J., Hall, R.L., Johnson, R.G., and Hearn, P.P. 1987. Age relations between soils and geology in the Coastal Plain of Maryland and Virginia. U.S. Geol. Surv. Bull. 1589-A, 34 p.

Marsella, K.A., Bierman, P.R., Davis, P.T., and Caffee, M.W. 1996. Stage II "big ice" on Baffin Island. Geol. Soc. Am. Abstr. with Programs 28, No. 7, A-433–A-434.

Martinson, D.G., Pisias, N.G., Hays, J.D., Imbrie, J., Moore, T.C., Jr., and Shackleton, N.J. 1987. Age dating and the orbital theory of the ice ages: Development of a high resolution 0 to 300,000-year chronostratigraphy. Quaternary Res. 27, 1–29.

Mayer, L., McFadden, L.D., and Harden, J.W. 1988. Distribution of calcium carbonate in desert soils: A model. Geology 16, 303–306.

McFadden, L.D. 1982. The impacts of temporal and spatial climatic changes on alluvial soils genesis in Southern California. Ph.D. thesis, University of Arizona, Tucson, 430 p.

McFadden, L.D. 1988. Climatic influences on rates and processes of soil development in Quaternary deposits of southern California. Geol. Soc. Am. Spec. Pap. 216, 153–177.

McFadden, L.D., and Hendricks, D.M. 1985. Changes in the content and composition of pedogenic iron oxyhydroxides in a chronosequence of soils in southern California. Quaternary Res. 23, 189–204.

McFadden, L.D., and Tinsley, J.C. 1985. Rate and depth of pedogenic-carbonate accumulation in soils. Formulation and testing of a compartment model. Geol. Soc. Am. Spec. Pap. 203, 23–41.

McFadden, L.D., and Weldon, R.J., III. 1987. Rates and processes of soil development on Quaternary terraces in Cajon Pass, California. Geol. Soc. Am. Bull. 98, 280–293.

McFadden, L.D., Wells, S.G., and Dohrenwend, J.C. 1986. Influences of Quaternary climatic changes on processes of soil development on desert loess deposits of the Cima volcanic field, California. Catena 13, 361–389.

McFadden, L.D., Wells, S.G., and Jercinovich, M.J. 1987. Influences of eolian and pedogenic processes on the origin and evolution of desert pavements. Geology 15, 504–508.

McFadden, L.D., Amundson, R.G., and Chadwick, O.A. 1991. Numerical modeling, chemical, and isotopic studies of carbonate accumulation in soils of arid regions. Soil Sci. Soc. Am. Spec. Publ. 26, 17–35.

McKeague, J.A., Ross, G.J., and Gamble, D.S. 1978. Properties, criteria of classification, and concepts of genesis of podzolic soils in Canada. In W.C. Mahaney, ed., Quaternary soils, p. 27–60. Geo. Abstracts, Univ. East Anglia, Norwich, England.

McTainsh, G.H. 1989. Quaternary aeolian dust processes and sediments in the Australian region. Quaternary Sci. Rev. 8, 235–253.

Mears, B., Jr. 1981. Periglacial wedges and the late Pleistocene environment of Wyoming's intermontane basins. Quaternary Res. 15, 171–198.

Meierding, T.C. 1981. Marble tombstone weathering rates: A transect of the United States. Phys. Geogr. 2, 1–18.

Meierding, T.C. 1993. Inscription legibility method for estimating rock weathering rates. Geomorphology 6, 273–286.

Mellor, A. 1985. Soil chronosequences on neoglacial moraine ridges, Jostedalsbreen and Jotunheimen, southern Norway: A quantitative pedogenic approach. In K.S. Richards, R.R. Arnett, and S. Ellis, eds., Geomorphology and soils, p. 289–308. George Allen & Unwin Ltd., London.

Merritts, D.J., Chadwick, O.A., and Hendricks, D.M. 1991. Rates and processes of soil evolution on uplifted marine terraces, northern California. Geoderma 51, 241–275.

Merritts, D.J., Chadwick, O.A., Hendricks, D.M., Brimhall, G.H., and Lewis, C.J. 1992. The mass balance of soil evolution on late Quaternary marine terraces, northern California. Geol. Soc. Am. Bull. 104, 1456–1470.

Messer, A.C. 1988. Regional variations in rates of pedogenesis and the influence of climate factors on moraine chronosequences, southern Norway. Arc. Alp. Res. 20, 31–39.

Miles, R.J., and Franzmeier, D.P. 1981. A lithochronosequence of soils formed in dune sand. Soil Sci. Soc. Am. J. 45, 362–367.

Miller, C.D. 1979. A statistical method for relative-age dating of moraines in the Sawatch Range, Colorado. Geol. Soc. Am. Bull. 90, 1153–1164.

Mokma, D.L., Jackson, M.L., Syers, J.K., and Stevens, R. 1973. Mineralogy of a chronosequence of soils from graywacke and mica-schist alluvium, Westland, New Zealand. New Zealand J. Sci. 16, 769–797.

Morrison, R.B. 1964. Lake Lahontan: Geology of the southern Carson Desert. U.S. Geol. Surv. Prof. Pap. 401, 156 p.

Muhs, D.R. 1982. A soil chronosequence on Quaternary marine terraces, San Clemente Island, California. Geoderma 28, 257–283.

Muhs, D.R. 1983. Airborne dust fall on the California Channel Islands, U.S.A. J. Arid Environ. 6, 222–238.

Muhs, D.R. 1984. Intrinsic thresholds in soil systems. Phys. Geogr. 5, 99–110.

Muhs, D.R. In preparation. Origin of soils and paleosols on carbonate islands and implications for stratigraphy and sea level history.

Muhs, D.R., Crittenden, R.C., Rosholt, J.N., Bush, C.A., and Stewart, K.C. 1987. Genesis of marine terrace soils, Barbados, West Indies: Evidence from mineralogy and geochemistry. Earth Surface Process. Landforms 12, 605–618.

Muhs, D.R., Bush, C.A., Stewart, K.C., Rowland, T.R., and Crittdendon, R.C. 1990. Geochemical evidence for Saharan dust parent material for soils developed on Quaternary limestones of Caribbean and western Atlantic islands. Quaternary Res. 33, 157–177.

Nikiforoff, C.C. 1949. Weathering and soil evolution. Soil Sci. 67, 219–230.

Nishiizumi, K., Winterer, E.L., Kohl, C.P., Klein, J., Middleton, R., Lal, D., and Arnold, J.R. 1989. Cosmic ray production rates of ^{10}Be and ^{26}Al in quartz from glacially polished rocks. J. Geophys. Res., No. B12, 17,907–17,915.

Noller, J.S. 1993. Late Cenozoic stratigraphy and soil geomorphology of the Peruvian Desert, 3–18 S: A long-term record of hyperaridity and El Nino. Ph.D. thesis, Univ. Colorado, Boulder, 279 p.

Ollier, C.D. 1969. Weathering. Oliver and Boyd, Edinburgh, 304 p.

Olson, J.S. 1958. Rates of succession and soil changes on southern Lake Michigan sand dunes. Bot. Gazette 119, 125–170.

Pedro, G., Jamagne, M., and Begon, J.C. 1978. Two routes in genesis of strongly differentiated acid soils under humid, cool-temperate conditions. Geoderma 20, 173–189.

Peterson, F.F. 1980. Holocene desert soil formation under sodium salt influence in a playa-margin environment. Quaternary Res. 13, 172–186.

Peterson, F.F. 1981. Landforms of the basin & range province, defined for soil survey. Nevada Agri. Exp. Sta. Tech. Bull. 28, 52 p.

Porter, S.C. 1989. Some geological implications of average Quaternary glacial conditions. Quaternary Res. 32, 245–261.

Porter, S.C. 1975. Weathering rinds as a relative-age criterion: Application to subdivision of glacial deposits in the Cascade Range. Geology 3, 101–104.

Rahn, P.H. 1971. The weathering of tombstones and its relationship to the topography of New England. J. Geol. Educ. 19, 112–118.

Reheis, M.C. 1987a. Soils in granitic alluvium and humid and semiarid climates along Rock Creek, Carbon County, Montana. U.S. Geol. Surv. Bull. 1590-D, 71 p.

Reheis, M.C. 1987b. Gypsic soils on the Kane alluvial fans, Big Horn County, Wyoming. U.S. Geol. Surv. Bull. 1590-C, 39 p.

Reheis, M.C. 1990. Influence of climate and eolian dust on the major-element chemistry and clay mineralogy of soils in the northern Bighorn Basin, U.S.A. Catena 17, 219–248.

Reheis, M.C., and Kihl, R. 1995. Dust deposition in southern Nevada and California 1984–1989: Relations to climate, source area, and source lithology. J. Geophys. Res. 100, No. D5, 8893–8918.

Reheis, M.C., Harden, J.W., McFadden, L.D., and Shroba, R.R. 1989. Development rates of late Quaternary soils, Silver Lake Playa, California. Soil Sci. Soc. Am. J. 53, 1127–1140.

Reheis, M.C., Sowers, J.M., Taylor, E.M., McFadden, L.D., and Harden, J.W. 1992. Morphology and genesis of carbonate soils on the Kyle Canyon fan, Nevada, U.S.A. Geoderma 52, 303–342.

Reheis, M.C., Goodmacher, J.C., Harden, J.W., McFadden, L.D., Rockwell, T.K., Shroba, R.R., Sowers, J.M., and Taylor, E.M. 1995. Quaternary soils and dust deposition in southern Nevada. Geol. Soc. Am. Bull. 107, 1003–1022.

Reider, R.G., Kuniansky, N.J., Stiller, D.M., and Uhl, P.J. 1974. Preliminary investigation of comparative soil development on Pleistocene and Holocene geomorphic surfaces of the Laramie Basin, Wyoming. In M.Wilson, ed., Applied geology and archeology: The Holocene of Wyoming, p. 27–33. Geol. Surv. Wyoming, Report of Investigations 10.

Richmond, G.M. 1962. Quaternary stratigraphy of the La Sal Mountains, Utah. U.S. Geol. Surv. Prof. Pap, 324, 135 p.

Richmond, G.M., and Fullerton, D.S. 1985. Introduction to Quaternary glaciations in the United States of America. In V. Šibrava, D.Q. Bowen, and G.M. Richmond, eds., Quaternary glaciations of the northern hemisphere, p. 3–10. Quaternary Science Rev. 5.

Rodbell, D.T. 1990. Soil-age relationships on late Quaternary moraines, Arrowsmith Range, Southern Alps, New Zealand. Arc. Alp. Res. 22, 355–365.

Rodbell, D.T. 1993. Subdivision of late Pleistocene moraines in the Cordillera Blanca, Peru, based on rock-weathering features, soils, and radiocarbon dates. Quaternary Res. 39, 133–143.

Rose, J., Boardman, J., Kemp, R.A., and Whiteman, C.A. 1985. Paleosols and the interpretation of the British Quaternary stratigraphy. In K.S. Richards, R.R. Arnett, and S. Ellis, eds., Geomorphology and soils, p. 348–375. George Allen & Unwin, London.

Ross, C. W., Mew, G., and Searle, P.L. 1977. Soil sequences on two terrace systems in the North Westland area, New Zealand. New Zealand J. Sci. 20, 231–244.

Ruhe, R.V. 1956. Geomorphic surfaces and the nature of soils. Soil Sci. 82, 441–455.

Ruhe, R.V. 1965. Quaternary paleopedology. In H.E. Wright, Jr. and D.G. Frey, eds., The Quaternary of the United States, p. 755–764. Princeton Univ. Press, Princeton, 922 p.

Ruhe, R.V. 1967. Geomorphology of parts of the Greenfield quadrangle, Adair County, Iowa. U.S. Dept. Agri. Tech. Bull. 1349, 93–161.

Ruhe, R.V. 1968. Identification of paleosols in loess deposits in the United States. In C.B. Schultz and J.C. Frye, eds., Loess and related eolian deposits of the world, p. 49–65. Int. Assoc. Quaternary Res., VII Cong., Proc. vol. 12, 369 p.

Ruhe, R.V. 1969. Quaternary landscapes in Iowa. Iowa St. Univ. Press, Ames, 225 p.

Ruhe, R.V. 1983. Aspects of Holocene pedology in the United States. In H.E. Wright, Jr., ed., Late-Quaternary environments of the United States, Vol. 2, p. 12–25. The Holocene. Univ. of Minnesota Press, Minneapolis.

Schaetzl, R.J. 1990. Effects of treethrow microtopography on the characteristics and genesis of Spodosols, Michigan, USA. Catena 17, 111–126.

Schaetzl, R.J., and Follmer, L.R. 1990. Longevity of treethrow microtopography: Implications for mass wasting. Geomorphology 3, 113–123.

Schaetzl, R.J., Johnson, D.J., Burns, S.F., and Small, T.W. 1989. Tree uprooting: review of terminology, process, and environmental implications. Can. J. Forest Res. 19, 1–11.

Schaetzl, R.J., Barrett, L.R., and Winkler, J.A. 1994. Choosing models for soil chronofunctions and fitting them to data. Eur. J. Soil Sci. 45, 219–232.

Schlesinger, W.H. 1990. Evidence from chronosequence studies for a low carbon-storage potential of soils. Nature (London) 348, 232–234.

Schmidt, D.L., and Mackin, J.H. 1970. Quaternary geology of the Long and Bear valleys, west-central Idaho. U.S. Geol. Surv. Bull. 1311-A, 22 p.

Scott, W.E., McCoy, W.C., Shroba, R.R., and Rubin, M. 1983. Reinterpretation of the exposed record of the last two cycles of Lake Bonneville, western United States. Quaternary Res. 20, 261–285.

Shackleton, N.J., and Opdyke, N.D. 1973. Oxygen-isotope and paleomagnetic stratigraphy of equatorial Pacific core V28-238: Oxygen isotope temperatures and ice volumes on a 10^3 year to 10^6 year scale. Quaternary Res. 3, 39–55.

Sharp, R.P., and Birman, J.H. 1963. Additions to classical sequence of Pleistocene glaciations, Sierra Nevada, California. Geol. Soc. Am. Bull. 74, 1079–1086.

Shroba, R.R. 1977. Soil development in Quaternary tills, rock-glacier deposits, and taluses, southern and central Rocky Mountains. Ph.D. thesis, Univ. Colorado, Boulder, 424 p.

Shroba, R.R. 1984. Secondary clay content and estimated rates of clay accumulation in soil B horizons in tills of Holocene and Pleistocene age, southern and central Rocky Mountains. Am. Quaternary Assoc., 8th Biennial Meeting, Program and Abstracts, 117.

Shroba, R.R., and Birkeland, P.W. 1983. Trends in the late-Quaternary soil development in the Rocky Mountains and Sierra Nevada of the western United States. In S.C. Porter, ed., Late-Quaternary environments of the United States, Vol. 1. The late Pleistocene, p. 145–156. Univ. of Minnesota Press, Minneapolis.

Singer, A. 1995. The mineral composition of hot and cold desert soils. Adv. GeoEcol. 28, 13–28.

Slate, J.L., Bull, W.B., Ku, T-L., Shafiqullah, M., Lynch, D.J., and Huang, Y-P. 1991. Soil-carbonate genesis in the Pinacate volcanic field, northwestern Sonora, Mexico. Quaternary Res. 35, 400–416.

Smith, S.M., and Lee, W.G. 1984. Vegetation and soil development on a Holocene river terrace sequence, Arawata Valley, south Westland, New Zealand. New Zealand J. Sci. 27, 187–196.

Soons, J.M. 1968. Erosion by needle ice in the Southern Alps, New Zealand. In H.E. Wright, Jr. and W.H. Osburn, eds., Arctic and alpine environments, p. 217–227. Indiana Univ. Press, Bloomington.

Soons, J.M., and Greenland, D.E. 1970. Observations on the growth of needle ice. Water Resources Res. 6, 579–593.

Stevens, P.R., and Walker, T.W. 1970. The chronosequence concept and soil formation. Quat. Rev. Biol. 45, 333–350.

Swanson, D.K. 1984. Soil catenas on Pinedale and Bull Lake moraines, Willow Lake, Wind River Mountains, Wyoming. M.Sc. thesis, Univ. Colorado, Boulder, 147 p.

Swanson, D.K. 1985. Soil catenas on Pinedale and Bull Lake moraines, Willow Lake, Wind River Mountains, Wyoming. Catena 12, 329–342.

Switzer, P., Harden, J.W., and Mark, R.K. 1988. A statistical method for estimating rates of soil development and ages of geologic deposits—a design for soil chronosequence studies. Math. Geol. 20, 49–61.

Tardy, Y., and Roquin, C. 1992. Geochemistry and evolution of lateritic landscapes. In I.P. Martini and W. Chesworth, eds., Weathering, soils & paleosols, p. 407–443. Elsevier, Amsterdam, 618 p.

Taylor, A., and Blum, J.D. 1995. Relation between soil age and silicate weathering rates determined from the chemical evolution of a glacial chronosequence. Geology 23, 979–982.

Taylor, A., and Blum, J.D. 1997. Relation between soil age and silicate weathering rates determined from the chemical evolution of a glacial chronosequence: Reply. Geology 25, 382–383.

Taylor, E.M. 1986. Impact of time and climate on Quaternary soils in the Yucca Mountain area of the Nevada Test Site. M.Sc. thesis, Univ. Colorado, Boulder, 217 p.

Thompson, C.H. 1983. Development and weathering of large parabolic dune systems along the subtropical coast of eastern Australia. Zeit. Geomorphol. Suppl.-Bd. 45, 205–225.

Thompson, C.H., and Bowman, G.M. 1984. Subaerial denudation and weathering of vegetated coastal dunes in eastern Australia. In B.G. Thom, ed., Coastal geomorphology in Australia, p. 263–290. Academic Press Australia, Sydney, 349 p.

Thorp, J. 1968. The soil—a reflection of Quaternary environments in Illinois. In R.E. Bergstrom, ed., The Quaternary of Illinois, p. 48–55. Univ. Illinois, College of Agri., Spec. Pub. 14, 179 p.

Tonkin, P.J., and Basher, L.R. 1990. Soil-stratigraphic techniques in the study of soil and landform evolution across the Southern Alps, New Zealand. Geomorphology 3, 547–575.

Torrent, J., and Neddleton, W.D. 1978. Feedback processes in soil genesis. Geoderma 20, 281–287.

Ugolini, F.C. 1986. Pedogenic zonation in the well-drained soils of the arctic regions. Quaternary Res. 26, 100–120.

Ulery, A.L., and Graham, R.C. 1993. Forest fire effects on soil color and texture. Soil Sci. Soc. Am. J. 57, 135–140.

Ulery, A.L., Graham, R.C., and Bowen, L.H. 1996. Forest fire effects on soil phyllosilicates in California. Soil Sci. Soc. Am. J. 60, 309–315.

Vidic, N.J. 1994. Pedogenesis and soil-age relationships of soils on glacial outwash terraces in the Ljubljana Basin. Ph.D. thesis, Univ. of Colorado, Boulder, 229 p.

Vidic, N.J., and Lobnik, F. 1997. Rates of soil development of the chronosequence in the Ljubljana Basin, Slovenia. Geoderma 76, 35–64.

Vreeken, W.J. 1975. Principal kinds of chronosequences and their significance in soil history. J. Soil Sci. 26, 378–394.

Wada, K. 1989. Allophane and imogolite. In J.B. Dixon and S.B. Weed, eds., Minerals in soil environments, p. 1052–1087. Soil Sci. Soc. Am., No. 1 in Book Series, Madison, WI.

Wagner, R.J., and Nelson, R.E. 1961. Soil Survey of the San Mateo area, California. U.S. Dept. Agri., Soil Surv. Series 1954, No. 13, 111 p.

Wahrhaftig, C. 1965. Stepped topography of the southern Sierra Nevada, California. Geol. Soc. Am. Bull. 76, 1165–1190.

Watanabe, T. 1990. Relative dating methods mainly applied to glacial and periglacial deposits (in Japanese). Quaternary Record 29, 49–77.

Wells, S.G., McFadden, L.D., and Dohrenwend, J.C. 1987. Influence of late Quaternary climate changes on geomorphic and pedogenic processes on a desert piedmont, eastern Mojave Desert, California. Quaternary Res. 27, 130–146.

Wells, S.G., McFadden, L.D., and Schultz, J.D. 1990. Eolian landscape evolution and soil formation in the Chaco dune field, southern Colorado Plateau, New Mexico. Geomorphology 3, 517–546.

Wells, W.G., II. 1987. The effects of fire on the generation of debris flows in southern California. Geol. Soc. Am., Rev. Eng. Geol. 7, 105–114.

Wilson, M. 1989. Igneous petrogenesis. Unwin Hyman, Boston, 466 p.

Whiteman, C.A., and Kemp, R.A. 1990. Pleistocene sediments, soils and landscape evolution at Stebbing, Essex. J. Quaternary Sci. 5, 145–161.

Willman, H.B., Glass, H.D., and Frye, J.C. 1966. Mineralogy of glacial tills and their weathering profiles in Illinois. Part II. Weathering profiles. Illinois State Geol. Surv. Circ. 400, 76 p.

Winkler, E.M. 1975. Stone. Properties, durability in man's environment. Springer-Verlag, New York, 230 p.

Wright, H.E., Jr. 1984. Sensitivity and response time of natural systems to climatic change in the late Quaternary. Quaternary Sci. Rev. 3, 91–131.

Yaalon, D.H. 1983. Climate, time and soil development. In L.P. Wilding, N.E. Smeck, and G.F. Hall, eds., Pedogenesis and Soil Taxonomy I. Concepts and interactions, p. 233–251. Elsevier, Amsterdam, 303 p.

Young, F.J. 1989. Soil survey of the islands of Aguijan, Rota, Saipan, and Tinian, Commonwealth of the Northern Mariana Islands. U.S. Dept. Agri., Soil Cons. Serv., 166 p.

■ Chapter 9

Al-Janabi, A.M., and Drew, J.V. 1967. Characterization and genesis of a Sharpsburg-Wymore soil sequence in southeastern Nebraska. Soil Sci. Soc. Am. Proc. 31, 238–244.

Bell, M., and Boardman, J., eds. 1992. Past and present erosion, archaeological and geographical perspectives. Oxbow Monograph 22, Oxbow Books, Oxford, England, 250 p.

Berry, M.E. 1987. Morphological and chemical characteristics of soil catenas on Pinedale and Bull Lake moraine slopes in the Salmon River Mountains, Idaho. Quaternary Res. 28, 210–225.

Berry, M.E. 1994. Soil-geomorphic analysis of late-Pleistocene glacial sequences in the McGee, Pine, and Bishop creek drainages, east-central Sierra Nevada, California. Quaternary Res. 41, 160–175.

Birkeland, P.W. 1994. Variation in soil-catena characteristics of moraines with time and climate, South Island, New Zealand. Quaternary Res. 42, 49–59.

Birkeland, P.W., and Burke, R.M. 1988. Soil catena chronosequences on eastern Sierra Nevada moraines, California, U.S.A. Arc. Alp. Res. 20, 473–484.

Birkeland, P.W., and Gerson, R. 1991. Soil-catena development with time in a hot desert, southern Israel—field data and salt distribution. J. Arid Environ. 21, 267–281.

Birkeland, P.W., Berry, M.E., and Swanson, D.K. 1991. Use of soil catena field data for estimating relative ages of moraines. Geology 19, 281–283.

Bravard, S., and Righi, D. 1989. Geochemical differences in an Oxisol-Spodosol toposequence in Amazonia (Brazil). Geoderma 44, 29–42.

Bravard, S., and Righi, D. 1990. Podzols in Amazonia. Catena 17, 461–475.

Bull, W.B. 1991. Geomorphic responses to climatic change. Oxford Univ. Press, New York, 326 p.

Burke, R.M., and Birkeland, P.W. 1979. Reevaluation of multiparameter relative dating techniques and their application to the glacial sequence along the eastern escarpment of the Sierra Nevada, California. Quaternary Res. 11, 21–51.

Burns, S.F., and Tonkin, P.J. 1982. Soil-geomorphic models and the spatial distribution and development of alpine soils. In C.E. Thorn, ed., Space and time in geomorphology, p. 25–43. Allen & Unwin, London.

Butler, B.E. 1959. Periodic phenomena in landscapes as a basis for soil studies. CSIRO Australia Soil Publ. No. 14.

Campbell, I.B., and Claridge, G.G.C. 1987. Antarctica: soils, weathering processes and environment. Elsevier, Amsterdam, 368 pp.

Conacher, A.J., and Dalrymple, J.B. 1977. The nine unit landsurface model: An approach to pedogeomorphic research. Geoderma 18, 1–154.

Coultas, C.L., Clewell, A.F., and Taylor, E.M., Jr. 1979. An aberrant toposequence of soils through a titi swamp. Soil Sci. Soc. Am. J. 43, 377–383.

Coventry, R.J. 1982. The distribution of red, yellow and grey earths in the Torrens Creek area, central north Queensland. Aust. J. Soil Res. 20, 1–14.

Coventry, R.J., Taylor, R.M., and Fitzpatrick, R.W. 1983. Pedological significance of the gravels in some red and grey earths of central north Queensland. Aust. J. Soil Sci. 21, 219–240.

Curi, N., and Franzmeier, D.P. 1984. Toposequence of Oxisols from the Central Plateau of Brazil. Soil Sci. Soc. Am. J. 48, 341–346.

Dahms, D.E. 1994. Mid-Holocene erosion of soil catenas on moraines near the type Pinedale till, Wind River Range, Wyoming. Quaternary Res. 42, 41–48.

Dan, J., Yaalon, D.H., and Koyumdjisky, H. 1968. Catenary soil relationships in Israel, 1. The Netanya catena on coastal dunes of the Sharon. Geoderma 2, 95–120.

Daniels, R.B., and Gamble, E.E. 1967. The edge effect in some Ultisols in the North Carolina Coastal Plain. Geoderma 1, 117–124.

Daniels, R.B., Gamble, E.E., and Nelson, L.A. 1967. Relation between A2 horizon characteristics and drainage in some fine loamy Ultisols. Soil Sci. 104, 364–369.

Daniels, R.B., Gamble, E.E., and Bartelli, L.J. 1968. Eluvial bodies in B horizons of some Ultisols. Soil Sci. 106, 200–206.

Daniels, R.B., Gamble, E.E., and Cady, J.G. 1971. The relation between geomorphology and soil morphology and genesis. Adv. Agron. 23, 51–88.

Division of Soils, CSIRO. 1983. Soils, an Australian viewpoint. CSIRO, Melbourne/Academic Press, London, 928 p.

Evans, C.V., and Franzmeier, D.P. 1986. Saturation, aeration, and color patterns in a toposequence of soils in north-central Indiana. Soil Sci. Soc. Am. J. 50, 975–980.

Fairbridge, F.W., ed. 1968. The encyclopedia of geomorphology. Reinhold, New York, 1295 p.

Finney, H.R., Holowaychuk, N., and Heddleson, M.R. 1962. The influence of microclimate on the morphology of certain soils of the Allegheny Plateau of Ohio. Soil Sci. Soc. Am. Proc. 26, 287–292.

Follmer, L.R. 1982. The geomorphology of the Sangamon surface: Its spatial and temporal attributes. In C.E. Thorn, ed., Space and time in geomorphology, p. 117–146. Allen & Unwin, London.

Forman, S.L., and Miller, G.H. 1984. Time dependent Soil morphologies and pedogenic processes on raised beaches, Bröggerhalvöya, Spitsbergen, Svaldbard Archipelago: Arc. Alp. Res. 16, 381–394.

Frolking, T.A. 1989. Forest soil uniformity along toposequences in the loess-mantled Driftless Area of Wisconsin. Soil Sci. Soc. Am. J. 53, 1168–1172.

Frye, J.C., Shaffer, P.R., Willman, H.B., and Ekblaw, G.E. 1960a. Accretion gley and the gumbotil dilemma. Am. J. Sci. 258, 185–190.

Frye, J.C., Willman, H.B., and Glass, H.D. 1960b. Gumbotil, accretion-gley, and the weathering profile. Ill. State Geol. Surv. Circ. 295, 39 p.

Gerrard, A.J. 1990. Soil variations on hillslopes in humid temperate climates: Geomorphology. 3, 225–244.

Gerrard, A.J. 1992. Soil geomorphology: An integration of pedology and geomorphology. Chapman & Hall, London, 269 p.

Gile, L.H., Hawley, J.W., and Grossman, R.B. 1981. Soils and geomorphology in the Basin and Range area of southern New Mexico—Guidebook to the Desert Project. New Mexico Bur. Mines and Mineral Resources Mem. 39, 222 p.

Glazovskaya, M.A. 1968. Geochemical landscapes and types of geochemical soil sequences. Trans. 9th Int. Cong. Soil Sci. 4, 303–312.

Graham, R.C., and Buol, S.W. 1990. Soil-geomorphic relations on the Blue Ridge front: II. Soil characteristics and pedogenesis. Soil Sci. Soc. Am. J. 54, 1367–1377.

Graham, R.C., Daniels, R.B., and Buol, S.W. 1990. Soil-geomorphic relations on the Blue Ridge front: I. regolith types and slope processes. Soil Sci. Soc. Am. J. 54, 1362–1367.

Hall, G.F. 1983. Pedology and geomorphology. In L.P. Wilding, N.E. Smeck, and G.F. Hall, eds., Pedogenesis and soil taxonomy, I. Concepts and interactions, p. 117–140. Elsevier, Amsterdam.

Honeycutt, C.W., Heil, R.D., and Cole, C.V. 1990a. Climatic and topographic relations of three Great Plains soils: I. Soil morphology. Soil Sci. Soc. Am. J. 54, 469–475.

Honeycutt, C.W., Heil, R.D., and Cole, C.V. 1990b. Climatic and topographic relations of three Great Plains soils: II. Carbon, nitrogen, and phosphorus. Soil Sci. Soc. Am. J. 54, 476–483.

Huggett, R.J. 1975. Soil landscape systems: A model of soil genesis. Geoderma 13, 1–22.

Huggett, R.J. 1976. Lateral translocation of soil plasma through a small valley basin in the Northaw Great Wood, Hertfordshire. Earth Surf. Proc. Landforms 1, 99–109.

Hunckler R.V., and Schaetzl, R.J. 1997. Spodosol development as affected by geomorphic aspect, Baraga County, Michigan. Soil Sci. Soc. Am. J. 61, 1105–1115.

Hussain, M.S., and Swindale, L.D. 1974. The physical and chemical properties of the gray hydromorphic soils of the Hawaiian Islands. Soil Sci. Soc. Am. Proc. 38, 935–941.

Jenny, H. 1980. The soil resource. Springer-Verlag, New York, 377 p.

Kantor, W., and Schwertmann, U. 1974. Mineralogy and genesis of clays in red-black soil toposequences on basic igneous rocks in Kenya. J. Soil Sci. 25, 67–78.

Kay, G.F. 1916. Gumbotil, a new term in Pleistocene geology. Science 44, 637–638.

Kay, G.F. 1931. Classification and duration of the Pleistocene period. Geol. Soc. Am. Bull. 42, 425–466.

Kay, G.F., and Pearce, J.N. 1920. The origin of gumbotil. J. Geol. 28, 89–125.

Krause, H.H., Rieger, S., and Wilde, S.A. 1959. Soils and forest growth on different aspects in the Tanana watershed of interior Alaska. Ecology 40, 492–495.

Leighton, M.M., and MacClintock, P. 1962. The weathered mantle of glacial tills beneath original surfaces in north-central United States. J. Geol. 70, 267–293.

Litaor, M.I. 1992. Aluminum mobility along a geochemical catena in an alpine watershed, Front Range, Colorado. Catena 19, 1–16.

Lotspeich, F.B., and Smith, H.W. 1953. Soils of the Palouse loess: I. The Palouse catena. Soil Sci. 76, 467–480.

Macedo, J., and Bryant, R.B. 1987. Morphology, mineralogy, and genesis of a hydrosequence of Oxisols in Brazil. Soil Sci. Soc. Am. J. 51, 690–698.

Machette, M.N. 1985. Calcic soils of the southwestern United States. Geol. Soc. Am. Spec. Pap. 203, 1–21.

Mahaney, W.C., and Sanmugadas, K. 1983. Early-Holocene soil catena in Titcomb Basin, Wind River Mountains, Wyoming. Zeit. Geomorphol. 27, 265–281.

Mahaney, W.C., and Sanmugadas, K. 1989. Late-Holocene soil toposequence in Stroud Basin, central Wind River Mountains, western Wyoming. Geografisk Tidsskrift 89, 58–65.

Marron, D.C., and Popenoe, J.H. 1986. A soil catena on schist in northwestern California. Geoderma 37, 307–324.

McCalpin, J.P., and Berry, M.E. 1996. Soil catenas to estimate ages of movement on normal fault scarps, with an example from the Wasatch Fault Zone, Utah, USA. Catena 27, 265–286.

McDaniel, P.A., Bathke, G.R., Buol, S.W., Cassel, D.K., and Falen, A.L. 1992. Secondary manganese/iron ratios as pedochemical indicators of field-scale throughflow water movement: Soil Sci. Soc. Am. J. 56, 1211–1217.

McKeague, J.A. 1965. Properties and genesis of three members of the uplands catena. Can. J. Soil Sci. 45, 63–77.

McMillan, M.E. 1990. Soil development on a Pinedale moraine and erosion ten years after a burn, Rocky Mountain National Park, Colorado. M.S. Thesis, Univ. of Colorado, Boulder.

Mears, B., Jr. 1981. Periglacial wedges and the late Pleistocene environment of Wyoming's intermontane basins: Quaternary Res. 171–198.

Meierding, T.C. 1984. Correlation of Rocky Mountain Pleistocene deposits by relative dating methods— A perspective. In W.C. Mahaney, ed., Correlation of Quaternary chronologies, p. 455–477. Geo Books, Norwich, England.

Menges, C.M. 1990. Soils and geomorphic evolution of bedrock facets on a tectonically active mountain front, western Sangre de Cristo Mountains, New Mexico. Geomorphology 3, 301–332.

Miller, C.D., and Birkeland, P.W. 1992. Soil catena variation along an alpine climatic transect, northern Peruvian Andes. Geoderma 55, 211–223.

Milne, G. 1935a. Some suggested units for classification and mapping, particularly for East African soils. Soil Res., Berlin 4, 183–198.

Milne, G. 1935b. Composite units for the mapping of complex soil associations. Trans. 3rd Int. Cong. Soil Sci. 1, 345–347.

Mohr, E.C.J., van Baren, F.A., and van Schuylenborgh, J. 1972. Tropical soils. Mouton-Ichtiar Baru-Van Hoeve, The Hague, Netherlands, 481 p.

Moreland, D.C., and Moreland, R.E. 1975. Soil survey of Boulder County area, Colorado. U.S. Dept. Agri., Soil Conservation Service.

Muhs, D.R. 1982. The influence of topography on the spatial variability of soils in Mediterranean climates. In C.E. Thorn, ed., Space and time in geomorphology, p. 269–284. Allen & Unwin, London.

Nettleton, W.D., Flach, K.W., and Borst, G. 1968. A toposequence of soils on tonalite grus in the southern California Peninsular Range. U.S. Dept. Agri., Soil Cons. Serv., Soil Surv. Invest. Rep. No. 21, 41 p.

Nye, P.H. 1954. Some soil-forming processes in the humid tropics. I. A field study of a catena in the west African forest. J. Soil Sci. 5, 7–21.

Ollier, C.D. 1973. Catenas in different climates. In E. Derbyshire, ed., Geomorphology and climate, p. 137–169. John Wiley & Sons, New York.

Panabokke, C.R. 1959. A study of some soils in the dry zone of Ceylon. Soil Sci. 87, 67–74.

Paton, T.R., Humphreys, G.S., and Mitchell, P.B. 1995. Soils, a new global view. Yale Univ. Press, New Haven, 213 p.

Pennock, D.J., and de Jong, E. 1990. Regional and catenary variations in properties of Borolls of southern Saskatchewan, Canada. Soil Sci. Soc. Am. J. 54, 1697–1701.

Pierce, K.L., and Colman, S.M. 1986. Effect of height and orientation (microclimate) on geomorphic degradation rates and processes, late-glacial terrace scarps in central Idaho. Geol. Soc. Am. Bull. 97, 869–885.

Radwanski, S.A., and Ollier, C.D. 1959. A study of an east African soil catena. Jour. Soil Sci. 10, 149–168.

Ruhe, R.V. 1956. Geomorphic surfaces and the nature of soils. Soil Sci. 82, 441–455.

Ruhe, R.V. 1965. Paleopedology. In H.E. Wright, Jr. and D.G. Frey, eds., The Quaternary of the United States, p. 755–764. Princeton Univ. Press, Princeton, 922 p.

Ruhe, R.V., and Walker, P.H. 1968. Hillslope models and soil formation. I. Open systems. Trans. 9th Int. Cong. Soil Sci. 4, 551–560.

Schumacher, B.A., Miller, B.J., and Day, W.J. 1987. A chronotoposequence of soils developed in loess in central Louisiana. Soil Sci. Soc. Am. J. 51, 1005–1010.

Sommer, M., and Schlichting, E. 1997. Archetypes of catenas in respect to matter—a concept for structuring and grouping catenas. Geoderma 76, 1–33.

Summerfield, M.A. 1991. Global geomorphology. Longman Scientific & Technical, England, 537 p.

Swanson, D.K. 1984. Soil catenas on Pinedale and Bull Lake moraines, Willow Lake, Wind River Mountains, Wyoming. M.S. Thesis, Univ. of Colorado, Boulder.

Swanson, D.K. 1985. Soil catenas on Pinedale and Bull Lake moraines, Willow Lake, Wind River Mountains, Wyoming. Catena 12, 329–342.

Swanson, D.K. 1995. Landscape ecosystems of the Kobuk Preserve Unit, Gates of the Arctic National Park, Alaska. U.S. National Park Service,

Technical Report NPS/ARRNR/NRTR-95-22, 169 p.

Swanson, D.K. 1996a. Soil geomorphology on bedrock and colluvial terrain with permafrost in central Alaska, USA. Geoderma 71, 157–172.

Swanson, D.K. 1996b. Susceptibility of permafrost soils to deep thaw after forest fires in interior Alaska, U.S.A., and some ecological implications. Arct. Alp. Res. 28, 217–227.

Tardy, Y., and Roquin, C. 1992. Geochemistry and evolution of lateritic landscapes. In I.P. Martini and W. Chesworth, eds., Weathering, soils & paleosols, p. 407–443. Elsevier, Amsterdam.

Tardy, Y., Bocquier, G., Paquet, H., and Millot, G. 1973. Formation of clay from granite and its distribution in relation to climate and topography. Geoderma 10, 271–284.

Tonkin, P.J. 1994. Principles of soil-landscape modelling and their application in the study of soil-landform relationships within drainage basins. In T.H. Webb, ed., Soil-landscape modelling in New Zealand, p. 20–37. Manaaki Whenua Press, Lincoln, Canterbury, New Zealand.

Tonkin, P.J., and Basher, L.R. 1990. Soil-stratigraphic techniques in the study of soil and landform evolution across the Southern Alps, New Zealand. Geomorphology 3, 547–575.

Tonkin, P.J., Young A.W., McKie, D.A., and Campbell, A.S. 1977. Conceptual models of soil development and soil distribution in hill country, central South Island, New Zealand. Part I. The analysis of the changes in soil pattern. New Zealand Soc. Soil Sci., 25th Jubilee Conf., Summaries of presented papers, 25–27.

Trowbridge, A.C. 1961. Discussion: Accretion-gley and the gumbotil dilemma. Am. J. Sci. 259, 154–157.

Valentine, K.W.G., and Dalrymple, J.B. 1975. The identification, lateral variation, and chronology of two buried paleocatenas at Woodhall Spa and West Runton, England. Quaternary Res. 5, 551–590.

Vreeken, W.J. 1984. Soil-landscape chronograms for pedochronological analysis. Geoderma 34, 149–164.

Walker, M.D., Everett, K.R., Walker, D.A., and Birkeland, P.W. 1996. Soil development as an indicator of relative pingo age, northern Alaska, U.S.A. Arct. Alp. Res. 28, 352–362.

Walker, P.R. 1966. Postglacial environments in relation to landscape and soils on the Cary drift, Iowa. Iowa State Univ., Agri. and Home Econ. Exp. Sta. Res. Bull. 549, 835–875.

Walker, P.R., and Ruhe, R.V. 1968. Hillslope models and soil formation. II. Closed systems. Trans. 9th Int. Cong. Soil Sci. 4, 561–568.

Watson, J.P. 1964a. A soil catena on granite in southern Rhodesia. I. Field observations. J. Soil Sci. 15, 238–250.

Watson, J.P. 1964b. A soil catena on granite in southern Rhodesia. II. Analytical data. J. Soil Sci. 15, 251–257.

Weider, M., Yair, A., and Arzi, A. 1985. Catenary soil relationships on arid hillslopes. Catena Suppl. 6, 41–57.

Whipkey, R.Z., and Kirby, M.J. 1978. Flow within the soil. In M.J. Kirby, ed., Hillslope hydrology, p. 121–144. John Wiley & Sons, Chichester.

Yaalon, D.H. 1975. Conceptual models in pedogenesis: Can the soil-forming functions be solved? Geoderma 14, 189–205.

Young, A. 1976. Tropical soils and soil survey. Cambridge University Press, Cambridge, 468 p.

Young, A.W. 1988. A catena of soils on Bealey Spur, Canterbury, New Zealand. Ph.D. thesis, Lincoln College, Univ. of Canterbury, New Zealand, 251 p.

Young, A.W., Tonkin, P.J., McKie, D.A., and Campbell, A.S. 1977. Conceptual models of soil development and soil distribution in hill country, central South Island, New Zealand. Part II. Chemical and mineralogical properties. New Zealand Soc. Soil Sci., 25th Jubilee Conf., Summaries of presented papers, 28–30.

Zaslavsky, D., and Rogowski, A.S. 1969. Hydrologic and morphologic implications of anisotropy and infiltration in soil profile development. Soil Sci. Soc. Am. Proc. 33, 594–599.

■ Chapter 10

Alexander, E.B., Mallory, J.I., and Colwell, W.L. 1993. Soil-elevation relationships on a volcanic plateau in the southern Cascade Range, northern California, U.S.A. Catena 20, 113–128.

Allen, B.L., and Hajek, B.F. 1989. Mineral occurrence in soil environments. In J.B. Dixon and S.B. Weed, eds., Minerals in soil environments, p. 199–278. No. 1 in Soil Sci. Soc. Am. Book Series, Madison, WI.

Amit, R., Gerson, R., and Yaalon, D.H. 1993. Stages and rate of the gravel shattering process by salts in desert Reg soils. Geoderma 57, 295–324.

Amundson, R.G., Chadwick, O.A., Sowers, J.A., and Doner, H.E. 1989a. Soil evolution along an elevational transect in the eastern Mojave Desert of Nevada, U.S.A. Geoderma 43, 349–371.

Amundson, R.G., Chadwick, O.A., Sowers, J.M., and Doner, H.E. 1989b. The stable isotope chemistry of pedogenic carbonates at Kyle Canyon, Nevada. Soil Sci. Soc. Am. J. 53, 201–210.

Amundson, R.G., Chadwick, O.A., Kendall, C., Wang, Y., and DeNiro, M. 1996. Isotopic evidence for shifts in atmospheric circulation patterns during the late Quaternary in mid-North America. Geology 24, 23–26.

Arkley, R.J. 1963. Calculation of carbonate and water movement in soil from climatic data. Soil Sci. 96, 239–248.

Arkley, R.J. 1967. Climates of some Great Soil Groups of the western United States. Soil Sci. 103, 389–400.

Avery, B.W. 1985. Argillic horizons and their significance in England and Wales. In J. Boardmen, ed., Soils and Quaternary landscape evolution, p. 69–86. John Wiley & Sons, New York.

Bachman, G.O., and Machette, M.N. 1977. Calcic soils and calcretes in the southwestern United States. U.S. Geol. Surv. Open-File Rept. 77–794, 163 p.

Barshad, I. 1966. The effect of a variation in precipitation on the nature of clay mineral formation in soils from acid and basic igneous rocks. 1966 Int. Clay Conf. (Jerusalem), Proc. 1, 167–173.

Bigham, J.M., Jaynes, W.F., and Allen, B.L. 1980. Pedogenic degradation of sepiolite and palygorskite in the Texas High Plains. Soil Sci. Soc. Am. J. 44, 159–167.

Birkeland, P.W. 1968. Correlation of Quaternary stratigraphy of the Sierra Nevada with that of the Lake Lahontan area. In R.B. Morrison and H.E. Wright, Jr., eds., Means of correlation of Quaternary successions, p. 469–500. Int. Assoc. Quaternary Res., VII Cong., Proc. 8, 631 p.

Birkeland, P.W. 1969. Quaternary paleoclimatic implications of soil clay mineral distribution in a Sierra Nevada-Great Basin transect. J. Geol. 77, 289–302.

Birkeland, P.W. 1984. Soils and geomorphology. Oxford Univ. Press, New York, 372 p.

Birkeland, P.W., and Janda, R.J. 1971. Clay mineralogy of soils developed from Quaternary deposits of the eastern Sierra Nevada, California. Geol. Soc. Am. Bull. 82, 2495–2514.

Birkeland, P.W., and Shroba, R.R. 1974. The status of the concept of Quaternary soil-forming intervals in the western United States. In W.C. Mahaney, ed.,

Quaternary environments, p. 241–276. York Univ. Geog. Mono. No. 5, Toronto, Canada.

Birkeland, P.W., Burke, R.M., and Walker, A.L. 1980. Soils and subsurface rock-weathering features of Sherwin and pre-Sherwin glacial deposits, eastern Sierra Nevada, California. Geol. Soc. Am. Bull. 91, 238–244.

Birkeland, P.W., Burke, R.M., and Benedict, J.J. 1989. Pedogenic gradients for iron and aluminum accumulation and phosphorus depletion in arctic and alpine soils as a function of time and climate. Quaternary Res. 32, 193–204.

Birkeland, P.W., and Gerson, R. 1991. Soil-catena development with time in a hot desert, southern Israel—field data and salt distribution. J. Arid Environ. 21, 267–281.

Blodgett, R.H., Crabaugh, J.P., and McBride, E.F. 1993. The color of red beds—a geologic perspective. Soil Sci. Soc. Am. Spec. Publ. No. 31, 127–159.

Bown, T.M., Kraus, M.J., Wing, S.L., Fleagle, J.G., Tiffney, B.H., Simons, E.L., and Vondra, C.F. 1982. The Fayum primate forest revisited. J. Human Evol. 11, 603–632.

Bretz, J.H., and Horberg, L. 1949. Caliche in southeastern New Mexico. J. Geol. 57, 491–511.

Brook, G.A., Folkoff, M.E., and Box, E.O. 1983. A world model of soil carbon dioxide. Earth Surface Proc. Landforms 8, 79–88.

Bryan, K., and Albritton, C.C., Jr. 1943. Soil phenomena as evidence of climatic changes. Am. J. Sci. 241, 469–490.

Busacca, A.J. 1989. Long Quaternary record in eastern Washington, U.S.A., interpreted from multiple buried paleosols in loess. Geoderma 45, 105–122.

Cerling, T.E. 1984. The stable isotope composition of modern soil carbonate and its relationship to climate. Earth Planetary Sci. Lett. 71, 229–240.

Cerling, T.E., and Hay, R.L. 1986. An isotopic study of paleosol carbonates from Olduvai Gorge. Quaternary Res. 25, 63–78.

Cerling, T.E., and Quade, J. 1992. Carbon isotopes in modern soils. Encyclopedia Earth Syst. Sci. 1 423–429.

Cerling, T.E., and Quade, J. 1993. Stable carbon and oxygen isotopes in soil carbonates. In P. Swart, J.A. McKenzie, and K.C. Lohman, eds., Climate change in continental isotopic records, p. 217–231. Geophysical Mon. p. 78.

Chadwick, O.A., and Davis, J.O. 1990. Soil-forming intervals caused by eolian sediment pulses in the Lahontan basin, northwestern Nevada. Geology 18, 243–246.

Chadwick, O.A., Nettleton, W.D., and Staidl, G.J. 1995. Soil polygenesis as a function of Quaternary climate change, northern Great Basin, USA. Geoderma 68, 1–26.

Colman, S.M., and Pierce, K.L. 1986. Glacial sequence near McCall, Idaho: Weathering rinds, soil development, morphology, and other relative-age criteria. Quaternary Res. 25, 25–42.

Cremaschi, M. 1987. Paleosols and vetusols in the central Po plain (northern Italy): A study in Quaternary geology and soil development. Studi e ricerche sul territorio, 28, Edizioni Unicopli, Milano, 306 p.

Dan, J., Yaalon, D.H., Moshe, R., and Nissim, S. 1982. Evolution of Reg soils in southern Israel and Sinai. Geoderma 28, 173–202.

Dethier, D.P. 1988. The soil chronosequence along the Cowlitz River, Washington. U.S. Geol. Surv. Bull. 1590-F, 47 p.

Division of Soils, CSIRO. 1983. Soils: An Australian viewpoint. CSIRO, Melbourne and Academic Press, London, 928 p.

Dunne, T., and Leopold, L.B. 1979. Water in environmental planning. W.H. Freeman, New York, 818 p.

Eswaran, H., and Gong, Z. 1991. Properties, genesis, classification, and distribution of soils with gypsum. Soil Sci. Soc. Am. Spec. Pap. 26, 89–119.

Eswaran, H., and Tavernier, R. 1980. Classification and genesis of Oxisols. In B.K.G. Theng, ed., Soils with variable charge, p. 427–442. New Zealand Society of Soil Science, Lower Hutt.

Eswaran, H., Van Den Berg, E., and Reich, P. 1993. Organic carbon in soils of the world. Soil Sci. Soc. Am. J. 57, 192–194.

Eswaran, H., Van den Berg, E., Reich, P., and Kimble, J. 1995. Global soil carbon resources. In R. Lal, J. Kimble, E. Levine, and B.A. Stewart, eds., Soils and global change, p. 27–43. CRC Press, Boca Raton, FL, 440 p.

Folkoff, M.E., and Meentemeyer, V. 1987. Climatic controls of the geography of clay minerals genesis. Ann. Assoc. Am. Geogr. 77, 635–650.

Forsyth, J.L. 1965. Age of the buried soil in the Sidney, Ohio, area. Am. J. Sci. 263, 571–597.

Foscolos, A.E., Rutter, N.W., and Hughes, O.L. 1977. The use of pedological studies in interpreting the Quaternary history of central Yukon Territory. Geol. Surv. Can. Bull. 271, 48 p.

Gerson, R. 1981. Geomorphic aspects of the Elat Mountains. In J. Dan, R. Gerson, H. Koyumdjisky, and D.H. Yaalon, eds., Aridic soils of Israel, p. 279–296. Agri. Res. Organization (Israel) Spec. Publ. No. 190.

Gerson, R., and Amit, R. 1987. Rates and modes of dust accretion and deposition in an arid region—the Negev, Israel. In L. Frostick, and I. Reid, eds., Desert sediments: Ancient and modern, p. 157–169. Geol. Soc. Spec. Publ. No. 35.

Gile, L.H., Peterson, F.F., and Grossman, R.B. 1966. Morphological and genetic sequences of carbonate accumulation in desert soils. Soil Sci. 101, 347–360.

Gile, L.H., Hawley, J.W., and Grossman, R.B. 1981. Soils and geomorphology in the Basin and Range area of southern New Mexico—Guidebook to the desert project. New Mexico Bur. Mines Mineral Resources Mem. 39, 222 p.

Goodfriend, G.A., and Magaritz, M. 1988. Palaeosols and late Pleistocene rainfall fluctuations in the Negev Desert. Nature (London) 332, 144–146.

Harden, J.W., and Taylor, E.M. 1983. A quantitative comparison of soil development in four climatic regimes. Quaternary Res. 10, 342–359.

Harradine, F., and Jenny, H. 1958. Influence of parent material and climate on texture and nitrogen and carbon contents of virgin California soils. I. Texture and nitrogen contents of soils. Soil Sci. 85, 235–243.

Hay, R.L., and Jones, B.F. 1972. Weathering of basaltic tephra on the island of Hawaii. Geol. Soc. Am. Bull. 83, 317–332.

Holliday, V.T. 1987. A reexamination of late-Pleistocene boreal forest reconstructions for the Southern High Plains. Quaternary Res. 28, 238–244.

Hunt, C.B., and Sokoloff, V.P. 1950. Pre-Wisconsin soil in the Rocky Mountain Region, a progress report. U.S. Geol. Surv. Prof. Pap. 221-G, 109–123.

Janda, R.J., and Croft, M.G. 1967. The stratigraphic significance of a sequence of Noncalcic Brown soils formed on the Quaternary alluvium of the northeastern San Joaquin Valley, California. In R.B. Morrison and H.E. Wright, Jr., eds., Quaternary soils, p. 157–190. Int. Assoc. Quaternary Res., VII Cong., Proc. 9, 338 p.

Jenny, H. 1935. The clay content of the soil as related to climatic factors, particularly temperature. Soil Sci. 40, 111–128.

Jenny, H. 1941a. Factors of soil formation. McGraw-Hill, New York, 281 p.

Jenny, H. 1941b. Calcium in the soil: III. Pedologic relations. Soil Sci. Soc. Am. Proc. 6, 27–37.

Jenny, H. 1950. Causes of the high nitrogen and organic matter content of certain tropical forest soils. Soil Sci. 69, 63–69.

Jenny, H. 1961. Comparison of soil nitrogen and carbon in tropical and temperate regions. Missouri Agri. Exp. Sta. Res. Bull. 765, 5–31.

Jenny, H. 1980. The soil resource. Springer-Verlag, New York, 377 p.

Jenny, H., and Leonard, C.D. 1934. Functional relationships between soil properties and rainfall. Soil Sci. 38, 363–381.

Kämpf, N., and Schwertmann, U. 1983. Goethite and hematite in a climosequence in southern Brazil and their application in classification of kaolinite soils. Geoderma 29, 27–39.

Karlstrom, E.T. 1991. Paleoclimatic significance of late Cenozoic paleosols east of Waterton-Glacier Parks, Alberta and Montana. Palaeogeogr., Palaeoclim., Palaeoecol. 85, 71–100.

Kemp, R.A. 1985. The Valley Farm Soil in southern East Anglia. In J. Boardman, ed., Soils and Quaternary landscape evolution, p. 179–196. John Wiley & Sons, New York.

Kemp, R.A., Whiteman, C.A., and Rose, J. 1993. Palaeoenvironmental and stratigraphic significance of the Valley Farm and Barham soils in eastern England. Quaternary Sci. Rev. 12, 833–848.

Kern, J.S. 1994. Spatial patterns of soil organic carbon in the contiguous United States. Soil Sci. Soc. Am. J. 58, 439–455.

Kukla, J. 1970. Correlation between loesses and deep-sea sediments. Geologiska Foreningens i Stockholm Förhandlingar 92, 148–180.

Kukla, J. 1977. Pleistocene land-sea correlations. 1. Europe. Earth-Sci. Rev. 13, 307–374.

Levine, E.L., and Ciolkosz, E.J. 1983. Soil development in till of various ages in northeastern Pennsylvania. Quaternary Res. 19, 85–99.

Locke, W.W., III. 1979. Etching of hornblende grains in arctic soils: An indicator of relative age and paleoclimate. Quaternary Res. 11, 197–212.

Machette, M.N. 1985. Calcic soils of the southwestern United States. Geol. Soc. Am. Spec. Pap. 203, p. 1–21.

Malloy, L. 1988. Soils in the New Zealand landscape, the living mantle. Mallinson Rendel Publ. Ltd., Wellington, New Zealand, 239 p.

Mather, J.R., ed. 1964. Average climatic water balance data of the continents. Part VIII United States. C.W. Thornthwaite Assoc., Laboratory of Climatology, Publ. in Climatology, XVII, No. 3, Centerton, NJ.

Mayer, L., McFadden, L.D., and Harden, J.W. 1988. Distribution of calcium carbonate in desert soils: A model. Geology 16, 303–306.

McCoy, W.D. 1981. Quaternary aminostratigraphy of the Bonneville and Lahontan basins, western United States, with paleoclimatic interpretations. Ph.D. thesis, Univ. of Colorado, Boulder, 603 p.

McFadden, L.D. 1982. The impacts of temporal and spatial climatic changes on alluvial soils genesis in Southern California. Ph.D. thesis, Univ. of Arizona, Tucson, 430 p.

McFadden, L.D. 1988. Climatic influences on rates and processes of soil development in Quaternary deposits of southern California. Geol. Soc. Am. Spec. Pap. 216, p. 153–177.

McFadden, L.D., and Tinsley, J.C. 1985. Rate and depth of pedogenic-carbonate accumulation in soils: Formulation and testing of a compartment model. Geol. Soc. Am. Spec. Pap. 203, p. 23–41.

McFadden, L.D., and Weldon, R.J., II. 1987. Rates and processes of soil development on Quaternary terraces in Cajon Pass, California. Geol. Soc. Am. Bull. 98, 280–293.

Mears, B., Jr. 1981. Periglacial wedges and the late Pleistocene environment of Wyoming's intermontane basins. Quaternary Res. 15, 171–198.

Miles, R.J., and Franzmeier, D.P. 1981. A lithosequence of soils formed in dune sand. Soil Sci. Soc. Am. J. 45, 362–367.

Morrison, R.B. 1991. Quaternary stratigraphic, hydrologic, and climatic history of the Great Basin, with emphasis on Lakes Lahontan, Bonneville, and Tekopa. In R.B. Morrison, ed., Quaternary nonglacial geology: Conterminous U.S., p. 283–320. Geol. Soc. Am., Boulder, Colorado, The Geology of North America, v. K-2.

Morrison, R.B., and Frye, J.C. 1965. Correlation of the middle and late Quaternary successions of the Lake Lahontan, Lake Bonneville, Rocky Mountain (Wasatch Range), southern Great Plains, and eastern mid-west areas. Nevada Bur. Mines Rept. 9, 45 p.

Muhs, D.R. 1984. Intrinsic thresholds in soil systems. Phys. Geogr. 5, 99–110.

Mulcahy, M.J. 1967. Landscapes, laterites, and soils in southwestern Australia. In J.N. Jennings and J.A. Mabbutt, eds., Landform Studies from Australia and New Guinea, p. 211–230. Australian National University Press, Canberra.

Mulcahy, M.J., and Hingston, F.J. 1961. The development and distribution of the soils of the York-Quairading area, west Australia, in relation to landscape evolution. CSIRO Soil Publ. 17, 43 p.

Nettleton, W.D., Witty, J.E., Nelson, R.E., and Hawley, J.W. 1975. Genesis of argillic horizons in soils of desert areas of the southwestern United States. Soil Sci. Soc. Am. Proc. 39, 919–926.

Nettleton, W.D., Nelson, R.E., Brasher, B.R., and Derr, P.S. 1982. Gypsiferous soils in the western United States. Soil Sci. Soc. Am. Spec. Publ. 10, 147–168.

Page, W.D. 1972. The geological setting of the archaeological site at Oued El Akarit and the paleoclimatic significance of gypsum soils, southern Tunisia. Ph.D. thesis, Univ. of Colorado, Boulder, 111 p.

Pawluk, S. 1978. The pedogenic profile in the stratigraphic section. In W.C. Mahaney, ed., Quaternary soils, p. 61–76. Geo Abstracts Ltd., Norwich, England.

Pedro, G., Jamagne, M., and Bejon, J.C. 1969. Mineral interactions and transformations in relation to pedogenesis during the Quaternary. Soil Sci. 107, 462–469.

Post, W.M., Pastor, J., Zinke, P.J., and Stangenberger, A.G. 1985. Global pattern of soil nitrogen storage. Nature (London) 317, 613–616.

Quade, J., Cerling, T.E., and Bowman, J.R. 1989a. Systematic variations in the carbon and oxygen isotopic composition of pedogenic carbonate along elevation transects in the southern Great Basin, United States. Geol. Soc. Am. Bull. 101, 464–475.

Quade, J., Cerling, T.E., and Bowman, J.R. 1989b. Development of Asian monsoon revealed by marked ecological shift during the latest Miocene in northern Pakistan Nature (London) 342, 163–166.

Quade, J., Cater, J.M.L., Ojha, T.P., Adam, J., and Harrison, T.M. 1995. Late Miocene environmental change in Nepal and the northern Indian subcontinent: Stable isotope evidence from paleosols. Geol. Soc. Am. Bull. 107, 1381–1397.

Reheis, M.C. 1987a. Soils in granitic alluvium in humid and semiarid climates along Rock Creek, Carbon County, Montana. U.S. Geol. Surv. Bull. 1590-D, 71 p.

Reheis, M.C. 1987b. Gypsic soils on the Kane alluvial fans, Big Horn County, Wyoming. U.S. Geol. Surv. Bull. 1590-C, 39 p.

Reheis, M.C. 1987c. Climatic implications of alternating clay and carbonate formation in semiarid soils of south-central Montana. Quaternary Res. 27, 270–282.

Retallack, G.J. 1994. The environmental factor approach to the interpretation of paleosols. Soil Sci. Soc. Am. Spec. Publ. No. 33, 31–64.

Richmond, G.M. 1962. Quaternary Stratigraphy of the La Sal Mountains, Utah. U.S. Geol. Surv. Prof. Pap. 324, 135 p.

Richmond, G.M. 1965. Glaciation of the Rocky Mountains. In H.E. Wright, Jr. and D.G. Frey, eds., The

Quaternary of the United States, p. 217–230. Princeton Univ. Press, Princeton, N.J, 922 p.

Richmond, G.M. 1972. Appraisal of the future climate of the Holocene in the Rocky Mountains. Quaternary Res. 2, 315–322.

Rose, J., Allen, P., Kemp, R.A., Whiteman, C.A., and Owen, N. 1985. The early Anglian Barham Soil of eastern England. In J. Boardman, ed., Soils and Quaternary landscape evolution, p. 197–299. John Wiley & Sons, New York.

Ross, G.J. 1980. The mineralogy of Spodosols. In B.K.G. Theng, ed., Soils with variable charge, p. 127–143. New Zealand Soc. Soil Sci., Lower Hutt.

Ruhe, R.V. 1984a. Loess-derived soils, Mississippi Valley region: I. Soil-sedimentation system. Soil Sci. Soc. Am. J. 48, 859–863.

Ruhe, R.V. 1984b. Loess-derived soils of the Mississippi Valley region: II. Soil-climate system. Soil Sci. Soc. Am. J. 48, 864–867.

Ruhe, R.V. 1984c. Soil-climate system across the prairies in midwestern U.S.A. Geoderma 34, 201–219.

Schaetzl, R.J., and Isard, S.A. 1996. Regional-scale relationships between climate and strength of podzolization in the Great Lakes region, North America. Catena 28, 47–69.

Schaetzl, R.J., and Mokma, D.L. 1988. A numerical index of podzol and podzolic soil development. Phys. Geogr. 9, 232–246.

Schlesinger, W.H. 1990. Evidence from chronosequence studies for a low carbon-storage potential of soils. Nature (London) 348, 232–234.

Schwertmann, U. 1993. Relations between iron oxides, soil color, and soil formation. Soil Sci. Soc. Am. Spec. Publ. 31, 51–69.

Schwertmann, U., Murad, E., and Schulze, D.G. 1982. Is there Holocene reddening (hematite formation) in soils of axeric temperate areas? Geoderma 27, 209–223.

Sherman, G.D. 1952. The genesis and morphology of the alumina-rich laterite clays. In Problems of clay and laterite genesis, p. 154–161. Am. Inst. Mining and Metallurgical Engr., New York.

Shroba, R.R. 1977. Soil development in Quaternary tills, rock-glacier deposits, and taluses, southern and central Rocky Mountains. Ph.D. thesis, Univ. of Colorado, Boulder, 424 p.

Singer, A. 1975. A Cretaceous laterite in the Negev Desert, southern Israel. Geol. Mag. 112, 151–162.

Singer, A. 1979/1980. The paleoclimatic interpretation of clay minerals in soils and weathering profiles. Earth-Sci. Rev. 15, 303–326.

Singer, A. 1989. Palygorskite and sepiolite group minerals. In J.B. Dixon and S.B. Weed, eds., Minerals in soil environments, 2nd ed., p. 829–872. No. 1 in Soil Sci. Soc. Am. Book Series, Madison, WI.

Singer, M.J., and Nkedi-Kizza, P. 1980. Properties and history of an exhumed Tertiary Oxisol in California. Soil Sci. Soc. Am. J. 44, 587–590.

Soil Survey Staff. 1975. Soil taxonomy. U.S. Dept. Agri., Agri. Handbook No. 436, 754 p.

Smeck, N.E., and Ciolkosz, E.J., eds. 1989. Fragipans: Their occurrence, classification, and genesis. Soil Sci. Soc. Am. Spec. Publ. No. 24, 153 p.

Strakhov, N.M. 1967. Principles of lithogenesis, vol. 1. Oliver and Boyd Ltd., Edinburgh, 245 p.

Tedrow, J.C.F. 1977. Soils of the polar landscapes: Rutgers Univ. Press., NJ, 638 p.

The Times of London. 1967. The Times atlas of the world. Houghton Mifflin, Boston.

Thornthwaite, C.W., and Mather, J.R. 1957. Instructions and tables for computing potential evapotranspiration and the water balance. Drexel Inst. Tech. Lab. of Climatology, Pubs. in Climatology 10, No. 3, 311 p.

Torrent, J., and Nettleton, W.D. 1979. Feedback processes in soil genesis. Geoderma 20, 281–287.

Veevers, J.J., ed. 1984. Phanerozoic earth history of Australia. Oxford University Press, New York, 418 p.

Waltman, W.J., Ciolkosz, E.J., Mausbach, M.J., Svoboda, M.D., Miller, D.A., and Kolb, P.J. 1997. Soil climate regimes of Pennsylvania. Penn. State Univ. Agri. Res. Sta. Bull. 873, 235 p.

Webb, T.H., Campbell, A.S., and Fox, F.B. 1986. Effect of rainfall on pedogenesis in a climosequence of soils near Lake Pukaki, New Zealand. New Zealand J. Geol. Geophys. 29, 323–334.

Wells, S.G., McFadden, L.D., and Dohrenwend, J.C. 1987. Influence of late Quaternary climatic changes on geomorphic and pedogenic processes on a desert piedmont, eastern Mojave Desert, California. Quaternary Res. 27, 130–146.

■ Chapter 11

Allen, J.R.L. 1989. Alluvial paleosols: Implications for architecture. In J.R.L. Allen and V.P. Wright, eds., Paleosols in siliclastic sequences, p. 49–69. Postgraduate Res. Inst. for Sedimentology, Univ. Reading, Reading, 98 p.

Alloway, B.J. 1990. Heavy metals in soils. Blackie and Son Ltd., Glasgow, Scotland, 339 p.

Amit. R., Harrison, J.B.J., and Enzel, Y. 1995. Use of soils and colluvial deposits in analysing tectonic events—the southern Arava Rift, Israel. Geomorphology, 12, 91–107.

Amit, R., Harrison, J.B.J., Enzel, Y., and Porat, N. 1996. Soils as a tool for estimating ages of Quaternary fault scarps in a hyperarid environment—the southern Arava valley, the Dead Sea Rift, Israel. Catena 28, 21–45.

An, Z., Kukla, G.J., Porter, S.C., and Xiao, J. 1991. Magnetic susceptibility evidence of monsoon variation on the Loess Plateau of central China during the last 130,000 years. Quaternary Res. 36, 29–36.

April, R., Newton, R., and coles, L.T. 1986. Chemical weathering in two Adirondack watersheds: Past and present-day rates. Geol. Soc. Am. Bull. 97, 1232–1238.

Banerjee, S.K. 1995. Chasing the paleomonsoon over China: Its magnetic proxy record. GSA Today, May issue, 93–97.

Barrientos, X., and Selverstone, J. 1987. Metamorphosed soils as stratigraphic indicators in deformed terrains: An example from the eastern Alps. Geology, 15, 841–844.

Basher, L.R., Tonkin, P.J., and McSaveney, M.J. 1988. Geomorphic history of a rapidly uplifting area on a compressional plate boundary: Cropp River, New Zealand. Zeit. Geomorphol, 69, 117–131.

Bell, M., and Boardman, J., eds. 1992. Past and present erosion: archaeological and geographical perspectives. Oxbow Monograph 22, Oxbow Books, Oxford, England, 250 p.

Benedict, J.B. 1981. The Fourth of July Valley. Center for Mountain Archaeology Report. No. 2, 139 p.

Benedict, J.B. 1985. Arapaho Pass: Glacial geology and archeology at the crest of the Colorado Front Range: Center for Mountain Archeology, Research Report No. 3, Ward, Colorado, 197 p.

Berry, M.E. 1990. Soil catena development on fault scarps of different ages, eastern escarpment of the Sierra Nevada, California. Geomorphology, 3, 333–350.

Berry, M.E. 1994. Soil-geomorphic analysis of late-Pleistocene glacial sequences in the McGee, Pine, and Bishop Creek drainages, east-central Sierra Nevada, California. Quaternary Res. 41, 160–175.

Bethell, P., and Mate, I. 1989. The use of phosphate analysis in archaeology. In J. Henderson, ed., Scientific analysis in archaeology and its interpretation, p. 1–29. UCLA Inst. of Archaeology, Archaeological Res. Tools 5.

Bettis, E.A., III, 1992, Soil morphologic properties and weathering sone characteristics as age indicators in Hollocene alluvium in the Upper Midwest. In V.T. Holliday, ed., Soils in archaeology, P. 119–144. Smithsonian Institution Press, Washington, D.C., 254 p.

Bettis, E.A., III, ed. 1995. Archaeological geology of the Archaic Period in North America. Geol. Soc. Am. Spec. Pap. 297, 154 p.

Billard, A. 1993. Is a middle Pleistocene climatic optimum recorded in the loess-palaeosol sequences of Eurasia? Quaternary Int. 17, 87–94.

Birkeland, P.W. 1968. Correlation of Quaternary stratigraphy of the Sierra Nevada with that of the Lake Lahontan area. In R.B. Morrison and H.E. Wright, Jr., eds., Means of correlation of Quaternary successions, p. 469–500. Int. Assoc. Quaternary Res., VII Cong., Proc. vol. 8.

Birkeland, P.W., and Shroba, R.R. 1974. The status of the concept of Quaternary soil-forming intervals in the western United States. In W.C. Mahaney, ed., Quaternary environments, p. 241–276. York University (Toronto, Canada) Geogr. Mono. No. 5.

Birkeland, P.W., Colman, S.M., Burke, R.M., Shroba, R.R., and Meierding, T.C. 1979. Nomenclature of alpine glacial deposits, or, what's in a name? Geology 7, 532–536.

Birkeland, P.W., Burke, R.M., and Walker, A.L. 1980. Soils and subsurface rock-weathering features of Sherwin and pre-Sherwin glacial deposits, eastern Sierra Nevada, California. Geol. Soc. Am. Bull. 91, 238–244.

Birkeland, P.W., Machette, M.N., and Haller, K.H. 1991. Soils as a tool for applied Quaternary research. Utah Geol. Mineral Surv. Mscl. Publ. 91–3, 63 p.

Blodgett, R.H., Crabough, J.P., and McBride, E.F. 1993. The color of red beds—a geologic perspective. Soil Sci. Soc. Am. Spec. Publ. 31, 127–159.

Bockheim, J.G., and Koerner, D. 1997. Pedogenesis in alpine ecosystems of the eastern Uinta Mountains, Utah, U.S.A. Arc. Alp. Res. 29, 164–172.

Borchardt, G., Rice, S., and Taylor, G. 1980a. Paleosols overlying the Foothills fault system near Auburn, California. California Division of Mines and Geology Special Report 149, 38 p.

Borchardt, G., Taylor, G., and Rice, S. 1980b. Fault features in soils of the Mehrten Formation, Auburn damsite, California. California Division of Mines and Geology Special Report 141, 45 p.

Bown, T.M., and Kraus, M.J. 1981. Lower Eocene alluvial paleosols (Willwood Formation, northwest Wyoming, U.S.A.) and their significance for paleoecology, paleoclimatology, and basin analysis. Palaeogeogr. Palaeoclimatol. Palaeoecol. 34, 1–30.

Bown, T.M., and Kraus, M.J. 1987. Integration of channel and floodplain suites, 1. Development sequence and lateral relations of alluvial paleosols. J. Sedimentary Petrol. 57, 587–601.

Bown, T.M., and Kraus, M.J. 1993. Time-stratigraphic reconstruction and integration of paleopedologic, sedimentologic, and biotic events (Willwood Formation, lower Eocene, northwest Wyoming, U.S.A.). Palaios 8, 68–80.

Bronger, A., and Heinkele, T. 1989. Paleosol sequences as witnesses of Pleistocene climatic history. Catena Suppl. 16, 163–186.

Bull, W.B. 1991. Geomorphic responses to climatic change. Oxford Univ. Press, New York, 326 p.

Burke, R.M., and Birkeland, P.W. 1979. Re-evaluation of multiparameter relative dating techniques and their application to the glacial sequence along the eastern escarpment of the Sierra Nevada, California. Quaternary Res. 11, 21–51.

Burns, S.F., Thompson, R.H., Beck, J.N., and Meriwether, J.R. 1991. Thorium, uranium and cesium-137 in Louisiana soils: Migration trends in a soil catena near Dubach, Louisiana, USA. Radiochim. Acta 52/53, 241–247.

Busacca, A.J. 1989. Long Quaternary record in eastern Washington, U.S.A., interpreted from multiple buried paleosols in loess. Geoderma 45, 105–122.

Carver, G.A., and McCalpin, J.P. 1996. Paleoseismicity of compressional tectonic environments. In J.P. McCalpin, ed., Paleoseismology, p. 183–270. Academic Press, New York.

Catt, J.A. 1986. Soils and Quaternary geology: A handbook for field scientists. Oxford Univ. Press, Oxford, 267 p.

Catt, J.A., ed. 1990. Paleopedology manual. Quaternary Int. 6, 1–95.

Catt, J.A., and Weir, A.H. 1976. The study of archeologically important sediments by petrographic techniques. In D.A. Davidson and M.L. Shackley, eds., Geoarchaeology, p. 65–91. G. Duckworth, London.

Chadwick, O.A., and Davis, J.O. 1990. Soil-forming intervals caused by eolian sediment pulses in the Lahontan basin, northwestern Nevada. Geology 18, 243–246.

Christenson, G.E., and Purcell, C. 1985. Correlation and age of Quaternary alluvial-fan sequences, Basin and Range Province, southwestern United States. Geol. Soc. Am. Spec. Pap. 203, 115–122.

Clayton, J.L., Kennedy, D.A., and Nagel, T. 1991a. Soil response to acid deposition, Wind River Mountains, Wyoming: I. Soil properties. Soil Sci. Soc. Am. J. 55, 1427–1433.

Clayton, J.L., Kennedy, D.A., and Nagel, T. 1991b. Soil response to acid deposition, Wind River Mountains, Wyoming: II. Column leaching studies. Soil Sci. Soc. Am. J. 55, 1433–1439.

Collins, M.E., Carter, B.J., Gladfelter, B.G., and Southard, R.J. eds. 1995. Pedological perspectives in archaeological research. Soil Sci. Soc. Am. Spec. Pub. No. 44, 157 p.

Colman, S.M., Choquette, A.F., and Hawkins, F.F. 1988. Physical, soil, and paleomagnetic stratigraphy of the upper Cenozoic sediments in Fisher Valley, southeastern Utah. U.S. Geol. Surv. Bull. 1686, 33 p.

Costa, J.E., and Baker, V.R. 1981. Surficial geology: Building with the earth. John Wiley & Sons, New York, 498 p.

Courty, M.A., Goldberg, P., and MacPhail, R. 1989. Soils and micromorphology in archaeology. Cambridge Univ. Press, Cambridge, England, 344 p.

Cremaschi, M. 1987. Paleosols and vetusols in the central Po plain (northern Italy): A study in Quaternary geology and soil development. Studi e ricerche sul territorio, 28, Edizioni Unicopli, Milano, 306 p.

Crone, A.J., Machette, M.N., and Bowman, J.R. 1992. Geologic investigations of the 1988 Tennant Creek, Australia, earthquakes—implications for paleoseismicity in stable continental interiors. U.S. Geol. Surv. Bull. 2032-A, 51 p.

Dahms, D.E., and Holliday, V.T. In press. Paleosols as environmental indicators: A critical review. Quaternary Int.

Derbyshire, E. 1983. On the morphology, sediments and origin of the Loess Plateau of central China. In R. Gardner and H. Scoging, eds., Mega-geomorphology, p. 172–194. Oxford Univ. Press, New York, 240 p.

Douglas, L.A. 1980. The use of soils in estimating the time of last movement of faults. Soil Sci. 129, 345–352.

Eberl, D.D. 1984. Clay mineral formation and transformation in rocks and soils. Phil. Trans. Royal Soc. London A 311, 241–257.

Eidt, R.C. 1985. Theoretical and practical considerations in the analysis of Anthrosols. In G.A. Rapp, Jr. and J.A. Gifford, eds., Archaeological geology, p. 155–190. Yale Univ. Press, New Haven, 435 p.

Environmental Laboratory, Dept. of the Army. 1987. Corps of Engineers wetlands delineation manual. Corps of Engineers, Waterways Experimental Station, Technical Report Y-87-1, Vicksburg, Miss., 100 p.

Farrand, W.R. 1975. Sediment analysis of a prehistoric rockshelter: The Abri Pataud. Quaternary Res. 5, 1–26.

Fastovsky, D.E., McSweeney, K., and Morton, L.D. 1989. Pedogenic development at the Cretaceous-Tertiary boundary, Garfield County, Montana. J. Sedimentary Petrol. 59, 758–767.

Fergusson, J.E. 1990. The heavy metals: Chemistry, environmental impact and health effects. Pergamon Press, Oxford, England, 614 p.

Ferring, C.R. 1992. Alluvial pedology and geoarchaeological research. In V.T. Holliday, ed., Soils in archaeology, p. 1–39. Smithsonian Institution Press, Washington, D.C., 254 p.

Follmer, L.R. 1978. The Sangamon Soil in its type area—a review. In W.C. Mahaney, ed., Quaternary soils, p. 125–165. Geo Abstracts, Ltd., Norwich, England, 508 p.

Follmer, L.R. 1983. Sangamon and Wisconsinan pedogenesis in the midwestern United States. In S.C. Porter, ed., Late Quaternary environments of the United States, vol. 1, The late Pleistocene, p. 138–144. Univ. Minnesota Press, Minneapolis, 407 p.

Follmer, L.R. 1996. Loess studies in central United states: Evolution of concepts. Eng. Geol. 45, 287–304.

Follmer, L.R., McKay, E.D., Lineback, J.A., and Gross, D.L. 1979. Wisconsinan, Sangamonian, and Illinoian stratigraphy in central Illinois. Ill. State Geol. Surv. Guidebook 13, 139 p.

Gerson, R., Grossman, S., Amit, R., and Greenbaum, N. 1993. Indicators of faulting events and periods of quiescence in desert alluvial fans. Earth Surf. Process. Landforms 18, 181–202.

Gill, J.D., West, M.W., Noe, D.C., Olsen, H.W., and McCarty, D.K. 1996. Geologic control of severe expansive clay damage to a subdivision in the Pierre Shale, southwest Denver metropolitan area, Colorado. Clays Clay Minerals, 44, 530–539.

Gillespie, A., and Molnar, P. 1995. Asynchronous maximum advances of mountain and continental glaciers. Rev. Geophys. 33, No. 3, 311–364.

Gillespie, A.R., Clark, D.H., and Bierman, P.R. 1996. New exposure ages support asynchronism between alpine glaciation and sea-level fluctuations, Sierra

Nevada, California. Geol. Soc. Am. Abstr. with Programs 28, No. 7, A-233.

Goldberg, P. 1992. Micromorphology, soils and archaeological sites. In V.T. Holliday, ed., Soils in archaeology, p. 145–167. Smithsonian Institution Press, Washington, D.C., 254 p.

Hall, R.D., and Shroba, R.R. 1993. Soils developed in the glacial deposits of the type areas of the Pinedale and Bull Lake glaciations, Wind River Range, Wyoming, U.S.A. Arc. Alp. Res. 25, 373–386.

Hall, R.D., and Shroba, R.R. 1995. Soil evidence for a glaciation intermediate between the Bull Lake and Pinedale glaciations at Fremont Lake, Wind River Range, Wyoming, U.S.A. Arc. Alp. Res. 27, 89–98.

Hallberg, G.R. 1980. Pleistocene stratigraphy in east-central Iowa. Iowa Geol. Surv. Tech. Info. Ser. No. 10, 168 p.

Hallet, B., and Putkonen, J. 1994. Surface dating of dynamic landforms: Young boulders on aging moraines. Science 265, 937–940.

Hasiotis, S.T., Aslan, A., and Bown, T.M. 1993. Origin, architecture, and paleoecology of the early Eocene continental ichnofossil *Scaphichnium hamatum*—integration of ichnology and paleopedology. Ichnos 3, 1–9.

Hawley, J.W., Bachman, G.O., and Manley, K. 1976. Quaternary stratigraphy in the Basin and Range and Great Plains Provinces, New Mexico and western Texas. In W.C. Mahaney, ed., Quaternary stratigraphy of North America, p. 235–274. Dowden, Hutchinson, and Ross, Stroudsberg, PA.

Hay, R.L. 1976. Geology of the Olduvai Gorge. University of California Press, Berkeley, 203 p.

Haynes, C.V., Jr. 1968. Chronology of late-Quaternary lluvium. In R.B. Morrison and H.E. Wright, Jr., eds., Means of correlation of Quaternary successions, p. 591–631. University of Utah Press, Salt Lake City.

Holliday, V.T. 1985a. Archaeological geology of the Lubbock Lake site, Southern High Plains of Texas. Geol. Soc. Am. Bull. 96, 1483–1492.

Holliday, V.T. 1985b. Early and middle Holocene soils at the Lubbock Lake archeological site, Texas. Catena 12, 61–78.

Holliday, V.T. 1985c. Morphology of late Holocene soils at the Lubbock Lake archeological site, Texas. Soil Sci. Soc. Am. J. 49, 938–946.

Holliday, V.T. 1988. Genesis of the late-Holocene soil chronosequence at the Lubbock Lake Archaeolog-ical site, Texas. Ann. Assoc. Am. Geogr. 78, 594–610.

Holliday, V.T. 1989. Paleopedology in archeology. Catena Suppl. 16, 187–206.

Holliday, V.T., ed. 1990. Pedology in archaeology. In N.P. Lasca and J. Donahue, eds., Archaeological geology of North America, p. 525–540. Geol. Soc. Am., Centennial Special Vol. 4, 633 p.

Holliday, V.T. 1992. Soils in archaeology. Smithsonian Institution Press, Washington, D.C., 254 p.

Holliday, V.T. 1994. The "state factor" approach in geoarchaeology. Soil Sci. Soc. Am. Spec. Publ. 33, 65–86.

Holliday, V.T. 1995. Stratigraphy and paleoenvironments of late Quaternary valley fills on the Southern High Plains. Geol. Soc. Am. Memoir 186, 136 p.

Hovan, S.A., Rea, D.K., Pisias, N.G., and Shackleton, N.J. 1989. A direct link between the China loess and marine O-18 records: aeolian flux to the North Pacific. Nature (London) 340, 296–298.

Hughes, J., and Heathwaite, L. 1995. Hydrology and hydrochemistry of British wetlands. John Wiley & Sons, Ltd., Chichester, England, 486 p.

Jacobs, P.M., Knox, J.C., and Mason, J.A. 1997. Preservation and recognition of middle and early Pleistocene loess in the Driftless Area, Wisconsin. Quaternary Res. 47, 147–154.

Johnson, D.L., and Watson-Stegner, D. 1990. The soil-evolution model as a framework for evaluating pedoturbation in archaeological site formation. In N.P. Lasca and J. Donahue, eds., Archaeological geology of North America, p. 541–560. Geol. Soc. Am., Centennial Special Vol. 4, 633 p.

Johnson, N.M. 1984. Acid rain neutralization by geologic materials. In O.P. Bricker, ed., Geological aspects of acid deposition, p. 37–53. Butterworth, Boston, 143 p.

Johnson, W.H., and six others, 1997. Late Quaternary temporal and event classifications, Great Lakes region, North America. Quaternary Res. 47, 1–12.

Jorgensen, D.W. 1992. Use of soils to differentiate dune age and to document spatial variation in eolian activity, northeast Colorado, U.S.A. J. Arid Environ. 23, 19–34.

Karlstrom, E.T. 1990. Relict periglacial features east of Waterton-Glacier Parks, Alberta and Montana, and their palaeoclimatic significance. Permafrost Periglacial Process. 1, 221–234.

Kennedy, I.R. 1992. Acid soil and acid rain. Research Studies Press Ltd., Somerset, England, 254 p.

Kirkham, R.M. 1977. Quaternary movements on the Golden fault, Colorado. Geology 5, 689–692.

Kover, R.W. 1967. Soils of the Malibu area, California. Soil Conservation Service, U.S. Dept. Agri., 89 p.

Kraus, M.J., and Bown, T.M. 1986. Paleosols and time resolution in alluvial stratigraphy. In V.P. Wright, ed., Paleosols, their recognition and interpretation, p. 180–207. Blackwell Sci. Publ., Oxford, 315 p.

Kraus, M.J., and Bown, T.M. 1988. Pedofacies analysis; a new approach to reconstructing ancient fluvial sequences. Geol. Soc. Am. Spec. Pap. 216, 143–152.

Kukla, J. 1970. Correlation between loesses and deep-sea sediments. Geolo. Förenin. Stolkholm För-hand. 92, 148–180.

Kukla, J. 1977. Pleistocene land-sea correlations. 1. Europe. Earth-Sci. Rev. 13, 307–374.

Kukla, G. 1987. Loess stratigraphy in central China. Quaternary Sci. Rev. 6, 191–219.

Lal, R., Kimble, J., Levine, E., and Stewart, B.A. 1995. Soils and global change. CRC Press, Boca Raton, FL, 440 p.

Langmuir, D. 1997. Aqueous environmental geochemistry. Prentice-Hall, Englewood Cliffs, NJ, 600 p.

Lasca, N.P., and Donahue, J., eds. 1990. Archaeological geology of North America. Geol. Soc. Am., Centennial Special Vol. 4, 633 p.

Leamy, M.L., Milne, J.D.G., Pullar, W.A., and Bruce, J.G. 1973. Paleopedology and soil stratigraphy in the New Zealand Quaternary succession. New Zealand J. Geol. Geophys. 16, 723–744.

Lehman, T.M. 1990. Upper Cretaceous (Maastrichtian) paleosols in Trans-Pecos Texas: reply. Geol. Soc. Am. Bull. 101, 845–846.

Lineback, J.A., compiler. 1979. Quaternary deposits of Illinois. Ill. State Geol. Surv.

Litaor, M.I. 1987. The influence of eolian deposition on the genesis of alpine soils in the Front Range, Colorado, U.S.A. Soil Sci. Soc. Am. J. 51, 142–147.

Litaor, M.I., and Ibrahim, S.A. 1996. Plutonium association with selected solid phases in soils of Rocky Flats, Colorado, using sequential extraction technique. J. Environ. Qual. 25, 1144–1152.

Litaor, M.I., and Thurman, E.M. 1988. Acid neutralizing processes in an alpine watershed, Front Range, Colorado, U.S.A.—1: Buffering capacity of dissolved organic carbon in soil solutions. Appl. Geochem. 3, 645–652.

Litaor, M.I., Thompson, M.L., Barth, G.R., and Molzer, P.C. 1994. Plutonium-239 + 240 and americium-241 in soils east of Rocky Flats, Colorado. J. Environ. Qual. 23, 1231–1239.

Liu, T. 1988. Loess in China. Springer-Verlag, Berlin, 224 p.

Machette, M.N. 1977. Geologic map of the Lafayette Quadrangle, Adams, Boulder, and Jefferson Counties, Colorado. U.S. Geol. Surv. Map GQ-1392.

Machette, M.N. 1978. Dating Quaternary faults in the southwestern United States by using buried calcic paleosols. U.S. Geol. Surv. J. Res. 6, 369–381.

Machette, M.N. 1985. Calcic soils of the southwestern United states. Geol. Soc. Am. Spec. Pap. 203, 1–21.

Machette, M.N. 1988. Quaternary movement along the La Jencia fault, central New Mexico. U.S. Geol. Surv. Prof. Pap. 1440, 82 p.

Mack, G.H., James, W.C., and Monger, H.C. 1993. Classification of paleosols. Geol. Soc. Am. Bull. 105, 129–136.

Mahaney, W.C. 1990. Ice on the Equator: Quaternary geology of Mount Kenya, East Africa. Wm Caxton Ltd., Sister Bay, Wisconsin, 386 p.

Mahaney, W.C. 1993. Moraine and soil stratigraphy, Mammoth Basin and upper Green River areas, Wind River Mountains, western Wyoming, U.S.A. Zeit. Gletscherk. Glazialgeol. 29, 103–118.

Mahaney, W.C., Hancock, R.G.V., Aufreiter, S., and Huffman, M.A. 1996. Geochemistry and clay mineralogy of termite mound soil and the role of geophagy in chimpanzees of the Mahale Mountains, Tanzania. Primates 37, 121–134.

Mandel, R.D. 1992. Soils and Holocene landscape evolution in central and southwestern Kansas: implications for archaeological research. In V.T. Holliday, ed., Soils in archaeology, p. 41–100. Smithsonian Institution Press, Washington, D.C., 254 p.

Martini, I.P., and Chesworth, W., eds. 1992. Weathering, soils & paleosols. Elsevier, Amsterdam, 618 p.

Mayberry, J.O. 1972a. Map showing landslide deposits and areas of potential landsliding in the Parker Quadrangle, Arapahoe and Douglas Counties, Colorado. U.S. Geol. Surv. Map I-770-E.

Mayberry, J.O. 1972b. Map showing relative swelling-pressure potential of geologic materials in the Parker Quadrangle, Arapahoe and Douglas Counties, Colorado. U.S.Geol. Surv. Map I-770-D.

Mayberry, J.O., and Lindvall, R.M. 1972. Geologic map of the Parker Quadrangle, Arapahoe and Douglas Counties, Colorado. U.S. Geol. Surv. Map I-770-A.

McBride, M.B. 1994. Environmental chemistry of soils. Oxford Univ. Press, New York, 406 p.

McCalpin, J.P. 1982. Quaternary geology and neotectonics of the west flank of the northern Sangre de

Cristo Mountains, south-central Colorado. Colo. School Mines Quart. 77, No. 3, 97 p.

McCalpin, J.P., ed. 1996. Paleoseismology. Academic Press, San Diego, 583 p.

McCalpin, J.P., and Berry, M.E. 1996. Soil catenas to estimate ages of movements on normal fault scarps, with an example from the Wasatch fault zone, Utah, USA. Catena 27, 265–286.

Meriwether, J.R., Burns, S.F., Thompson, R.H., and Beck, J.N. 1995. Evaluation of soil radioactivities using pedologically based sampling techniques. Health Phys. 69, 406–409.

Milne, J.D.G. 1973. Map and sections of river terraces in the Rangitikei Basin, North Island, New Zealand. New Zealand Soil Surv. Report 4.

Monger, H.C., and Adams, H.P. 1996. Micromorphology of calcite-silica deposits, Yucca Mountain, Nevada. Soil Sci. Soc. Am. J. 60, 519–530.

Morrison, R.B. 1964. Lake Lahontan: Geology of the southern Carson Desert, Nevada. U.S. Geol. Surv. Prof. Pap. 401, 156 p.

Morrison, R.B. 1965. Lake Bonneville: Quaternary stratigraphy of eastern Jordan Valley, south of Salt Lake City, Utah. U.S. Geol. Surv. Prof. Pap. 477, 80 p.

Morrison, R.B. 1967. Principles of Quaternary stratigraphy. p. 1–69. In R.B. Morrison and H.E. Wright, Jr., eds., Quaternary soils. Int. Assoc. Quaternary Res., VII Cong., Proc. vol. 9.

Morrison, R.B. 1968. Means of time-stratigraphic division and long-distance correlation of Quaternary successions. In R.B. Morrison and H.E. Wright, Jr., eds., Means of correlation of Quaternary successions, p. 1–113. Int. Assoc. Quaternary Res., VII Cong., Proc. vol. 8.

Morrison, R.B. 1978. Quaternary soil stratigraphy—concepts, methods, and problems. In W.C. Mahaney, ed., Quaternary soils, p. 77–108. Geo Abstracts, Norwich, England.

Morrison, R.B. 1991. Quaternary stratigraphic, hydrologic, and climatic history of the Great Basin, with emphasis on Lakes Lahontan, Bonneville and Tekopa. In R.B. Morrison, ed., Quaternary nonglacial geology; Conterminous U.S., p. 283–320. Geol. Soc. Am., The geology of North America, vol. K-2.

Morrison, R.B., and Frye, J.C. 1965. Correlation of the middle and late Quaternary successions of the Lake Lahontan, Lake Bonneville, Rocky Mountain (Wasatch Range), southern Great Plains, and eastern midwest areas: Nevada Bur. Mines Report 9, 45 p.

Mueller, K.J., and Rockwell, T.K. 1995. Late Quaternary activity of the Laguna Salida fault in northern Baja California, Mexico. Geol. Soc. Am. Bull. 107, 8–18.

Muhs, D.R., Stafford, T.W., Cowherd, S.D., Mahan, S.A., Kihl, R., Maat, P.B., Bush, C.A., and Nehring, J. 1996. Origin of the late Quaternary dune fields of northeastern Colorado. Geomorphology 17, 129–149.

Mulamoottil, G., Warner, G., and McBean, E.A., eds. 1996. Wetlands: Environmental gradients, boundaries, and buffers. CRC Press, Boca Raton, FL, 298 p.

Mulcahy, M.J., and Churchward, H.M. 1973. Quaternary environments and soils in Australia. Soil Sci. 116, 156–169.

Murray, R.C., and Tedrow, J.C.F. 1975. Forensic geology. Rutgers Univ. Press, New Brunswick, NJ, 217 p.

Nelson, A.R. 1992. Lithofacies analysis of colluvial sediments—an aid in interpreting the recent history of Quaternary normal faults in the Basin and Range Province, western United States. J. Sedimentary Petrol. 62, 607–621.

Nesbitt, H.W. 1992. Diagenesis and metasomatism of weathering profiles, with emphasis on Precambrian paleosols. In I.P. Martini and W. Chesworth, eds., Weathering, soils & paleosols, p. 127–152. Elsevier, Amsterdam, 618 p.

Nishiizumi, K., Winterer, E.L., Kohl, C.P., Klein, J., Middleton, R., Lal, D., and Arnold, J.R. 1989. Cosmic ray production rates of ^{10}Be and ^{26}Al in quartz from glacially polished rocks. J. Geophys. Res. 94, No. B12, 17,907–17,915.

North American Commission on Stratigraphic Nomenclature. 1983. North American stratigraphic code. Amer. Assoc. Petrol. Geol. 67, 841–875.

Olson, C.G. 1989. Soil geomorphic research and the importance of paleosol stratigraphy to Quaternary investigations, midwestern USA. Catena Suppl. 16, 129–142.

Ostenaa, D.A., Levish, D.R., and O'Connell, D.R.H. 1996. Paleoflood study for Bradbury Dam, Cachuma project, California. U.S. Bureau Reclamation, Seismotectonic Report 96-3, Denver, Colorado, 86 p.

Patterson, P.E. 1990. Differentiation between the effects of diagenesis and pedogenesis in the origin of color banding in the Wind River Formation (lower Eocene), Wind River Basin, Wyoming. Ph.D. thesis, Univ. Colorado, Boulder, 301 p.

Patterson, P.E., Birkeland, P.W., and Larson, E.E. 1990. Upper Cretaceous (Maastrichtian) paleosols in

Trans-Pecos Texas: Discussion. Geol. Soc. Am. Bull. 102, 844–845.

Phillips, F.M., Zreda, M.G., Smith, S.S., Elmore, D., Kubik, P.W., and Sharma, P. 1990. Cosmogenic chlorine-36 chronology for glacial deposits at Bloody Canyon, eastern Sierra Nevada. Science 248, 1529–1532.

Pierce, K.L., and Schmidt, P.W. 1975. Reconnaissance map showing relative amounts of soil and bedrock in the mountainous part of the Eldorado Springs Quadrangle, Boulder and Jefferson Counties, Colorado. U.S. Geol. Surv. Map MF-695.

Pierzynski, G.M., Sims, J.T., and Vance, G.F. 1994. Soils and environmental quality. CRC Press, Boca Raton, FL, 313 p.

Piety, L.A. 1996. Compilation of known or suspected Quaternary faults within 100 km of Yucca Mountain, Nevada and California. U.S. Geol. Surv. Open-File Report 94-112.

Piety, L.A., Sullivan, J.T., and Anders, M.H. 1992. Segmentation and paleoseismicity of the Grand Valley fault, southeastern Idaho and western Wyoming. Geol. Soc. Am. Memoir 179, 155–182.

Pimentel, N.L., Wright, V.P., and Azevedo, T.M. 1996. Distinguishing early groundwater alteration effects from pedogenesis in ancient alluvial basins: examples from the Palaeogene of southern Portugal. Sedimentary Geol. 105, 1–10.

Ponti, D.J. 1985. The Quaternary alluvial sequence of the Antelope Valley, California. Geol. Soc. Am. Spec. Pap. 203, 79–96.

Porter, S.C., An, Z. and Hongbo, Z. 1992. Cyclic Quaternary alluviation and terracing in a nonglaciated drainage on the north flank of the Qinling Shan, central China. Quaternary Res. 38, 157–169.

Price, A.B., and Amen, A.E. 1984. Soil survey of Golden area, Colorado. Soil Conservation Service, U.S. Dept. Agri., 405 p.

Pye, K. 1987. Aeolian dust and dust deposits. Academic Press Inc. Ltd., London, 334 p.

Quade, J., and Cerling, T.E. 1990. Stable isotope evidence for a pedogenic origin of carbonates in trench 14 near Yucca Mountain, Nevada. Science 250, 1549–1552.

Quade, J., Cater, J.M.L., Ojha, T.P., Adam, J., and Harrison, T.M. 1995. Late Miocene environmental change in Nepal and the northern Indian subcontinent: stable isotope evidence from paleosols. Geol. Soc. Am. Bull. 107, 1381–1397.

Reheis, M.C. 1985. Evidence for Quaternary tectonism in the northern Bighorn Basin, Wyoming and Montana. Geology 13, 364–367.

Reheis, M.C., Sawyer, T.L., Slate, J.L., and Gillespie, A.R. 1993. Geologic map of the late Cenozoic deposits and faults in the southern part of the Davis Mountain 15′ Quadrangle, Esmeralda County, Nevada. U.S. Geol. Surv. Map I-2342.

Reider, R.G. 1990. Late Pleistocene and Holocene pedogenic and environmental trends at archaeological sites in the plains and mountains areas of Colorado and Wyoming. In N.P. Lasca, and J. Donahue, eds., Archaeological geology of North america, p. 335–360. Geol. Soc. Am., Centennial Special Vol. 4, 633 p.

Reinhardt, J., and Sigleo, W.R. 1988. Paleosols and weathering through geologic time: Principles and applications. Geol. Soc. Am. Spec. Pap. 216, 181 p.

Retallack G.J. 1988. Field recognition of paleosols. Geol. Soc. Am. Spec. Pap. 216, 1–20.

Retallack, G.J. 1983. Late Eocene and Oligocene paleosols from Badlands National Park, South Dakota. Geol. Soc. Am. Spec. Pap. 193.

Retallack, G.J. 1990. Soils of the past: An introduction to paleopedology. Unwin Hyman, Boston, 520 p.

Retallack, G.J. 1991. Untangling the effects of burial alteration and ancient soil formation. Ann. Rev. Earth Planetary Sci. 19, 183–206.

Retallack, G.J. 1993. Classification of paleosols: Discussion. Geol. Soc. Am. Bull. 105, 1635–1636.

Retallack, G.J. 1994. The environmental factor approach to the interpretation of paleosols. Soil Sci. Soc. Am. Spec. Publ. 33, 31–64.

Reynolds R.L., and King, J.W. 1995. Magnetic records of climate change. Rev. Geophys. Suppl. 101–110.

Richmond, G.M. 1962. Quaternary stratigraphy of the La Sal Mountains, Utah. U.S. Geol. Surv. Prof. Pap. 324, 135 p.

Richmond, G.M. 1965. Glaciation of the Rocky Mountains. In H.E. Wright, Jr. and D.G. Frey, eds., The Quaternary of the United States, p. 217–230. Princeton Univ. Press, Princeton, NJ, 922 p.

Rockwell, T. 1988. Neotectonics of the San Cayetano fault, Transverse Ranges, California. Geol. Soc. Am. Bull. 100, 500–513.

Rodbell, D.T., and Schweig, E.S., III. 1993. The record of seismically induced liquifaction on late Quaternary terraces in northwestern Tennessee. Seis. Soc. Am. Bull. 83, 269–278.

Rodbell, D.T., Forman, S.L., Pierson, J., and Lynn, W.C. 1997. The stratigraphy and chronology of Mississippi Valley loess in western Tennessee. Geol. Soc. Am. Bull. 109, 1134–1148.

Ruhe, R.V. 1965. Quaternary paleopedology. In H.E. Wright, Jr. and D.G. Frey, eds., The Quaternary of the United States, p. 755–764. Princeton Univ. Press, Princeton, NJ, 922 p.

Ruhe, R.V. 1969. Quaternary landscapes in Iowa: Iowa State Univ. Press, Ames, 255 p.

Ruhe, R.V., Hall, R.D., and Canepa, A.P. 1974. Sangamon paleosols of southwestern Indiana, U.S.A. Geoderma 12, 191–200.

Sandor, J.A. 1992. Long-term effects of prehistoric agriculture on soils: Examples from New Mexico and Peru. In V.T. Holliday, ed., Soils in Archaeology, p. 217–245. Smithsonian Institution Press, Washington, D.C., 254 p.

Scharpenseel, H.W., Schomaker, M., and Ayoub, A., eds. 1990. Soils on a warmer earth. Elsevier, Amsterdam, 274 p.

Schwochow, S.D., Shroba, R.R., and Wicklein, P.C. 1974. Atlas of sand, gravel, and quarry aggregate resources, Colorado Front Range counties. Colorado Geol. Surv., Spec. Publ. 5-B.

Scott, G.R. 1963. Quaternary geology and geomorphic history of the Kassler Quadrangle, Colorado. U.S. Geol. Surv. Prof. Pap. 421-A, 70 p.

Scott, W.E., McCoy, W.D., Shroba, R.R., and Rubin, M. 1983. Reinterpretation of the exposed record of the last two cycles of Lake Bonneville, western United States. Quaternary Res. 20, 261–285.

Sharp, R.P. 1968. Sherwin Till-Bishop Tuff geological relationships, Sierra Nevada, California. Geol. Soc. Am. Bull. 79, 351–364.

Sharp, R.P., and Birman, J.H. 1963. Additions to the classical sequence of Pleistocene glaciations, Sierra Nevada, California: Geol. Soc. Am. Bull. 74, 1079–1086.

Schwertmann, U., and Taylor, R.M. 1989. Iron oxides. In J.B. Dixon and S.B. Weed, eds., Minerals in soil environments, p. 379–437. Soil Sci. Soc. Am. Book Series No. 1, 1244 p.

Shlemon, R.J. 1978. Late Quaternary evolution of the Camp Pendleton-San Onofre State Beach coastal area, northwestern San Diego County, California. Unpublished report for Southern California Edison Co. and San Diego Gas and Electric Co., 123 p.

Shlemon, R.J. 1986. Late Quaternary stratigraphy, South Platte River, Two Forks area, east-central Front Range, Colorado. Colorado Geol. Surv. Spec. Publ. No. 28, 282–294.

Shlemon, R.J., and Budinger, F.E., Jr. 1990. The archaeological geology of the Calico Site, Mojave Desert, California. In N.P. Lasca and J. Donahue, eds., Archaeological geology of North America, p.

301–313. Geol. Soc. Am., Centennial Special Vol. 4, 633 p.

Shroba, R.R. 1980. Geologic map and physical properties of the surficial and bedrock units of the Englewood Quadrangle, Denver, Arapahoe, and Adams Counties, Colorado. U.S. Geol. Surv. Map GQ-1524.

Sigleo, W., and Reinhardt, J. 1988. Paleosols from some Cretaceous environments in the southeastern United States. Geol. Soc. Am. Spec. Pap. 216, 123–142.

Singer, M.J., and Warkentin, B.P. 1996. Soils in an environmental context: An American perspective. Catena 27, 179–189.

Skully, R.W., and Arnold, R.W. 1981. Holocene alluvial stratigraphy in the upper Susquehanna River basin, New York. Quaternary Res. 15, 327–344.

Stein, J.K., and Farrand, W.R., eds. 1985. Archaeological sediments in context. Center for the Study of Early Man, Inst. Quaternary Studies, Univ. Maine, Orono, 147 p.

Stewart, G.H., and Harrison, J.B.J. 1987. Physical influences on forest types and deer habitat, northern Fiordland, New Zealand. New Zealand J. Ecol. 10, 1–10.

Suggate, R.P. 1965. Late Pleistocene geology of the northern part of the South Island, New Zealand. New Zealand Geol. Surv. Bull. 77, 91 p.

Swan, F.H., III, Schwartz, D.P., and Cluff, L.S. 1980. Recurrence of moderate to large magnitude earthquakes produced by surface faulting on the Wasatch fault zone, Utah. Bull. Seis. Soc. Am. 70, 1431–1462.

Taylor, E.M. 1986. Impact of time and climate on Quaternary soils in the Yucca Mountain area of the Nevada Test Site. M.S. thesis, Univ. Colorado, Boulder, 217 p.

Taylor, E.M., and Huckins, H.E. 1995. Lithology, fault displacement, and origin of secondary calcium carbonate and opaline silica at trenches 14 and 14D on the Bow Ridge fault at Exile Hill, Nye County, Nevada. U.S. Geol. Surv. Open-File Report 93-477, 38 p.

Thorson, R.M., and Holliday, V.T. 1990. Just what is geoarchaeology? Geotimes 35 (July), 19–20.

Tonkin, P.J., and Basher, L.R. 1990. Soil-stratigraphic techniques in the study of soil and landform evolution across the Southern Alps, New Zealand. Geomorphology 3, 547–575.

Wahlstrom, E.E. 1948. Pre-fountain and recent weathering on Flagstaff Mountain near Boulder, Colorado. Geol. Soc. Am. Bull. 59, 1173–1190.

Walker, T.R. 1967. Formation of red beds in modern and ancient deserts. Geol. Soc. Am. Bull. 78, 353–368.

Walker, T.R. 1975. Red beds in the western interior of the United States. U.S. Geol. Surv. Prof. Pap. 853, part II, 49–56.

Walker, T.R., Waugh, B., and Crone, A.J. 1978. Diagenesis of first-cycle desert alluvium of Cenozoic age, southwestern United States and northwestern Mexico. Geol. Soc. Am. Bull. 89, 19–32.

Waltman, W.J., Cunningham, R.L., and Ciolkosz, E.J. 1990. Stratigraphy and parent material relationships of red substratum soils on the Allegheny Plateau. Soil Sci. Soc. Am. J. 54, 1049–1057.

Wells, S.G., McFadden, L.D., and Schultz, J.D. 1990a. Eolian landscape evolution and soil formation in the Chaco dune field, southern Colorado Plateau, New Mexico. Geomorphology 3, 517–546.

Wells, S.G., McFadden, L.D., Renault, C.E., and Crowe, B.M. 1990b. Geomorphic assessment of late Quaternary volcanism in the Yucca Mountain area, southern Nevada: Implications for the proposed high-level radioactive waste repository. Geology 18, 549–553.

Whitney, J.W., and Taylor, E.M., eds. In press. Quaternary paleoseismology and stratigraphy of the Yucca Mountain site area, Nevada and California. U.S. Geol. Surv. Open-File report, CD Rom.

Willman, H.B., and Frye, J.C. 1970. Pleistocene stratigraphy of Illinois. Illinois State Geol. Surv. Bull. 94, 204 p.

Woodward, J.C., Macklin, M.G., and Lewin, J. 1994. Pedogenic weathering and relative-age dating of Quaternary alluvial sediments in the Pindus Mountains of northwest Greece. In D.A. Robinson and R.B.G. Williams, eds., Rock weathering and landform evolution, p. 259–283. John Wiley & Sons Ltd., London.

Wright, V.P. 1986. Paleosols, their recognition and interpretation. Blackwell Sci. Publ., Oxford, 315 p.

Wright, V.P. 1992. Paleopedology: stratigraphic relationships and empirical models. In I.P. Martini and W. Chesworth, eds., Weathering, soils & paleosols, p. 475–499. Elsevier, Amsterdam, 618 p.

Yaalon, D.H. 1971. Soil-forming processes in time and space. In D.H. Yaalon, ed., Paleopedology, p. 29–39. Israel Univ. Press, Jerusalem, 350 p.

Young, F.J. 1989. Soil survey of the islands of Aguijan, Rota, Saipan, and Tinian, Commonwealth of the Northern Mariana Islands. Soil Conservation Service, U.S. Dept. Agri., 166 p.

■ Appendix 1

Ahmad, N. 1983. Vertisols. In L.P. Wilding, N.E. Smeck, and G.F. Hall, eds., Pedogenesis and soil taxonomy, II. Soil orders, p. 91–123. Elsevier, Amsterdam, 410 p.

Berry, M.E. 1987. Morphological and chemical characteristics of soil catenas on Pinedale and Bull Lake moraine slopes in the Salmon River Mountains, Idaho. Quaternary Res. 28, 210–225.

Forman, S.L., and Miller, G.H. 1984. Time dependent soil morphologies and pedogenic processes on raised beaches, Bröggerhalvöya, Spitsbergen, Svaldbard Archipelago. Arc. Alp. Res. 16, 381–394.

Foss, J.E., Wright, W.R., and Coles, R.H. 1975. Testing the accuracy of field textures. Soil Sci. Soc. Am. Proc. 29, 800–802.

Gile, L.H., Peterson, F.F., and Grossman, R.B. 1966. Morphological and genetic sequences of carbonate accumulation in desert soils: Soil Sci. 101, 347–360.

Guthrie, R.L., and Witty, J.E. 1982. New designations for soil horizons and layers and the new Soil Survey Manual. Soil Sci. Soc. Am. J. 46, 443–444.

Harden, J.W., Taylor, E.M., McFadden, L.D., and Reheis, M.C. 1991. Calcic, gypsic, and siliceous soil chronosequences in arid and semiarid environments. Soil Sci. Soc. Am. Spec. Publ. No. 26, 1–16.

Holliday, V.T. 1982. Morphological and chemical trends in Holocene soils at the Lubbock Lake archeological site, Texas. Ph.D. thesis, Univ. of Colorado, Boulder, 285 p.

Lattman, L.H. 1973. Calcium carbonate cementation of alluvial fans in southern Nevada. Geol. Soc. Am. Bull. 84, 3013–3028.

Machette, M.N. 1985. Calcic soils of the southwestern United States. Geol. Soc. Am. Spec. Pap. 203, 1–21.

Ruhe, R.V. 1975. Geomorphology. Houghton Mifflin, Boston, 246 p.

Soil Survey Division Staff. 1993. Soil Survey Manual. U.S. Dept. Agri. Handbook No. 18, 437 p.

Soil Survey Staff. 1975. Soil taxonomy. U.S. Dept. Argi., Agri. Handbook No. 436, 754 p.

Vepraskas, M.J. 1994. Redoximorphic features for identifying aquic features. N. Carolina Agri. Res. Service, Tech. Bull. 301, 33 p.

Walker, T.R., Waugh, B., and Crone, A.J. 1978. Diagenesis in first-cycle desert alluvium of Cenozoic age, southwestern United States and northwestern Mexico. Geol. Soc. Am. Bull. 89, 19–32.

Yaalon, D.H. 1966. Chart for quantifying estimation of mottling and nodules in soil profiles. Soil Sci. 102, 212–213.

■ Appendix 2

Harden, J.W. 1982. A quantitative index of soil development from field descriptions—examples from a chronosequence in central California. Geoderma 28, 1–28.

Harden, J.W., and Taylor, E.M. 1983. A quantitative comparison of soil development in four climatic regimes. Quaternary Res. 20, 342–359.

Knuepfer, P.L.K. 1988. Estimating ages of late Quaternary stream terraces from analysis of weathering rinds and soils. Geol. Soc. Am. Bull. 100, 1224–1236.

Machette, M.N. 1985. Calcic soils of the southwestern United States. Geol. Soc. Am. Spec. Pap. 203, p. 1–21.

Reheis, M.C., Harden, J.W., McFadden, L.D., and Shroba, R.R. 1989. Development rates of late Quaternary soils, Silver Lake playa, California. Soil Sci. Soc. Am. J. 53, 1127–1140.

Sowers, J.M., and nine others. 1988. Geomorphology and pedology on the Kyle Canyon alluvial fan, southern Nevada. In D.L. Weide and M.L. Faber, eds., This extended land—geological journeys in the southern Basin and Range, p. 137–157. Dept. Geoscience, Univ. of Nevada, Las Vegas.

Taylor, E.M. 1988. Instructions for the soil development index template—Lotus 1-2-3. U.S. Geol. Surv. Open-File Report 88-233, 23 p.

Index

A

A horizon, formation of, 107
  age of, 108
  radiocarbon dating, 137
Abrasion pH, 62, 67, 74
Accretion gley, 257
Accumulation index, 14, 187
Acid rain and soils, 338
Acid sulphate soils, 136
Africa, 127, 305, 319, 337
Ages using soils, 229
Alaska, 168, 235, 242
Alfisol, 46
Alloformation, 309
Allophane, 90
Allostratigraphic units, 309
Alpine soils, 160, 185, 337
Alps, European, 289
Aluminum, compounds, 90
  effect on pH, 18
  extracts (treatments), 91
  mobility, 95
  oxyhydroxide test, 92
  translocation, 108
Amazon Basin, 81
Andes, 82
Andisol, 43, 96, 159
Antarctica, 183, 202, 241
Applications, soils, to other studies, 307–46
Archeological studies and soils, 333–37
Arctic, 184, 242
Aridisol, 45
Arizona, 210
Aspect vs. soils, 232
Atlantic islands, 158
Australia, 195, 259, 295
Available water-holding capacity, 16

B

Bs horizon, process of formation, 108
Bt horizon
  clay with time, 217
  color with time, 220
  iron with time, 216
  process of formation, 112
Backslope, 231
Baffin Island, 187
Basalt weathering, 65
Base saturation, 18
  vs. pH, 19
Beidellite, 88
Bilzi-Ciolkosz index (profile development), 21
Bioclimate, 268
Biotite-induced shattering of rocks, 75
Bioturbation, 105
Boehmite, 90, 198
Boundary, calcic-noncalcic soils, 290
Brazil, 258
Bulk density, 13
Buntley-Westin color index, 20
Buried soils, 8, 24–28. *See also* Paleosols.
  horizon nomenclature, 7
  persistence of buried horizons, 26

C

Ca^{2+} in precipitation, 131
Calcium carbonate. *See* Carbonate horizon
Calcrete. *See* K horizon
Calcic soils, 127, 203
Caliche. *See* K horizon
California, 33, 74, 166, 192, 201, 208, 234, 245, 247, 287, 292, 294, 330
Canada, 77, 251

Capillary rise of water, 17
Carbon, global, 280
Carbonate
  atmosphere origin, 143
  biogenic origin, 128
  build up, climate, 206
  capillary rise, 132
  chemistry and mineralogy, 130
  climate, 203, 290
  equilibria, 128
  groundwater origin, 132
  gypsum origin, 132
  horizon morphology stages, 129, 206, 356
  horizon, origins, 127
  leaching, 204, 214
  radiocarbon dating, 138
  source, 131, 204
  time to form stages, 206
  U-series dates, 140
Carbon dioxide partial pressures, 60, 128
Catena, 235–67
  alpine, 243
  Antarctic, 241
  arctic, 242
  calcic, 261
  clay-minerals, 241, 249
  dry climates, 245
  faults, association with, 327
  fire-induced erosion, 251
  geochemistry, 240, 244
  gypsic, 263
  humid, 251
  kinds of, 239
  model, 235, 265
  PDI, 250
  polar, 241
  synthetic, 240
  water movement in, 236
  warm humid, 257
Cation exchange capacity, 18, 92
  vs. clay minerals, 93
  vs. organics, 92
Cementation, 357
Chelation, 65, 110
Chelating complexes, 65
Chemical analysis, weathering, 66, 69, 190, 193, 198
  eluvial-illuvial coefficient, 72
  molar ratios, 68
Chemical denudation, 78, 81
Chemical extracts (treatments), 91
Chemical weathering, 59
  amount, 66
  classes, 75
  global, 80

  mass balance, 72
  rate of present day, 76
Chile, 75
China loess, 323
Chlorite, 88
Chronogram, 256
Chronofunctions. *See* Chronosequences
Chronosequences, 178–229
  A horizon, 215
  alpine, 183
  aridic soils, 202
  Atlantic islands, 196
  calcic, 203
  classification, 178
  clay comparisons (Bt), 216
  clay mineralogy, 221
  desert soil summary, 214
  diagnostic horizons, 225
  gypsic, 210
  iron comparisons, 216
  loess on basalt, 208
  marine terraces, 192
  noncalcic-calcic transects, 201
  orders, 225
  Pacific islands, 198
  polar, 183
  red color, 220
  surface and buried soils, 209
  tills, 190
  western USA, 190
  wet climates, 194
  river terraces, 191
  tropical islands, 196
Chronostratigraphic unit, 309
Clast disintegration index, 172
Clay
  accumulation index, 14, 194, 210
  bands (lamellae), 113
  content, climate, 188, 282
  eolian origin, 157, 161. *See also* eolian additions
  films, 113, 355
  film destruction, 115
  influence of original texture, 155
  mass balance, 117
  translocation, 112
Clay mineralogy
  climate, 286
  climatic change, 299
  global distribution, 104
  soil orders, 286
  stages (Jackson), 222
Clay minerals, 85–104
  alteration, 100
  change with time, 72, 198

classification, 86
chemical composition, 88
distribution with depth, 102
families vs. soil orders, 286
formation, 96
formation, parent material influence, 154
inheritance, 97
neoformation, 97
percentages, 90
phase diagrams, 98
Si/Al oxide ratios, 89
transformation, 97
Climate
  carbonate, 290
  clay, 282
  clay mineralogy, 285
  factor, 268–306
  gradient, 182
  iron, aluminum, phosphorus, 292
  iron minerals, 288
  parameters, 269
  PDI, 293
  redness, 288
  soil orders, 278
  soil relationships, 268–306
Climatic change, soil evidence, 293–306
Climatic parameters, 269
C:N ratio, 107, 281
Coastal Plain, eastern USA, 194, 259
Color, 9, 349
Colorado, 160, 185, 243, 251, 261, 264, 312
Color indices, 20
Columbia Plateau, 295
Composite soils, 27
Congruent dissolution, 59
Consistence, 353
Cordilleran, 174, 203, 206, 247, 282, 315. *See also* Rocky
  Mountains, Sierra Nevada
Correlation of deposits using soils, 312, 322
Cosmogenic isotopes, 315
Cumulative curve, particle size, 161
Cumulative profile, 8, 165, 167
Cutan. *See* Clay films

D

Dehydration, 66
Depletion index, 14, 187
Development of red color, 220, 288
Diagnostic horizons, 36
Dioctahedral structure, 85
Dispersion of clay, 118, 181
Dissolution (congruent, incongruent), 59

Distribution of soil orders and suborders
  in the United States, 48
  in the world, 49
Dithionite-citrate extract (treatment), 91
Duripan, 37
Dust-trap data, 205,

E

Earth chemistry, 105
Egypt, 296
Eh-pH diagram, 135
Elevation-soil transects, 292
El Niño, 212
Eluvial, 4
England, 298
Entisols, 46
Environmental studies and soils, 337
Eolian additions and influx (carbonate, fines), 199, 204,
  211, 284, 298
  in laterite formation, 126
Epipedon, 37
  histic, 36
  melanic, 159
  mollic, 36
  ochric, 36
  umbric, 36
Equilibrium, clay minerals, 98
Europe, 289, 294, 319
  loess, 323
Exchangeable cations, 18
Exhumed soils, 8
Extracts of Fe, Al and P, 91, 137

F

Factors of soil formation, 141–47
  biotic, 143, 268
  climate, 143, 268–306
  parent material, 143, 148–70
  time, 143, 171–229
  topography, 143, 230–67
Fe activity ratio, 91
Ferrihydrite, 9, 91
Ferrolysis, 136, 228
Fibrous clay minerals, 88, 102
Field capacity, 15
Fire weathering, 56
Flocculation of clay, 118, 181
Florida, 260
Footslope, 231

Fragipan, 37, 121, 290
Freeze-thaw, 54
Fulvic acid, 110
Fundamental soil formation equation, 141

G

Gelic materials, 42
Gelisol, 42
Geology maps from soils and soil maps, 31–39, 311
Geophagy, 339
Geosols, 308
Gibbsite, 90, 123, 198
Glaebular zone (laterites), 122
Gley, accretion, 258
Gley, 134
Goethite, 9, 91, 123
Granite
   grus, 75
   weathering stages, 75
Great groups, 46
Ground water carbonate, 132, 332
Grus, 75
Gumbotil, 256
Gypsum
   accumulation morphology stages, 357
   Antarctica, 133
   origin in soils, 132
   Israel, 203, 210, 263
   Peru, 211
   USA, 211

H

H_2O^+, H_2O^-, 66
Halite, 132
Halloysite, 87
Hawaii, 286, 287
Hematite, 9, 91, 123
High Plains, 334
High water table. *See also* Horizon, redoximorphic
   and salts, 136
Himalaya, 185
Histosol, 43
Holocene, 171
Horizon,
   A, 4
   albic, 36
   argillic, 36
   boundaries, 7, 356
   B, all kinds, 5

C, all kinds, 5
calcic, 36
cambic, 36
duripan, 37
E, 6
fragipan, 37, 121
gypsic, 37
K, 7, 129, 356
kandic, 36
master, 4
natric, 36
nomenclature, field, 5–6
O, 4
old soil-horizon nomenclature, 347
oxic, 36
petrocalcic, 37
petrogypsic, 37
redoximorphic, 134
salic, 37
spodic, 36
subdivision of, 7
subordinate nomenclature, 6
Hornblende, stability of, 176
Humic acids, 111
Hurst index (color), 20
Hydration, 66
Hydrolysis, 61
Hypersthene, stability of, 177

I

Idaho, 70, 79, 247
Illinois, 168, 176, 312
Illite, 88
Illuvial, 4
Imogolite, 90
Inceptisols, 46
Incongruent dissolution, 60
Index of profile anisotropy (IPA), 22
Indiana, 191
Influx limited, 205
Insolation weathering, 58
Ion exchange, 66, 101
Ion mobilities, 62, 93
Ionic potential, 66, 93
Iowa, 32, 168, 176, 255
Iron
   extract ratios, 91
   extracts (treatments), 91
   mobility, 95
   translocation, 108
Iron-bearing pedogenic minerals, formation of, 96
Israel, 169, 203, 210, 246, 263, 305

K

K cycle model, 238
K horizon, chemical analyses, 130
ka, 171
Kaolinite, 87
Kenya, 319
Krotovinas, 105
ky, 171

L

Lake Lahontan, 207, 298, 312
Lamellae. *See* clay bands
La Sal Mountains, Utah, 291, 312
Laterite, morphology and origin, 121
    chemistry, 123
    mineralogy, 123
    and plate tectonics, 127
Leaching index, 273
Leaching of original carbonate, 214
Lepidocrocite, 91
Lichen weathering, 65
Limestone, weathering of, 60, 153
Limestone soils, 156, 196
Loess, 167, 200, 294, 322
Loess transect, 167
Louisiana, 255
Lower case letter (for horizon nomenclature), 5
Lysimeter, 77

M

Ma, 171
Maghemite, 91
Magnetic susceptibility, 164, 323
Maryland, 78, 195
Mass accumulation, 14
Mean residence time (radiocarbon dating), 137
meq, 92
Mexico, 208
Mica half life, 185
Michigan, 235, 275
Micromorphology, 23
Midcontinent, 223, 253, 275, 284. *See also* individual
    states
Midcontinent loess, 324
Mineral
    depletion, 176
    etching, 177
    ratios, 176
    in rock, Si/Al oxide ratio, 90

stability, 148, 164
    weathering, 172
mIPA. *See* Index of profile anisotropy
Missouri, 177, 330
Mixed-layer minerals, 88
Models of pedogenesis, 145
Moisture limited, 205
Moisture regimes, soil,
    aquic, 42
    aridic, 42
    perudic, 42
    udic, 42
    ustic, 42
    xeric, 42
Molar ratios
    of clay minerals, 89
    of rocks, 151
    of rock-forming minerals, 90
    of weathered materials, 67
Mollisol, 46
Monogenetic soils, 144
Montana, 201, 297
Montmorillonite, 88
Mottled color pattern, 9, 134, 349
Mud rain, 120
My, 171

N

Na influence on clay movement, 119
Nebraska, 253
Needle ice, influence on soil erosion, 227
Neoformation of clay minerals, 96
Nevada, 207, 292, 299, 332
New Hampshire, 79
New Mexico, 203, 262, 328
New York, 153
New Zealand, 22, 73, 82, 92, 164, 185, 195, 251, 275,
    290, 319
Nitrogen, global, 281
Non-crystalline materials, 90
Non-cumulative profile, 8
Nontronite, 88
North Carolina, 235
Norway, 245

O

Ohio, 189, 232
Octahedral sheet, 85
Ooids, 129
Orders, 42
    vs. process, 107

Oregon, 176, 194
Organic carbon, radiocarbon dates, 137
Organic matter, 11
  content and climate, 279
Oxalate extract (treatment), 91
Oxic horizon origin, 121
Oxidation, 64
Oxide molar ratios. *See* molar ratios
Oxidizing conditions. *See* redox
Oxisol, 44, 201
Oxygen-isotope data, 164
Oxygen-isotope stages and ages, 171

P

Pacific islands, 156
Pakistan, 305
Paleoclimatic interpretation, 293–306
  Bt horizon properties, 297
  carbonate, 300
  clay mineralogy, 299
  eolian, 298
  gypsum and halite, 303
  isotopic data, 304
  mineral etching, 300
  morphology, 294
  periglacial, 298
  poor correlation with soils, 293
  salt-shattered clasts, 303
  water movement, 297
Paleosols, 339–46
  alluvial, 345
  classification, 341
  criticism, 342
  defined, 339
  diagenesis, 342
  examples, 344
  isotopes, 346
  laboratory analyses, 340
  K-T boundary, 346
  metamorphosed, 345
  recognition, 340
Palygorskite, 88, 131, 287
Parent material
  factor, 141–70
  layering, 7, 159
Parkers weathering index, 70
Particle-size classes. *See* texture classes
Pedoderm, 308
Pedogenic gradient, 273
Pedogenic processes, 105–40
Pedology, 4
Pedon, 2

Peds, 12
Pennsylvania, 194
Percent base saturation, 18
Permanent-wilting point, 15
Peru, 175, 211, 245
pH, 18, 354
  vs. base saturation, 19
  NaF pH, 92, 159
Phase diagrams, clay minerals, 98
Phosphorus transformation, 137
Phyllosilicates, 85
Physical weathering, 53[n]59
Pingo soils, 242
Pisolith, 123, 129
Pleistocene, 171
Plinthite, 6, 127
Podzolization, 108, 156
  index, 21
Polygenetic soil, 143
Porosity, 13
Precipitation chemistry (Australia), 292
Process, soil formation, 105–40
  vs. soil orders, 107
Profile development index (PDI), 21, 218, 250, 293
  calculation of, 360
Progressive pedogenesis, 146
Proto-imogolite, 111
Pyrophosphate extract, 91

R

Radionuclides, 338
Radiocarbon dating
  inorganic carbon of $CaCO_3$, 138
  organic carbon, 137
Rare-earth elements, 200
Rates of organic matter decomposition, 279
Ratios
  resistant mineral to non-resistant mineral, 176
  resistant mineral to resistant mineral, 164
Red color, 9
Redness, iron-bearing minerals, 9, 91
Redox conditions, 134
Reducing conditions, 134
Regional rates of chemical weathering, 76
Regressive pedogenesis, 146, 179, 226
Reiche's product index, 70
Reiche's weathering potential index, 70
Relict soils, 8
Rhodesia, 258
Rock chemical composition, 150
  Si/Al oxide ratios, 151
Rock weathering, 151, 171

Rocky Mountains, 190, 298, 303. *See also* Cordilleran
Rubification index, 362

S

Salt
  catenas, 263
  climate, 203, 290
  climatic change, 303
  morphology stages, 258
  weathering, 172
  weathering, climatic change, 304
Sand dune soils, 113, 191, 326
Sangamon soil, 257, 325
Saprolite, 3, 123
Scanning electron microscope, 114
Sepiolite, 88, 131, 287
Shoulder, 231
Si/Al molar ratio, 68
Sierra Nevada, 175, 177, 187, 190, 287, 299, 315. *See also* California
Silica accumulation, 132
  stages of, 357
Silcrete, 132
Silicon tetrahedron, 85
Silt
  cap stages, 358
  translocation, 120
Slope orientation and soils, 232–35
Smectite, 88
Soda niter, 133
Soil
  classification, 29–52
  date colluvium, 320
  date faulting, 326
  date floods, 321
  date folds, 331
  date landslides, 319
  date sand blows, 330
  defined, 2
  development defined, 19
  development with time, 180–229
  facies, 308
  formation, time of, 309
  formation, vs. landscape evolution, 179
  forming factors, 141–47
  forming factors, criticism, 144
  forming interval, 309
  geomorphology, 1
  geophagy, 339
  great groups, 48
  horizon destruction, 226
  horizon nomenclature, 5, 347

  horizon symbol, 5, 347
  horizons, 3
  horizons, persistence with burial, 26
  indices, 19
  landscape analysis, 49
  on limestone, 156
  maps, 30, 50
  moisture, 15–18
  moisture regimes, 42
  orders, 37, 40–48
  orders and time, 45, 72
  phases, 29
  plant nutrition association, 319
  profile, 2
  profile vs. weathering profile, 3
  processes. *See* Process, soil formation
  photographing, 347
  properties, 348
  series, 29
  stratigraphic units (studies), 308, 311
  stratigraphy, 307
  suborder, 37, 40, 43
  taxonomy, 32
  taxonomy, distribution USA, 48
  taxonomy, distribution world, 49
  taxonomy vs. process, 107
  temperature regimes, 42
  volcanic parent material, 158
  water content, movement, 15
  welding, 27
Soil-elevation transects, 292
Soil-forming processes, 104–40
  vs. orders, 107
South America, 127. *See also* Amazon, Brazil, Chile, Peru
Southeastern USA, 177
Spitzbergen, 184
Spodosol, 43, 275
  in tropics, 260
Sri Lanka, 258
Stability fields for clay minerals, 98
Stages in the build up
  of carbonate, 129, 356
  of gypsum, 358
  of silica, 359
  of silt, 358
Steady state, 146, 226
Stone line, 27, 126
Structure, 12, 352
Subdivision and correlation of Quaternary deposits, 307–26
Suborder nomenclature, 40
Summit, 231
Surface area per unit volume, 10
Surface-boulder frequency, 175

T

Taiwan, 8
Temperature regimes, soil, 42
    cryic, 42
    frigid, 42
    hyperthermic, 42
    mesic, 42
    pergelic, 42
    thermic, 42
Tetrahedral sheet, 191
Texas, 210, 334
Texture, 10, 354
    class, 10, 354
    influence on soil formation, 155
Termites, 259
Thin sections, 23, 114
Threshold, 181, 195, 293
Throughflow water, 236
Time factor, 171–229
Time scale, 171
Toeslope, 231
Tombstone weathering, 172
Topographic factor (toposequence, topofunction),
    230–67
Transformation, clay minerals, 97
Transport-limited slopes, 239
Translocation
    clay, 112
    silt, 120
Tree windthrow, 228
Trioctahedral structure, 85
Tropical soils, 121, 196, 257

U

Ultisol, 45
Uniformity of parent material, 159
Uranium-series dating, 140
Utah, 331

V

Vegetation-soil relationships, 268
Vermiculite, 88
Vertisol, 44
    and gravel movement, 161
Vesicular structure, 13
Virginia, 83, 195

Volcanic airfall deposits
    soils, 158
    weathering of, 96

W

Washington, 194, 234, 295
    loess, 323
Water
    balance, 269
    chemistry, 77
    movement, 15, 271
    movement vs clay, carbonate, 272
    movement on slopes, 231
Weathered boulders, percent of, 175
Weathered minerals, 176
Weathering, 53–84
    alpine, 75
    chemical, 59–65
    elevation, with, 58
    equations, 62
    fire, 56
    indices, 68
    minerals, 63, 79
    molar ratios, 68
    physical, 53–59
    pits, 58
    profile, 3
    rinds, 152, 173
    of rocks, 58, 65, 151, 172
    salt, 55, 172
    subaerial vs. subsurface of rocks, 174
Weathering-limited slopes, 239
Weighted mean percent, 14
Welded soils, 27
Wetland soils, 337
Wind erosion, 227
Wyoming, 185, 190, 211, 247, 306, 315. *See also*
    Cordilleran, Rocky Mountains

Y

Yellowstone, 71, 174
Yucca Mountain, 332

Z

Z-score, 178
Zeolite, 98